Die Grundlehren der mathematischen Wissenschaften

in Einzeldarstellungen
mit besonderer Berücksichtigung
der Anwendungsgebiete

Band 149

Herausgegeben von

J. L. Doob · E. Heinz · F. Hirzebruch · E. Hopf · H. Hopf
W. Maak · S. MacLane · W. Magnus · D. Mumford
M. M. Postnikov · F. K. Schmidt · D. S. Scott · K. Stein

Geschäftsführende Herausgeber

B. Eckmann und B. L. van der Waerden

L. Sario · K. Oikawa

Capacity Functions

Springer-Verlag New York Inc. 1969

Prof. Leo Sario
University of California, Los Angeles

Prof. Kôtaro Oikawa
University of Tokyo

Geschäftsführende Herausgeber:

Prof. Dr. B. Eckmann
Eidgenössische Technische Hochschule Zürich

Prof. Dr. B. L. van der Waerden
Mathematisches Institut der Universität Zürich

ISBN-13: 978-3-642-46183-5 e-ISBN-13: 978-3-642-46181-1
DOI: 10.1007/978-3-642-46181-1

 Library of Congress Catalog Card Number 69-20300.
Softcover reprint of the hardcover 1st edition 1969

Preface

Capacity functions were born out of geometric necessity, a decade and a half ago. Plane regions had been found of arbitrarily small area, yet with a totally disconnected boundary. Such regions seemed to defy the very spirit of Riemann's mapping theorem. They could be mapped conformally and univalently into a disk, with the single boundary point at infinity being stretched into a circle.

The plausible explanation of the mystery is, of course, as follows. Under a mapping of the punctured sphere onto a disk, an area element near the punctured point would have to stretch more in the circular direction than in the radial direction, and the conformality would be destroyed. But if there is a sufficiently heavy accumulation of other boundary components, these can take over the distortion, and the mapping of the region itself remains conformal.

Such phenomena made it an important problem to characterize pointlike boundary components which were *unstable*, i.e., hid in themselves this power of stretching into proper continua. Standard tools such as mass distributions, potentials, and transfinite diameters could not be used here, as they were subject to the vagaries of the other components. The characterization had to be intrinsic, depending only on the region itself, in a conformally invariant manner.

This goal was achieved in the following fashion (SARIO [10, 13]). Given a region W and a boundary component γ, choose a point $\zeta \in W$, a local parameter z at ζ, and a regular region Ω containing ζ. Let γ_Ω, $\beta_{i\Omega}$ $(i=1, \ldots, n_\Omega)$ be the components of $\partial\Omega$, with γ_Ω separating γ from ζ. Take the harmonic function $p_{1\gamma\Omega}$ on $\bar{\Omega}$ with singularity $\log|z-\zeta|$ such that $p_{1\gamma\Omega}(z)-\log|z-\zeta| \to 0$ as $z \to \zeta$, and with constant boundary values $k_{1\gamma\Omega}$ and $k_{1i\Omega}$ on γ_Ω and $\beta_{i\Omega}$. The constants are here so chosen that the flux $\int * dp_{1\gamma\Omega}$ is 2π and 0 across γ_Ω and each $\beta_{i\Omega}$, respectively. As the Ω exhaust W, the $c_{1\gamma\Omega}=e^{-k_{1\gamma\Omega}}$ decrease to a limit $c_{1\gamma} \geq 0$, which was defined as the *capacity of* γ. The normal family $\{p_{1\gamma\Omega}\}$ has a limiting function $p_{1\gamma}$, unique if $c_{1\gamma}>0$; it was called the *capacity function of* γ.

Our problem was herewith solved: $c_{1\gamma}>0$ turned out to be necessary and sufficient for a point boundary component to be unstable.

The solution of the problem had immediate applications to classification theory. A boundary component γ was called *weak*, if it was a point under all univalent mappings of the region, i.e., if $c_{1\gamma}=0$. A boundary was said to be *absolutely disconnected* if all its components were weak. The class O_γ of regions with absolutely disconnected boundaries was obviously identical with the class O_{SB} of regions without univalent bounded functions. This in turn had been shown by AHLFORS and BEURLING [1] to coincide with the class O_{SD} of regions without univalent Dirichlet-finite functions. In particular, the identity map of the aforementioned region of finite area was in SD, hence the instability of the component at the point at infinity.

The concept of a weak boundary component γ was, however, not restricted to plane regions. For an arbitrary open Riemann surface W, the function $p_{1\Omega}$ on a regular subregion Ω could be formed as before, except that γ_Ω and the $\beta_{i\Omega}$ had to be borders of components of $W-\Omega$. Using the terminology of the then recently developed theory of normal operators, $p_{1\Omega}$ was to have the L_1-behavior on $\partial\Omega-\gamma_\Omega$ corresponding to the canonical partition. As in the case of plane regions, the $c_{1\gamma\Omega}$ decreased, the family $\{p_{1\gamma\Omega}\}$ was normal, and the limits $c_{1\gamma}$ and $p_{1\gamma}$ were the capacity and the capacity function of γ. An important class was again the class O_γ of Riemann surfaces whose boundary was, in analogy with plane regions, called absolutely disconnected. A later result of JURCHESCU [3,4] was illuminating: a necessary and sufficient condition for a Riemann surface not to be essentially extendable is that every regularly imbedded planar subregion V whose relative boundary components are connected by the ideal boundary have its double $\hat{V}$ about ∂V in the class O_γ.

The thought was now near to replace the L_1-behavior on $\partial\Omega-\gamma_\Omega$ by the L_0-behavior. This meant that the normal derivative of the function, now denoted by $p_{0\Omega}$ on $\bar{\Omega}$, had to vanish on $\partial\Omega-\gamma_\Omega$. On γ_Ω the function was again to be a constant, $k_{0\gamma\Omega}$, and one set $c_{0\gamma\Omega}=e^{-k_{0\gamma\Omega}}$. For $\Omega\to W$, the limits $c_{0\gamma}$ and $p_{0\gamma}$ gave a new kind of capacity and capacity function of γ. The immediate relation $c_{1\gamma}\geq c_{0\gamma}$ gave rise to five (not mutually exclusive) types of boundary components: (1) $c_{1\gamma}=0$, (2) $c_{0\gamma}=0$, (3) $c_{0\gamma}=0$, $c_{1\gamma}>0$, (4) $c_{1\gamma}>0$, (5) $c_{0\gamma}>0$. Several classes of Riemann surfaces are obtained by requiring that all boundary components, or none of them be of one of these five types. An interesting problem, largely unexplored thus far, is to find geometric properties of boundary components of types (1)–(5) under compactifications.

For plane regions, such properties gave a direct three-way classification: in addition to weak and unstable components, one considered *strong* ones, which were continua under all univalent mappings of the region. It having been established that a component γ is weak if and only if $c_{1\gamma}=0$, it was tempting to assume that γ is unstable if and only if $c_{0\gamma}=0$, $c_{1\gamma}>0$, and consequently strong if and only if $c_{0\gamma}>0$. The correctness of this conjecture remains, however, an open question, perhaps the most outstanding one in the theory. All that is known is that $c_{0\gamma}>0$ implies the strength of γ.

One then encountered the following surprising phenomenon on parallel slit mappings (SARIO [14]). Take a function P_φ with singularity $1/(z-\zeta)$, which maps the given region onto a plane region bounded by a set of parallel slits with inclination φ relative to the positive x-axis. As φ varies from 0 to π, a given boundary component is either always a point or always a continuum or else – and this is the essence – it is a point *for exactly one* φ.

Although $c_{0\gamma}$ and $p_{0\gamma}$ have thus far failed to solve the problem of instability, they have proved to possess interesting mapping properties. STREBEL [2] and REICH [2] succeeded in showing that $P_{0\gamma}=\exp(p_{0\gamma}+i\,p_{0\gamma}^*)$ gives a mapping onto an incised radial slit disk of radius $c_{0\gamma}^{-1}$. By this is meant a region whose outer boundary, the *image* of γ, is a circle of radius $c_{0\gamma}^{-1}$ from which radial slits emanate into the interior of the disk, the other boundary components being radial slits. It is, however, not known how to characterize the points of γ which go to the incisions, i.e., whose image has a modulus $<c_{0\gamma}^{-1}$.

Whereas, in symbolic notation, $\min|P_{0\gamma}(\gamma)|$ can thus be $<c_{0\gamma}^{-1}$, other univalent functions P_γ with the same normalization $1/(z-\zeta)$ and the same flux conditions can bring $\min|P_\gamma(\gamma)|$ as close to $c_{0\gamma}^{-1}$ as one wishes. This is the striking result of RENGEL [1]. If this minimum actually reaches $c_{0\gamma}^{-1}$, then the mapping function is $P_{0\gamma}$.

The original capacity function $P_{1\gamma}=\exp(p_{1\gamma}+i\,p_{1\gamma}^*)$ has less sophisticated mapping properties: it always gives a disk of radius $c_{1\gamma}^{-1}$ less circular slits, and it minimizes $\max|P_\gamma(\gamma)|$. However, for the component γ with the smallest minmax $|P_\gamma(\gamma)|$, REICH and WARSCHAWSKI [1] obtained the elegant result that every circular slit subtends a central angle less than π.

In the definition of $p_{1\gamma}$, the component γ can be replaced by any subboundary of the given Riemann surface. If γ is the entire ideal boundary β, then $p_{1\gamma}$, $c_{1\gamma}$ are denoted by p_β, c_β. In contrast with the Green's function, the capacity function p_β exists on every open Riemann surface.

For plane regions, we have here the connection with classical concepts: the capacity function p_β is the restriction to the region of the equilibrium potential of the complement E. The capacity c_β of the boundary β of the region coincides with the classical capacity of E and has therefore the well-known relation to the transfinite diameter.

These, then, were the beginnings, with some later results indicated, of the theory of capacity functions. During the intervening decade and a half, the theory of these functions, of weak, unstable, and strong boundary components, and of related slit mappings has developed by leaps and bounds in several directions. The main results were achieved by AHLFORS, AKAZA, BEURLING, CONSTANTINESCU, FUJIIA, JURCHESCU, KURAMOCHI, MARDEN, NAKAI, OHTSUKA, REICH, RODIN, SAKAI, SAVAGE, STREBEL, SUITA, and WARSCHAWSKI (cf. Author Index). New problems have been opened up by these writers, and important tools, in particular that of extremal length, have been brought into the theory.

It is the purpose of the present first monograph in the field to give a systematic account of its present status, much of it developed by the junior author (see Bibliography). No attempt, however, is made at comprehensiveness. We make uninhibited use of all authors' right to choose topics for a coherent unity, close to their own interests. Thus many results are omitted, not because of their being less significant, but because they did not fall within the general plan.

The book is divided into three parts. In Part One we develop the analytic tools. If the reader's patience is stretched in the jungle of formulas, we hope he will find the effort rewarded in Part Two, where the analytic tools are put to concrete geometric use. In a sense, Part Two, with its stress on univalent mappings, is the most essential part of the book. Both the analytic theory of Part One and the geometric theory of Part Two will be needed in Part Three, where we develop the role played by capacities in classification theory.

Since capacity functions are limits of functions $p_{0\gamma\Omega}$, $p_{1\gamma\Omega}$, both Part One and Part Two open with the theory of principal functions. The presentation in each of these parts then proceeds to capacity functions and to modulus functions. The latter are the counterparts of capacity functions in boundary neighborhoods. They are used to define moduli and to give mappings into annular regions.

Part One closes with an account of interrelations of the fundamental functions, and Part Two with the theory of extremal slit regions, i.e., regions which can be images under mappings related to principal, capacity, or modulus functions.

Part Three consists of two chapters. The first is concerned with characterization and properties of null classes related to capacity functions, and with their significance in the general classification theory. The second is devoted to practical tests, e.g., for a boundary component to be weak.

In Appendices I and II, we have compiled those parts of the theories of extremal length and conductor potentials that are used in the book.

At the end of the book we have included some exercises to give the interested reader a chance for active participation in the getting acquainted process. A list of unsolved problems is also included to challenge the cognoscenti.

The number of results contained in the book is quite large, and the presentation concise. In the detailed Table of Contents, the chapter and section headings give the general framework of the theory, the actual contents lying somewhat hidden in the subsection headings. Before a systematic study, the reader may wish to thumb through the theorems, perhaps starting with those of concrete geometric meaning. The following sampling should be useful for this purpose: Chapter VI, 3 F, 4 B, 4 D, 4 F, 5 B; Chapter VII, 2 B, 2 D–2 G, 3 B–3 D, 3 H, 3 I, 4 A, 4 B, 5 A–5 C, 5 F, 5 G, 6 D, 6 G, 6 I, 6 L; Chapter VIII, 2 C, 2 G, 2 I; Chapter IX, 2 A, 2 B, 3 G, 4 F, 4 G; Chapter X, 2 C, 2 K, 2 L, 3 A, 3 C, 3 D, 4 C, 4 F, 5 D; Chapter XI, 1 A, 1 E, 1 K, 2 E, 3 B, 3 D, 3 F, 3 H, 3 L, 4 A, 4 E, 5 A, 5 B, 5 D, 5 J.

In some proofs, we have not hesitated to use auxiliary results established in later parts of the book, when clarity and adherence to the general plan have so demanded.

It is in the nature of a first monograph in the field that the overlapping with existing books is almost nil. In those few instances, where making the book self-contained necessitated background information covered in another book, a new approach was developed, to serve our particular needs.

For prerequisites a standard graduate curriculum should be sufficient. Beyond this, we have on occasion taken for granted some very basic concepts and results in the monograph "Riemann surfaces" by AHLFORS and SARIO, and some generally known theorems in advanced complex analysis; an exact reference is then given, and the reader should find no difficulty on this score.

A lucid general orientation regarding the problem of weak, strong, and unstable boundary components is found in a lecture by M. H. STONE [2].

Research on the theory developed in this book was sponsored by the U.S. Army Research Office–Durham. We are deeply grateful to Drs. J. DAWSON, A. S. GALBRAITH, and G. PARRISH for their excellent cooperation throughout the 6-year work and for their liberal policy of allowing us to make and break proposals according as the project wound through its unpredictable course.

Without the administrative good will of Vice President A. TAYLOR, Vice Chancellor F. SHERWOOD, and Dean L. PAIGE, we would not have had the time and the favorable circumstances to carry out the many-phased research work.

We are indebted to S. COUNCILMAN, Y. KWON, and I. LIN who helped untiringly in checking the entire manuscript, and to M. NAKAI, N. SUITA,

G. WEILL, and J. CHANG, who read various portions of it and gave their valued advice.

Our sincere appreciation goes to Professor B. ECKMANN for the kind interest he has shown in our work and for including our book in this series of challenging tradition.

The renowned expertness of the Springer-Verlag in both technical and aesthetic matters made the entire publishing process a pleasant experience.

Our thanks are due to Mrs. ELAINE BARTH, whose skilled typists processed several versions of the manuscript, and to Miss ELLEN COLE, whose staff transcribed our voluminous trans-Pacific communication with unfailing promptness and accuracy.

Los Angeles and Tokyo
March 15, 1969

LEO SARIO
KÔTARO OIKAWA

Contents

Part I · Analytic Theory

Chapter I · Normal Operators

Chapter II · Principal Functions

Chapter VII · Mappings Related to Capacity Functions

Chapter XI · Practical Tests

Appendices

PART ONE

Analytic Theory

Chapter I

Normal Operators

The main tool used in this book is the normal operator method. In 1, we introduce normal operators and employ them to construct, on a Riemann surface, harmonic functions with a given behavior at the ideal boundary. In 2–4, are given the definition and basic properties of the important special operators L_0 and L_1, which will be used throughout the book. Another operator H, not necessarily normal, will also be discussed; it will be introduced in 5.

1. Fundamentals of the Normal Operator Method

1 A. End. We shall make free use of concepts defined in AHLFORS-SARIO [1, Chs. I, II, and Ch. V, §1]. New ones will be introduced as needed.

The boundary of a set E on a Riemann surface W is denoted by ∂E. To distinguish it from the ideal boundary, ∂E will often be called the *relative* boundary of E. In particular, if E is a regularly imbedded region of W, then ∂E is the border of the bordered Riemann surface $E \cup \partial E$. In this case, ∂E will also stand for the oriented border.

By an *end* of a Riemann surface W, we mean the union of a finite number of disjoint regularly imbedded regions with compact relative boundaries. We use this term in a less restricted sense than HEINS [3], in that we do not require an end to be connected and we allow it to be relatively compact. With respect to an end U, the symbol ∂U will designate its relative boundary (as a set) as well as its oriented border (as a cycle).

1 B. Subboundary. The ideal boundary of an open Riemann surface W will be denoted by β. It can be realized as a set under the Kerékjártó-Stoïlow compactification (AHLFORS-SARIO [1, pp. 81 ff.], and CONSTANTINESCU-CORNEA [1]). However, we shall seldom make use of this realization of β.

By a subboundary γ of W we shall mean a closed subset of β under the above compactification. Equivalently, γ is a part of a regular partition (Ahlfors-Sario [1, pp. 87ff.]). It may also be introduced directly as follows:

Definition. *A subboundary γ is a mapping from the family of all regular subregions Ω of W into the family of ends such that* (a) *the image U_Ω of Ω under γ is a union of some components of $W-\bar{\Omega}$ and* (b) *if $\bar{\Omega}_1 \subset \Omega_2$, then every component of U_{Ω_1} contains a component of U_{Ω_2}.*

The proof of the equivalence of these three definitions will be left to the reader.

With respect to a subboundary γ, each U_Ω will be called a *boundary neighborhood of* γ. If there is no ambiguity, we shall use the simpler term *neighborhood of* γ.

The ideal boundary β may be regarded as a subboundary determined by $U_\Omega = W-\bar{\Omega}$. A neighborhood of β is therefore an end with compact complement.

A subboundary such that U_Ω is connected for every Ω will be called an *ideal boundary component*.

Let U be a relatively non-compact end of W. We denote by β_U the subboundary of W determined by U as follows: for every regular subregion Ω of W, the image U_Ω under β_U is the union of all components R of $W-\bar{\Omega}$ such that $R \cap U$ is not void and is relatively non-compact.

Sometimes we shall consider a connected end U itself as an open Riemann surface. In this case, the ideal boundary of U is to be strictly distinguished from β_U; the former is the "union" of β_U and ∂U.

Now suppose an open Riemann surface W is a proper subregion of another (closed or open) Riemann surface W^*. Denote by β_0 the relative boundary of W in W^*. Let $\gamma: \Omega \to U_\Omega$ be a subboundary of W. Take the closure $\bar{U}_\Omega$ in W^* and consider the set $E_\gamma = \bigcap \bar{U}_\Omega$. It is a closed subset of β_0. It may be void. If not, we shall say that *a part of the subboundary γ is realized as the set E_γ*. If γ is a boundary component, then E_γ is connected and it is a component of β_0.

If, in particular, γ has the property that there exists a U_Ω which is relatively compact in W^*, then E_γ is a compact subset of W^*. In this case, we shall say that the *subboundary γ is realized as the set E_γ*.

For example, let W be the interior of a bordered Riemann surface $\bar{W}$. The latter can be imbedded in a compact Riemann surface W^*, and every subboundary of W is realized as the union of some contours on the border of $\bar{W}$.

1C. Definition of Normal Operator. Given an end U of a Riemann surface W, consider the family $C^\omega(\alpha)$ of real-valued real-analytic functions f on $\alpha=\partial U$. Let $H(\bar{U})$ be the family of real-valued functions harmonic on $\bar{U}=U\cup\alpha$ (more precisely, on some open sets containing $\bar{U}$).

DEFINITION. *A normal operator L is a mapping from $C^\omega(\alpha)$ into $H(\bar{U})$ satisfying the following conditions:*

$$Lf=f \quad \text{on } \alpha, \tag{1}$$

$$L(c_1 f_1+c_2 f_2)=c_1 Lf_1+c_2 Lf_2, \tag{2}$$

$$L1=1, \tag{3}$$

$$Lf\geq 0 \quad \text{if } f\geq 0, \tag{4}$$

$$\int_\alpha * dLf=0. \tag{5}$$

Here c_1, c_2 are real constants, and the asterisk stands for the conjugate differential. In general (AHLFORS-SARIO [1, p. 270]),

$$*\omega=-b\,dx+a\,dy \quad \text{if} \quad \omega=a\,dx+b\,dy.$$

The conjugate of a harmonic function u will be written u^*; thus $*du=d(u^*)$. Here and in the sequel (cf. 2 A) we shall view U as a surface although it may not be connected.

We shall say, for the sake of brevity, that the operator L *acts from* α *into* U.

If U is relatively compact, we can solve the Dirichlet problem to obtain a uniquely determined function u in $H(\bar{U})$ such that $u=f$ on α. It is immediately seen that the correspondence $f\to u$ is a normal operator, and the only one in this case. On an arbitrary end U, there are several normal operators. Those discussed in 2, 3 are basic for the development of our theory.

The normal operator method was introduced in SARIO [3, 5].

1 D. Maximum Principle. It is easy to see that, under assumptions (1) and (2), conditions (3) and (4) are equivalent to the following *maximum principle for normal operators:*

$$\min_\alpha f\leq Lf\leq \max_\alpha f \quad \text{on } U.$$

As its direct application we obtain the *continuity of normal operators:* if $\lim_{n\to\infty} f_n=f$ uniformly on α, then $\lim_{n\to\infty} Lf_n=Lf$ uniformly on U.

1 E. Extension of Domains of Definition. A normal operator L defined on $C^\omega(\alpha)$ can be extended to a larger family of functions. First, if f is continuous on α, then we take a sequence of functions $f_n\in C^\omega(\alpha)$ which converge uniformly to f. By the continuity of L, we see that $\lim Lf_n$

exists and is independent of the choice of the sequence f_n. We define Lf as $\lim Lf_n$. For this extended L, relations (1)–(4) continue to hold, and so does (5) in the following slightly modified form:

$$\int_{\alpha'} *dLf=0 \qquad \text{for } \alpha'\subset U,\ \alpha'\sim\alpha,$$

where "$\sim$" stands for homology.

Second, for a complex-valued continuous function f on α, we define Lf by $L(\operatorname{Re} f)+iL(\operatorname{Im} f)$. For this L the linearity (2) is valid even for complex coefficients. We also have

$$L\bar{f}=\overline{Lf}.$$

Next, suppose an end U is decomposed into a finite number of disjoint ends $U_1, \dots, U_n$. If normal operators L_k acting from ∂U_k into U_k, $k=1, \dots, n$, are given, we define, for every $f\in C^\omega(\alpha)$, the function $u\in H(\bar{U})$ in such a way that its restriction to each U_k is $L_k f_k$. Here f_k is the restriction of f to ∂U_k. The correspondence $f\to u$ is readily seen to be a normal operator, called the *direct sum* of $L_1, \dots, L_n$.

We remark that the reverse process, *decomposition*, is not always possible. In fact, the restriction of Lf to U_k does not necessarily satisfy (5) with respect to ∂U_k.

1 F. Existence Theorem for Harmonic Functions. On an arbitrary (closed or open) Riemann surface W, consider an end U with compact complement. Given a normal operator L acting from $\alpha=\partial U$ into U, and a harmonic function (complex-valued in general) s on $\bar{U}$, we are going to construct a harmonic function p on W which behaves like s on U modulo L. More precisely, $L(p-s|\alpha)=p|\bar{U}-s$, or, as we simply write, $L(p-s)=p-s$ on U.

If p exists, then we have necessarily $\int_\alpha *ds=0$. In fact, $\int_\alpha *dp=0$ by the compactness of $W-U$, and $\int_\alpha *dL(p-s)=0$ by (5). We shall prove, conversely, that the vanishing of the flux of s is sufficient for the existence of p:

THEOREM. *If*

$$\int_\alpha *ds=0,$$

then there exists a harmonic function p on W satisfying

$$L(p-s)=p-s \qquad \text{on } U.$$

It is determined uniquely up to an additive constant.

The existence of p is assured even if W is the union of a finite number of Riemann surfaces (cf. 1 A). The uniqueness means, explicitly, that the difference of two functions p is constant on each component of W.

Remark. In SARIO [5] and RODIN-SARIO [1], the function p is called the "principal function" corresponding to s and L. In our book, however, we shall use this term in a more restricted sense (see II. 1 A).

From the uniqueness we deduce at once:

(a) $p=\text{const}$ if and only if $Ls=s$ on U.

(b) If s is real-valued, so is p (with a suitable constant added).

(c) If p_1 and p_2 correspond to $\operatorname{Re} s$ and $\operatorname{Im} s$ respectively, then $p=p_1+i\,p_2+\text{const}$.

The proof of the theorem will be given in 1 G – 1 J. A portion of it has its roots in the classical method used by SCHWARZ [1] on closed surfaces. The essence of our approach is that it gives the existence on arbitrary surfaces.

1 G. Uniqueness. We begin by establishing the uniqueness. For two solutions p, p', we have $L(p-p')=p-p'$ on U, so that $\operatorname{Re}(p-p')\leq \max_\alpha \operatorname{Re}(p-p')$, $\operatorname{Im}(p-p')\leq \max_\alpha \operatorname{Im}(p-p')$ on U. These inequalities are trivially satisfied on $W-U$ as well. Thus the harmonic functions $\operatorname{Re}(p-p')$ and $\operatorname{Im}(p-p')$ attain their maxima on W at points on α and, therefore, $p-p'=\text{const}$.

1 H. Construction. For the proof of the existence of p we may assume without loss of generality that s is real-valued. Take a regular region $\Omega\subset W$ with $W-U\subset\Omega$, and denote by $\beta(\Omega)$ its border. Let L' be the (uniquely determined) normal operator acting from $\beta(\Omega)$ into Ω.

Suppose, for a moment, that we have the function p. Then $L'p=p$ holds on α. We substitute this relation into

$$L(p-s)=p-s \tag{6}$$

to obtain $p-LL'p=s-Ls$ on U and, in particular, on $\beta(\Omega)$. Conversely, suppose there exists a continuous function h on $\beta(\Omega)$ satisfying

$$h-LL'h=s-Ls. \tag{7}$$

Then the harmonic functions $L'h$ (on Ω) and $LL'h+s-Ls$ (on U) are harmonic continuations of each other. Indeed, $L'h=LL'h+s-Ls$ holds on α trivially, and on $\beta(\Omega)$ by (7), hence on $U\cap\Omega$. The function

$$p=\begin{cases} L'h & \text{on } \Omega, \\ LL'h+s-Ls & \text{on } U \end{cases} \tag{8}$$

is thus defined on W and satisfies (6). It is the desired function.

We have reduced the existence proof to showing that there exists an h which satisfies (7) on $\beta(\Omega)$. Let us introduce the notation $\Lambda=LL'$ and

$s_0 = s - Ls$ to simplify (7):

$$h - \Lambda h = s_0. \tag{9}$$

Λ is a linear operator from the space of continuous functions on $\beta(\Omega)$ into itself. The formal solution of (9) is given by

$$h = \sum_{n=0}^{\infty} \Lambda^n s_0. \tag{10}$$

As soon as we have shown its uniform convergence on $\beta(\Omega)$, we shall conclude that (10) gives a solution of (9). In fact, by the continuity of operators (see 1 D), the interchange of Λ and Σ is legitimate, so that $\Lambda h = \Sigma_1^\infty \Lambda^n s_0 = \Sigma_0^\infty \Lambda^n s_0 - s_0 = h - s_0$.

1 I. The *q*-Lemma. In order to show the convergence of (10) we need a property of harmonic functions on Ω.

LEMMA. *There exists a positive constant* $q < 1$ *such that*

$$|u| \leq q \cdot \sup_{\Omega} |u| \qquad \text{on } \alpha$$

for any real-valued harmonic function u defined on Ω *which changes sign on* α.

For the proof, we may normalize u by $\sup_\Omega |u| = 1$. If there were no q such that $|u| \leq q$ on α for every normalized u, we would have a sequence of functions u_n and a sequence of points $z_n \in \alpha$ such that $\lim_{n\to\infty} u_n(z_n) = 1$. On taking subsequences, again denoted by $\{u_n\}$ and $\{z_n\}$, we find a harmonic function u on Ω with $u = \lim u_n$ uniformly on α, and a point $z_0 \in \alpha$ with $|u(z_0)| = 1$. By the maximum principle, $u \equiv 1$ or $u \equiv -1$. This means that, for sufficiently large n, the function u_n does not change its sign on α, a contradiction. The proof of the lemma is herewith complete.

Note that Ω need not be a regular region, nor α an analytic arc. Only the compactness of α is used.

1 J. Convergence. We begin the proof of the uniform convergence of (10) by showing that $L' \Lambda^n s_0$ changes sign on α for each $n = 0, 1, \ldots$. With respect to the harmonic function v on $\bar{\Omega} \cap \bar{U}$ which is equal to 1 on α and 0 on $\beta(\Omega)$, we have

$$\int_{\alpha+\beta(\Omega)} (\Lambda^n s_0 - L' \Lambda^n s_0) * dv = \int_{\alpha+\beta(\Omega)} v * d(\Lambda^n s_0 - L' \Lambda^n s_0).$$

Since $\Lambda^n s_0 = L' \Lambda^n s_0$ on $\beta(\Omega)$,

$$\int_\alpha (\Lambda^n s_0 - L' \Lambda^n s_0) * dv = \int_\alpha * d(\Lambda^n s_0 - L' \Lambda^n s_0).$$

The right-hand side vanishes. In fact, for $n \geq 1$, this is a property of the normal operators L and L'; for $n=0$, $\int_\alpha * ds_0 = 0$ by the assumption of the theorem, and $\int_\alpha * dL' s_0 = 0$ is again a property of L'. Thus

$$\int_\alpha (\Lambda^n s_0) * dv = \int_\alpha (L' \Lambda^n s_0) * dv. \tag{11}$$

The right-hand side is equal to $\int_\alpha (\Lambda^{n+1} s_0) * dv$ for $n=0, 1, \ldots$. Hence

$$\int_\alpha (\Lambda^n s_0) * dv = \int_\alpha s_0 * dv = \int_\alpha (s - L s) * dv = 0$$

and, again by (11),

$$\int_\alpha (L' \Lambda^n s_0) * dv = 0.$$

Since $* dv > 0$ along α, this relation implies that $L' \Lambda^n s_0$ changes sign on α for each $n=0, 1, \ldots$.

An application of the lemma to $L' \Lambda^{n-1} s_0$ gives $\max_\alpha |L' \Lambda^{n-1} s_0| \leq q \cdot \max_{\beta(\Omega)} |L' \Lambda^{n-1} s_0|$. Clearly $|\Lambda^n s_0| = |L L' \Lambda^{n-1} s_0| \leq \max_\alpha |L' \Lambda^{n-1} s_0|$ on $\beta(\Omega)$. Consequently, $|\Lambda^n s_0| \leq q \cdot \max_{\beta(\Omega)} |\Lambda^{n-1} s_0|$ and, therefore,

$$|\Lambda^n s_0| \leq q^n \cdot \max_{\beta(\Omega)} |s_0| \quad \text{on } \beta(\Omega)$$

for $n=1, 2, \ldots$. Since $q<1$ this estimate guarantees the uniform convergence of (10).

2. Operators L_0 and L_1

2A. Case of Compact Bordered Surfaces. Suppose first that W is the interior of a compact bordered Riemann surface $\overline{W}$. An end U of W is also the interior of some bordered Riemann surface (cf. remark in 1C on connectedness), whose border consists of $\alpha = \partial U$ (the relative boundary of U in W) and β_U (the subboundary of W determined by U). Set $\overline{U} = U \cup \alpha$.

Given a function $f \in C^\omega(\alpha)$, consider the family $H(\overline{U}, f) = \{u \mid u \in H(\overline{U}), u = f \text{ on } \alpha\}$. There exist uniquely determined functions

$$u_0, u_1 \in H(\overline{U}, f),$$

harmonically extendable to $\overline{U} \cup \beta_U$, which satisfy the following additional conditions:

$$* du_0 = 0 \quad \text{along } \beta_U, \tag{1}$$

$$u_1 = \text{const} \quad \text{on } \beta_U, \tag{2}$$

$$\int_\alpha * du_1 = 0. \tag{3}$$

Condition (1) means that the normal derivative of u_0 vanishes on β_U, and (3) is equivalent to $\int_{\beta_U} * du_1 = 0$.

The uniqueness is verified by Green's formula

$$D_\Omega[u] = \int_{\partial\Omega} u * d\bar{u}$$

for the Dirichlet integral

$$D_\Omega[u] = \iint_\Omega \left(\left|\frac{\partial u}{\partial x}\right|^2 + \left|\frac{\partial u}{\partial y}\right|^2 \right) dx\, dy$$

of a harmonic function u, complex-valued in general. In the present case, if u_0' and u_1' also satisfy our conditions, then $D_U[u_0 - u_0'] = \int_{\alpha+\beta_U} (u_0 - u_0') * d(u_0 - u_0') = 0$ and $D_U[u_1 - u_1'] = \int_{\alpha+\beta_U} (u_1 - u_1') * d(u_1 - u_1') = 0$, so that $u_0 = u_0'$ and $u_1 = u_1'$. The function u_0 is obtained as the restriction to $\bar{U}$ of the function $\hat{u}_0$ which is harmonic on the double $\hat{U}$ of $\bar{U} \cup \beta_U$ about β_U, symmetric about β_U, and coincides with f on α. To form u_1, let u and v be harmonic on $\bar{U} \cup \beta_U$ with $u = 0$ on β_U, $u = f$ on α, and $v = 1$ on β_U, $v = 0$ on α. Denote by c the constant $(\int_{\beta_U} * du)(\int_{\beta_U} * dv)^{-1}$, where $\int_{\beta_U} * dv = D_U[v] \neq 0$. Then $u_1 = u - c\, v$. The auxiliary functions $\hat{u}_0, u, v$ are obtained by solving the Dirichlet problem on compact bordered Riemann surfaces (cf. AHLFORS-SARIO [1, pp. 138ff.]).

The correspondences $f \to u_0$ and $f \to u_1$ will be denoted by L_0 and L_1:

$$u_0 = L_0 f, \quad u_1 = L_1 f.$$

It is easily seen that L_0 and L_1 are normal operators. Here we shall only verify 1 C.(4). If $f \geq 0$, then $\hat{u}_0 \geq 0$ is trivial and $u_0 \geq 0$. As for u_1, it suffices to show that its constant value k on β_U is non-negative. If $k < 0$, then k is the minimum value of u_1 on $\bar{U} \cup \beta_U$. Accordingly, $* du_1 \leq 0$ along β_U, which implies, together with condition (3), that $* du_1 = 0$ along β_U. Then u_1 is identical with u_0, which was seen to be non-negative, in violation of $k < 0$.

2 B. Arbitrary Ends. We proceed to extend the definitions of L_0 and L_1 to an arbitrary end. This will be suggested by properties (4) and (5) below, satisfied by u_0 and u_1 in the compact case (cf. OIKAWA [3]).

Under the assumption of 2 A on u, let h be a function of class C^1 on $\bar{U} \cup \beta_U$. Then the mixed Dirichlet integral

$$D_U[u_0, h] = \iint_U \left(\frac{\partial u_0}{\partial x} \frac{\partial \bar{h}}{\partial x} + \frac{\partial u_0}{\partial y} \frac{\partial \bar{h}}{\partial y} \right) dx\, dy$$

over U is equal to the line integral $\int \bar{h} * du_0$ along $\alpha + \beta_U$. In the present case the functions are real-valued, and $* du_0 = 0$ along β_U, so that

$$D_U[u_0, h] = \int_\alpha h * du_0. \tag{4}$$

Let ω be a closed C^1-differential on $\bar{U} \cup \beta_U$ with

$$\int_\alpha \omega = 0.$$

Then $\int_U du_1 \wedge \omega = \int_{\alpha+\beta_U} u_1 \omega$. Here the symbol $\wedge$ expresses the (exterior) product of differentials (AHLFORS-SARIO [1, p. 267]):

$$\omega_1 \wedge \omega_2 = (a_1 b_2 - a_2 b_1)\, dx\, dy$$

for $\omega_i = a_i\, dx + b_i\, dy$, $i = 1, 2$. Since $u_1 = \text{const} = k$ on β_U, we see that $\int_{\beta_U} u_1 \omega = k \int_{\beta_U} \omega = k \int_\alpha \omega = 0$. Consequently,

$$\int_U du_1 \wedge \omega = \int_\alpha u_1 \omega. \tag{5}$$

Now let U be an arbitrary end of an arbitrary Riemann surface W. Set $\alpha = \partial U$, $\bar{U} = U \cup \alpha$, and

$$HD(\bar{U}) = \{u \mid u \in H(\bar{U}),\ D_U[u] < \infty\}.$$

Let $\Gamma_c^1(\bar{U})$ be the family of closed C^1-differentials ω on $\bar{U}$ (more precisely, on some open sets containing $\bar{U}$) which have finite norm $\|\omega\| = (\int_U \omega \wedge * \bar{\omega})^{\frac{1}{2}}$.

THEOREM. *Given an $f \in C^\omega(\alpha)$, there exist uniquely determined functions u_0 and u_1 in the family $HD(\bar{U}, f) = \{u \mid u \in HD(\bar{U}),\ u = f \text{ on } \alpha\}$ such that*

$$D_U[u_0, h] = \int_\alpha h * du_0 \tag{6}$$

for every $h \in C^1(\bar{U})$ with $D_U[h] < \infty$,

$$\int_U du_1 \wedge \omega = \int_\alpha u_1 \omega \tag{7}$$

for every $\omega \in \Gamma_c^1(\bar{U})$ with $\int_\alpha \omega = 0$, and

$$\int_\alpha * du_1 = 0. \tag{8}$$

The correspondences $f \to u_0$ and $f \to u_1$ are normal operators. If U satisfies the assumption of 2A, then the functions u_0 and u_1 coincide with those constructed there.

The proof will be given in 2C–2E.

DEFINITION. *L_0 and L_1 are the normal operators $f \to u_0$ and $f \to u_1$.*

The operator L_1 will be further generalized in 3. For reasons to be given in 3A, the above L_1 will also be referred to as the *operator L_1 for* **I**, the identity partition of the ideal boundary.

2C. Construction of u_1. For the proof of the theorem it will be convenient to construct u_1 first. Exhaust the Riemann surface W by regular subregions Ω with $\alpha \subset \Omega$. The border of $U \cap \Omega$ consists of α and $U \cap \partial\Omega$, both oriented positively with respect to $U \cap \Omega$. We recall from 2A that there exists a function $u_{1\Omega}$ harmonic on $\overline{U \cap \Omega}$ with $u_{1\Omega}=f$ on α, $\int_\alpha * du_{1\Omega}=0$, and $u_{1\Omega}=\text{const}$ on $U \cap \partial\Omega$.

Let $\bar{\Omega} \subset \Omega'$. Apply relation (5) to $U \cap \Omega$ and $U \cap \Omega'$ with $\omega = * du_{1\Omega'}$:

$$D_{U\cap\Omega}[u_{1\Omega}, u_{1\Omega'}] = \int_\alpha u_{1\Omega} * du_{1\Omega'}, \qquad D_{U\cap\Omega'}[u_{1\Omega'}] = \int_\alpha u_{1\Omega'} * du_{1\Omega'}.$$

The integrals are equal, for $u_{1\Omega}=u_{1\Omega'}=f$ on α. We have

$$D_{U\cap\Omega}[u_{1\Omega}-u_{1\Omega'}] \le D_{U\cap\Omega}[u_{1\Omega}] - D_{U\cap\Omega'}[u_{1\Omega'}], \tag{9}$$

since $D_{U\cap\Omega}[u_{1\Omega'}] \le D_{U\cap\Omega'}[u_{1\Omega'}]$. As a consequence,

$$0 \le D_{U\cap\Omega'}[u_{1\Omega'}] \le D_{U\cap\Omega}[u_{1\Omega}]$$

for $\bar{\Omega} \subset \Omega'$, which implies the existence of $\lim_{\Omega\to W} D_{U\cap\Omega}[u_{1\Omega}]$. Therefore, again by (9),

$$\lim_{\substack{\Omega, \Omega' \to W \\ \bar{\Omega} \subset \Omega'}} D_{U\cap\Omega}[u_{1\Omega'} - u_{1\Omega}] = 0.$$

By a standard argument (e.g. AHLFORS-SARIO [1, p. 147]), we see that there exists a harmonic function u_1 on U with $D[u_1] < \infty$ and

$$\lim_{\Omega\to W} D_{U\cap\Omega}[u_{1\Omega} - u_1] = 0. \tag{10}$$

We use reflection about α and observe that $u_{1\Omega}$ converges to u_1 uniformly on every compact set in $\bar{U}$. Consequently, u_1 is in $HD(\bar{U}, f)$ and satisfies (7) since, by (5) and (10),

$$\int_U du_1 \wedge \omega = \lim \int_{U\cap\Omega} du_{1\Omega} \wedge \omega = \lim \int_\alpha u_{1\Omega}\, \omega = \int_\alpha u_1\, \omega.$$

Equality (8) is similarly verified, and the construction of u_1 is complete.

The function u_1 with (7) and (8) is uniquely determined. In fact, if u_1' is another such function, an application of (7) to $\omega = *d(u_1 - u_1')$ shows that $D[u_1 - u_1'] = 0$ and, therefore, $u_1 = u_1'$.

Since the correspondence $f \to u_{1\Omega}$ satisfies 1C.(1)–(5), and these properties are preserved under the limiting process, we conclude that $f \to u_1$ is a normal operator.

2D. Construction of u_0. We use the same exhaustion $\Omega \to W$ as before and consider the function $u_{0\Omega}$ harmonic on $\overline{U \cap \Omega}$ with $u_{0\Omega}=f$ on α, $* du_{0\Omega}=0$ along $U \cap \partial\Omega$. Let $\bar{\Omega} \subset \Omega'$ and apply (4) to $U \cap \Omega$ with

$u_0 = u_{0\Omega}$ and $h = u_{0\Omega'}$:

$$D_{U\cap\Omega}[u_{0\Omega'}, u_{0\Omega}] = \int_\alpha u_{0\Omega'} * du_{0\Omega}, \qquad D_{U\cap\Omega}[u_{0\Omega}] = \int_\alpha u_{0\Omega} * du_{0\Omega}. \tag{11}$$

We obtain as before

$$D_{U\cap\Omega}[u_{0\Omega'} - u_{0\Omega}] \le D_{U\cap\Omega'}[u_{0\Omega'}] - D_{U\cap\Omega}[u_{0\Omega}]. \tag{12}$$

Now $D_{U\cap\Omega}[u_{0\Omega}]$ increases with Ω. To guarantee the existence of $\lim D_{U\cap\Omega}[u_{0\Omega}]$, we have to show its boundedness. To this end we apply (4) to $U\cap\Omega$ with $h = u_1$:

$$D_{U\cap\Omega}[u_1, u_{0\Omega}] = \int_\alpha u_1 * du_{0\Omega}.$$

This is equal to $D_{U\cap\Omega}[u_{0\Omega}]$ by the second identity (11) and, therefore, $D_{U\cap\Omega}[u_1 - u_{0\Omega}] = D_{U\cap\Omega}[u_1] - D_{U\cap\Omega}[u_{0\Omega}]$. The desired boundedness follows:

$$D_{U\cap\Omega}[u_{0\Omega}] \le D_U[u_1] < \infty.$$

The existence of $\lim D_{U\cap\Omega}[u_{0\Omega}]$ assured, the reasoning parallels that in the case of u_1. We obtain a uniquely determined $u_0 \in HD(\bar{U}, f)$ with the properties

$$\lim_{\Omega\to W} D_{U\cap\Omega}[u_{0\Omega} - u_0] = 0 \tag{13}$$

and (6). The correspondence $f \to u_0$ is a normal operator.

2 E. Identity with u_0 and u_1 of 2 A. Suppose U satisfies the assumption of 2 A. We shall show that the above u_0 and u_1 coincide with those in 2 A. It suffices to verify that the functions u_0 and u_1 in $H(\bar{U}, f)$ which are harmonic on $\bar{U} \cup \beta_U$ and satisfy (1), (2), and (3) have the properties (6), (7), and (8).

First consider u_0. Every h of Theorem 2 B is easily deformed near β_U into C^1-functions h_n on $\bar{U} \cup \beta_U$ such that $D_U[h - h_n] \to 0$ and $h_n = h$ on α. By (4), identity (6) is true for every h_n, hence also for h.

Concerning u_1, every ω of Theorem 2 B can be deformed near β_U to give $\omega_n \in \Gamma_c^1(\bar{U} \cup \beta_U)$ with $\|\omega - \omega_n\| \to 0$ as $n \to \infty$ and $\omega = \omega_n$ off the vicinity of β_U. Since (7) is valid for every ω_n by (5), it also holds for ω.

The reasoning in 2 E was communicated to the authors by M. NAKAI.

3. Operator L_1 for the Canonical Partition

3 A. Definition on Bordered Surfaces. As in 2 A, let U be a relatively non-compact end of the interior W of a compact bordered Riemann surface $\bar{W}$. We decompose the subboundary β_U, realized as a part of the border of $\overline{W}$, into its contours:

$$\beta_U = \gamma_1 \cup \cdots \cup \gamma_k. \tag{1}$$

For any $f \in C^\omega(\alpha)$, there exists a function $u_1 \in H(\overline{U}, f)$ which is harmonic on $\overline{U} \cup \beta_U$ and such that

$$u_1 = \text{const} \quad \text{on } \gamma_i, \tag{2}$$

$$\int_{\gamma_i} * du_1 = 0, \tag{3}$$

for $i = 1, \ldots, k$. The constant generally depends on i.

The function can be obtained as a linear combination of harmonic measures and the function u_1 of 2 A. But we shall postpone this approach until V.1 K and give here another proof, based on Theorem 1 F.

Consider disjoint ends $U'_0, \ldots, U'_k$ of the surface U, where U'_0 is a neighborhood of α and U'_i is that of γ_i, $i = 1, \ldots, k$. Let L be the direct sum on $U' = \bigcup_{i=0}^k U'_i$ of the operators L_1 acting from $\alpha_i = \partial U'_i$ (considered in U) into U'_i, $i = 0, \ldots, k$. Denote by s the function on U' defined as follows: $s = 0$ on U'_i, $i = 1, \ldots, k$, and $s = L_1 f$ on U'_0, where L_1 acts from α into U'_0. Theorem 1 F is applicable on U to these L, U', and s, since $\int_{\partial U'} * ds = 0$. We obtain a function p harmonic on U and such that $L(p-s) = p - s$ on U'. It is harmonic on $\overline{U} \cup \beta_U$, $p = f + \text{const}$ on α, $p =$ const on γ_i, and $\int_{\gamma_i} * dp = 0$, $i = 1, \ldots, k$. On subtracting the first constant from p we obtain a function $u_1 \in H(\overline{U}, f)$ satisfying (2) and (3).

As in 2 A, it is seen that the function u_1 is uniquely determined and the correspondence $f \to u_1$ is a normal operator.

Instead of (1), we can consider, more generally, an arbitrary partition **P** of β_U (AHLFORS-SARIO [1, p. 87]). In exactly the same way, we obtain a function u_1 and a normal operator. The special case of u_1 and L_1 discussed above corresponds to the canonical partition **Q**, and the case in 2 A to the identity partition **I**.

Our next task is to generalize the operator L_1 for **Q** on an arbitrary end U.

3 B. Dividing Cycles in an End. In an arbitrary end U of a Riemann surface W, we mean by a *dividing cycle relative to* α $(=\partial U)$ a cycle in U which is, for any compact $K \subset U \cup \alpha$, homologous to a cycle in $U - K$. A closed differential on U will be called *semiexact relative to* α if it has vanishing periods along all dividing cycles relative to α. We denote by $KD(\overline{U})$ the family of harmonic functions $u \in HD(\overline{U})$ such that $*du$ is semiexact relative to α.

3 C. Operator L_1 for Q on Arbitrary Ends. Let U be an arbitrary end of an arbitrary Riemann surface.

THEOREM. *Given an $f \in C^\omega(\alpha)$, there exists a uniquely determined function u_1 in $KD(\bar{U}, f) = \{u \mid u \in KD(\bar{U}),\ u = f \text{ on } \alpha\}$ such that*

$$\int_U du_1 \wedge \omega = \int_\alpha u_1 \omega \tag{4}$$

for every $\omega \in \Gamma_c^1(\bar{U})$ semiexact relative to α.

The correspondence $f \to u_1$ is a normal operator. If U satisfies the assumption of 3 A, *then u_1 coincides with the one constructed there.*

Note that u_1 has the property $\int_\alpha * du_1 = 0$.

The proof of the theorem is completely parallel to that in 2 C and 2 E, except that one now uses canonical subregions Ω. The details are left to the reader.

DEFINITION. *L_1 for* **Q** *is the normal operator $f \to u_1$.*

4. Basic Properties of L_0 and L_1

4 A. Decomposition. For the operators L_0 and L_1 for **Q**, the decomposition in the sense of 1 E is always possible. This is a special case of (a) below.

4 B. Consistency. Let U and U' be ends with $U' \subset U$, and let $\alpha = \partial U$, $\alpha' = \partial U'$. Consider the operators $L = L_0$, L_1 for **I**, and L_1 for **Q** on both U and U'. The dependence on the end will be indicated by a subscript. We claim:

(a) *$L_U f = L_{U'}(L_U f)$ on U' for every $f \in C^\omega(\alpha)$. In the case $L = L_1$ for* **I** *it is here assumed that $\alpha \sim \alpha'$.*

(b) *If $\alpha \sim \alpha'$ and if $u \in H(\bar{U})$ satisfies $u = L_{U'} u$ on U', then $u = L_U u$ on U.*

Proof. We shall establish (a) and (b) for $L = L_0$; the other cases are analogous.

To prove (a), set $u = L_U f$. We are to show that $u = L_{U'} u$ on U'. Every $h \in C^1(\bar{U}')$ with $D_{U'}[h] < \infty$ can be extended to a C^1-function on $\bar{U}$ which vanishes identically on a neighborhood of the subboundary $\beta_{U-U'}$ determined by the end $U - \bar{U}'$. Then $D_U[h] < \infty$ and

$$D_{U-U'}[u, h] = \int_\alpha h * du - \int_{\alpha'} h * du. \tag{1}$$

By the assumption, we have $D_U[u, h] = \int_\alpha h * du$, so that $D_{U'}[u, h] = \int_{\alpha'} h * du$. Since this is true for every h, we conclude that $u = L_{U'} u$ on U'.

As for (b), we start from an $h \in C^1(\bar{U})$ with a finite $D_U[h]$. Since $\alpha \sim \alpha'$, we have (1), and on reversing the above reasoning we obtain $u = L_U u$ on U.

4 C. Semiexactness of $*dL_0 f$. We maintain that

$$L_0 f \in KD(\bar{U})$$

for every $f \in C^\omega(\alpha)$.

For the proof, let $c \subset U$ be a dividing cycle relative to α. Exhaust W by canonical subregions Ω with $c \subset \Omega$. Since c is homologous to a cycle c' on $\partial\Omega \cap U$, we have $\int_c *dL_{0\Omega} f = \int_{c'} *dL_{0\Omega} f = 0$ and, on letting $\Omega \to W$, $\int_c *dL_0 f = 0$.

4 D. Construction by Exhaustion. We exhaust an open Riemann surface W by arbitrary subregions V which are not necessarily regular or even relatively compact.

On an end U, consider the operators $L = L_0$, L_1 for **I**, and L_1 for **Q**. If $\alpha = \partial U$ is contained in V, the same operators acting from α into $U \cap V$ will be denoted by L_V. With respect to L_1 for **Q** we require that V have the following property: a cycle in V divides W whenever it divides V. For instance, a canonical subregion satisfies this condition.

For every $f \in C^\omega(\alpha)$,

(a) $\lim_{V \to W} L_V f = L f$ *uniformly on every compact set in* $\bar{U}$,

(b) $\lim_{V \to W} D_{U \cap V}[L_V f - L f] = 0$,

(c) $\lim_{V \to W} D_{U \cap V}[L_V f] = D_U[L f]$; *as V increases, $D_{U \cap V}[L_V f]$ increases if $L = L_0$ and decreases if $L = L_1$ for* **I** *or* **Q**, *provided f is real-valued.*

Note that if every V is a regular subregion, the result is contained in 2 C and 2 D.

For an arbitrary V, the proof is parallel to that in 2 C and 2 D. We shall give its outline only in the case $L = L_1$ for **I**, and leave the details to the reader.

We use 2 B.(7) instead of 2 B.(5) employed in 2 C. For $V \subset V'$, we obtain in the same manner

$$D_{U \cap V}[u_V - u_{V'}] \le D_{U \cap V}[u_V] - D_{U \cap V'}[u_{V'}],$$

where $u_V = L_V f$ and $u_{V'} = L_{V'} f$. This implies the existence of u with $\lim_{V \to W} D_{U \cap V}[u_V - u] = 0$. It satisfies 2 B.(7) and 2 B.(8) and we have $u = L f$. For this u, we obtain (a), (b), and (c).

4 E. Behavior on the Border. We shall slightly generalize the results in 2 A and 3 A. Let W be the interior of a bordered Riemann surface $\bar{W}$, which is not necessarily compact. A part of the ideal boundary β of W is realized as the border β_0 of $\bar{W}$ (cf. 1 B). If U is an end of W, then a part of β_U is realized as a part of β_0. We denote it by δ.

For every $f \in C^\omega(\alpha)$, the functions $L_0 f$ and $L_1 f$ for **Q** *are harmonic on $\bar{U} \cup \delta$ and such that*

$$*dL_0 f=0, \quad dL_1 f=0$$

along δ.

The latter property may be restated as follows: $L_1 f = \text{const}$ on each component of δ.

With respect to L_1 for **I**, we also conclude that $L_1 f$ is harmonic on $\bar{U} \cup \delta$ and constant on δ. We shall, however, consider a more general situation in 5 F.

Proof. Let $\hat{W}$ be the double of W about β_0, and let $\hat{U}$ be that of U about δ. Denote by $\hat{\alpha}$ the relative boundary of $\hat{U}$ in $\hat{W}$; it is the double of α about δ. Take a point $z_0 \in \delta$ and a neighborhood $N \subset \hat{U}$ of z_0 symmetric about $\delta \cap N$.

As for L_0, consider a regular subregion Ω of $\hat{W}$ which contains $N \cup \hat{\alpha}$ and is symmetric about $\delta \cap N$. The set $\Omega \cap W$ is not necessarily a regular subregion of W. However, by 4 D,

$$L_0 f = \lim_{\Omega \to \hat{W}} L_{0\Omega} f$$

at every point of U, with $L_{0\Omega}$ acting from α into $\Omega \cap U$. Strictly speaking, $\overline{\Omega \cap U}$ is not a bordered surface, since the "border" may have vertices. But it is not difficult to see that the reasoning in 2 E applies, so that the result in 2 A extends to the present case. Denoting by $\hat{L}_{0\Omega}$ the operator L_0 which acts from $\hat{\alpha}$ into $\Omega \cap \hat{U}$ and by $\hat{f}$ the symmetric extension of f onto $\hat{\alpha}$, we see that $\hat{L}_{0\Omega} \hat{f} = L_{0\Omega} f$ on $\Omega \cap U$ and $*d\hat{L}_{0\Omega} f = 0$ along $\delta \cap N$. If $L_{0\hat{U}}$ stands for L_0 acting from $\hat{\alpha}$ into $\hat{U}$, then

$$L_{0\hat{U}} \hat{f} = \lim_{\Omega \to \hat{W}} \hat{L}_{0\Omega} \hat{f}$$

uniformly in N. We conclude that $L_0 f$ is harmonic on $U \cup (\delta \cap N)$ and that $*dL_0 f = 0$ along $\delta \cap N$.

For the operator L_1, we must use a different technique, since we do not know in advance the values of $L_1 f$ on δ. The reasoning, which makes use of Weyl's lemma, follows closely that in Ahlfors-Sario [1, p. 290] (also see Accola [1, p. 158]).

Extend the differential $\omega_1 = dL_1 f$ antisymmetrically to $\hat{U} - \delta$; leave it undefined on δ. For every $g \in C^2(N)$ with compact support, it will be shown that

$$\int_N \omega_1 \wedge dg = \int_N \omega_1 \wedge *dg = 0. \tag{2}$$

We assume this for a moment and continue our proof. Weyl's lemma guarantees the existence on N of a harmonic differential equal to ω_1

almost everywhere. Since ω_1 is harmonic on $N \cap U$ and was extended antisymmetrically, we conclude that it can be extended harmonically to N and that the extension vanishes along $\delta \cap N$.

To prove (2), consider the function $\tilde{g}$ which is symmetric to g about $\delta \cap N$. Then $\omega_1 \wedge d(g-\tilde{g})$ and $\omega_1 \wedge * d(g+\tilde{g})$ are symmetric and satisfy

$$\int_N \omega_1 \wedge dg = \int_{N \cap U} \omega_1 \wedge d(g-\tilde{g}), \qquad \int_N \omega_1 \wedge * dg = \int_{N \cap U} \omega_1 \wedge * d(g+\tilde{g}).$$

Extend the differentials $d(g-\tilde{g})$ and $*d(g+\tilde{g})$ to $\bar{U}$ by letting them vanish on $\bar{U}-N$. Then by the definition of L_1 in 3 C,

$$\int_{N \cap U} \omega_1 \wedge d(g-\tilde{g}) = \int_U \omega_1 \wedge d(g-\tilde{g}) = \int_\alpha f d(g-\tilde{g}) = 0,$$

$$\int_{N \cap U} \omega_1 \wedge * d(g+\tilde{g}) = \int_U \omega_1 \wedge * d(g+\tilde{g}) = \int_\alpha f * d(g+\tilde{g}) = 0.$$

5. Operator H

5 A. Operator H on a Bordered Surface. The operator H that we shall consider is, in general, *not* a normal operator in the sense of 1 C. It has been used and generalized by several authors, e. g., NEVANLINNA [2, p. 329] and CONSTANTINESCU-CORNEA [1, pp. 21 ff.].

First, as in 2 A, let U be an end of the interior W of a compact bordered Riemann surface $\bar{W}$. For any $f \in C^\omega(\alpha)$, there exists a function $u \in H(\bar{U})$ harmonic on $\bar{U} \cup \beta_U$ and such that

$$u = f \quad \text{on } \alpha, \qquad u = 0 \quad \text{on } \beta_U.$$

The function is unique, and the correspondence $f \to u$ gives an operator H from $C^\omega(\alpha)$ into $H(\bar{U})$. It has the following easily verified properties:

(a) $Hf = f$ on α,

(b) $H(c_1 f_1 + c_2 f_2) = c_1 Hf_1 + c_2 Hf_2$,

(c) $Hf \geq 0$ if $f \geq 0$,

where $f, f_1, f_2 \in C^\omega(\alpha)$ and c_1, c_2 are real constants.

We do not, in general, have $H1 = 1$, but the *maximum principle* holds in the following form:

(d) If $m_1 \leq f \leq m_2$ on α with $m_1 \leq 0 \leq m_2$, then

$$m_1 \leq Hf \leq m_2 \quad \text{on } U;$$

in particular,

$$|Hf| \leq \max_\alpha |f| \quad \text{on } U.$$

5 B. Operator H on an Arbitrary End. Now let U be an arbitrary end of an arbitrary Riemann surface, and let $\alpha = \partial U$.

THEOREM. *For every $f \in C^{\omega}(\alpha)$, there exists a uniquely determined function $u \in HD(\bar{U}, f)$ such that*

$$\int_U du \wedge \omega = \int_\alpha u\,\omega \tag{1}$$

for every $\omega \in \Gamma_c^1(\bar{U})$.

The function coincides with the one in 5 A *if U satisfies the assumption stated there.*

The proof is analogous to that for the normal operator L_1 in 2 C and 2 E; for exhausting regular subregions, we may use properties (a)–(d) instead of 1 C.(1)–(4).

DEFINITION. *The operator H is the correspondence $f \to u$.*

5 C. Basic Properties of H. We claim:

(a) *H has properties* 5 A.(a)–(d).

(b) *If* $\lim f_n = f$ *uniformly on α, then* $\lim H f_n = H f$ *uniformly on $\bar{U}$.*

(c) *The domain of definition can be extended as in* 1 E.

(d) *The decomposition is possible as in* 4 A.

(e) *The consistency condition $H_U f = H_{U'}(H_U f)$ holds as in* 4 B.(a) *without the assumption $\alpha \sim \alpha'$.*

(f) *The following consistency is valid in analogy with* 4 B.(b)*: if $\alpha \sim \alpha'$ and if $u \in H(\bar{U})$ satisfies $u = H_{U'} u$ on U', then $u = H_U u$ on U.*

(g) *As in* 4 D, $\lim H_V f = Hf$ *as $V \to W$; $D_{V \cap U}[H_V f]$ decreases as V increases.*

(h) *If $u \in H(\bar{U})$ is non-negative, then*

$$Hu \le u \quad \text{on } U.$$

Here (a) is immediate and (b), the continuity property of H, is implied by 5 A.(d). Properties (c)–(g) are proved in exactly the same manner as those of L_1 for **I**, and property (h) is easily established.

5 D. Relation between H and L_1. The function

$$u_\gamma = 1 - H1$$

will be called the *harmonic measure* of $\gamma = \beta_U$ with respect to U. From the definition of H (Theorem 5 B) we at once have $u_\gamma \in HD(\bar{U})$ and

$$\int_U du_\gamma \wedge \omega = -\int_\alpha \omega \tag{2}$$

for every $\omega \in \Gamma_c^1(\overline{U})$. In particular,

$$D[u_\gamma] = -\int_\alpha * du_\gamma. \tag{3}$$

We shall return to the properties of u_γ in IV.4 B.

In terms of the harmonic measure, a natural relation exists between H and L_1 for **I**:

THEOREM. *For every $f \in C^\omega(\alpha)$, $\alpha = \partial U$,*

$$L_1 f = Hf \quad \textit{if} \ u_\gamma = 0,$$
$$L_1 f = Hf - \Big(\int_\alpha * dHf\Big)\Big(\int_\alpha * du_\gamma\Big)^{-1} u_\gamma \quad \textit{if} \ u_\gamma \neq 0.$$

Proof. If $u_\gamma = 0$, the left-hand side of (2) vanishes. This means that every $\omega \in \Gamma_c^1(\overline{U})$ has a vanishing integral over α. Therefore, conditions (1) and 2 B.(7) are equivalent. For the same reason, $\int_\alpha * dHf = 0$, that is, condition 2 B.(8) is satisfied, and we have $H = L_1$. If $u_\gamma \neq 0$, we exhaust W by regular regions Ω. Since, as pointed out in 2 A, the relation is true on Ω, we obtain its general validity by a limiting process; note that (3) implies $\int_\alpha * du_\gamma \neq 0$.

5 E. Boundary Behavior of Hf. An analogue of 4 E will be discussed in a less restricted situation.

Let W be a proper subregion of a Riemann surface W^*. Then a part of its ideal boundary is realized as its relative boundary with respect to W^*. Consider an end U of W such that a part of β_U is realized as a subset δ of W^*.

THEOREM. *If $z_0 \in \delta$ is a regular point with respect to the Dirichlet problem on U as a subregion of W^*, then*

$$\lim_{z \to z_0,\, z \in U} Hf(z) = 0$$

for every $f \in C^\omega(\alpha)$.

Proof. The regular point z_0 possesses a barrier. In the present case there exists a neighborhood N of z_0 (in W^*) and a subharmonic function b on U satisfying the following conditions (cf. AHLFORS-SARIO [1, p. 139]): $b \equiv -1$ on $U - N$, $b < 0$ on $N \cap U$, $\lim_{z \to z_0} b(z) = 0$ ($z \in U$), and $\overline{\lim}_{z \to \zeta} b(z) < 0$ ($z \in U$) at every boundary point ζ of U (in W^*) with $\zeta \neq z_0$, $\zeta \in N$.

Exhaust W by regular subregions Ω with $\alpha \subset \Omega$ and denote by H_Ω the operator H acting from α into $\Omega \cap U$. We have

$$Hf = \lim_{\Omega \to W} H_\Omega f.$$

Suppose $f \geq 0$, and let $A = \max_\alpha f$. The function $H_\Omega f + Ab$ is non-positive at every boundary point of $U \cap \Omega$ and a fortiori at every point of $U \cap \Omega$. On letting $\Omega \to W$, we obtain

$$Hf + Ab \leq 0$$

and $\overline{\lim}\, Hf \leq 0$ as $z \to z_0$. On the other hand, $Hf \geq 0$ on U gives $\underline{\lim}_{z \to z_0} Hf \geq 0$ and, therefore, $\lim_{z \to z_0} Hf = 0$.

For an arbitrary $f \in C^\omega(\alpha)$, we set $f = f^+ - f^-$, where f^+ and f^- are non-negative continuous functions on α. On using the above result and 5 C.(c), we infer that $\lim Hf^+ = \lim Hf^- = 0$ as $z \to z_0$, that is, $\lim_{z \to z_0} Hf = 0$.

5 F. Boundary Behavior of $L_1 f$. In the same situation as above, the result in 5 D permits us to draw an analogous conclusion with respect to the operator L_1 for **I**: for every $f \in C^\omega(\alpha)$, there exists a constant c_f such that

$$\lim_{z \to z_0,\, z \in U} L_1 f(z) = c_f$$

at every regular point $z_0 \in \delta$. The constant is given by

$$c_f = -\Big(\int_\alpha * dHf\Big)\Big(\int_\alpha * du_\gamma\Big)^{-1}.$$

Note that if there is a regular point, then necessarily $u_\gamma = 1 - H1 \neq 0$ by Theorem 5 E.

Remark. Under the most general circumstances, the boundary behavior of Hf has been discussed, e.g., in terms of the Royden compactification of W, in Constantinescu-Cornea [1] and Nakai-Sario [2]. In the latter, the operator L_1 for **I** was also considered, and it was shown that behavior 2 A.(2) at the Royden boundary in turn characterizes the functions in question. The corresponding problems concerning the operators L_0 and L_1 for **Q** remain open (cf. also Kusunoki [2]).

Chapter II

Principal Functions

The purpose of the present chapter is to give some fundamentals on principal functions, to be compared later with capacity functions, our main subject. We shall again be concerned only with the operators L_0 and L_1. In 1, we consider principal functions with arbitrary singularities. A more detailed investigation will be carried out in 2 with respect to a number of special singularities. In 3, we discuss analytic differentials derived from principal functions. Further detailed study on plane regions will be postponed until Chapter VI.

1. Principal Functions Corresponding to L_0 and L_1

1 A. Principal Functions in General. Given a Riemann surface W, closed or open, a *singularity* shall mean a triple (s, E, N), where E is a compact set in W, N is the union of a finite number of regular subregions of W with disjoint closures and $E \subset N$, and s is a real-valued harmonic function on $\bar{N} - E$. Note that $W - \bar{N}$ is an end which is not necessarily connected. If no ambiguity arises, we shall simply write s for (s, E, N). For economy of notation we shall also speak of s on $N - E$, although we take integrals along ∂N.

Given a singularity, a neighborhood U of the ideal boundary β such that $\bar{U} \cap \bar{N} = \emptyset$, and a normal operator L acting from ∂U into U, we wish to construct a harmonic function p on W with singularity s such that $Lp = p$ on U. More precisely, p is to be harmonic on $W - E$, $p - s$ harmonically extendable to N, and $Lp = p$ on U. If such a function exists, then we have necessarily

$$\int_{\partial N} * ds = 0. \tag{1}$$

Indeed, $\int_{\partial N} * ds = \int_{\partial N} * dp = -\int_{\partial U} * dp = -\int_{\partial U} * dLp = 0$.

Conversely, we shall show that this is sufficient for the existence of p.

THEOREM. *If condition* (1) *is satisfied, there exists a real-valued harmonic function on* W *with singularity* s *and such that*

$$Lp=p \quad \text{on } U.$$

It is determined uniquely up to an additive constant.

Proof. We apply Theorem I.1 F to the Riemann surface $W'=W-E$. The function s is extended to $U'=U\cup(N-E)$, an end of W', by choosing $s=0$ in U. Clearly $\int_{\partial U'} * ds=0$. Let L' be the direct sum on U' of the given L on U and (the restriction to $N-E$ of) the uniquely determined normal operator acting from ∂N into N. The harmonic function p on W' such that $L'(p-s)=p-s$ on U' is what we set out to find. If there is another such function, p', then $p-p'$ is harmonic on W and satisfies $L(p-p')=p-p'$ on U. Hence the maximum of $p-p'$ over W is attained at a point on ∂U, so that $p-p'=\text{const}$.

DEFINITION. *The function* p *determined up to an additive constant is the principal function with respect to the given singularity (with vanishing flux) and the given normal operator.*

Note that the above definition is more restrictive than that in SARIO [5] and RODIN-SARIO [1] (cf. I.1 F).

We shall be interested only in the cases $L=L_0$ and $L=L_1$ for **I** or **Q**.

1 B. Remarks. We can actually say more about the uniqueness of principal functions than was stated in the theorem. Consider two singularities, (s, E, N) and (s', E, N'), such that $s-s'$ is harmonically extendable to $N\cap N'$. Then the principal functions with respect to these singularities and a normal operator on U with $\bar{U}\cap\bar{N}=\bar{U}\cap\bar{N}'=\emptyset$ coincide up to an additive constant. The proof is the same as above.

If W is a closed surface, the end U is relatively compact and, therefore, $Lp=p$ is no restriction. The principal function is characterized up to an additive constant as the harmonic function on W having the singularity s.

We can easily see that Theorem 1 A can be extended to complex-valued principal functions with a complex-valued singularity. In this case, if p_1 and p_2 correspond to $\operatorname{Re} s$ and $\operatorname{Im} s$, respectively, then p corresponding to s is $p_1+i\,p_2$, up to an additive constant. In particular, the (complex-valued) principal function with respect to a real-valued singularity is real-valued if the additive constant is suitably chosen. In our book, however, we shall seldom have occasion to use complex-valued principal functions.

1 C. Principal Functions with Respect to L_0 and L_1. In the cases $L=L_0$ or L_1, the choice of the neighborhood U of β for which the operator is given is immaterial. In fact, because of the consistency (I.4 B), the validity of $Lp=p$ on some U implies it on every U. For this reason we shall simply say that

$$Lp=p \quad \text{on } \beta.$$

Given a singularity (s, E, N), the principal functions with respect to $L=L_0$ and L_1 will be denoted by p_0 and p_1, with additive constants disregarded. The principal functions corresponding to L_1 for **I** and **Q** will be referred to as p_1 for **I** and p_1 for **Q**, or simply $p_1(\mathbf{I})$ and $p_1(\mathbf{Q})$.

Later we shall determine the additive constants *in casu*, so that the principal functions are uniquely determined.

If, in particular, W is the interior of a bordered Riemann surface $\overline{W}$, then p_0 and p_1 are the harmonic functions on W with the given singularity and with the following behavior on the border β of $\overline{W}$:

$$*dp_0=0 \quad \text{along } \beta,$$

$$p_1(\mathbf{I})=\text{const} \quad \text{on } \beta,$$

$$dp_1(\mathbf{Q})=0 \quad \text{along } \beta,$$

$$\int_\gamma *dp_1(\mathbf{Q})=0$$

for every component γ of β.

Remark. The introduction of principal functions (1 A) and the properties discussed in the sequel (Theorems 1 D, 1 F, 1 G) are due to SARIO [3, 5–12]. Historically, special cases of principal functions might be traced back to Abelian integrals in the theory of algebraic functions (cf. 2 A and 3). If one stresses the boundary behavior on open surfaces, other cases are found in the potential-theoretic approach to the theory of conformal mappings (cf. Chapter VI) by SCHOTTKY [1, 2], CECIONI [1], HILBERT [1], KOEBE [1, 4, 8, 9], and COURANT [1]. The mapping theory was further developed by GRUNSKY [1], who obtained Theorems 1 D, 1 F, 1 G with respect to particular singularities on plane regions (cf. 2 C, 2 G, and Chapter VI); see also SCHIFFER [6]. For such regions, a different approach was later introduced by GRÖTZSCH [3–5, 8] and DE POSSEL [3, 4] (cf. Chapter VI).

The importance of the theory of principal functions lies in that it gives simultaneously the existence and properties of a great variety of functions and differentials which earlier needed separate, often quite involved, treatment. This versatility is due to the fact that, within the necessary and sufficient condition (1), one can prescribe E, N, s, U, and L at will. The sets N and U need not even be connected, and L can be chosen differently in their various components.

1 D. Extremal Property. Given a (real-valued) singularity $s=(s, E, N)$ satisfying (1), let

$$HD_s = HD_s(W), \qquad KD_s = KD_s(W)$$

be the families of real-valued harmonic functions on W with singularity s, whose restrictions to $W-N$ belong to $HD(W-N)$ and $KD(W-N)$, respectively.

The principal functions p_0 and p_1 satisfy $L_\nu p_\nu = p_\nu$, $\nu=0,1$ on $W-N$. Therefore

$$p_1(\mathbf{I}) \in HD_s, \qquad p_1(\mathbf{Q}) \in KD_s, \qquad p_0 \in KD_s. \tag{2}$$

In view of I.2 B and I.3 C, we obtain

$$D_{W-N}[u, p_0] = -\int_{\partial N} u * dp_0 \tag{3}$$

for every $u \in HD(W-N)$, and

$$D_{W-N}[u, p_1] = -\int_{\partial N} p_1 * du \tag{4}$$

for every $u \in HD(W-N)$ with $\int_{\partial N} * du = 0$ if $p_1 = p_1(\mathbf{I})$, and for every $u \in KD(W-N)$ if $p_1 = p_1(\mathbf{Q})$.

This observation leads us to some extremal properties of p_0 and p_1. They will be stated in terms of auxiliary quantities which we now introduce.

Given a real-valued harmonic function u whose domain contains the closure of a neighborhood U of the ideal boundary β, we define $\int_\beta u * du$ by

$$\int_\beta u * du = D_U[u] - \int_{\partial U} u * du.$$

As is easily verified, this definition is independent of the choice of U. If W is exhausted by regular subregions Ω with $\partial U \subset \Omega$, then

$$\int_\beta u * du = \lim_{\Omega \to W} \int_{\beta(\Omega)} u * du, \tag{5}$$

where $\beta(\Omega)$ is the border of Ω.

Given a singularity (s, E, N) and a harmonic function u on $\bar{N}-E$, we set

$$C_s[u] = \int_{\partial N} (u * ds - s * du).$$

The following easily verified relation will be useful later: if $p, p' \in HD_s(W)$, then

$$C_s[p] - C_s[p'] = \int_{\partial N} (p * dp' - p' * dp). \tag{6}$$

We can now state the extremal properties of p_0 and p_1:

THEOREM. *The principal function p_0 minimizes*

$$C_s[p]+\int_\beta p*dp$$

in HD_s, as well as in KD_s, and is the unique minimizing function, up to an additive constant.

*The principal function p_1 for **I** (or **Q**) maximizes*

$$C_s[p]-\int_\beta p*dp$$

in HD_s (or KD_s, resp.), and is the unique maximizing function, up to an additive constant.

Furthermore,

$$\int_\beta p_0*dp_0=\int_\beta p_1*dp_1=0.$$

The proof, to be given in 1 E, amounts to establishing the relations

$$D_W[p-p_0]=\int_\beta p*dp+C_s[p]-C_s[p_0], \tag{7}$$

$$D_W[p-p_1]=\int_\beta p*dp-C_s[p]+C_s[p_1] \tag{8}$$

for all competing functions p.

1 E. Proof of the Theorem. If (7) and (8) have been demonstrated, then the theorem is obtained as follows. For $p=p_0$ and p_1, we have $\int_\beta p_0*dp_0=\int_\beta p_1*dp_1=0$. Moreover, $D_W[p-p_\nu]\geq 0$, and the equality holds if and only if $p-p_\nu=\text{const}$ for $\nu=0,1$.

We shall prove (8) for $p_1=p_1(\mathbf{I})$; the other cases are treated in a similar manner. For an arbitrary $p\in HD_s$, we may apply (4) with $u=p-p_1$ since $\int_{\partial N}*d(p-p_1)=0$:

$$D_{W-N}[p-p_1,p_1]=-\int_{\partial N}(p_1*dp-p_1*dp_1).$$

Then

$$\begin{aligned}D_{W-N}[p-p_1]&=D_{W-N}[p]-D_{W-N}[p_1]-2D_{W-N}[p-p_1,p_1]\\&=\int_\beta p*dp-\int_\beta p_1*dp_1-\int_{\partial N}(p*dp_1-p_1*dp)-\int_{\partial N}(p-p_1)*d(p-p_1)\\&=\int_\beta p*dp-\int_\beta p_1*dp_1-C_s[p]+C_s[p_1]-D_N[p-p_1],\end{aligned}$$

so that

$$D_W[p-p_1]=\int_\beta p*dp-\int_\beta p_1*dp_1-C_s[p]+C_s[p_1].$$

Here $\int_\beta p_1 * dp_1 = 0$ must be shown directly. In (5) for $u = p_1$, we know that $L_1 p_1 = p_1$ on $W - \Omega$, and $\int_{\beta(\Omega)} p_1 * dp_1 = -D_{W-\Omega}[p_1]$ by I.2 B.(7), so that

$$\int_\beta p_1 * dp_1 = -\lim_{\Omega \to W} D_{W-\Omega}[p_1] = 0.$$

This completes the proof of (8).

1 F. The Function $\frac{1}{2}(p_0 + p_1)$. Given the same singularity s, consider the function

$$p_2 = \tfrac{1}{2}(p_0 + p_1).$$

Corresponding to $p_1(\mathbf{I})$ and $p_1(\mathbf{Q})$, we have two functions p_2, which will be referred to as p_2 for $\mathbf{I}$ and p_2 for $\mathbf{Q}$, or simply $p_2(\mathbf{I})$ and $p_2(\mathbf{Q})$.

Clearly

$$p_2(\mathbf{I}) \in HD_s, \qquad p_2(\mathbf{Q}) \in KD_s. \tag{9}$$

From (3) and (4) we have

$$D_{W-N}[u, p_2] = -\tfrac{1}{2} \int_{\partial N} (p_1 * du + u * dp_0) \tag{10}$$

for every $u \in HD(W-N)$ with $\int_{\partial N} * du = 0$ if $p_2 = p_2(\mathbf{I})$, and for every $u \in KD(W-N)$ if $p_2 = p_2(\mathbf{Q})$. The right-hand side is equal to

$$-\tfrac{1}{2} \int_{\partial N} ((p_1 - s) * du + u * d(p_0 - s)) - \tfrac{1}{2} \int_{\partial N} (s * du + u * ds).$$

Thus

$$D_{W-N}[u, p_2] = -\tfrac{1}{2} D_N[p_1 - s, u] - \tfrac{1}{2} D_N[p_0 - s, u] - \tfrac{1}{2} \int_{\partial N} (s * du + u * ds) \tag{11}$$

for $u \in HD(W)$ if $p_2 = p_2(\mathbf{I})$, and for $u \in KD(W)$ if $p_2 = p_2(\mathbf{Q})$. Here the family $HD(W)$ consists, by definition, of harmonic functions with finite Dirichlet integrals over W, and the family $KD(W)$ of functions $u \in HD(W)$ with semiexact $* du$.

THEOREM. *The function p_2 for $\mathbf{I}$ (or $\mathbf{Q}$) minimizes*

$$\int_\beta p * dp$$

in $HD_s(W)$ (or $KD_s(W)$, resp.), and is the unique minimizing function up to an additive constant. Furthermore,

$$\int_\beta p_2 * dp_2 = \tfrac{1}{2} C_s[\tfrac{1}{2}(p_0 - p_1)].$$

The first statement will be derived from

$$D_W[p - p_2] = \int_\beta p * dp - \int_\beta p_2 * dp_2 \tag{12}$$

for every competing function p.

Proof. We shall prove (12) only for the case $p_2=p_2(\mathbf{I})$. Apply (10) with $u=p-p_2$ to the last term of

$$D_{W-N}[p-p_2]=D_{W-N}[p]-D_{W-N}[p_2]-2D_{W-N}[p-p_2,p_2].$$

Then

$$\begin{aligned}D_{W-N}[p-p_2]&=\int_\beta p*dp-\int_\beta p_2*dp_2\\&\quad-\int_{\partial N}(p*dp-p_2*dp_2-p_1*dp+p_1*dp_2-p*dp_0+p_2*dp_0)\\&=\int_\beta p*dp-\int_\beta p_2*dp_2-\int_{\partial N}(p-p_2)*d(p-p_2)\\&\quad-\int_{\partial N}(p_3*d(p-p_1)-(p-p_1)*dp_3),\end{aligned}$$

where $p_3=\frac{1}{2}(p_0-p_1)$ is harmonic on N. It follows that

$$D_{W-N}[p-p_2]=\int_\beta p*dp-\int_\beta p_2*dp_2-D_N[p-p_2],$$

which gives (12).

To evaluate $\int_\beta p_2*dp_2$, exhaust W by regular regions Ω with $\bar{N}\subset\Omega$. We have

$$\int_\beta p_2*dp_2=\lim_{\Omega\to W}\int_{\beta(\Omega)}p_2*dp_2$$

and, by I.2 B.(6), (7), and II.1 D.(6),

$$\begin{aligned}\int_{\beta(\Omega)}p_2*dp_2&=\tfrac{1}{4}\int_{\beta(\Omega)}(p_1*dp_1+2p_1*dp_0+p_0*dp_0)+\tfrac{1}{4}\int_{\beta(\Omega)}(p_0*dp_1-p_1*dp_0)\\&=-\tfrac{1}{4}(D_{W-\Omega}[p_1]+2D_{W-\Omega}[p_1,p_0]+D_{W-\Omega}[p_0])\\&\quad+\tfrac{1}{4}(C_s[p_0]-C_s[p_1]).\end{aligned}$$

On letting $\Omega\to W$, we obtain

$$\int_\beta p_2*dp_2=\tfrac{1}{2}C_s[\tfrac{1}{2}(p_0-p_1)].$$

1 G. The Function $\frac{1}{2}(p_0-p_1)$. Given a singularity s, consider the function

$$p_3=\tfrac{1}{2}(p_0-p_1).$$

Corresponding to $p_1(\mathbf{I})$ and $p_1(\mathbf{Q})$, we have two functions p_3, which will be referred to as p_3 for $\mathbf{I}$ and p_3 for $\mathbf{Q}$, or simply $p_3(\mathbf{I})$ and $p_3(\mathbf{Q})$.

The singularities cancel, so that p_3 is harmonic throughout W. Moreover,

$$p_3(\mathbf{I})\in HD(W),\qquad p_3(\mathbf{Q})\in KD(W).$$

The function p_3 is characterized by an extremal property, the statement of which would be somewhat involved in the present general situation. Its discussion will be postponed until we consider special singularities (see Theorems 2 E and 2 I).

Here we give the following "reproducing property", whose meaning will become clear later when it is applied to special singularities:

THEOREM. *The function p_3 has the property*

$$D_W[u, p_3] = -\tfrac{1}{2} C_s[u]$$

for every $u \in HD(W)$ if $p_3 = p_3(\mathbf{I})$, and for every $u \in KD(W)$ if $p_3 = p_3(\mathbf{Q})$.

In particular, by Theorem 1 F,

$$D_W[p_3] = -\tfrac{1}{2} C_s[p_3] = -\int_\beta p_2 * dp_2. \tag{13}$$

Proof. We shall consider only the case $p_3 = p_3(\mathbf{I})$. By 1 D.(3) and (4),

$$\begin{aligned} D_{W-N}[u, p_3] &= \tfrac{1}{2} D_{W-N}[u, p_0] - \tfrac{1}{2} D_{W-N}[u, p_1] = \tfrac{1}{2} \int_{\partial N} (p_1 * du - u * dp_0) \\ &= \tfrac{1}{2} \int_{\partial N} (p_1 - p_0) * du + \tfrac{1}{2} \int_{\partial N} ((p_0 - s) * du - u * d(p_0 - s)) \\ &\quad - \tfrac{1}{2} \int_{\partial N} (u * ds - s * du). \end{aligned}$$

The integral in the middle vanishes, so that

$$D_{W-N}[u, p_3] = -D_N[u, p_3] - \tfrac{1}{2} C_s[u].$$

1 H. Construction by Exhaustion. Let (s, E, N) be a singularity on an open Riemann surface W and consider the principal function p for s and the operator $L = L_0$, L_1 for $\mathbf{I}$, or L_1 for $\mathbf{Q}$. If V is a subregion of W with $\bar{N} \subset V$, let p_V be the principal function on V for the same singularity s and the same operator L. In the case of L_1 for $\mathbf{Q}$ we assume that V has the following property: a cycle in V divides W whenever it divides V. Note that V is not necessarily relatively compact nor regularly imbedded.

THEOREM.

$$\lim_{V \to W} D_V[p_V - p] = 0.$$

Since we are considering Dirichlet integrals, additive constants of p_V and p are arbitrary. If they are so chosen that $\lim (p_V - p) = 0$ at some point, the same is true on each compact set (see AHLFORS-SARIO [1, p. 147]).

Proof. The reasoning is analogous to that in I.2 C, I.2 D, and I.4 D.

Let $p=p_1(\mathbf{I})$ and suppose $V\subset V'$. Apply (8) to V with $p=p_{V'}$:

$$D_V[p_{V'}-p_V]=\int_{\beta(V)} p_{V'}*dp_{V'}-C_s[p_{V'}]+C_s[p_V],$$

where $\beta(V)$ stands for the ideal boundary of V (not the relative boundary in W). Since $\int_{\beta(V)} p_{V'}*dp_{V'}\leq\int_{\beta(V')} p_{V'}*dp_{V'}=0$,

$$D_V[p_{V'}-p_V]\leq C_s[p_V]-C_s[p_{V'}]. \tag{14}$$

Thus $C_s[p_V]$ decreases as V increases. To show that it is bounded from below, apply (8) again:

$$D_V[p-p_V]\leq C_s[p_V]-C_s[p].$$

Then $C_s[p_V]\geq C_s[p]>-\infty$, so that $\lim C_s[p_V]$ exists. In view of (14) we obtain a function $q\in HD_s(W)$ with

$$\lim_{V\to W} D_V[p_V-q]=0.$$

In the proof that $q=p+\text{const}$, we make use of the relation

$$Lf=\lim L_V f,$$

which was established in I.4 D. On $V-N$,

$$\begin{aligned}|q-L_V q|&\leq|q-L_V p_V|+|L_V p_V-L_V q|\\&=|q-p_V|+|L_V(p_V-q)|\leq|q-p_V|+\max_{\partial N}|p_V-q|\to 0\end{aligned}$$

as $V\to W$. Thus $q=Lq$, i.e., q is a principal function. We conclude that $q=p+\text{const}$, and therefore $D_V[p_V-p]\to 0$.

The same proof applies to $p=p_1(\mathbf{Q})$. In the case $p=p_0$ we have similarly

$$D_V[p_{V'}-p_V]\leq C_s[p_{V'}]-C_s[p_V]$$

instead of (14). From this the conclusion follows in an analogous manner.

2. Special Singularities

2 A. Integrals with Discontinuities Across a Cycle. On a Riemann surface W, closed or open, consider an analytic Jordan curve c. Take a doubly connected (planar) regular region N such that $c\subset N$ and the components of ∂N are separated by c. Denote by N_l and N_r the components of $N-c$ adjacent to the left and right edges of c, respectively. Let (s, c, N) be a singularity with

$$s=\begin{cases}1 & \text{on } N_l,\\ 0 & \text{on } N_r.\end{cases}$$

It satisfies $\int_{\partial N} * ds = 0$. In terms of the principal function for this s and L_1, we define the differentials

$$\sigma(c) = dp_1(\mathbf{I}), \qquad \tilde{\sigma}(c) = dp_1(\mathbf{Q}).$$

The additive constants vanish, and the differentials depend only on c, not on N.

The construction reveals that, *if c is not a dividing cycle*, the differentials have the following properties, which in turn characterize them:

(a) $\sigma(c)$ and $\tilde{\sigma}(c)$ are harmonic on W.

(b) $\sigma(c)$ and $\tilde{\sigma}(c)$ are exact on $W - c$, and the value of the indefinite integral on the left edge of c exceeds that on the right edge by 1.

(c) $L_1(\int \sigma(c)) = \int \sigma(c)$ on β, where L_1 is for **I**.
$L_1(\int \tilde{\sigma}(c)) = \int \tilde{\sigma}(c)$ on β, where L_1 is for **Q**.

Condition (b) may be replaced by the following:

(b′) $$\int_{c'} \sigma(c) = c \times c', \qquad \int_{c'} \tilde{\sigma}(c) = c \times c'$$

for every cycle $c' \subset W$. Here the cross stands for the intersection number, with $c \times c' = 1$ if c' crosses c from the left shore of c to the right (see, e.g. AHLFORS-SARIO [1, pp. 67 ff.]).

On a closed surface W, these properties are classical (see, e.g., WEYL [1]); condition (c) is void, and $\sigma(c) = \tilde{\sigma}(c)$.

The differential $\sigma(c)$ is exact if and only if c is a dividing cycle. Indeed, if W_l, W_r are the components of $W - c$ with $W_l \cap N_l = \emptyset$, $W_r \cap N_r = \emptyset$, then

$$u = \begin{cases} p_1 & \text{on } W_l, \\ p_1 + 1 & \text{on } W_r \end{cases}$$

is harmonic on W and satisfies $du = \sigma(c)$. By contrast, for a nondividing c, there exists a c' with $\int_{c'} \sigma(c) = c \times c' \neq 0$, which shows that $\sigma(c)$ is not exact. The above function u will be discussed further in V.1 H.

$\tilde{\sigma}(c) = 0$ for a dividing cycle c, as is apparent from Theorem 2 B below. If c is not a dividing cycle, $\tilde{\sigma}(c)$ is not exact; the proof is the same as for $\sigma(c)$ above.

If we use condition (b′) instead of (b), we can speak of $\sigma(c)$ and $\tilde{\sigma}(c)$ for any cycle c, not necessarily a closed analytic curve. In such a general case, however, the construction will be more cumbersome (cf. 3 D, 3 E).

2 B. Reproducing Property. As in AHLFORS-SARIO [1], denote by $\Gamma_h = \Gamma_h(W)$ the family of harmonic differentials ω with $\|\omega\| < \infty$, and by $\Gamma_{hse} = \Gamma_{hse}(W)$ the family of semiexact ω in Γ_h. The inner product is defined by $(\omega_1, \omega_2) = \int_W \omega_1 \wedge * \bar{\omega}_2$. We shall later make use of the basic relation $(*\omega_1, *\omega_2) = (\omega_1, \omega_2)$.

THEOREM. *The differentials $\sigma(c)$ and $\tilde{\sigma}(c)$ have the reproducing properties*

$$\sigma(c)\in\Gamma_h \quad \text{and} \quad (\omega, *\sigma(c))=\int_c \omega \quad \text{for all } \omega\in\Gamma_h,$$

$$*\tilde{\sigma}(c)\in\Gamma_{hse} \quad \text{and} \quad (\omega, *\tilde{\sigma}(c))=\int_c \omega \quad \text{for all } \omega\in\Gamma_{hse}.$$

They are uniquely determined by these properties.

Proof. By I.2 B.(7),

$$(\omega, *\sigma(c))_{W-N}=\int_{W-N} dp_1\wedge\omega=-\int_{\partial N} p_1\,\omega$$

$$=-\int_{\partial N}(p_1-s)\,\omega-\int_{\partial N} s\,\omega=-\int_N d(p_1-s)\wedge\omega+\int_c\omega.$$

We let N shrink to c and have $(\omega, *\sigma(c))=\int_c \omega$. To see the uniqueness, suppose σ' has the above property. On applying the relation to $\omega=*(\sigma(c)-\sigma')$, we obtain $\|\sigma(c)-\sigma'\|^2=(\omega, \sigma(c)-\sigma')=\int_c \omega-\int_c \omega=0$ and, therefore, $\sigma'=\sigma(c)$. The proof for $\tilde{\sigma}(c)$ is similar.

Remark. On open Riemann surfaces, the differentials $\sigma(c)$ and $\tilde{\sigma}(c)$ with the above property were introduced by the method of orthogonal projection in AHLFORS [3] (see also AHLFORS-SARIO [1, pp. 288 and 315] and PFLUGER [2, pp. 180ff.]). The purpose of the present section is to present the constructive approach by the normal operator method, adopted by RODIN [1] and M. MORI [1]. For a detailed discussion of the properties and applications of these differentials, the reader is referred to AHLFORS-SARIO [1] and RODIN-SARIO [1]. The existence of $\sigma(c)$ on arbitrary Riemann surfaces was first proved in SARIO [4].

2 C. Singularity $\mathrm{Re}(z-\zeta)^{-m-1}$. Let ζ be an arbitrary point of a Riemann surface and let t be a local parameter with $t(\zeta)=0$ defined in a neighborhood N of ζ. For a non-negative integer m, we consider the singularity $(s, \{\zeta\}, N)$ with

$$s(t)=\mathrm{Re}\,\frac{1}{t^{m+1}}.$$

We shall refer to this simply as *the singularity* $\mathrm{Re}(z-\zeta)^{-m-1}$.

Since $\int_{\partial N} *ds=0$, the principal functions p_0 and p_1 with this singularity exist. To achieve uniqueness, we normalize them by

$$\lim_{t\to 0}\left(p_\nu(t)-\mathrm{Re}\,\frac{1}{t^{m+1}}\right)=0, \qquad \nu=0,1. \tag{1}$$

We know from 1 B that they do not depend on N. Note, however, that they do depend on the choice of the parameter t about ζ. Whenever it is necessary to specify the point ζ, the functions p_ν will be denoted by

$$p_{\nu m}(z;\zeta), \qquad \nu=0,1.$$

In the case $\nu=1$, the distinction between **I** and **Q** will sometimes be indicated.

Let $HD_{\zeta m}=HD_{\zeta m}(W)$ and $KD_{\zeta m}=KD_{\zeta m}(W)$ be the families of functions p in HD_s and KD_s, respectively, subject to the normalization (1). For an arbitrary $p\in HD_{\zeta m}$, we obtain

$$C_s[p]=-2\pi\,\mathrm{Re}\,a_m, \tag{2}$$

where a_m is a coefficient in the expansion

$$dp+i*dp=\left(-\frac{m+1}{t^{m+2}}+\sum_{k=0}^{\infty}a_k t^k\right)dt$$

about ζ. In fact,

$$\int_{\partial N}(p*ds-s*dp)=-\int_{\partial N}(s^*\,dp+s*dp)=-\mathrm{Im}\int_{\partial N}(s+i\,s^*)\,d(p+i\,p^*)$$

$$=-\mathrm{Im}\int_{\partial N}\frac{1}{t^{m+1}}\left(\frac{-m-1}{t^{m+2}}+a_0+a_1\,t+\cdots\right)dt$$

$$=-2\pi\,\mathrm{Re}\,a_m.$$

The coefficient a_m will also be denoted by $a_m[p]$.

By Theorem 1 D, we obtain the following characterization of $p_{\nu m}(z;\zeta)$ by its extremal property:

The principal function p_0 minimizes

$$-2\pi\,\mathrm{Re}\,a_m+\int_\beta p*dp$$

in $HD_{\zeta m}$, as well as in $KD_{\zeta m}$, and is the unique minimizing function. The principal function p_1 for **I** *(or* **Q***) minimizes*

$$2\pi\,\mathrm{Re}\,a_m+\int_\beta p*dp$$

in $HD_{\zeta m}$ (or $KD_{\zeta m}$, resp.), and is the unique minimizing function.

2 D. Reproducing Properties of p_2 and p_3. Let us examine relation 1 F.(11) for the singularity $\mathrm{Re}(z-\zeta)^{-m-1}$:

$$D_{W-N}[u,p_2]=-\tfrac{1}{2}D_N[p_1-s,u]-\tfrac{1}{2}D_N[p_0-s,u]-\tfrac{1}{2}\int_{\partial N}(s*du+u*ds).$$

For

$$du+i*du=\left(\sum_{k=0}^{\infty}c_k\,t^k\right)dt$$

at ζ we have

$$\tfrac{1}{2}\int\limits_{\partial N}(s*du+u*ds)=-\tfrac{1}{2}\int\limits_{\partial N}(s^*\,du-s*du)$$

$$=\tfrac{1}{2}\operatorname{Im}\int\limits_{\partial N}(s-i\,s^*)(du+i*du)=\tfrac{1}{2}\operatorname{Im}\int\limits_{\partial N}\frac{c_0+c_1\,t+\cdots}{\bar{t}^{m+1}}\,dt.$$

If N is the disk $|t|<r$, then this integral has the value

$$\int\limits_{|t|=r}\frac{t^{m+1}}{r^{2m+2}}(c_0+c_1\,t+\cdots)\,dt=0.$$

Consequently,

$$\lim_{r\to 0} D_{W-N}[u,p_2]=0.$$

The left-hand side is the Cauchy principal value of the integral, with the disk about ζ taken with respect to the preassigned local parameter. We express this by writing

$$\underset{\zeta}{p.v.}\, D_W[u,p_2]=0.$$

Conversely, the function p_2 for **I** is characterized by $p\in HD_{\zeta m}$ and the validity of the above relation for every $u\in HD$. In fact, if there is another such function, say q, we obtain $D_W[p_2-q]=\lim_{r\to 0} D_{W-N}[p_2-q]=0$ on choosing $u=p_2-q$, so that $p_2=q$.

Next consider Theorem 1 G. For $u\in HD$, we again use the above expansion of $du+i*du$ about ζ.

The coefficient of t^m has the value $c_m=(m!)^{-1}(d^{m+1}(u+i\,u^*)/dt^{m+1})_{t=0}$ and, if $t=\xi+i\eta$,

$$\operatorname{Re} c_m=\frac{1}{m!}\left(\frac{\partial^{m+1}u}{\partial\xi^{m+1}}\right)_{t=0}.$$

We shall use for this the expression

$$\operatorname{Re} c_m=\frac{1}{m!}\left(\frac{\partial^{m+1}u}{\partial x^{m+1}}\right)_{z=\zeta}.$$

Then Theorem 1 G takes the form

$$D_W[u,p_3]=-\tfrac{1}{2}\,C_s[u]=\tfrac{1}{2}\int\limits_{\partial N}(s*du-u*ds)$$

$$=\tfrac{1}{2}\operatorname{Im}\int\limits_{\partial N}\frac{c_0+c_1\,t+\cdots}{t^{m+1}}\,dt=\pi\operatorname{Re} c_m.$$

Conversely, we see as before that this property characterizes p_3.

In summary, we have the following properties of $p_2=\frac{1}{2}(p_0+p_1)$ and $p_3=\frac{1}{2}(p_0-p_1)$ for the singularity $\operatorname{Re}(z-\zeta)^{-m-1}$:

THEOREM. *The function p_2 for* **I** *(or* **Q***) is characterized by*

$$p_2\in HD_{\zeta m} \ (or\ KD_{\zeta m},\ resp.), \qquad \underset{\zeta}{p.v.}\, D_W[u, p_2]=0$$

for every $u\in HD$ (or KD, resp.).

The function p_3 for **I** *(or* **Q***) is characterized by*

$$p_3\in HD \ (or\ KD,\ resp.), \qquad p_3(\zeta)=0,$$

$$D_W[u, p_3]=\frac{\pi}{m!}\left(\frac{\partial^{m+1} u}{\partial x^{m+1}}\right)_{z=\zeta}$$

for every $u\in HD$ (or KD, resp.).

2 E. Extremal Properties of p_2 and p_3. We shall refer to the quantity

$$S_m=S_m(\zeta)=\frac{m!}{\pi} D_W[p_3]$$

as the *span*. According as p_3 is for **I** or **Q**, it will be called the *H-span* or the *K-span*. On substituting $u=p_3$ in the last relation of the above theorem, we observe that the span can also be written as

$$S_m=\left(\frac{\partial^{m+1} p_3}{\partial x^{m+1}}\right)_{z=\zeta}=m!\operatorname{Re}\tfrac{1}{2}(a_m[p_0]-a_m[p_1]).$$

From the second half of the above theorem, we obtain:

The H-span vanishes if and only if $HD(\zeta, m)=\emptyset$.

The K-span vanishes if and only if $KD(\zeta, m)=\emptyset$.

Here $HD(\zeta, m)$ is the family of functions $u\in HD(W)$ with $u(\zeta)=0$, $(\partial^{m+1} u/\partial x^{m+1})_{z=\zeta}=1$ and $KD(\zeta, m)$ stands for $KD(W)\cap HD(\zeta, m)$.

If $S_m>0$, the function p_3/S_m belongs to these families and satisfies $D[u, p_3/S_m]=\pi/m!\, S_m=D[p_3/S_m]$ for every u in the same family. Thus

$$D\left[u-\frac{p_3}{S_m}\right]=D[u]-D\left[\frac{p_3}{S_m}\right].$$

This shows that p_3/S_m is the unique minimizing function of the Dirichlet integral in the family.

We remark that by (2), the extreme value of the functional in Theorem 1 F is now $-\pi S_m/m!$.

In summary, we have the following properties of $p_2=\frac{1}{2}(p_0+p_1)$ and $p_3=\frac{1}{2}(p_0-p_1)$ for the singularity $\operatorname{Re}(z-\zeta)^{-m-1}$:

THEOREM. *The function p_2 for* **I** *(or* **Q***) is the unique function in $HD_{\zeta m}$ (or $KD_{\zeta m}$, resp.) minimizing $\int_\beta p*dp$. The minimum value is $-\pi S_m(\zeta)/m!$.*

If the H-span (or the K-span) $S_m(\zeta)$ is positive, then $p_3/S_m(\zeta)$ for **I** *(or* **Q***) is the unique function in $HD(\zeta, m)$ (or $KD(\zeta, m)$, resp.) minimizing $D[u]$. The minimum value is $\pi/m!\, S_m(\zeta)$.*

The H-span (or the K-span) vanishes if and only if the family $HD(\zeta, m)$ (or $KD(\zeta, m)$, resp.) is void.

The following properties of the span are immediate consequences of this theorem:

(a) *The K-span is dominated by the H-span.*

(b) *If $\zeta\in V\subset W$, then the span $S_{mV}(\zeta)$ considered on V dominates $S_m(\zeta)$. With respect to the K-span, the region V is subject to the condition that every cycle in V dividing V divides W.*

(c) *$HD(W)$ (or $KD(W)$) consists entirely of constants if and only if the H-span (or the K-span, resp.) for $m=0$ vanishes at every point $\zeta\in W$ and with respect to every local parameter about ζ.*

(d) *The product of the extreme values in the theorem is a constant independent of S_m, ζ, and W:*

$$\left(\min \int_\beta p*dp\right)\cdot\left(\min D(u)\right) = -\left(\frac{\pi}{m!}\right)^2.$$

Here the extrema are taken in $HD_{\zeta m}$ and $HD(\zeta, m)$, or else in $KD_{\zeta m}$ and $KD(\zeta, m)$, respectively.

Remark 1. The span was introduced for plane regions in SCHIFFER [2] and generalized to Riemann surfaces and to *HD*- and *KD*-classes in SARIO [6].

Remark 2. In VI.2 D, we shall prove that on a planar region, *KD* consists entirely of constants if and only if the *K*-span for $m=0$ vanishes at *some* point with respect to *some* local parameter about it. It is not known to the authors if this remains true on an arbitrary Riemann surface, or for the *H*-span.

2 F. Conformally Invariant Metric. We restrict our attention to the case $m=0$ and write $S=S(\zeta)$ for $S_0(\zeta)$.

$S(\zeta)$ depends not only on the point ζ, but also on the parameter t about ζ.

In addition to the principal functions p_0 and p_1 with the singularity $\operatorname{Re} t^{-1}$, consider the principal functions p_0' and p_1' with the singularity

$\operatorname{Im} t^{-1}$. The functions $p_\nu + i\,p'_\nu$, $\nu = 0, 1$, are principal functions with the complex-valued singularity t^{-1}. Normalize them by

$$\lim_{t\to 0}\left(p_\nu + i\,p'_\nu - \frac{1}{t}\right) = 0.$$

If the parameter t is changed to $\tilde{t}(\tilde{t}(\zeta)=0)$, then the function with respect to $\tilde{t}$ is readily seen to be equal to

$$(p_\nu + i\,p'_\nu)\left(\frac{dt}{d\tilde{t}}\right)_{\tilde{t}=0}.$$

Let us express this fact symbolically by saying that $(p_\nu + i\,p'_\nu)\,d\zeta$ *is invariant*.

The functions p'_0 and p'_1 can also be regarded as the principal functions p_0 and p_1 with respect to the parameter *it* about the point ζ. An application of the second half of Theorem 2 D gives

$$D[u, p'_3] = -\pi\left(\frac{\partial u}{\partial y}\right)_{z=\zeta},$$

where the right-hand side is *considered with respect to t*. Thus

$$D[u, p_3 - i\,p'_3] = \pi\left(\frac{\partial u}{\partial x} - i\frac{\partial u}{\partial y}\right)_{z=\zeta} \tag{3}$$

and, in particular,

$$D[p_3 - i\,p'_3] = \pi\left(\frac{\partial p_3}{\partial x} - \frac{\partial p'_3}{\partial y}\right)_{z=\zeta}, \tag{4}$$

where we made use of the fact that the left-hand side is real.

In addition to the span $S(\zeta)$, we consider $S'(\zeta)$, the span with respect to the parameter *it* about the point ζ. The right-hand side of (4) is equal to $\pi(S+S')$. We write

$$\Sigma(\zeta) = \tfrac{1}{2}(S(\zeta) + S'(\zeta))$$

and call $\Sigma(\zeta)$ the *modified* (H- or K-)*span*. From the left-hand side of (4), it is not difficult to see that $\Sigma(\zeta)\,|d\zeta|^2$ is invariant, or $\sqrt{\Sigma(\zeta)}\,|d\zeta|$ is a conformal metric.

The validity of $\Sigma(\zeta)=0$ depends only on the point ζ, and *not on the parameter about it*. Moreover, as is easily seen from (3) and (4), it is equivalent to the fact that $HD(W)$ (if Σ is the modified H-span) or $KD(W)$ (if Σ is the modified K-span) consists entirely of constants. Compare this conclusion with 2 E. (c).

Remark. In VI.2 D, we shall see that, if W is a plane region, then the K-span and the modified K-span coincide, and vanish identically whenever they vanish at a point ζ.

2 G. Singularity $\log|(z-\zeta)/(z-\zeta')|$. Let ζ and ζ' be distinct points of a Riemann surface W. Let t and t' be local parameters with $t(\zeta)=0$, $t'(\zeta')=0$ defined on neighborhoods N_0 and N_0' of ζ and ζ', respectively. Assume $\bar{N}_0 \cap \bar{N}_0' = \emptyset$, and consider the singularity $(s, \{\zeta, \zeta'\}, N_0 \cup N_0')$ with

$$s = \begin{cases} \log|t| & \text{in } N_0 - \{\zeta\}, \\ -\log|t'| & \text{in } N_0' - \{\zeta'\}. \end{cases}$$

For the sake of brevity, we shall call it *the singularity* $\log|(z-\zeta)/(z-\zeta')|$.

Since $\int_{\partial N} * ds = 0$, $N = N_0 \cup N_0'$, we have principal functions p_0 and p_1 with this singularity. We normalize them at ζ by

$$\lim_{t \to 0} (p_\nu(t) - \log|t|) = 0, \qquad \nu = 0, 1, \tag{5}$$

so that they are uniquely determined. They do not depend on N or the parameter t'. They do depend on the parameter t, but a change of this parameter results in an additive constant only. The functions will also be denoted by

$$p_\nu(z; \zeta, \zeta'), \qquad \nu = 0, 1.$$

A distinction between **I** and **Q** will occasionally be necessary.

Concerning an interchange of the reference points, we note the following easily obtained relations:

$$p_\nu(z; \zeta, \zeta') = -p_\nu(z; \zeta', \zeta) + \text{const}, \tag{6}$$

$$p_\nu(z; \zeta, \zeta') = p_\nu(z; \zeta, \zeta'') - p_\nu(z; \zeta', \zeta'') + \text{const}. \tag{7}$$

By $HD_{\zeta\zeta'} = HD_{\zeta\zeta'}(W)$ and $KD_{\zeta\zeta'} = KD_{\zeta\zeta'}(W)$ we mean the families of functions in HD_s and KD_s, respectively, normalized by (5). For an arbitrary $p \in HD_{\zeta\zeta'}$, we have

$$C_s[p] = -2\pi\, p\langle\zeta'\rangle, \tag{8}$$

where

$$p\langle\zeta'\rangle = \lim_{t' \to 0} (p(t') + \log|t'|). \tag{9}$$

In fact,

$$\int_{\partial N} (p * ds - s * dp)$$

$$= \int_{\partial N_0} (p\, d \arg t - \log|t| * dp) - \int_{\partial N_0'} (p\, d \arg t' - \log|t'| * dp),$$

which in turn is equal to $-2\pi\, p\langle\zeta'\rangle$.

From Theorem 1 D, we obtain the following characterization of p_0 and p_1 by extremal properties:

The principal function p_0 minimizes

$$-2\pi\, p\langle\zeta'\rangle + \int_\beta p * dp$$

in $HD_{\zeta\zeta'}(W)$, as well as in $KD_{\zeta\zeta'}(W)$, and is the unique minimizing function. The principal function p_1 for **I** *(or* **Q***) minimizes*

$$2\pi\, p\langle\zeta'\rangle + \int_\beta p * dp$$

in $HD_{\zeta\zeta'}(W)$ (or $KD_{\zeta\zeta'}(W)$, resp.), and is the unique minimizing function.

2 H. Reproducing Properties of p_2 and p_3. To obtain the counterpart of 2 D, we remark that the present p_0 and p_1 are absolutely integrable, so that we need not consider the Cauchy principal value of the integral. We obtain

$$\int_{\partial N} (s * du + u * ds) = 2\pi\big(u(\zeta) - u(\zeta')\big),$$

$$C_s[u] = 2\pi\big(u(\zeta) - u(\zeta')\big).$$

Regarding the functions $p_2 = \frac{1}{2}(p_0 + p_1)$ and $p_3 = \frac{1}{2}(p_0 - p_1)$ for the singularity $\log|(z-\zeta)/(z-\zeta')|$, we have:

THEOREM. *The function p_2 for* **I** *(or* **Q***) is characterized by*

$$p_2 \in HD_{\zeta\zeta'} \ (\text{or } KD_{\zeta\zeta'}, \text{ resp.}), \qquad D_W[u, p_2] = \pi\big(u(\zeta') - u(\zeta)\big)$$

for every $u \in HD$ (or KD, resp.).

The function p_3 for **I** *(or* **Q***) is characterized by*

$$p_3 \in HD \ (\text{or } KD, \text{ resp.}), \qquad p_3(\zeta) = 0, \qquad D_W[u, p_3] = \pi\big(u(\zeta') - u(\zeta)\big)$$

for every $u \in HD$ (or KD, resp.).

2 I. Extremal Properties and the Span. We introduce the quantity

$$S(\zeta, \zeta') = \frac{1}{\pi} D[p_3] = p_3(\zeta') = \frac{1}{2}\big(p_0\langle\zeta'\rangle - p_1\langle\zeta'\rangle\big).$$

It does not depend on the choice of the parameters about ζ and ζ'. It is symmetric:

$$S(\zeta, \zeta') = S(\zeta', \zeta).$$

$HD(W)$ (or $KD(W)$) consists only of constants if and only if $S(\zeta, \zeta')$ for **I** *(or* **Q***, resp.) vanishes for every pair ζ, ζ'.*

Again it will be shown in VI.2 D that, for a plane region, the above is valid if $S(\zeta, \zeta')=0$ for a single pair ζ, ζ'.

The functions $p_2=\frac{1}{2}(p_0+p_1)$ and $p_3=\frac{1}{2}(p_0-p_1)$ for the singularity $\log|(z-\zeta)/(z-\zeta')|$ have the following extremal properties:

THEOREM. *p_2 for* **I** *(or* **Q***) is the unique function in $HD_{\zeta\zeta'}$ (or $KD_{\zeta\zeta'}$, resp.) minimizing $\int_\beta p * dp$; the minimum value is $-\pi S(\zeta, \zeta')$.*

If $S(\zeta, \zeta')$ for **I** *(or* **Q***) is positive, then $p_3/S(\zeta, \zeta')$ for* **I** *(or* **Q***, resp.) is the unique function in $HD(\zeta, \zeta')$ (or $KD(\zeta, \zeta')$, resp.) minimizing $D[u]$; the minimum value is $\pi/S(\zeta, \zeta')$.*

Here $HD(\zeta, \zeta')=\{u | u \in HD(W),\ u(\zeta)=0,\ u(\zeta')=1\}$ and $KD(\zeta, \zeta')=KD(W) \cap HD(\zeta, \zeta')$. The proof is analogous to that of Theorem 2 E.

3. Reproducing Analytic Differentials

3 A. Preliminaries. We shall consider a singularity (s, E, N) such that s is real-valued and s^* is single-valued in $N-E$. Since $\int_{\partial N} * ds=0$ and $\int_{\partial N} * ds^*=0$, we can consider principal functions p_1 and p_1' with singularities s and s^*, respectively, and, for both functions, the same normal operator L_1 for **I** or **Q**. (The function p_1' occurred already in 2 F.)

The functions

$$p_1+i p_1', \quad p_1-i p_1'$$

are complex-valued principal functions with complex-valued singularities $(s+i s^*, E, N)$ and $(s-i s^*, E, N)$, respectively.

Take the complex derivatives $\partial/\partial z=\frac{1}{2}(\partial/\partial x-i\, \partial/\partial y)$ to form the differentials

$$\varphi=\frac{\partial}{\partial z}(p_1+i p_1')\, dz, \quad \psi=\frac{\partial}{\partial z}(p_1-i p_1')\, dz.$$

The former is a meromorphic differential on W with the analytic singularity $d(s+i s^*)$ and with finite norm on $W-N$. The latter has no singularity and belongs to $\Gamma_a=\Gamma_a(W)$, the family of analytic differentials of finite norm. For any harmonic function p we have $(\partial p/\partial z)\, dz=\frac{1}{2}(dp+i * dp)$.

If the operator L_1 is for **I**, the differentials satisfy the following relations for every $\omega \in \Gamma_a$:

$$(\omega, \varphi)_{W-N}=i \int_N d((p_1-s)-i(p_1'-s^*)) \wedge \omega+i \int_{\partial N}(s-i s^*)\, \omega, \tag{1}$$

$$(\omega, \psi)_{W-N}=i \int_N d((p_1-s)+i(p_1'-s^*)) \wedge \omega+i \int_{\partial N}(s+i s^*)\, \omega. \tag{2}$$

In fact, in view of $*\varphi=-i\varphi$ and $*\omega=-i\omega$, we have

$$(\omega,\varphi)_{W-N}$$

$$=\int_{W-N}\omega\wedge*\bar{\varphi}=-i\int_{W-N}(\operatorname{Re}\varphi)\wedge\omega-\int_{W-N}*(\operatorname{Re}\varphi)\wedge\omega=-2i\int_{W-N}(\operatorname{Re}\varphi)\wedge\omega$$

$$=-i\int_{W-N}(dp_1\wedge\omega+dp_1'\wedge*\omega)=-i\int_{W-N}d(p_1-i\,p_1')\wedge\omega$$

and, by I.2 B.(5),

$$\int_{W-N}d(p_1-i\,p_1')\wedge\omega=-\int_{\partial N}(p_1-i\,p_1')\,\omega$$

$$=-\int_{\partial N}((p_1-i\,p_1')-(s-i\,s^*))\,\omega-\int_{\partial N}(s-i\,s^*)\,\omega$$

$$=-\int_{N}d((p_1-i\,p_1')-(s-i\,s^*))\wedge\omega-\int_{\partial N}(s-i\,s^*)\,\omega.$$

Relation (2) is proved similarly.

If L_1 is for **Q**, then (1) and (2) are satisfied for every $\omega\in\Gamma_a$ which is semiexact on $W-\bar{N}$ relative to ∂N.

3 B. Singularity $\operatorname{Re}(z-\zeta)^{-m-1}$. We shall apply the above results to the singularity $\operatorname{Re}(z-\zeta)^{-m-1}$ (see 2 C). It will be convenient to divide p_1 by $-(m+1)$ and consider the differentials

$$\tau_m=\frac{-1}{m+1}(dp_1(\mathbf{I})+i\,dp_1'(\mathbf{I})),\quad \tilde{\tau}_m=\frac{-1}{m+1}(dp_1(\mathbf{Q})+i\,dp_1'(\mathbf{Q})).$$

The construction reveals that they have the following properties, which in turn characterize them:

(a) τ_m and $\tilde{\tau}_m$ are harmonic and exact on $W-\{\zeta\}$.

(b) τ_m and $\tilde{\tau}_m$ have the singularity $t^{-m-2}\,dt$ at ζ, where t is the preassigned local parameter.

(c) $L_1(\int\tau_m)=\int\tau_m$ on β, where L_1 is for **I**,
$L_1(\int\tilde{\tau}_m)=\int\tilde{\tau}_m$ on β, where L_1 is for **Q**.

Set

$$\varphi_m=\frac{1}{2}(\tau_m+i*\tau_m)=\frac{-1}{m+1}\frac{\partial}{\partial z}(p_1+i\,p_1')\,dz,$$

$$\psi_m=\frac{1}{2}(\tilde{\tau}_m+i*\tilde{\tau}_m)=\frac{-1}{m+1}\frac{\partial}{\partial z}(p_1-i\,p_1')\,dz,$$

and similarly for $\tilde{\varphi}_m$, $\tilde{\psi}_m$. The differentials φ_m, $\tilde{\varphi}_m$ are meromorphic and possess the singularity $t^{-m-2}\,dt$ at ζ, whereas ψ_m, $\tilde{\psi}_m$ are analytic throughout W.

If W is closed, the above properties are classical (see, e.g., WEYL [1]); condition (c) is void, and $\varphi_m = \tilde{\varphi}_m$, $\psi_m = \tilde{\psi}_m$.

The differentials depend on the local parameter t about ζ. If t is changed to $\tilde{t}$, then φ_m with respect to the latter is equal to

$$\varphi_m \left(\frac{dt}{d\tilde{t}}\right)_{\tilde{t}=0}^{m+1}.$$

We shall express this simply by saying that $\varphi_m\, d\zeta^{m+1}$ is invariant. Similarly, $\psi_m\, d\bar{\zeta}^{m+1}$ is invariant. Regarding $\tilde{\varphi}_m$ and $\tilde{\psi}_m$, the manner of dependence is the same.

3 C. Reproducing Properties. The differential φ_m is analytic in $W - \{\zeta\}$, has finite norm on $W - N$, and possesses the singularity $t^{-m-2}\,dt$ at ζ, where t is the local parameter under consideration. In general, for a φ with these properties and for $\omega \in \Gamma_a$, we mean by $p.v.(\omega, \varphi)$ the Cauchy principal value $\lim_{r\to 0}(\varphi, \omega)_{W-N}$ with $N = \{t \,|\, |t| < r\}$.

We introduce the following notation. Given a point ζ, a local parameter t about ζ with $t(\zeta) = 0$, and a differential $\omega = a\,dz \in \Gamma_a$, the expression

$$\left(\frac{d^k a(z)}{dz^k}\right)_{z=\zeta} \quad \text{or simply} \quad \frac{d^k a(\zeta)}{d\zeta^k},$$

$k = 0, 1, 2, \ldots$, stands for the k-th derivative of $a(t)$ at the origion, where $a(t)$ is defined by $\omega = a(t)\,dt$ about ζ.

THEOREM. *φ_m is characterized as the unique differential with the above properties and such that*

$$\underset{\zeta}{p.\,v.}(\omega, \varphi_m) = 0$$

for every $\omega \in \Gamma_a$.

ψ_m is characterized as the unique differential in Γ_a such that

$$(\omega, \psi_m) = \frac{2\pi}{(m+1)!}\,\frac{d^m a(\zeta)}{d\zeta^m}$$

for every $\omega = a\,dz \in \Gamma_a$.

Corresponding characterizations of $\tilde{\varphi}_m$ and $\tilde{\psi}_m$ are valid with the additional requirements of semiexactness of $\tilde{\varphi}_m$ on $W - \{\zeta\}$ and of $\tilde{\psi}_m$ and ω on W.

Proof. Apply 3 A.(1) and (2) with $\varphi = -(m+1)\varphi_m$, $\psi = -(m+1)\psi_m$, and let N shrink to ζ. If N is a disk $|t|<r$ in the t-plane, then

$$i\int_{\partial N}(s-i\,s^*)\,\omega = i\int_{|t|=r}\frac{c_0+c_1\,t+\cdots}{\bar{t}^{m+1}}\,dt$$

$$=\frac{i}{r^{2m+2}}\int_{|t|=r}t^{m+1}(c_0+c_1\,t+\cdots)\,dt=0,$$

$$i\int_{\partial N}(s+i\,s^*)\,\omega = i\int_{|t|=r}\frac{c_0+c_1\,t+\cdots}{t^{m+1}}\,dt = -2\pi\,c_m.$$

The proof of uniqueness is similar to that in 2 B.

The differentials ψ_0 and φ_0 are the *Bergman kernel* (function) and the *adjoint kernel* (function) (BERGMAN [4]).

Remark. On open Riemann surfaces, differentials $\varphi_m, \psi_m, \tilde{\varphi}_m$, and $\tilde{\psi}_m$ with this property were constructed by the method of orthogonal projection by AHLFORS [3] (see AHLFORS-SARIO [1, pp. 301, 309, and 314]). They were also investigated by KUSUNOKI [1]. The present approach using the normal operator method was adopted by RODIN [1] (see also RODIN-SARIO [1]) and M. MORI [1]. The Bergman kernel was introduced by BERGMAN [1] and BOCHNER [1] by means of an orthogonal basis of the (function) space Γ_a. The importance of the adjoint kernel was pointed out by SCHIFFER [3].

In terms of $\psi_m = a_m\,dz$ and $\tilde{\psi}_m = \tilde{a}_m\,dz$, we introduce the quantities

$$M(\zeta)=\frac{1}{2\pi}\,\|\psi_0\|^2 = a_0(\zeta), \qquad \tilde{M}(\zeta)=\frac{1}{2\pi}\,\|\tilde{\psi}_0\|^2=\tilde{a}_0(\zeta).$$

They depend on the parameter about ζ in such a way that $M(\zeta)\,|d\zeta|^2$ and $\tilde{M}(\zeta)\,|d\zeta|^2$ are invariant. Here $(\pi^{-1}M(\zeta))^{\frac{1}{2}}\,|d\zeta|$ is the *Bergman metric* (BERGMAN [4]).

The validity of $M(\zeta)=0$ or $\tilde{M}(\zeta)=0$ is independent of the choice of the parameter about ζ. By the above theorem, $\Gamma_a=\{0\}$ if and only if $M(\zeta)=0$ at every $\zeta\in W$, and $\Gamma_{ase}=\{0\}$ if and only if $\tilde{M}(\zeta)=0$ at every $\zeta\in W$, where Γ_{ase} is the family of semiexact differentials in Γ_a.

The former result was sharpened by VIRTANEN [2] as follows:

$\Gamma_a=\{0\}$ *whenever* $M(\zeta)=0$ *at a point* ζ.

Proof. If $M(\zeta)=0$, then Theorem 3 C implies that every $\omega\in\Gamma_a$ vanishes at ζ and that $\psi_0=-(\partial(p_1-i\,p_1')/\partial z)\,dz$ with the reference point at ζ vanishes identically. The latter property means that the function $f=p_1+i\,p_1'$ is analytic. It has a simple pole at ζ and is bounded outside of a neighborhood of ζ. If there existed an $\omega\in\Gamma_a$ with $\omega\neq 0$, then one could find a k with $\omega=(c_k t^k+\cdots)\,dt$ and $c_k\neq 0$, so that $f^k\omega\in\Gamma_a$ would not vanish at ζ, a contradiction.

Remark. The authors do not know whether the corresponding fact is true for $\tilde{M}$. If W is plane region, we shall see in VI.2 D, 5 B that $\tilde{M}$, the K-span, and the modified K-span coincide and that $\Gamma_{ase}=\{0\}$ whenever they vanish at some ζ. It will also be seen in X.2 B, 2 G that $\Gamma_a=\{0\}$ or $\Gamma_{ase}=\{0\}$ necessarily imply that W is planar.

3 D. Differentials Associated with Chains. For the sake of simplicity, a chain c on W is assumed in the sequel to be a (closed or open) analytic arc. With a slight technical modification, however, the entire reasoning carries over to arbitrary singular chains.

Suppose first that c is a simple open arc contained in a parametric disk N, on which a local parameter t is given. Let $c=\overrightarrow{t_0' t_0}$; by this we mean that $t=t_0'$ is the initial point and t_0 is the terminal point of c. Then the singularity (s, c, N) with $s(t)=\log|(t-t_0)/(t-t_0')|$ has a single-valued conjugate s^* in $N-c$. Denote by p_{1c} and p'_{1c} the principal functions p_1 and p'_1 with respect to s and s^*, respectively, with L_1 for either **I** or **Q**.

Given an arbitrary curve c, we subdivide it into a finite number of arcs c_j satisfying the above assumption, and set

$$p_{1c}=\sum_j p_{1c_j}, \qquad p'_{1c}=\sum_j p'_{1c_j}.$$

These functions are, up to an additive constant, independent of the subdivision and local parameters. This is apparent from properties (a)–(d) below, which also show that

$$p_{1c}(z)=p_1(z;\zeta,\zeta')+\text{const}$$

if $c=\overrightarrow{\zeta'\zeta}$. Therefore, p_{1c} depends only on ζ and ζ'. The function p'_{1c}, however, depends on c joining ζ and ζ'.

Consider the differentials

$$\tau(c)=dp_{1c}(\mathbf{I})+i\,dp'_{1c}(\mathbf{I}), \qquad \tilde{\tau}(c)=dp_{1c}(\mathbf{Q})+i\,dp'_{1c}(\mathbf{Q}).$$

The additive constants vanish, so that the differentials depend only on c. They have the following properties, which in turn characterize them:

(a) *$\tau(c)$ and $\tilde{\tau}(c)$ are harmonic in $W-\{\zeta,\zeta'\}$, with $c=\overrightarrow{\zeta'\zeta}$.*

(b) *$\tau(c)$ and $\tilde{\tau}(c)$ have the singularities $-t^{-1}\,dt$ at ζ' and $t^{-1}\,dt$ at ζ.*

(c) *$\int_{c'}\tau(c)=\int_{c'}\tilde{\tau}(c)=2\pi i(c\times c')$ for every cycle $c'\subset W-\{\zeta,\zeta'\}$.*

(d) *$L_1(\int\tau(c))=\int\tau(c)$ on β, with L_1 for* **I**,
$L_1(\int\tilde{\tau}(c))=\int\tilde{\tau}(c)$ on β, with L_1 for **Q**.

We set

$$\varphi(c)=\tfrac{1}{2}(\tau(c)+i*\tau(c))=\frac{\partial}{\partial z}(p_{1c}+i\,p'_{1c})\,dz,$$

$$\psi(c)=\tfrac{1}{2}(\overline{\tau(c)}+i*\overline{\tau(c)})=\frac{\partial}{\partial z}(p_{1c}-i\,p'_{1c})\,dz,$$

and define $\tilde{\varphi}(c)$, $\tilde{\psi}(c)$ similarly. Here $\varphi(c)$ and $\tilde{\varphi}(c)$ are meromorphic, with simple poles at ζ and ζ', whereas $\psi(c)$ and $\tilde{\psi}(c)$ are analytic throughout W.

If W is a closed surface, these differentials are classical (cf. WEYL [1]); condition (d) is void and, therefore, $\varphi(c)=\tilde{\varphi}(c)$, $\psi(c)=\tilde{\psi}(c)$.

If c is a cycle, then $p_{1c}=$ const, and

$$\varphi(c)=-\psi(c),\quad \tilde{\varphi}(c)=-\tilde{\psi}(c).$$

If c is an analytic Jordan curve, then $\tau(c)$ and $\tilde{\tau}(c)$ coincide, up to a constant factor, with the differentials introduced in 2 A. In fact,

$$\tau(c)=2\pi\, i\,\sigma(c),\quad \tilde{\tau}(c)=2\pi\, i\,\tilde{\sigma}(c).$$

Finally, if c is a dividing cycle, then $\tilde{\varphi}(c)=\tilde{\psi}(c)=0$ (cf. 2 A).

3 E. Reproducing Property. The differential $\varphi(c)$ is analytic on $W-\{\zeta,\zeta'\}$ with finite norm on $W-N$ and has a simple pole with residue $1(-1)$ at ζ (ζ', resp.). In the present case we need not consider the Cauchy principal value for the absolutely integrable $\varphi(c)$ and $\tilde{\varphi}(c)$, and can state the reproducing property as follows:

THEOREM. *$\varphi(c)$ is characterized as the unique differential with the above properties and such that*

$$(\omega,\varphi(c))=-2\pi\int_c \omega$$

for every $\omega\in\Gamma_a$.

$\psi(c)$ is characterized as the unique differential in Γ_a such that

$$(\omega,\psi(c))=2\pi\int_c \omega$$

for every $\omega\in\Gamma_a$.

Similar characterizations of $\tilde{\varphi}(c)$ and $\tilde{\psi}(c)$ are valid under the additional assumptions that $\tilde{\varphi}(c)$ is semiexact on $W-c$ and that $\tilde{\psi}(c),\omega\in\Gamma_{ase}$.

By virtue of 3 A.(1) and (2), the proof is reduced to that of

$$i\int_{\partial N}(s-i\,s^*)\,\omega=-2\pi\int_c\omega,\qquad i\int_{\partial N}(s+i\,s^*)\,\omega=2\pi\int_c\omega,$$

where N is, as at the beginning of 3 D, a parametric disk containing $c=\overrightarrow{t'_0\,t_0}$. The details are left to the reader.

These differentials again coincide with those constructed by Ahlfors by means of the method of orthogonal projection (see the references in 3 C).

3 F. Harmonic Period Reproducer. We retain the notation of 3 D.

THEOREM. *The harmonic differential* $\omega(c) = \operatorname{Re}\psi(c) = \frac{1}{2}(dp_{1c} + *dp'_{1c})$ *is characterized by the properties*

$$\omega(c) \in \Gamma_h, \qquad (\omega, \omega(c)) = \pi \int_c \omega$$

for every $\omega \in \Gamma_h$.

Proof. Without loss of generality, ω may be assumed to be real-valued. Then $*\overline{\psi(c)} = *\omega(c) - i\,\omega(c)$ and we have

$$2\int \omega \wedge *\omega(c) + 2i \int *\omega \wedge *\omega(c) = \int (\omega + i*\omega) \wedge *\overline{\psi(c)} = 2\pi \int_c (\omega + i*\omega).$$

The real parts provide us with the desired result. The uniqueness is shown exactly as in 2 B.

The differential $\omega(c)$ was first considered by AHLFORS [2] (cf. V.1 G).

If c is an analytic Jordan curve, then

$$\omega(c) = \pi * \sigma(c),$$

so that the first half of Theorem 2 B is a special case of the above result.

We remark in passing that, in Theorem 2 H, the statement concerning p_3 for **I** may be written as follows:

$$-dp_3 \in \Gamma_{he}, \qquad (du, -dp_3) = \pi \int_{\zeta'}^{\zeta} du$$

for every du on Γ_{he}, the space of exact differentials in Γ_h. In other words, $-dp_3$ for **I** with respect to the singularity $\log|(z-\zeta)/(z-\zeta')|$ plays the same role as $\omega(c)$ in the space Γ_{he}.

Chapter III

Capacity Functions

Thus far we have discussed principal functions whose singularities have vanishing total flux. We now turn to the important case, the main object of the present book, where the flux does not vanish. On certain bordered Riemann surfaces, we still have harmonic functions with given singularities and a prescribed boundary behavior. By a limiting process, we obtain the corresponding functions on arbitrary open Riemann surfaces.

Functions with a logarithmic singularity will be called capacity functions. In contrast with Green's functions, they exist on arbitrary open Riemann surfaces. They permit us to define the capacity of a subboundary and, in particular, of a boundary component. They also yield conformal mappings of plane regions onto circular or radial slit disks.

In 1, we carry out the construction of capacity functions on bordered surfaces with compact borders. The convergence proof on an arbitrary surface is given in 2, and properties of the limiting function are discussed in 3 and 4. Uniqueness problems are considered in 5. Further study of the case of planar regions will be postponed until Chapter VII.

1. Capacity Functions on Bordered Surfaces

1 A. Subboundary. A subboundary of an open Riemann surface is a closed subset of the ideal boundary in the sense of the Kerékjártó-Stoïlow compactification. Accordingly, set-theoretic concepts like "intersection" and "union", and topological notions like "isolated subboundary" are well-defined.

Without reference to the compactification, we defined in I.1 B a subboundary γ as a correspondence $\Omega \to U_\Omega$. Two subboundaries γ^i: $\Omega \to U_\Omega^i$, $i=1,2$, are said to be *disjoint*, $\gamma^1 \cap \gamma^2 = \emptyset$, if $U_\Omega^1 \cap U_\Omega^2 = \emptyset$ for some Ω. A subboundary γ^1 is *contained in* γ^2, i.e., $\gamma^1 \subset \gamma^2$, if $U_\Omega^1 \subset U_\Omega^2$ for all Ω. A subboundary γ is said to be the *union* $\gamma = \bigcup_{\lambda \in \Lambda} \gamma^\lambda$ of subboundaries γ^λ ($\lambda \in \Lambda$) if $U_\Omega = \bigcup_{\lambda \in \Lambda} U_\Omega^\lambda$ for all Ω. If $\gamma = \gamma^1 \cup \gamma^2$ and $\gamma^1 \cap \gamma^2 = \emptyset$, we shall write $\gamma^2 = \gamma - \gamma^1$. For arbitrary subboundaries, we shall not consider $\bigcup \gamma^\lambda$ and $\gamma - \gamma^1$.

A subboundary γ is *isolated* if there exists an end U such that $\gamma = \beta_U$, where β_U is the subboundary of W determined by U (I. 1 B). For example, the ideal boundary β is an isolated subboundary. If a subboundary γ is the union of (an arbitrary number of) isolated subboundaries, then γ is isolated. If γ is an isolated subboundary, then $\beta - \gamma$ is well-defined and is an isolated subboundary.

1 B. Exhaustion towards γ. Let γ be a subboundary of an open Riemann surface W, and consider subregions V such that $W - \overline{V}$ is a neighborhood of γ. A nested sequence of subregions with this property which exhausts W will be called an *exhaustion towards* γ. We shall also (e.g. in 2 B) use this term for a family of such subregions even if it is not nested or not a sequence.

Denote by $\gamma(V)$ the relative boundary ∂V of V. Observe that $\gamma(V)$ is not necessarily connected even if γ is a boundary component. In this case, however, it is possible to find an exhaustion towards γ for which every $\gamma(V)$ is connected.

V is the interior of the bordered surface $\overline{V} = V \cup \gamma(V)$; if $\gamma \neq \beta$, then $\overline{V}$ is not compact. $\gamma(V)$ is its entire border and is compact. It is regarded in a natural manner as an isolated subboundary of V and, if connected, as a boundary component.

1 C. Capacity Functions. Let W be the interior of a bordered Riemann surface $\overline{W}$ with compact border γ. Here $\overline{W}$ is not assumed to be compact. Take a point $\zeta \in W$, a neighborhood N of ζ, with $\overline{N} \subset W$, and a local parameter t on N with $t(\zeta) = 0$. The singularity $(\log|t|, \{\zeta\}, N)$ will be referred to as *the singularity* $\log|z - \zeta|$.

If $\gamma \neq \beta$, we consider also a normal operator on $\beta - \gamma$, where β is the entire ideal boundary of W viewed itself as a Riemann surface (not as an end of some surface). More precisely, we take a normal operator L on an end U with $\beta - \gamma = \beta_U$ and $\overline{U} \cap \overline{N} = \emptyset$, acting from ∂U into U.

Under these circumstances we have:

THEOREM. *There exists a harmonic function p in $\overline{W} - \{\zeta\}$ with the singularity* $\log|z - \zeta|$ *such that*

(a) $p = \text{const}$ *on* γ,

(b) $L p = p$ *on* U.

It is uniquely determined up to an additive constant.

Proof. Choose a neighborhood U_γ of γ such that $\beta_{U_\gamma} = \gamma$ and $\overline{U}_\gamma \cap \overline{U} = \overline{U}_\gamma \cap \overline{N} = \emptyset$. Then the union of U, U_γ, and $N - \{\zeta\}$ is an end of the Riemann surface $W-\{\zeta\}$, and its complement is compact. On this end we consider the direct sum L^* of the following operators: the given L on U, L_1 for **I** on U_γ, and the restriction to $N-\{\zeta\}$ of the unique normal operator acting from ∂N into N. Define the function s on this end as $s=0$ on U, $s=\log|t|$ on $N-\{\zeta\}$, and $s=\text{const}\cdot u_\gamma$ on U_γ. Here u_γ is the harmonic measure in the sense of I.5 D defined on U_γ, so that $u_\gamma=1$ on γ, and the constant is $-2\pi(\int_{\partial U_\gamma} *du_\gamma)^{-1}$. The flux of s vanishes and Theorem I.1 F applies. The resulting function p with $L^*(p-s)=p-s$ is what we set out to find.

To verify uniqueness, let p' be another function with the required properties. Then $p-p'$ is harmonic on $\overline{W}$, constant on γ, and satisfies $L(p-p')=p-p'$ on U. Moreover, $\min_{\partial U}(p-p')\le p-p'\le \max_{\partial U}(p-p')$ on U, so that the maximum or minimum on W is attained at a point on ∂U and, therefore, $p-p'=\text{const}$.

For a general singularity (s, E, N), not necessarily $\log|z-\zeta|$, the corresponding function p is obtained in the same manner provided $\int_{\partial N} *ds\neq 0$. However, we shall not consider this case.

The function furnished by the theorem is called the *capacity function* with respect to L.

1 D. Capacities. In the sequel we shall be concerned only with the operators L_0 and L_1 for **Q**. We shall never use L_1 for **I**, even if we merely write L_1.

Because of the consistency property of the operators (I.4 B), the choice of the end U for which the operator is given is immaterial. Therefore, condition (b) of Theorem 1 C may be stated, using the convention in II.1 C,

$$L_\nu p=p \quad \text{on } \beta-\gamma,$$

with $\nu=0,1$. The choice of a neighborhood N about ζ affects only an additive constant. We normalize the capacity function by

$$\lim_{t\to 0}(p(t)-\log|t|)=0 \tag{1}$$

about ζ. The normalized functions corresponding to L_0 and L_1 are called the *capacity functions* $p_{0\gamma}$ *and* $p_{1\gamma}$ on a bordered surface with compact border γ. In reference to the pole ζ, we also use the notation $p_{\nu\gamma}=p_{\nu\gamma}(z;\zeta)$, $\nu=0,1$.

The constant values on γ, now uniquely determined, are denoted by $k_{0\gamma}=k_{0\gamma}(\zeta)$ and $k_{1\gamma}=k_{1\gamma}(\zeta)$. We introduce the *capacities* $c_{0\gamma}=c_{0\gamma}(\zeta)$ and $c_{1\gamma}=c_{1\gamma}(\zeta)$ of γ with respect to ζ: $c_{\nu\gamma}=e^{-k_{\nu\gamma}}$.

All quantities above depend on the choice of the local parameter t about ζ in such a way that $e^{-p_{\nu\gamma}}|d\zeta|$ and $c_{\nu\gamma}|d\zeta|$ are invariant. Accordingly, the function $k_{\nu\gamma}-p_{\nu\gamma}$ does not depend on the parameter.

In particular, if $\overline{W}$ is compact, the border γ is regarded as the ideal boundary β of W. Condition (b) of Theorem 1 C is void, and the distinction between $\nu=0$ and $\nu=1$ is redundant. In this case we write

$$p_\beta=p_\beta(z;\zeta),\quad k_\beta=k_\beta(\zeta),\quad c_\beta=c_\beta(\zeta)$$

for $p_{\nu\gamma}$, $k_{\nu\gamma}$, and $c_{\nu\gamma}$, respectively. The function

$$g=g(z;\zeta)=k_\beta(\zeta)-p_\beta(z;\zeta)$$

is the *Green's function*, independent of the choice of the parameter about the pole ζ. It is characterized by the following conditions: it is harmonic on $\overline{W}-\{\zeta\}$, has the singularity $-\log|z-\zeta|$, and vanishes on the border β. The quantity k_β is the *Robin constant.*

1 E. Basic Identities for $p_{1\gamma}$. Denote by $HD_\zeta(\overline{W})$ the family of real-valued functions p harmonic on $\overline{W}-\{\zeta\}$, with the singularity $\log|z-\zeta|$, normalized by (1), and with a finite Dirichlet integral over the complement on W of a neighborhood of ζ. We are interested in the subfamily $KD_{\zeta\gamma}(\overline{W})$ of functions $p\in HD_\zeta(\overline{W})$ such that $\int_c *dp=0$ for every dividing cycle $c\subset W-\{\zeta\}$ which does not separate γ from ζ. The latter condition means that every point on γ can be joined to ζ by an arc in $\overline{W}-c$.

The capacity function $p_{1\gamma}$ belongs to $KD_{\zeta\gamma}(\overline{W})$. It is readily seen to be the unique function in $KD_{\zeta\gamma}(\overline{W})$ (also in $HD_\zeta(\overline{W})$) such that $p_{1\gamma}=\text{const}$ on γ and $L_1\,p_{1\gamma}=p_{1\gamma}$ on $\beta-\gamma$.

LEMMA. *The capacity function $p_{1\gamma}$ satisfies the identities*

$$\int_\beta p_{1\gamma}*dp_{1\gamma}=\int_\gamma p_{1\gamma}*dp_{1\gamma}=2\pi\,k_{1\gamma},\tag{2}$$

$$D_W[p-p_{1\gamma}]=\int_\beta p*dp-\int_\beta p_{1\gamma}*dp_{1\gamma},\tag{3}$$

for every $p\in KD_{\zeta\gamma}(\overline{W})$.

Here the notation $\int_\beta$ is used in the sense of II.1 D.

Proof. We shall first prove that

$$D_W[u,p_{1\gamma}]=0\tag{4}$$

for every u in the family $KD(\overline{W})$ of functions $u\in HD(\overline{W})$ with $*du$ semiexact relative to γ (cf. I.3 B). Note that $|\operatorname{grad} p_{1\gamma}|$ is integrable. Let N be a neighborhood of ζ, and U a neighborhood of $\beta-\gamma$ disjoint from a neighborhood of γ and such that $\overline{U}\cap\overline{N}=\emptyset$. By property I.3 C.(4) of the operator L_1, we have

$$D_U[u,p_{1\gamma}]=\int_{\partial U}p_{1\gamma}*du$$

and, by Green's formula,

$$D_{W-U-N}[u, p_{1\gamma}] = \int_{\gamma-\partial U-\partial N} p_{1\gamma} * du .$$

Since $\int_\gamma p_{1\gamma} * du = k_{1\gamma} \int_\gamma * du = 0$ and $\int_{\partial N} p_{1\gamma} * du \to 0$ as $N \to \zeta$, we obtain (4).

We let $N \to \zeta$ in the identity

$$\begin{aligned} D_{W-N}[p-p_{1\gamma}] &= D_{W-N}[p] - D_{W-N}[p_{1\gamma}] - 2D_{W-N}[p-p_{1\gamma}, p_{1\gamma}] \\ &= \int_\beta p * dp - \int_\beta p_{1\gamma} * dp_{1\gamma} - \int_{\partial N} (p * dp - p_{1\gamma} * dp_{1\gamma}) - 2D_{W-N}[p-p_{1\gamma}, p_{1\gamma}]. \end{aligned}$$

The third integral converges to zero because of condition (1), while the last term converges to 0 by (4) with $u = p - p_{1\gamma}$. Thus (3) holds.

Next observe that $\int_\gamma p_{1\gamma} * dp_{1\gamma} = k_{1\gamma} \int_\gamma * dp_{1\gamma} = k_{1\gamma} \int_{\partial N + \partial U} * dp_{1\gamma}$, and $\int_{\partial N} * dp_{1\gamma} = 2\pi$. Since $L_1 p_{1\gamma} = p_{1\gamma}$ on U, we have $\int_{\partial U} * dp_{1\gamma} = 0$. Hence $\int_\gamma p_{1\gamma} * dp_{1\gamma} = 2\pi k_{1\gamma}$, which is the latter half of (2).

Finally, let U_γ be a neighborhood of γ such that $\bar{U}_\gamma \cap \bar{U} = \emptyset$ and $\bar{U}_\gamma \cap \bar{N} = \emptyset$. We recall the definition of $\int_\beta p_{1\gamma} * dp_{1\gamma}$ in II.1 D:

$$\int_\beta p_{1\gamma} * dp_{1\gamma} = D_U[p_{1\gamma}] + D_{U_\gamma}[p_{1\gamma}] - \int_{\partial U} p_{1\gamma} * dp_{1\gamma} - \int_{\partial U_\gamma} p_{1\gamma} * dp_{1\gamma}.$$

Since $p_{1\gamma}$ is harmonic on $U_\gamma \cup \gamma$,

$$D_{U_\gamma}[p_{1\gamma}] = \int_{\partial U_\gamma} p_{1\gamma} * dp_{1\gamma} + \int_\gamma p_{1\gamma} * dp_{1\gamma}$$

and, since $L_1 p_{1\gamma} = p_{1\gamma}$ on U,

$$D_U[p_{1\gamma}] = \int_{\partial U} p_{1\gamma} * dp_{1\gamma}$$

by I.3 C.(4). We conclude that

$$\int_\beta p_{1\gamma} * dp_{1\gamma} = \int_\gamma p_{1\gamma} * dp_{1\gamma}.$$

The proof of (2) is herewith complete.

1 F. Basic Identities for $p_{0\gamma}$. The capacity function $p_{0\gamma}$ is the unique function in the family $KD_{\zeta\gamma}(\bar{W})$ (and in $HD_\zeta(\bar{W})$) such that $p_{0\gamma} = \text{const}$ on γ and $L_0 p_{0\gamma} = p_{0\gamma}$ on $\beta - \gamma$.

Consider also the family

$$KD^0_{\zeta\gamma}(\bar{W}) = \{p \mid p \in HD_\zeta(\bar{W}),\ L_0 p = p \text{ on } \beta - \gamma\}.$$

It is a subfamily of $KD_{\zeta\gamma}(\bar{W})$. The function $p_{0\gamma}$ is readily seen to be uniquely determined in $KD^0_{\zeta\gamma}(\bar{W})$ by the property $p_{0\gamma} = \text{const}$ on γ.

LEMMA. *Every* $p^0 \in KD^0_{\zeta\gamma}(\overline{W})$ *satisfies*

$$\int_\beta p^0 * dp^0 = \int_\gamma p^0 * dp^0. \tag{5}$$

In particular,

$$\int_\beta p_{0\gamma} * dp_{0\gamma} = \int_\gamma p_{0\gamma} * dp_{0\gamma} = 2\pi\, k_{0\gamma}. \tag{5'}$$

Moreover,

$$D_W[p-p^0] = \int_\beta p * dp - 2\int_\gamma p * dp^0 + \int_\gamma p^0 * dp^0 \tag{6}$$

for $p \in HD_\zeta(\overline{W})$, $p^0 \in KD^0_{\zeta\gamma}(\overline{W})$, *and*

$$D_W[p-p_{0\gamma}] = \int_{\beta-\gamma} p * dp - \int_\gamma p * dp + \int_\gamma p_{0\gamma} * dp_{0\gamma} \tag{7}$$

for $p \in KD_{\zeta\gamma}(\overline{W})$ *with* $p = \text{const}$ *on* γ. *Finally,*

$$D_W[p-p_{0\gamma}] = \int_\beta p * dp - \int_\beta p_{0\gamma} * dp_{0\gamma} \tag{8}$$

for every $p \in KD^0_{\zeta\gamma}(\overline{W})$.

The integral $\int_{\beta-\gamma} p * dp$ in (7) has the same meaning as in II.1 D (cf. 3 E below).

Proof. First, for $p^0 \in KD^0_{\zeta\gamma}(\overline{W})$ and $u \in HD(\overline{W})$, we have

$$D_W[u, p^0] = -2\pi\, u(\zeta) + \int_\gamma u * dp^0. \tag{9}$$

In fact, we obtain as before

$$D_{W-N}[u, p^0] = -\int_{\partial N} u * dp^0 + \int_\gamma u * dp^0,$$

which yields (9) in the limit $N \to \zeta$.

Now we use the notation of 1 E. The proofs of (5) and (5′) are then exactly the same as that of (2).

To prove (6), we start as in 1 E with

$$D_{W-N}[p-p^0] = \int_\beta p * dp - \int_\beta p^0 * dp^0 - \int_{\partial N} (p * dp - p^0 * dp^0) - 2D_{W-N}[p-p^0, p^0].$$

As $N \to \zeta$, the third integral converges to 0, and by (9) with $u = p - p^0$, which vanishes at ζ, we have

$$\lim D_{W-N}[p-p^0, p^0] = \int_\gamma p * dp^0 - \int_\gamma p^0 * dp^0.$$

Hence (6) is established.

To obtain (7), substitute $p^0 = p_{0\gamma}$ in (6):

$$D_W[p - p_{0\gamma}] = \int_\beta p * dp - 2\int_\gamma p * dp_{0\gamma} + \int_\gamma p_{0\gamma} * dp_{0\gamma}. \tag{10}$$

Clearly $\int_{\partial U} * dp = \int_{\partial U} * dp_{0\gamma} = 0$ for $p \in KD_{\zeta\gamma}(\overline{W})$, so that $\int_\gamma * dp = \int_\gamma * dp_{0\gamma} = 2\pi$. Since p is constant on γ, this gives

$$\int_\gamma p * dp_{0\gamma} = \int_\gamma p * dp. \tag{11}$$

On the other hand, as in 1 E,

$$\int_\beta p * dp = D_U[p] + D_{U_\gamma}[p] - \int_{\partial U} p * dp - \int_{\partial U_\gamma} p * dp$$

and, therefore,

$$\int_\beta p * dp - \int_\gamma p * dp = D_U[p] - \int_{\partial U} p * dp,$$

which is equal to $\int_{\beta - \gamma} p * dp$ by definition (see also 3 E). A substitution of these quantities into (10) yields (7).

To obtain (8), set $p = p_{0\gamma}$ in (6):

$$D_W[p_{0\gamma} - p^0] = \int_\beta p_{0\gamma} * dp_{0\gamma} - 2\int_\gamma p_{0\gamma} * dp^0 + \int_\gamma p^0 * dp^0.$$

It is seen that $\int_\gamma p_{0\gamma} * dp^0 = \int_\gamma p_{0\gamma} * dp_{0\gamma}$, as in (11). Using (5), we conclude that $D[p_{0\gamma} - p^0] = \int_\beta p^0 * dp^0 - \int_\beta p_{0\gamma} * dp_{0\gamma}$, which is nothing but (8).

The proof of the lemma is herewith complete.

2. Capacity Functions on Arbitrary Surfaces

2 A. Preliminaries. Let γ be a subboundary of an arbitrary open Riemann surface W. Choose a point $\zeta \in W$ and a local parameter t about it. On every subregion V of W with $\zeta \in V$ and with the property in 1 B, consider $p_{\nu\gamma(V)}$ and $k_{\nu\gamma(V)}$, $\nu = 0, 1$, introduced in 1 D.

We claim that, for $\overline{V} \subset V'$,

$$D_V[p_{\nu\gamma(V')} - p_{\nu\gamma(V)}] \leq 2\pi(k_{\nu\gamma(V')} - k_{\nu\gamma(V)}). \tag{1}$$

For the proof in the case $\nu = 0$, observe that $p_{0\gamma(V')} \in KD^0_{\zeta\gamma(V)}(\overline{V})$, and apply 1 F.(8):

$$D_V[p_{0\gamma(V')} - p_{0\gamma(V)}] = \int_{\beta(V)} p_{0\gamma(V')} * dp_{0\gamma(V')} - 2\pi\, k_{0\gamma(V)}.$$

By 1 F.(5'),

$$2\pi\, k_{0\gamma(V')} = \int_{\gamma(V')} p_{0\gamma(V')} * dp_{0\gamma(V')}.$$

On the end $V'-\bar{V}$ of V', we have $L_0\, p_{0\gamma(V')} = p_{0\gamma(V')}$, so that

$$\int_{\gamma(V')-\gamma(V)} p_{0\gamma(V')} * dp_{0\gamma(V')} = D_{V'-V}[p_{0\gamma(V')}] \geq 0$$

by I.2 C.(6). Thus

$$2\pi\, k_{0\gamma(V')} \geq \int_{\gamma(V)} p_{0\gamma(V')} * dp_{0\gamma(V')}.$$

By virtue of 1 F.(5), the right-hand side is equal to

$$\int_{\beta(V)} p_{0\gamma(V')} * dp_{0\gamma(V')},$$

and we have (1).

The proof for $\nu=1$ is obtained in a similar manner by using 1 E.(2), (3). At the final step we invoke the equality

$$\int_{\gamma(V)} p_{1\gamma(V')} * dp_{1\gamma(V')} = \int_{\beta(V)} p_{1\gamma(V')} * dp_{1\gamma(V')},$$

which is proved in exactly the same fashion as 1 E.(2).

2 B. Definitions. Exhaust W towards γ. Since $k_{\nu\gamma(V)}$ increases with V (see (1)), the limit

$$k_{\nu\gamma} = k_{\nu\gamma}(\zeta) = \lim_{V\to W} k_{\nu\gamma(V)}(\zeta)$$

exists, the case $k_{\nu\gamma}=\infty$ not excluded. The non-negative quantities

$$c_{\nu\gamma} = c_{\nu\gamma}(\zeta) = e^{-k_{\nu\gamma}}$$

will be called the *capacities* $c_{0\gamma}$ and $c_{1\gamma}$ of γ, with the reference point ζ. We shall see in 2 D that, if W satisfies the assumption of 1 C, then these capacities are positive and coincide with those in 1 D.

If $c_{\nu\gamma}>0$ (i.e., $k_{\nu\gamma}<\infty$), then a standard argument shows that (1) implies the existence of the functions

$$p_{\nu\gamma} = p_{\nu\gamma}(z;\zeta) \quad \text{with} \quad \lim_{V\to W} D_V[p_{\nu\gamma(V)} - p_{\nu\gamma}] = 0 \tag{2}$$

and, therefore,

$$\lim_{V\to W} (p_{\nu\gamma(V)} - p_{\nu\gamma}) = 0, \tag{3}$$

uniformly on every compact set in W. The functions will be called the *capacity functions* $p_{0\gamma}$ *and* $p_{1\gamma}$. It will be seen in 2 D that, if W satisfies the assumption of 1 C, then these functions coincide with those introduced in 1 D. The case $c_{\nu\gamma}=0$ will be considered in 2 E.

The limits in (2) and (3) are directed limits. This implies that (2) and (3) are valid with respect to an arbitrary exhausting *sequence* towards γ, and that the limiting function is unique, i.e., independent of the choice of the exhausting sequence.

Capacities and capacity functions depend on the choice of the parameter about ζ in such a manner that $e^{-p_{\nu\gamma}}|d\zeta|$ and $c_{\nu\gamma}|d\zeta|$ are invariant. Accordingly, the function $k_{\nu\gamma}-p_{\nu\gamma}$ is independent of the parameter.

In the particular case $\beta=\gamma$, an exhaustion towards γ is an ordinary exhaustion by regular subregions. We need not refer to $\nu=0, 1$, so that we shall write, as in 1 D,

$$k_\beta=k_\beta(\zeta), \qquad c_\beta=c_\beta(\zeta), \qquad p_\beta=p_\beta(z;\zeta)$$

for $k_{\nu\gamma}$, $c_{\nu\gamma}$, $p_{\nu\gamma}$. The function

$$g=g(z;\zeta)=k_\beta(\zeta)-p_\beta(z;\zeta),$$

defined only for a finite k_β, is the *Green's function*. It is independent of the local parameter about ζ. The quantity k_β is the *Robin constant* and is equal to $\lim_{t\to 0}(g+\log|t|)$.

Remark. Capacity functions and capacities of the ideal boundary and of an ideal boundary component of a Riemann surface were introduced in SARIO [10]. The function $p_{0\gamma}$ was considered in STREBEL [2] and, under the assumption of 1 C, in SARIO [13]; the function $p_{1\gamma}$ was discussed in SARIO [10]. Recently, far-reaching generalizations of both $p_{0\gamma}$ and $p_{1\gamma}$ were introduced in MARDEN-RODIN [1] (see also RODIN-SARIO [1]). Historically, the starting point was the classical Green's function. But if stress is placed on distinguishing a subboundary γ and the boundary behavior on $\beta-\gamma$, the basic idea, like that of principal functions (Remark in II. 1 C), had its analogues in the potential-theoretic approach to the theory of conformal mappings. Such potential functions were considered by KOEBE [8] in connection with conformal mappings onto circular or radial slit disks, and later by GRUNSKY [1], SCHIFFER [3], and LOKKI [1] (see also SCHIFFER [6]). Other approaches were made by GRÖTZSCH [3, 5], RENGEL [1, 2], SCHIFFER [1], REICH-WARSCHAWSKI [1], and REICH [2] (cf. Chapter VII). The normalization condition in 1 D, adopted by SARIO [10] to define $p_{1\gamma}$, goes back to HEINS [2] and TSUJI [3], who considered the function p_β for $k_\beta\leq\infty$ (cf. Theorem 2 E in the case $\gamma=\beta$). The significance of $p_{0\gamma}$ and $p_{1\gamma}$ is in that they yield capacities $c_{0\gamma}$ and $c_{1\gamma}$ of a single boundary component, provide related conformal mappings, and lead to a classification of boundary components and Riemann surfaces (cf. Preface).

2 C. Elementary Properties. The capacity function $p_{\nu\gamma}$ for $c_{\nu\gamma}>0$, $\nu=0, 1$, has the following properties:

(a) $p_{\nu\gamma}$ *is real-valued and harmonic on* $W-\{\zeta\}$.

(b) $p_{\nu\gamma}$ *has the singularity* $\log|z-\zeta|$ *and the normalization*

$$\lim_{t\to 0}(p_{\nu\gamma}(t)-\log|t|)=0.$$

(c) $p_{\nu\gamma}$ *satisfies*

$$L_\nu p_{\nu\gamma}=p_{\nu\gamma}$$

on every end U *which is disjoint from a neighborhood of* γ *and such that* $\zeta\notin\bar{U}$.

For the sake of simplicity, we shall express this property by

$$L_\nu p_{\nu\gamma}=p_{\nu\gamma} \quad \text{on } \beta-\gamma$$

even in the case where $\beta-\gamma$ is not a subboundary (cf. 1 A).

(d) $p_{\nu\gamma}$ *satisfies*

$$\int_c * dp_{\nu\gamma}=0$$

for every dividing cycle c *in* $W-\{\zeta\}$ *not separating* γ *from* ζ.

More precisely, the requirement for c is as follows: for every (or equivalently some) regular subregion Ω with $c\subset\Omega$, every component of U_Ω (see the definition of γ in I.1 B) can be joined to ζ by a curve in $\bar{\Omega}-c$.

Properties (a) and (b) follow directly from (3), and (c) is shown as follows. We may assume that $U\subset V$. Then

$$\begin{aligned}|p_{\nu\gamma}-L_\nu p_{\nu\gamma}| &\le |p_{\nu\gamma}-L_\nu p_{\nu\gamma(V)}|+|L_\nu p_{\nu\gamma(V)}-L_\nu p_{\nu\gamma}|\\ &\le |p_{\nu\gamma}-p_{\nu\gamma(V)}|+\max_{\partial U}|p_{\nu\gamma(V)}-p_{\nu\gamma}|\to 0\end{aligned}$$

on U as $V\to W$. Property (d) is directly implied by (c).

2 D. Isolated γ. We shall use the operator H introduced in I.5 B.

THEOREM. *If* γ *is an isolated subboundary with* $c_{\nu\gamma}>0$, *then* $p_{\nu\gamma}$, $\nu=0, 1$, *is characterized by conditions* 2 C.(a)–(c) *and the existence of a constant* k *with*

$$H(p_{\nu\gamma}-k)=p_{\nu\gamma}-k \quad \text{on } \gamma.$$

The constant k *coincides with* $k_{\nu\gamma}$.

If, in particular, W *and* γ *satisfy the assumption in* 1 C, *then the constants* $k_{\nu\gamma}$ *in* 1 D *and in* 2 B *coincide. Moreover,* $c_{\nu\gamma}>0$, *and the functions* $p_{\nu\gamma}$ *in* 1 D *and in* 2 B *coincide.*

As a special case of the first half of the theorem, we have:

COROLLARY. *If* $c_\beta > 0$, *then the Green's function* g *is characterized by these properties: it is harmonic on* $W - \{\zeta\}$ *with the singularity* $-\log|z-\zeta|$, *and*

$$Hg = g \quad \text{on } \beta.$$

Proof of the Theorem. Let U' be an end of W with $\beta_{U'} = \gamma$ and $\zeta \notin \bar{U}'$. Exhaust W towards γ by V with $\partial U' \subset V$. Let H act from $\partial U'$ into U', and H_V act from $\partial U'$ into $U' \cap V$. For brevity, we write p and p_V for $p_{\nu\gamma}$ and $p_{\nu\gamma(V)}$; we use k and k_V similarly. Then

$$\begin{aligned} |H(p-k)-(p-k)| &\le |H(p-k) - H_V(p-k)| + |H_V(p-k) - H_V(p_V - k_V)| \\ &\quad + |(p_V - k_V) - (p-k)| \\ &\le |H(p-k) - H_V(p-k)| + \max_{\partial U'} |(p-k) - (p_V - k_V)| \\ &\quad + |(p-k) - (p_V - k_V)| \to 0 \end{aligned}$$

as $V \to W$, and therefore $H(p-k) = p-k$ on U'.

Next, to show the uniqueness, suppose there is another function p' satisfying the conditions with the constant k'. The function $u = (p_{\nu\gamma} - p') - (k_{\nu\gamma} - k')$ is harmonic on W with $L_\nu u = u$ on $\beta - \gamma$ and $Hu = u$ on U'. The maximum principle for L_ν gives $\min_{\partial U'} u \le u \le \max_{\partial U'} u$ on $W - U'$. If $u \not\equiv 0$, then either $\min_{\partial U'} u < 0$ or $\max_{\partial U'} u > 0$, so that, by the property of H, either $\min_{\partial U'} u \le u$ or $u \le \max_{\partial U'} u$ on U'. Hence $\min_W u$ or $\max_W u$ is attained at a point on $\partial U'$ and, therefore, $u \equiv$ const. Thus $p' = p_{\nu\gamma} +$ const and, in view of the normalization at ζ, $p' = p_{\nu\gamma}$.

For the proof of the latter half of the theorem, we use $p_{\nu\gamma}$ and $k_{\nu\gamma}$ *in the sense of* 1 D. It is sufficient to show the existence of an exhaustion $V \to W$ towards γ, with $\zeta \in V$, such that the restriction to V of $p_{\nu\gamma}$ coincides with $p_{\nu\gamma(V)}$ and that $\lim_{V \to W} k_{\nu\gamma(V)} = k_{\nu\gamma}$.

For each r with $0 < r$, let $V_r = \{z \,|\, p_{\nu\gamma}(z;\zeta) < k_{\nu\gamma} - r\}$. Take a neighborhood U' of γ such that $\beta_{U'} = \gamma$ and $\zeta \notin \bar{U}'$. By Lemma 2 F below, $m = \max_{\partial U'} p_{\nu\gamma}$ has the properties $m < k_{\nu\gamma}$, and $p_{\nu\gamma} < m$ on $W - \bar{U}'$. If $r < k_{\nu\gamma} - m$, then V_r satisfies the condition in 1 B. Since $p_{\nu\gamma} = k_{\nu\gamma} - r$ on $\gamma(V_r)$ and $L_\nu p_{\nu\gamma} = p_{\nu\gamma}$ on $\beta(V_r) - \gamma(V_r)$, we see that $p_{\nu\gamma} = p_{\nu\gamma(V_r)}$ and $k_{\nu\gamma(V_r)} = k_{\nu\gamma} - r$. The choice $V = V_r$ for $r \to 0$ gives the desired exhaustion. The proof of the theorem will be complete upon establishing Lemma 2 F.

2 E. Capacity Functions in the Case $c_{\nu\gamma} = 0$. By virtue of the normalization 1 D.(1), we obtain the following result for $\nu = 0, 1$:

THEOREM. *If* $c_{\nu\gamma} = 0$, *then, with respect to any exhaustion* $V_n \to W$ *towards* γ *with* $\zeta \in V_n$, *the sequence* $\{p_{\nu\gamma(V_n)}\}_{n=1}^{\infty}$ *contains a subsequence which converges to a harmonic function uniformly on every compact subset of* $W-\{\zeta\}$.

The proof will be given in 2 G.

The limiting function is not necessarily unique, as will be shown in 5 A. In the case $c_{\nu\gamma}>0$, it is always unique and equal to $p_{\nu\gamma}$ (see 2 B.(3)).

If no ambiguity arises, any limiting function will also be referred to as *a capacity function* $p_{\nu\gamma}=p_{\nu\gamma}(z;\zeta)$, or, in the case $\gamma=\beta$, as $p_\beta=p_\beta(z;\zeta)$. It is easily verified that they have properties 2 C.(a)–(d).

The question of the validity of $\lim_{n\to\infty} D_{V_n}[p_{\nu\gamma(V_n)}-p_{\nu\gamma}]=0$ will be discussed in 5 C.

2 F. Maximum Principle. We shall make use of the following auxiliary result, valid even in the case $c_{\nu\gamma}=0$:

LEMMA. *The capacity function* $p_{\nu\gamma}$ *has the property*

$$p_{\nu\gamma}<k_{\nu\gamma} \quad \text{on } W.$$

If Ω *is a regular region with* $\zeta\in\Omega$, *then*

$$p_{\nu\gamma}<\max_{\partial\Omega} p_{\nu\gamma} \quad \text{on } \Omega,$$

$$p_{\nu\gamma}>\min_{\partial\Omega} p_{\nu\gamma} \quad \text{on } W-\bar{\Omega}.$$

Proof. First assume that W satisfies the assumption of 1 C. Let U be a neighborhood of $\beta-\gamma$ disjoint from $\bar{\Omega}$ and from a neighborhood of γ. Then $\min_{\partial U} p_{\nu\gamma}\le p_{\nu\gamma}\le\max_{\partial U} p_{\nu\gamma}$ on $\bar{U}$.

On $W-U$, the maximum of $p_{\nu\gamma}$ is attained at a point of γ, since otherwise the maximum of $p_{\nu\gamma}$ on W would be taken at an interior point. Therefore, $p_{\nu\gamma}<k_{\nu\gamma}$ on W. On $W-U-\Omega$, the minimum is reached on $\partial\Omega$, for otherwise the minimum on $W-\Omega$ would be attained at an interior point. Thus $p_{\nu\gamma}>\min_{\partial\Omega} p_{\nu\gamma}$ on $W-\bar{\Omega}$. The inequality $p_{\nu\gamma}<\max_{\partial\Omega} p_{\nu\gamma}$ on Ω is evident.

By a limiting process (Theorem 2 E), we see that the lemma remains valid, for an arbitrary W. In the proof of Theorem 2 E, however, the above special case will suffice.

2 G. Proof of Theorem 2 E. For the sake of simplicity, write p_V for $p_{\nu\gamma(V)}$. Let N_0 and N be neighborhoods of ζ which, in terms of the local parameter t about ζ, are determined by the disks $|t|<e^{-1}$ and $|t|<1$. The proof of the theorem will be complete if we show that, for every compact set K such that $\zeta\notin K$ and $\partial N\subset K\subset W-\bar{N}_0$, the functions p_V with $V\supset K$ are uniformly bounded on K.

We set $m(V)=\min_{\partial N_0} p_V$ and have, by Lemma 2 F applied to $\Omega=N_0$,

$$p_V-m(V)>0 \quad \text{on } V-\bar{N}_0.$$

The maximum principle for the harmonic function $p_V-\log|t|$ on $\bar{N}$ implies that its minimum on ∂N is dominated by that on ∂N_0. Hence $\min_{\partial N} p_V \le m(V)+1$, that is,

$$\min_{\partial N}(p_V-m(V))\le 1.$$

These inequalities permit us to use Harnack's theorem on $W-\bar{N}_0$, and we see that the functions $p_V-m(V)$ are uniformly bounded on K:

$$|p_V-m(V)|\le A \quad \text{on } K. \tag{4}$$

Next, the harmonic function $p_V-\log|t|-m(V)$ on $\bar{N}$ attains its extrema on ∂N:

$$|p_V-\log|t|-m(V)|\le \max_{\partial N}|p_V-m(V)|\le A,$$

where we note that $\partial N\subset K$. For $t\to 0$, the normalization condition 1 D.(1) gives $|m(V)|\le A$. We substitute this into (4) to obtain

$$|p_V|\le 2A \quad \text{on } K,$$

that is, the p_V are uniformly bounded. This completes the proof.

2 H. Condition for $c_{\nu\gamma}=0$. By Lemma 2 F we easily see that $k_{\nu\gamma(V)}-p_{\nu\gamma(V)}$ increases with V. From the results in 2 B and 2 E, we conclude:

The capacity $c_{\nu\gamma}$ vanishes if and only if

$$\lim_{V\to W}(k_{\nu\gamma(V)}-p_{\nu\gamma(V)})=\infty.$$

In particular, if $\beta=\gamma$, then V is a regular subregion of W. The Green's function $g_V=k_{\beta(V)}-p_{\beta(V)}$ on V increases with V and, if $c_\beta>0$, we have $\lim_{V\to W} g_V=g$, the Green's function on W. The above result that $c_\beta=0$ if and only if $\lim_{V\to W} g_V=\infty$ is more commonly stated in the equivalent form: $c_\beta=0$ *if and only if W carries no Green's function.*

3. Extremal Properties

3 A. Functions $p_{0\gamma}$ and $p_{1\gamma}$. Let γ be a subboundary of an open Riemann surface W. Take a point $\zeta\in W$ together with a local parameter t about it. Denote by $HD_\zeta(W)$ the family of real-valued harmonic functions p on $W-\{\zeta\}$ which have the singularity $\log|z-\zeta|$, the normalization $\lim_{t\to 0}(p(t)-\log|t|)=0$, and a finite Dirichlet integral outside of a neighborhood of ζ. Let $KD_{\zeta\gamma}(W)$ be the family of functions p in $HD_\zeta(W)$

such that $\int_c *dp=0$ along every dividing cycle c in $W-\{\zeta\}$ not separating γ from ζ (see 2 C.(d)). Finally, consider the family $KD^0_{\zeta\gamma}(W)$ of functions in $HD_\zeta(W)$, such that $L_0\, p=p$ on $\beta-\gamma$ in the sense of 2 C.(c). Clearly $KD^0_{\zeta\gamma}\subset KD_{\zeta\gamma}$.

THEOREM. *If $c_{0\gamma}, c_{1\gamma}>0$, then the capacity functions $p_{0\gamma}$ and $p_{1\gamma}$ are the unique functions minimizing $\int_\beta p*dp$ in $KD^0_{\zeta\gamma}(W)$ and $KD_{\zeta\gamma}(W)$. The minimum values are:*

$$\min_{KD^0_{\zeta\gamma}} \int_\beta p*dp=2\pi\, k_{0\gamma}, \qquad \min_{KD_{\zeta\gamma}} \int_\beta p*dp=2\pi\, k_{1\gamma}.$$

In particular, if $c_\beta>0$, then p_β is the unique function minimizing $\int_\beta p*dp$ in $HD_\zeta(W)$, and the minimum value is $2\pi\, k_\beta$.

For plane regions W, an extremal property characterizing $p_{0\gamma}$ in $KD_{\zeta\gamma}$ will be discussed in VII.3 I.

Remark. The above result for $p_{1\gamma}$ was obtained in SARIO [10, 12]. For plane regions, it was found by GRUNSKY [1] and LOKKI [1]. The property of $p_{0\gamma}$ was given by OIKAWA [4] and MARDEN-RODIN [1].

Proof of the Theorem. It suffices to show that

$$p_{0\gamma}\in KD^0_{\zeta\gamma}, \qquad p_{1\gamma}\in KD_{\zeta\gamma}, \tag{1}$$

$$\int_\beta p_{\nu\gamma}*dp_{\nu\gamma}=2\pi\, k_{\nu\gamma}, \tag{2}$$

$$D_W[p-p_{0\gamma}]=\int_\beta p*dp-\int_\beta p_{0\gamma}*dp_{0\gamma} \tag{3}$$

for every $p\in KD^0_{\zeta\gamma}$, and

$$D_W[p-p_{1\gamma}]=\int_\beta p*dp-\int_\beta p_{1\gamma}*dp_{1\gamma} \tag{4}$$

for every $p\in KD_{\zeta\gamma}$.

Relation (1) is an immediate consequence of 2 B.(2) and the properties in 2 C.

For the proofs of (2) with $\nu=0$ and (3), observe that the restriction to V of a $p\in KD^0_{\zeta\gamma}$ belongs to $KD^0_{\zeta\gamma(V)}(\overline{V})$. By identities 1 F.(5′) and (8),

$$D_V[p-p_{0\gamma(V)}]=\int_{\beta(V)} p*dp-2\pi\, k_{0\gamma(V)}. \tag{5}$$

If $V\to W$, the left-hand side converges to $D_W[p-p_{0\gamma}]$, as is easily seen by 2 B.(2) and the triangle inequality. Consequently,

$$D_W[p-p_{0\gamma}]=\int_\beta p*dp-2\pi\, k_{0\gamma}, \tag{6}$$

and $p=p_{0\gamma}$ yields (2). On substituting (2) in (6), we have (3).

In a similar manner, using identities 1 E.(2), (3), we obtain (2) for $\nu=1$ and (4).

3 B. Condition for $c_{\nu\gamma}=0$. From (5), we have

$$\int_{\beta(V)} p * dp \geq 2\pi\, k_{0\gamma(V)}$$

for every $p \in KD^0_{\zeta\gamma}$. If $c_{0\gamma}=0$, then $\int_\beta p * dp$ cannot be finite. A similar conclusion is obtained in the case $\nu=1$, and we conclude:

THEOREM. $c_{0\gamma}=0$ *if and only if* $KD^0_{\zeta\gamma}=\emptyset$, *and* $c_{1\gamma}=0$ *if and only if* $KD_{\zeta\gamma}=\emptyset$.

In particular, if $c_{\nu\gamma}=0$, all capacity functions satisfy

$$\int_\beta p_{\nu\gamma} * dp_{\nu\gamma} = \infty,$$

that is, (2) continues to hold.

3 C. Relations between Capacities. The above extremal properties lead to the following results:

THEOREM. (a) *The capacities satisfy the inequality*

$$c_{0\gamma} \leq c_{1\gamma}.$$

In the case $c_{0\gamma}>0$, *the equality holds if and only if* $p_{0\gamma}=p_{1\gamma}$.

(b) *The capacity* $c_{1\gamma}$ *has the bound*

$$c_{1\gamma} \leq c_\beta.$$

In the case $c_{1\gamma}>0$, *the equality holds if and only if* $p_{1\gamma}=p_\beta$.

(c) *If* $\gamma \subset \gamma'$, *then*

$$c_{\nu\gamma} \leq c_{\nu\gamma'}, \qquad \nu=0,1.$$

In the case $c_{\nu\gamma}>0$, *the equality holds if and only if* $p_{\nu\gamma}=p_{\nu\gamma'}$.

Proof. (a) The case $c_{0\gamma}=0$ is trivial. Suppose $c_{0\gamma}>0$ and apply (4) with $p=p_{0\gamma} \in KD^0_{\zeta\gamma} \subset KD_{\zeta\gamma}$:

$$D_W[p_{0\gamma}-p_{1\gamma}] = 2\pi(k_{0\gamma}-k_{1\gamma}).$$

The conclusion is immediate.

(b) is a special case of (c), which is proved in the same manner as (a) using $KD_{\zeta\gamma} \subset KD_{\zeta\gamma'}$ and $KD^0_{\zeta\gamma} \subset KD^0_{\zeta\gamma'}$.

3 D. The Case of a Bordered Surface. As in 1 C, let W be the interior of a bordered surface $\overline{W}$, not necessarily compact, with compact border γ. To analyse (a) and (b) on W, consider the families $HD(\overline{W})$ and $KD(\overline{W})$

as in I.2 B and I.3 B, regarding W as an end of a surface and γ as its relative boundary. Set

$$H_0D(\overline{W})=\{u|u\in HD(\overline{W}),\ u=0 \text{ on } \gamma\},$$
$$K_0D(\overline{W})=KD(\overline{W})\cap H_0D(\overline{W}).$$

We claim that

$$D_W[u, p_{1\gamma}-p_{0\gamma}]=2\pi\, u(\zeta) \tag{7}$$

for $u\in K_0D(\overline{W})$, and

$$D_W[u, p_\beta-p_{0\gamma}]=2\pi\, u(\zeta) \tag{8}$$

for $u\in H_0D(\overline{W})$.

For the proof, let N be a neighborhood of ζ. Then for $u\in K_0D(\overline{W})$,

$$D_{W-N}[u, p_{1\gamma}]-D_{W-N}[u, p_{0\gamma}]=\int_{\gamma-\partial N} p_{1\gamma} * du - \int_{\gamma-\partial N} u * dp_{0\gamma}$$

by properties I.3 C.(4) and I.2 B.(6) of the operators L_1 and L_0. This quantity is equal to

$$-\int_{\partial N} p_{1\gamma} * du + \int_{\partial N} u * dp_{0\gamma},$$

which converges to $2\pi\, u(\zeta)$ as $N\to\zeta$, and we have (7). As for (8), we use the relation $\overline{H(p_\beta-k_\beta)}=p_\beta-k_\beta$ obtained in 2 D and note that $c_\beta\geq c_{1\gamma}>0$. For $u\in H_0D(\overline{W})$, we have

$$D_{W-N}[u, p_\beta]-D_{W-N}[u, p_{0\gamma}]=\int_{-\partial N} (p_\beta-k_\beta) * du - \int_{\gamma-\partial N} u * dp_{0\gamma},$$

where we invoked property I.5 B.(1) of H acting from ∂N into $W-N$. On letting $N\to\zeta$ we obtain (8).

Observe that $p_{1\gamma}-p_{0\gamma}-(k_{1\gamma}-k_{0\gamma})\in K_0D(\overline{W})$. By I.5 E we also see that $p_\beta-p_{0\gamma}-(k_\beta-k_{0\gamma})\in H_0D(\overline{W})$. On combining these with (7) and (8), we conclude:

THEOREM. (a) *The following conditions are equivalent:* $c_{1\gamma}=c_{0\gamma}$ *for every* ζ; $p_{1\gamma}=p_{0\gamma}$ *for every* ζ; *and* $K_0D(\overline{W})=\{0\}$.

(b) *The following conditions are equivalent:* $c_\beta=c_{1\gamma}=c_{0\gamma}$ *for every* ζ; $p_\beta=p_{1\gamma}=p_{0\gamma}$ *for every* ζ; *and* $H_0D(\overline{W})=\{0\}$.

3 E. Another Extremal Property of $p_{0\gamma}$. The quantity $\int_\beta u * du$ introduced in II.1 D is easily generalized to an isolated subboundary γ: given a $u\in H(\overline{U})$ with $\beta_U=\gamma$, we set

$$\int_\gamma u * du = D_U[u] - \int_{\partial U} u * du.$$

This is independent of the choice of U with the above property.

THEOREM. *If γ is an isolated subboundary with $c_{0\gamma}>0$, then $p_{0\gamma}$ is the unique function maximizing*

$$\int_{\gamma} p*dp - \int_{\beta-\gamma} p*dp$$

in the family of functions $p\in KD_{\zeta\gamma}$ such that $H(p-k)=p-k$ on γ, with k a constant (which may depend on p). The maximum value is $2\pi\, k_{0\gamma}$.

Remark. The theorem was obtained in SARIO [13] under a stronger assumption (cf. 1 F.(7)).

Proof. Observe that $\beta-\gamma$ is an isolated subboundary. On a neighborhood U of $\beta-\gamma$ with $\beta_U=\beta-\gamma$, the function $p_{0\gamma}$ satisfies $L_0\, p_{0\gamma}=p_{0\gamma}$, so that $D_U[p_{0\gamma}]=\int_{\partial U} p_{0\gamma}*dp_{0\gamma}$. Hence

$$\int_{\beta-\gamma} p_{0\gamma}*dp_{0\gamma}=D_U[p_{0\gamma}]-\int_{\partial U} p_{0\gamma}*dp_{0\gamma}=0.$$

The identity

$$\int_{\beta} p*dp=\int_{\gamma} p*dp+\int_{\beta-\gamma} p*dp$$

is an immediate consequence of the definition. From 3 A.(2) we obtain

$$\int_{\gamma} p_{0\gamma}*dp_{0\gamma}=2\pi\, k_{0\gamma}.$$

The proof of the theorem is thus reduced to showing that

$$D_W[p-p_{0\gamma}]=\int_{\beta-\gamma} p*dp-\int_{\gamma} p*dp+\int_{\gamma} p_{0\gamma}*dp_{0\gamma}, \tag{9}$$

a generalization of 1 F.(7).

To establish (9), take a neighborhood U_γ of γ such that $\beta_{U_\gamma}=\gamma$ and $V=W-\overline{U}_\gamma$ is connected. Assume $\zeta\in V$. Apply 1 F.(6) with $p^0=p_{0\gamma}$ to the first term of the right-hand side of

$$D_W[p-p_{0\gamma}]=D_V[p-p_{0\gamma}]+D_{U_\gamma}[p-p_{0\gamma}]$$

to obtain

$$D_V[p-p_{0\gamma}]=\int_{\beta-\gamma} p*dp+\int_{\gamma(V)} p*dp-2\int_{\gamma(V)} p*dp_{0\gamma}+\int_{\gamma(V)} p_{0\gamma}*dp_{0\gamma}.$$

Then

$$\begin{aligned} D_W[p-p_{0\gamma}]={}&\int_{\beta-\gamma} p*dp+\Big(D_{U_\gamma}[p]+\int_{\gamma(V)} p*dp\Big)\\ &-2\Big(D_{U_\gamma}[p,p_{0\gamma}]+\int_{\gamma(V)} p*dp_{0\gamma}\Big)+\Big(D_{U_\gamma}[p_{0\gamma}]+\int_{\gamma(V)} p_{0\gamma}*dp_{0\gamma}\Big). \end{aligned} \tag{10}$$

On U_γ, we have $H(p-k)=p-k$. Therefore, by property I.5 B.(1) of H,

$$D_{U_\gamma}[p,p_{0\gamma}]=-\int_{\gamma(V)} p*dp_{0\gamma}+k\int_{\gamma(V)} *dp_{0\gamma},$$

$$D_{U_\gamma}[p]=-\int_{\gamma(V)} p*dp+k\int_{\gamma(V)} *dp.$$

Since $\int_{\gamma(V)} *dp_{0\gamma} = \int_{\gamma(V)} *dp = 2\pi$, we have

$$D_{U_\gamma}[p, p_{0\gamma}] + \int_{\gamma(V)} p * dp_{0\gamma} = D_{U_\gamma}[p] + \int_{\gamma(V)} p * dp,$$

which is equal to $\int_\gamma p * dp$ by definition. For the same reason,

$$D_{U_\gamma}[p_{0\gamma}] + \int_{\gamma(V)} p_{0\gamma} * dp_{0\gamma} = \int_\gamma p_{0\gamma} * dp_{0\gamma}.$$

On substituting these into (10), we obtain (9).

3 F. Further Extremal Properties of $p_{\nu\gamma}$. Let γ be a subboundary of an open Riemann surface W, and ζ a given point of W. In this section we shall denote by $\mathscr{H}_\zeta$ the family of non-negative harmonic functions in $W - \{\zeta\}$ with the singularity $-\log|z-\zeta|$. No normalization corresponding to 1 D.(1) is used so that it is not necessary to specify a local parameter. Let $\mathscr{H}_{\zeta\gamma}^\nu$, $\nu = 0, 1$, be the family of functions $h \in \mathscr{H}_\zeta$ with $L_\nu h = h$ on $\beta - \gamma$ in the sense of 2 C.(c).

THEOREM. $c_{\nu\gamma} = 0$ *if and only if* $\mathscr{H}_{\zeta\gamma}^\nu = \emptyset$. *If* $c_{\nu\gamma} > 0$, *then*

$$k_{\nu\gamma}(\zeta) - p_{\nu\gamma}(z;\zeta) = \min_{h \in \mathscr{H}_{\zeta\gamma}^\nu} h(z).$$

In the case $\gamma = \beta$, we obtain the following well-known characterization of the Green's function: $c_\beta = 0$ *if and only if* $\mathscr{H}_\zeta = \emptyset$; *if* $c_\beta > 0$, *then*

$$g(z;\zeta) = \min_{h \in \mathscr{H}_\zeta} h(z).$$

Proof of the Theorem. Exhaust W towards γ by V with $\zeta \in V$. The relations

$$0 \le k_{\nu\gamma(V)} - p_{\nu\gamma(V)} \le h \quad \text{on } V$$

for $h \in \mathscr{H}_{\zeta\gamma}^\nu$ are verified exactly as in 2 F, and the theorem follows easily (cf. also Theorem 2 E).

COROLLARY. *For an arbitrary* $c_{\nu\gamma}$,

$$\sup_{z \in W} p_{\nu\gamma}(z;\zeta) = k_{\nu\gamma}(\zeta).$$

In particular, if $c_\beta > 0$, *then*

$$\inf_{z \in W} g(z;\zeta) = 0.$$

Proof. Suppose $\sup p_{\nu\gamma} = a_\nu$ is less than $k_{\nu\gamma}$. Then $a_\nu - p_{\nu\gamma}$ belongs to $\mathscr{H}_{\zeta\gamma}^\nu$. If $k_{\nu\gamma} = \infty$, then $\mathscr{H}_{\zeta\gamma}^\nu \neq \emptyset$, and if $k_{\nu\gamma} < \infty$, then $a_\nu - p_{\nu\gamma} < k_{\nu\gamma} - p_{\nu\gamma}$, a contradiction.

4. Construction by Exhaustion

4 A. Subboundary of a Subregion. We recall the definition of a subboundary γ of an open Riemann surface as a correspondence $\Omega \to U_\Omega$, where Ω is a regular subregion of W, and U_Ω is a union of components of $W - \bar{\Omega}$.

For every Ω,

$$\gamma(\Omega) = \bar{\Omega} \cap \bar{U}_\Omega$$

is a union of components of $\partial\Omega$, and may not be connected even if γ is an ideal boundary component. However, if Ω is a canonical region, then $\gamma(\Omega)$ is a component of $\partial\Omega$.

We next consider an arbitrary subregion V of W. For every regular subregion Ω of V, we denote by $U_\Omega(V)$ the union of those components R of $V \cap U_\Omega$ which satisfy the following condition: R meets $V \cap U_{\Omega'}$ for every Ω' such that $\bar{\Omega} \subset \Omega'$ and $\bar{\Omega}' \subset V$. It is not difficult to see that the correspondence $\Omega \to U_\Omega(V)$ is a subboundary of V. We shall denote it by $\gamma(V)$.

In general, if γ is an ideal boundary component, the same is not necessarily true of $\gamma(V)$.

If V is a regular subregion, then this $\gamma(V)$ is regarded in a natural manner as identical with $\bar{V} \cap \bar{U}_V$. More generally, if V satisfies the condition in 1 B, then the present $\gamma(V)$ can be viewed as the same as the one considered there.

If V contains a neighborhood of γ, then $\gamma(V)$ and γ are considered identical.

If γ is an isolated subboundary of W, then $\gamma(V)$ for a sufficiently large V is also isolated. Indeed, if V contains ∂U for a neighborhood U of γ with $\beta_U = \gamma$, then $V \cap U$ is an end of V and satisfies $\gamma(V) = \beta(V)_{U \cap V}$, where $\beta(V)$ is the ideal boundary of V regarded as an open Riemann surface.

4 B. Capacity Function $p_{1\gamma}$. Let γ be a subboundary of an open Riemann surface W, and let $\zeta \in W$. Consider $c_{1\gamma(V)}$ and $p_{1\gamma(V)}$ for subregions V such that $\zeta \in V$ and all dividing cycles of V which do not separate components of $\gamma(V)$ are dividing cycles of W. This condition is satisfied, for example, by a canonical subregion or by a V which meets the condition in 1 B.

THEOREM. *The capacity $c_{1\gamma(V)}$ decreases as V increases and*

$$c_{1\gamma} = \lim_{V \to W} c_{1\gamma(V)}.$$

If $c_{1\gamma} > 0$, then

$$\lim_{V \to W} D_V[p_{1\gamma(V)} - p_{1\gamma}] = 0.$$

In the special case where $V \to W$ is an exhaustion towards γ, the above is merely the definition in 2 B.

Proof. We write k_V and p_V for $k_{1\gamma(V)}$ and $p_{1\gamma(V)}$. If $V' \supset V$, the assumption for V implies that $p_{V'} \in KD_{\zeta\gamma(V)}(V)$. By 3 A.(4) and (2),

$$D_V[p_{V'} - p_V] \leq 2\pi(k_{V'} - k_V). \tag{1}$$

For $p_W = p_{1\gamma}$ and $k_W = k_{1\gamma}$, we have similarly

$$D_V[p_W - p_V] \leq 2\pi(k_W - k_V). \tag{2}$$

It follows from (1) that k_V increases with V, and from (2) that it is bounded by k_W.

If $\lim k_V = \infty$, then $k_W = \infty$, so that $k_W = \lim_{V \to W} k_V$.

If $\lim k_V < \infty$, then (1) shows the existence of a function q with

$$\lim_{V \to W} D_V[q - p_V] = 0.$$

Clearly $q \in KD_{\zeta\gamma}(W)$, so that $k_W < \infty$ by Theorem 3 B. For an arbitrary $p \in KD_{\zeta\gamma}(W)$, we have on V

$$D_V[p - p_V] = \int_{\beta(V)} p * dp - 2\pi k_V,$$

where $\beta(V)$ is the entire ideal boundary of V. Thus on W,

$$D_W[p - q] = \int_\beta p * dp - 2\pi \lim_{V \to W} k_V. \tag{3}$$

In particular, for $p = q$,

$$\int_\beta q * dq = 2\pi \lim_{V \to W} k_V. \tag{4}$$

From (3) and (4) we see that q is the minimizing function of $\int_\beta p * dp$ in $KD_{\zeta\gamma}$. By Theorem 3 A we conclude that $q = p_W$, hence $\lim D_V[p_W - p_V] = 0$ and $\lim k_V = k_W$.

4 C. Capacity Function $p_{0\gamma}$. In the case of $p_{0\gamma}$, exhaustion as above by general subregions V is not convenient, and we shall only use exhaustions by canonical subregions Ω with $\zeta \in \Omega$.

THEOREM. *The capacity $c_{0\gamma}$ and the capacity function $p_{0\gamma}$ have the following limit properties:*

(a) *For a subboundary γ,*

$$c_{0\gamma} = \overline{\lim_{\Omega \to W}}\, c_{0\gamma(\Omega)}$$

and, if $c_{0\gamma} > 0$,

$$\lim_{\Omega \to W} D_\Omega[p_{0\gamma(\Omega)} - p_{0\gamma}] = 0.$$

(b) *For an isolated* γ,

$$c_{0\gamma} = \lim_{\Omega \to W} c_{0\gamma(\Omega)}$$

and, if $c_{0\gamma} > 0$,

$$\lim_{\Omega \to W} D_\Omega [p_{0\gamma(\Omega)} - p_{0\gamma}] = 0.$$

(c) *For a non-isolated* γ,

$$\underline{\lim}_{\Omega \to W} c_{0\gamma(\Omega)} = 0$$

and, if $c_{0\gamma} > 0$,

$$\overline{\lim}_{\Omega \to W} D_\Omega [p_{0\gamma(\Omega)} - p_{0\gamma}] = \infty.$$

The result is due to Oikawa [2, 4].

The convergence of $D_\Omega [p_{0\gamma(\Omega)} - p_{0\gamma}]$ in the case $c_{0\gamma} = 0$ will be discussed in 5 C.

Proof. Let U_Ω be as in I.1 B. The region $V(\Omega) = W - \overline{U}_\Omega$ satisfies the condition in III.1 B with $\gamma(V(\Omega)) = \gamma(\Omega)$. As $\Omega \to W$, $V(\Omega)$ exhausts W towards γ, so that

$$c_{0\gamma} = \lim_{\Omega \to W} c_{0\gamma(V(\Omega))}.$$

The capacity $c_{0\gamma(V(\Omega))}$ dominates $c_{0\gamma(\Omega)}$ for Ω, as is seen from 3 E.(9) using a process similar to (a) of Theorem 3 C, or else from the expression in terms of extremal length in IV.3 F and IV.3 G. We conclude that

$$c_{0\gamma} \geq \overline{\lim}_{\Omega \to W} c_{0\gamma(\Omega)}.$$

Next, let $\zeta \in V_n \to W$ be an exhaustion towards γ. We have $c_{0\gamma} = \lim c_{0\gamma(V_n)}$ and, if $c_{0\gamma} > 0$, $\lim D_{V_n} [p_{0\gamma(V_n)} - p_{0\gamma}] = 0$. Consider an exhaustion $\zeta \in \Omega_n^m \to V_n$ $(m \to \infty)$ towards $\beta(V_n) - \gamma(V_n)$, where $\beta(V_n)$ is the entire ideal boundary of V_n and, therefore, $\beta(V_n) - \gamma(V_n)$ is a subboundary of V_n. Then $\gamma(V_n)$ is a subboundary of Ω_n^m for which the capacity $c_{0\gamma(V_n)}$ and the capacity function $p_{0\gamma(V_n)}$ will be denoted by c_n^m and p_n^m. Theorem 4 F below gives

$$c_{0\gamma(V_n)} = \lim_{m \to \infty} c_n^m > 0, \qquad \lim_{m \to \infty} D_{\Omega_n^m} [p_{0\gamma(V_n)} - p_n^m] = 0,$$

where $c_{0\gamma(V_n)}$ and $p_{0\gamma(V_n)}$ are for V_n. It is easy to find integers $m = m(n)$ such that for $\Omega_n = \Omega_n^{m(n)}$, $n = 1, 2, \ldots$, we have $\lim c_{0\gamma(\Omega_n)} = c_{0\gamma}$ and, if $c_{0\gamma} > 0$, $\lim D_{\Omega_n} [p_{0\gamma(\Omega_n)} - p_{0\gamma}] = 0$. We conclude that

$$\overline{\lim}_{\Omega \to W} c_{0\gamma(\Omega)} = c_{0\gamma}$$

and, if $c_{0\gamma} > 0$,

$$\underline{\lim}_{\Omega \to W} D_\Omega [p_{0\gamma(\Omega)} - p_{0\gamma}] = 0.$$

The proof of (a) is herewith complete.

4 D. Proof of (b). Suppose γ is isolated, and let U be a neighborhood of γ with $\zeta \notin \bar{U}$ and $\beta_U = \gamma$. In the present case, $\beta - \gamma$ is a subboundary of W. As $\Omega \to W$, the regions $U \cup \Omega$ exhaust W towards $\beta - \gamma$. But γ is a subboundary of $U \cup \Omega$ as well, for which the capacity $c_{0\gamma}$ will be denoted by $c_{0\gamma(U \cup \Omega)}$. We obtain as before $c_{0\gamma(U\cup\Omega)} \leq c_{0\gamma(\Omega)}$ and, by Theorem 4 F below, $\lim_{\Omega \to W} c_{0\gamma(U\cup\Omega)} = c_{0\gamma}$. Thus

$$c_{0\gamma} \leq \varliminf_{\Omega \to W} c_{0\gamma(\Omega)}.$$

On combining this with (a), we infer that $c_{0\gamma} = \lim c_{0\gamma(\Omega)}$ and, therefore, $k_{0\gamma} = \lim k_{0\gamma(\Omega)}$.

In the case $c_{0\gamma} > 0$, identities 1 F.(5′), (6) give

$$\begin{aligned} D_\Omega[p_{0\gamma(\Omega)} - p_{0\gamma}] &= \int_{\beta(\Omega)} p_{0\gamma} * dp_{0\gamma} - 2 \int_{\gamma(\Omega)} p_{0\gamma} * dp_{0\gamma(\Omega)} + 2\pi\, k_{0\gamma(\Omega)} \\ &\leq 2\pi\, k_{0\gamma} - 2 \int_{\gamma(\Omega)} p_{0\gamma} * dp_{0\gamma(\Omega)} + 2\pi\, k_{0\gamma(\Omega)}. \end{aligned}$$

Here

$$\begin{aligned} \int_{\gamma(\Omega)} p_{0\gamma} * dp_{0\gamma(\Omega)} &= \int_{\gamma(\Omega)} p_{0\gamma(\Omega)} * dp_{0\gamma} - \int_{\partial U} (p_{0\gamma} * dp_{0\gamma(\Omega)} - p_{0\gamma(\Omega)} * dp_{0\gamma}) \\ &= 2\pi\, k_{0\gamma(\Omega)} - \int_{\partial U} (p_{0\gamma} * dp_{0\gamma(\Omega)} - p_{0\gamma(\Omega)} * dp_{0\gamma}) \to 2\pi\, k_{0\gamma} \end{aligned}$$

as $\Omega \to W$. Consequently, $\lim D_\Omega[p_{0\gamma(\Omega)} - p_{0\gamma}] = 0$.

4 E. Proof of (c). Suppose γ is not isolated. There exists an exhaustion $\zeta \in \Omega_n \to W$ such that $\gamma(\Omega_n)$ and $\gamma(\Omega_{n-1})$ are not homologous, that is, $\partial(\Omega_n - \Omega_{n-1}) - \gamma(\Omega_n) - \gamma(\Omega_{n-1}) = \delta_n$ is not void. We replace δ_n by a curve in $\bar{\Omega}_n - \Omega_{n-1}$ as indicated in Fig. 1 and obtain an $\tilde{\Omega}_n$ with $\Omega_{n-1} \subset \tilde{\Omega}_n \subset \Omega_n$.

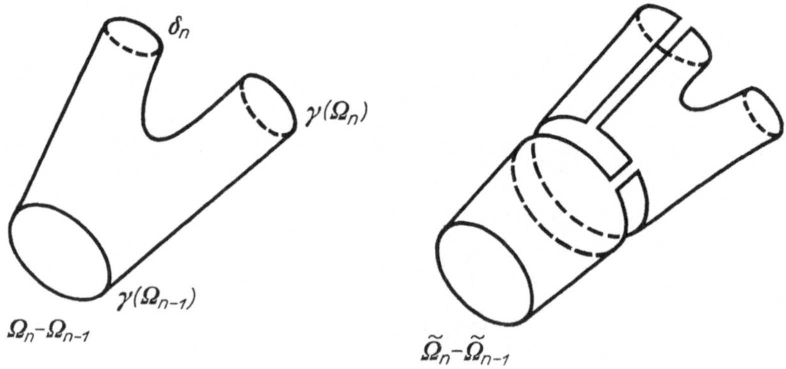

Fig. 1

By taking the unexcised portion of the band sufficiently narrow, we can make $k_{0\gamma(\tilde{\Omega}_n)}$ arbitrarily large, as is seen from the extremal length expression in IV.3 G. Thus we obtain an exhaustion $\tilde{\Omega}_n \to W$ with

$$\lim_{n\to\infty} k_{0\gamma(\tilde{\Omega}_n)} = \infty$$

and, therefore, $\underline{\lim}_{\Omega\to W}\, c_{0\gamma(\Omega)} = 0$.

Next suppose $c_{0\gamma} > 0$. We shall show that

$$\lim_{n\to\infty} D_{\Omega_n}[p_{0\gamma(\Omega_n)} - p_{0\gamma}] = \infty$$

whenever $\lim_{n\to\infty} k_{0\gamma(\Omega_n)} = \infty$. This will complete the proof of

$$\overline{\lim_{\Omega\to W}}\, D_\Omega[p_{0\gamma(\Omega)} - p_{0\gamma}] = \infty\,.$$

Write k_n and p_n for $k_{0\gamma(\Omega_n)}$ and $p_{0\gamma(\Omega_n)}$. By identity 1 F.(6),

$$D_{\Omega_n}[p_{0\gamma} - p_n] = \int_{\beta(\Omega_n)} p_{0\gamma} * dp_{0\gamma} - 2 \int_{\gamma(\Omega_n)} p_{0\gamma} * dp_n + \int_{\gamma(\Omega_n)} p_n * dp_n\,.$$

The first integral on the right dominates $\int_{\beta(\Omega_1)} p_{0\gamma} * dp_{0\gamma}$. As for the second integral, we know that $p_{0\gamma} < k_{0\gamma}$ on $\gamma(\Omega_n)$ and $*dp_n > 0$ along $\gamma(\Omega_n)$. Therefore,

$$D_{\Omega_n}[p_{0\gamma} - p_n] > \int_{\beta(\Omega_1)} p_{0\gamma} * dp_{0\gamma} - 4\pi\, k_{0\gamma} + 2\pi\, k_n \to \infty$$

as $n \to \infty$.

This completes the proof of Theorem 4 C.

4 F. Exhaustion towards $\beta - \gamma$. If γ is an isolated subboundary of W, then so is $\beta - \gamma$. We can speak of an exhaustion of W towards $\beta - \gamma$ by subregions V with $\zeta \in V$. Here γ is a subboundary of V as well; it is equal to $\gamma(V)$ in the sense of 4 A. For the region V, the capacity $c_{0\gamma}$ and the capacity function $p_{0\gamma}$ will be denoted by $c_{0\gamma(V)}$ and $p_{0\gamma(V)}$. Assume further that V has the following property: every dividing cycle in V which does not separate components of γ is a dividing cycle in W.

THEOREM. *For an exhaustion towards $\beta - \gamma$, the capacity $c_{0\gamma(V)}$ increases with V, and*

$$c_{0\gamma} = \lim_{V\to W} c_{0\gamma(V)}\,.$$

If $c_{0\gamma} > 0$, then

$$\lim_{V\to W} D_V[p_{0\gamma(V)} - p_{0\gamma}] = 0\,.$$

Proof. The inequality $c_{0\gamma} \geq c_{0\gamma(V)}$ is clear from Theorem 3 E applied to V. Indeed,

$$2\pi\, k_{0\gamma(V)} \geq \int_{\gamma(V)} p_{0\gamma} * dp_{0\gamma} - \int_{\beta(V)-\gamma(V)} p_{0\gamma} * dp_{0\gamma}$$
$$\geq \int_{\gamma} p_{0\gamma} * dp_{0\gamma} - \int_{\beta-\gamma} p_{0\gamma} * dp_{0\gamma} = 2\pi\, k_{0\gamma}.$$

From this we obtain

$$c_{0\gamma} \geq \overline{\lim_{V \to W}}\, c_{0\gamma(V)},$$

which completes the proof in the case $c_{0\gamma}=0$.

Now assume that $c_{0\gamma}>0$. Let U be a neighborhood of γ with $\zeta \notin \overline{U}$ and $\beta_U=\gamma$, and write k_V and p_V for $k_{0\gamma(V)}$ and $p_{0\gamma(V)}$.

Suppose $V \subset V'$. Apply identity 3 E.(9) to V with $p=p_{V'}$:

$$D_V[p_{V'}-p_V] = \int_{\beta(V)-\gamma} p_{V'} * dp_{V'} - \int_{\gamma} p_{V'} * dp_{V'} + \int_{\gamma} p_V * dp_V$$
$$\leq \int_{\beta(V')-\gamma} p_{V'} * dp_{V'} - \int_{\gamma} p_{V'} * dp_{V'} + \int_{\gamma} p_V * dp_V$$
$$= 2\pi (k_V - k_{V'}).$$

Similarly,

$$D_V[p_{0\gamma}-p_V] \leq 2\pi (k_V - k_{0\gamma}).$$

We see that k_V decreases as V increases, and k_V is bounded from below. Thus

$$\tilde{k} = \lim_{V \to W} k_V$$

is defined, and there exists a function $\tilde{p}$ such that

$$\lim_{V \to W} D_V[p_V - \tilde{p}] = 0.$$

Since $H(p_V-k_V)=p_V-k_V$ on γ, we obtain $H(\tilde{p}-\tilde{k})=\tilde{p}-\tilde{k}$ on γ. Since $L_0\, p_V=p_V$ on $\beta(V)-\gamma$, we have $L_0\, \tilde{p}=\tilde{p}$ on $\beta-\gamma$. From this together with Theorem 2 D, we infer that $\tilde{p}=p_{0\gamma}$ and, therefore,

$$\lim_{V \to W} D_V[p_V - p_{0\gamma}] = 0.$$

We know that $\int_\gamma p_V * dp_V = 2\pi\, k_V$ and $\int_\gamma p_{0\gamma} * dp_{0\gamma} = 2\pi\, k_{0\gamma}$. By definition,

$$\int_{\gamma} p_V * dp_V = D_U[p_V] - \int_{\partial U} p_V * dp_V,$$

$$\int_{\gamma} p_{0\gamma} * dp_{0\gamma} = D_U[p_{0\gamma}] - \int_{\partial U} p_{0\gamma} * dp_{0\gamma}.$$

From the convergence of p_V to $p_{0\gamma}$ in the Dirichlet norm and its resulting uniform convergence on ∂U, we conclude that

$$k_{0\gamma} = \lim_{V \to W} k_V.$$

4 G. Approximation by Isolated Subboundaries. Let γ again be an arbitrary subboundary of W, defined in I.1 B as a correspondence $\Omega \to U_\Omega$, and set $\gamma_\Omega = \beta_{U_\Omega}$. Since γ_Ω is an isolated subboundary of W, the characterization of the capacity function $p_{v\gamma_\Omega}$ is essentially simpler than in the case of an arbitrary subboundary (e.g., Theorem 2 D).

We see that

$$\gamma = \bigcap_\Omega \gamma_\Omega$$

in the sense of 1 A. Following CONSTANTINESCU [2], we state:

THEOREM. *If* $c_{v\gamma} > 0$, *then*

$$\lim_{\Omega \to W} D_W[p_{v\gamma_\Omega} - p_{v\gamma}] = 0.$$

Proof. Clearly $p_{1\gamma} \in KD_{\zeta\gamma_\Omega}(W)$ and $p_{0\gamma} \in KD^0_{\zeta\gamma_\Omega}(W)$. Hence by 3 A.(2) to (4) applied to $p_{v\gamma_\Omega}$,

$$D_W[p_{v\gamma} - p_{v\gamma_\Omega}] = 2\pi(k_{v\gamma} - k_{v\gamma_\Omega}).$$

The sets $V_\Omega = W - U_\Omega$ exhaust W towards γ as $\Omega \to W$, and we have

$$0 \le k_{v\gamma} - k_{v\gamma_\Omega} \le k_{v\gamma(V_\Omega)} \to 0.$$

5. Uniqueness Problem

5 A. Example of Non-Uniqueness. Let γ be a subboundary of an open Riemann surface W. Suppose $c_{v\gamma} = 0$, and consider the capacity function $p_{v\gamma}$ in the sense of 2 E. In contrast with the case of positive capacity, the function $p_{v\gamma}$ is not necessarily unique. The following theorem, which is modelled after an example communicated to the authors by R. ACCOLA, gives an example:

THEOREM. *If an isolated subboundary* γ *with* $c_{v\gamma} = 0$ *consists of more than one ideal boundary component, then the capacity function* $p_{v\gamma}$ *is not uniquely determined.*

The proof will be given in 5 D and 5 E.

There do exist cases of some interest in which the functions are unique. An example is the following result, which will be established in VII.2 H: *If* W *is a plane region and if* γ *is a boundary component with* $c_{1\gamma} = 0$, *then* $p_{1\gamma}$ *is uniquely determined.* Another example is given below.

5 B. Capacity Functions and Principal Functions. Let $\tilde{W}$ be an arbitrary Riemann surface, and ζ, ζ' distinct points of $\tilde{W}$. Consider the Riemann surface $W = \tilde{W} - \{\zeta'\}$, which possesses a boundary component γ realized as the point ζ'.

THEOREM. *The capacity $c_{\nu\gamma}$ corresponding to W vanishes. The capacity function $p_{\nu\gamma}$ coincides with the restriction to W of the principal function $p_\nu(z;\zeta,\zeta')$ for* $\mathbf{Q}$ *on $\tilde{W}$ (cf.* II.2 G*) and, therefore, is uniquely determined.*

Proof. $p_{\nu\gamma}$ has the singularity $\log|z-\zeta|$ with the normalization II.2 G.(5) and satisfies $L_\nu p_{\nu\gamma}=p_{\nu\gamma}$ on the ideal boundary of $\tilde{W}$. About the point ζ', it is bounded from below (see 2 F) and has flux -2π, so that it possesses the singularity $-\log|z-\zeta'|$. Thus $p_{\nu\gamma}=p_\nu(z;\zeta,\zeta')$. Since this function is not bounded from above, we obtain $c_{\nu\gamma}=0$ by Corollary 3 F. One can also prove that $c_{\nu\gamma}=0$ by using extremal length.

Further relations between capacity and principal functions will be given in Chapter V.

5 C. Convergence in Dirichlet Norm. In this context we consider the question posed in 2 E of whether

$$\lim_{V\to W} D_V[p_{\nu\gamma(V)}-p_{\nu\gamma}]=0$$

holds even if $c_{\nu\gamma}=0$.

We shall show that this depends on the exhaustion:

On the surface W of Theorem 5 B, *there exists an exhaustion $V_n\to W$ towards γ with $\zeta\in V_n$ and*

$$\lim_{n\to\infty} D_{V_n}[p_{\nu n}-p_{\nu\gamma}]=0, \tag{1}$$

as well as one with

$$\lim_{n\to\infty} D_{V_n}[p_{\nu n}-p_{\nu\gamma}]=\infty, \tag{2}$$

where $p_{\nu n}=p_{\nu\gamma(V_n)}$.

For the proof, we retain the notation of 5 B. Since $p_{\nu\gamma}=p_\nu(z;\zeta,\zeta')$ on W, the sequence of regions $V_n=\{z\,|\,p_{\nu\gamma}<n\}$, $n=1,2,\dots$, forms an exhaustion towards γ. The function $p_{\nu n}$ coincides with the restriction to V_n of $p_{\nu\gamma}$, so that (1) holds.

Next, to obtain an exhaustion with (2) for $\nu=1$, let U be a neighborhood of γ in W such that $\tilde{U}=U\cup\{\zeta'\}$ is a neighborhood of ζ' in $\tilde{W}$. Suppose a parameter t with $t(\zeta')=0$, $|t|<1$, is defined on $\tilde{U}$. Consider the set $\sigma_n\subset\tilde{U}$, $n=2,3,\dots$, corresponding to the line segment $[0,1/n]$ of the t-plane. The subregion $W_n=W-\sigma_n$ has a boundary component realized as σ_n with a finite $k_{1\sigma_n}$, as is easily seen by the expression in terms of extremal length (IV.3 F, 3 G). On the other hand, σ_n has vanishing area, so that (see II.1 D)

$$\int_{\beta(W_n)} p_{1\gamma}*dp_{1\gamma}=D_U[p_{1\gamma}]-\int_{\partial U} p_{1\gamma}*dp_{1\gamma}=\infty.$$

For an exhaustion $\zeta \in V_n^m \to W_n$ $(m \to \infty)$ towards σ_n,

$$\lim_{m\to\infty}\Big(\int_{\beta(V_n^m)} p_{1\gamma} * dp_{1\gamma} - 2\pi\, k_{1\gamma(V_n^m)}\Big) = \int_{\beta(W_n)} p_{1\gamma} * dp_{1\gamma} - 2\pi\, k_{1\gamma(W_n)} = \infty.$$

On taking a suitable $m = m(n)$, we obtain an exhaustion $\zeta \in V_n = V_n^{m(n)} \to W$ towards γ. It satisfies

$$D_{V_n}[p_{1\gamma(V_n)} - p_{1\gamma}] = \int_{\beta(V_n)} p_{1\gamma} * dp_{1\gamma} - 2\pi\, k_{1\gamma(V_n)}$$

because of identities 1 E.(2) and (3) on V_n for $p = p_{1\gamma}$, and we conclude that (2) holds.

The same proof is obtained for $p_{0\gamma}$ if we use identities 1 F.(5′) and (8).

5 D. Proof of Theorem 5 A. We proceed to establish Theorem 5 A for $\nu = 0$. The proof for $\nu = 1$ is exactly the same.

We shall make use of the following result of M. NAKAI:

If U is an end of a parabolic Riemann surface (X.1 B), *then there exists a harmonic function v on $\bar{U}$ such that*

$$v = 0 \quad \text{on } \partial U, \qquad \int_{\partial U} * dv = -2\pi, \qquad \lim_{n\to\infty} v(z_n) = \infty$$

for any sequence $\{z_n\}$ in $\bar{U}$ having no accumulation points.

For a complete proof, we refer to NAKAI [1], SARIO-NOSHIRO [1, pp. 98–115], and NAKAI-SARIO [2]. The result is a generalization to Riemann surfaces of the Evans-Selberg theorem for plane regions (see VII.6 K, 6L).

Now the hypothesis of Theorem 5 A implies the existence of isolated and disjoint subboundaries $\gamma^i \subset \gamma$, $i = 1, 2$. Let U^i be a neighborhood of γ^i with $\beta_{U^i} = \gamma^i$, $\zeta \notin \bar{U}^i$, and $\bar{U}^1 \cap \bar{U}^2 = \emptyset$. By X.1 A, each U^i satisfies the assumption of the above result of NAKAI. Therefore, we obtain a function v^i on U^i with the corresponding properties.

Using the technique of 1 C, we construct a function q^i which is harmonic on $W - \{\zeta\}$, has singularity $\log|z - \zeta|$, is normalized by 1 D.(1), and satisfies

$$L_1(q^i - v^i) = q^i - v^i \qquad \text{on } \gamma^i,$$

$$L_0\, q^i = q^i \qquad \text{on } \beta - \gamma^i.$$

These properties imply that q^i is not bounded from above on U^i but is bounded from above on $W - U^i$. Accordingly, $q^1 \neq q^2$. To complete the proof, we must show that both q^1 and q^2 are capacity functions $p_{0\gamma}$.

5 E. Capacity Functions q^1, q^2. It suffices to consider q^1, since the proof for q^2 is the same. We write q for q^1.

Let

$$U_n = \{z \mid q(z) > \log n\}$$

for $n=1, 2, \ldots$. Since $\lim q(z_k) = \infty$ for every sequence of points $z_k \in \overline{U}^1$ without accumulation points, U_n is a neighborhood of γ^1. It is contained in U^1 if n is sufficiently large.

Take a neighborhood U of the subboundary $\beta - \gamma^1$ such that $\overline{U} \cap \overline{U}_n = \emptyset$, $n=1, 2, \ldots$, and $\zeta \notin \overline{U}$. Consider a neighborhood U^1 of the subboundary $\gamma - \gamma^1$ such that $U^1 \subset U$. By 3 C.(c), $\gamma - \gamma^1$ has vanishing capacity. Our later study of the harmonic measure of $\gamma - \gamma^1$ (IV.4 A) will show that we can choose neighborhoods $U_n' \subset U'$ of $\gamma - \gamma^1$, $n=1, 2, \ldots$, in such a way that the $V_n = W - \overline{U}_n - \overline{U}_n'$ exhaust W towards γ, and on $\partial U'$,

$$0 < u_n \leq \frac{1}{n}. \tag{3}$$

Here u_n is harmonic on V_n; 0 on ∂U_n; 1 on $\partial U_n'$; and $L_0 u_n = u_n$ on $\beta - \gamma^1 - \beta_{U_n'}$.

Set $\tilde{u}_n = L_0\, q$, with L_0 acting from ∂V_n into V_n. Then

$$q + \tilde{u}_n(\zeta) - \tilde{u}_n$$

evidently coincides with the capacity function $p_{0\gamma(V_n)}$ on V_n.

In terms of $A = \sup_U |q|$, we have

$$\tilde{u}_n = (1 - u_n) \log n \quad \text{on } \partial U_n, \quad |\tilde{u}_n| \leq A \quad \text{on } \partial U_n'$$

and, therefore,

$$-u_n A + (1 - u_n) \log n \leq \tilde{u}_n \leq u_n A + (1 - u_n) \log n$$

on V_n. By (3),

$$-\frac{A}{n} + \left(1 - \frac{1}{n}\right) \log n \leq \tilde{u}_n \leq \frac{A}{n} + \log n \tag{4}$$

on $\partial U'$. The same is true on ∂U_n, since $\tilde{u}_n = q = \log n$ there. We deduce that (4) holds on $\overline{V}_n - U'$ and, in particular, at ζ, i.e.,

$$-\frac{A}{n} - \log n \leq -\tilde{u}_n(\zeta) \leq \frac{A}{n} - \left(1 - \frac{1}{n}\right) \log n. \tag{5}$$

From (4) and (5) we obtain

$$|\tilde{u}_n - \tilde{u}_n(\zeta)| \leq \frac{2A}{n} + \frac{\log n}{n}$$

on $\overline{V}_n - U'$. Consequently,

$$|p_{0\gamma(V_n)} - q| \leq \frac{2A}{n} + \frac{\log n}{n} \to 0$$

as $n \to \infty$ on $W - U'$, so that q is a capacity function $p_{0\gamma}$.

Chapter IV

Modulus Functions

Modulus functions are closely related to capacity functions. The classical examples are harmonic measures and conformal mappings of plane regions onto circular or radial slit annuli. Accordingly, we shall discuss in 1 topics parallel to those in Chapter III. In 2, we express moduli in terms of extremal length, which, together with the normal operator method, is our main tool. The results will be applied in 3 to capacity functions as well. In 4, we shall give an extension of the classical harmonic measure. Functions of a different nature than multiples of modulus functions will also be considered.

1. Modulus Functions

1 A. Modulus Functions on Bordered Surfaces. Let W be the interior of a bordered Riemann surface $\overline{W}$ with disconnected compact border. Partition the border into two parts γ and γ', which are subboundaries of W. If the entire ideal boundary β of W does not coincide with $\gamma \cup \gamma'$, then we shall consider, on an end U of W with $\beta_U = \beta - (\gamma \cup \gamma')$, the normal operator L_0 or L_1 for $\mathbf{Q}$ acting from ∂U into U. Actually, we could consider an arbitrary normal operator L and derive the theorem below, but we shall not need such generality.

THEOREM. *There exists a harmonic function q_ν on $\overline{W}$ for $\nu = 0, 1$ with the following properties:*

(a) *q_ν is constant on γ and constant on γ',*

(b) *$L_\nu q_\nu = q_\nu$ on $\beta - (\gamma \cup \gamma')$,*

(c) $\int_\gamma * dq_\nu \neq 0$.

The function q_ν is uniquely determined up to an integral transformation $q_\nu \to a\, q_\nu + b$, $a \neq 0$.

The meaning of (b) is that $L_\nu q_\nu = q_\nu$ on some (or equivalently every) end U with $\beta_U = \beta - (\gamma \cup \gamma')$ (cf. II.1 C).

It is to be noted that the *constant values of q_ν on γ and γ' are different.* If they were equal, then by the maximum principle of normal operators (I.1 D) we could easily conclude that q_ν is constant, contrary to condition (c).

Proof of the Theorem. Take an end U with $\beta_U=\beta-(\gamma\cup\gamma')$. Let U_γ and $U_{\gamma'}$ be neighborhoods of γ and γ' such that $\bar U\cap\bar U_\gamma=\bar U\cap\bar U_{\gamma'}=\bar U_\gamma\cap\bar U_{\gamma'}=\emptyset$. Consider the harmonic measures u_γ and $u_{\gamma'}$ (I.5 D) on U_γ and $U_{\gamma'}$. Define the function s on $\bar U\cup\bar U_\gamma\cup\bar U_{\gamma'}$ as follows: $s=0$ on $\bar U$, $s=(\int_{\partial U_{\gamma'}}*du_{\gamma'})\,u_\gamma$ on $\bar U_\gamma$, $s=-(\int_{\partial U_\gamma}*du_\gamma)\,u_{\gamma'}$ on $\bar U_{\gamma'}$. Let L_ν^* be the direct sum of the given operator L_ν on U, L_1 for $\mathbf{I}$ on U_γ, and L_1 for $\mathbf{I}$ on $U_{\gamma'}$. Since s has vanishing flux, we can apply Theorem I.1 F to obtain a function q_ν on W such that $L_\nu^*(q_\nu-s)=q_\nu-s$ on $U\cup U_\gamma\cup U_{\gamma'}$. Clearly conditions (a) and (b) are satisfied, and so is (c), since $\int_\gamma *dq_\nu=(\int_{\partial U_\gamma}*du_\gamma)(\int_{\partial U_{\gamma'}}*du_{\gamma'})\neq 0$. Thus q_ν is the desired function.

To prove uniqueness, let q'_ν be another function with the required properties. Both q_ν and q'_ν have different constant values on γ and γ'. Hence it is possible to take constants a and b in such a way that q'_ν and $a\,q_\nu+b$ coincide on $\gamma\cup\gamma'$. The function $u=q'_\nu-(a\,q_\nu+b)$ vanishes on $\gamma\cup\gamma'$ and satisfies $L_\nu u=u$ on U. We conclude easily that u is identically zero.

On multiplying by a non-zero constant, if necessary, we can find a function q_ν satisfying the additional condition

$$\int_\gamma *dq_\nu=2\pi. \tag{1}$$

The additive constant still remains, but it will be convenient *not* to normalize further. We shall call q_0 and q_1, determined uniquely up to additive constants, the *modulus functions q_0 and q_1* of W with respect to γ and γ'. We shall also use the notation

$$q_0(z;\gamma,\gamma'),\quad q_1(z;\gamma,\gamma').$$

Let $k_\gamma[q_\nu]$ and $k_{\gamma'}[q_\nu]$ be the constant values of q_ν on γ and γ'. Under assumption (1), the former is greater than the latter. In fact, by properties of the normal operators L_0 and L_1, we have $\int_\gamma *dq_\nu=-\int_{\gamma'}*dq_\nu=2\pi$ and

$$0<D_W[q_\nu]=\int_{\gamma+\gamma'} q_\nu *dq_\nu=2\pi(k_\gamma[q_\nu]-k_{\gamma'}[q_\nu]).$$

The positive numbers

$$l_\nu=l_\nu(\gamma,\gamma')=k_\gamma[q_\nu]-k_{\gamma'}[q_\nu]$$

are uniquely determined, independently of the additive constant of the function q_ν. The quantities

$$\mu_\nu=\mu_\nu(\gamma,\gamma')=e^{l_\nu}>1$$

will be called the *moduli μ_0 and μ_1* of W with respect to γ and γ'.

As is easily seen, the interchange of γ and γ' has only the following effect:

$$q_\nu(z;\gamma',\gamma) = -q_\nu(z;\gamma,\gamma') + \text{const},$$

$$\mu_\nu(\gamma',\gamma) = \mu_\nu(\gamma,\gamma').$$

1 B. Harmonic Measure on a Bordered Surface. We use the modulus function q_ν which vanishes on γ' and consider for $\nu=0,1$ the harmonic function

$$u_\nu = u_\nu(z;\gamma,\gamma') = \frac{q_\nu(z;\gamma,\gamma')}{\log \mu_\nu(\gamma,\gamma')}$$

on $\overline{W}$. It has the properties

(a) $u_\nu = 1$ on γ, $u_\nu = 0$ on γ',

(b) $L_\nu u_\nu = u_\nu$ on $\beta - (\gamma \cup \gamma')$.

Conversely, these properties together with the harmonicity on $\overline{W}$ characterize u_ν.

We remark that the modulus μ_ν is determined by the function u_ν as follows:

$$\log \mu_\nu = \frac{2\pi}{D_W[u_\nu]}.$$

In particular, if $\beta = \gamma \cup \gamma'$ and $\overline{W}$ is thus a compact bordered surface, then u_0, q_0, and μ_0 coincide with u_1, q_1, and μ_1, so that we may omit the subindex. The function

$$u_\gamma(z) = \frac{q(z;\gamma,\gamma')}{\log \mu(\gamma,\gamma')}, \qquad \gamma' = \beta - \gamma,$$

coincides with the *harmonic measure* in the sense of I.5 D.

1 C. Basic Identities for q_0 and q_1. For later use, we shall give here the analogues of the basic identities for capacity functions in III.1 E, 1 F.

As before, let W be the interior of a (not necessarily compact) bordered Riemann surface $\overline{W}$ with disconnected compact border, and let $\gamma \cup \gamma'$ with $\gamma \cap \gamma' = \emptyset$ be a partition of the border. The families $HD(\overline{W})$ and $KD(\overline{W})$ are considered, as in I.2 B and I.3 B, with W as an end of some surface: $HD(\overline{W})$ is the family of harmonic functions on $\overline{W}$ with finite Dirichlet integrals, and $KD(\overline{W})$ is the subfamily of functions q with the additional property $\int * dq = 0$ along all dividing cycles in W which do not separate components of $\gamma \cup \gamma'$. We now introduce the families

$$HD_{\gamma'\gamma}(\overline{W}) = \{q \mid q \in HD(\overline{W}),\ \int_\gamma * dq = 2\pi\},$$

$$KD_{\gamma'\gamma}(\overline{W}) = HD_{\gamma'\gamma}(\overline{W}) \cap KD(\overline{W}),$$

$$KD^0_{\gamma'\gamma}(\overline{W}) = \{q \mid q \in HD_{\gamma'\gamma}(\overline{W}),\ L_0\, q = q \text{ on } \beta - (\gamma \cup \gamma')\}.$$

The following properties are verified in the same manner as those in III.1 E and 1 F, if we observe that

$$\int_\beta q * dq = D_W[q]$$

for every $q \in HD(\overline{W})$.

(a) $D_W[q_\nu] = \int_{\gamma+\gamma'} q_\nu * dq_\nu = 2\pi\, l_\nu$, *and* $D_W[q^0] = \int_{\gamma+\gamma'} q^0 * dq^0$ *for every* $q^0 \in KD^0_{\gamma'\gamma}(\overline{W})$.

(b) $D_W[q - q_1] = D_W[q] - D_W[q_1]$ *for every* $q \in KD_{\gamma'\gamma}(\overline{W})$.

(c) *Every* $q^0 \in KD^0_{\gamma'\gamma}(\overline{W})$ *satisfies*

$$D_W[q - q^0] = D_W[q] - 2 \int_{\gamma+\gamma'} q * dq^0 + D_W[q^0]$$

for every $q \in HD_{\gamma'\gamma}(\overline{W})$.

(d) $D_W[q - q_0] = \int_{\beta-\gamma-\gamma'} q * dq - \int_{\gamma+\gamma'} q * dq + \int_{\gamma+\gamma'} q_0 * dq_0$ *for every* $q \in KD_{\gamma'\gamma}(\overline{W})$ *with* q *constant on* γ *and constant on* γ'.

(e) $D_W[q - q_0] = D_W[q] - D_W[q_0]$ *for every* $q \in KD^0_{\gamma'\gamma}(\overline{W})$.

(f) $D_W[u - u_1] = D_W[u] - D_W[u_1]$ *for every* $u \in KD(\overline{W})$ *with* $\int_\gamma * du = 2\pi/l_1$.

(g) $D_W[u - u_0] = D_W[u] - D_W[u_0]$ *for every* $u \in HD(\overline{W})$ *with* $u = 1$ *on* γ, $u = 0$ *on* γ'.

Here properties (f) and (g) of u_0 and u_1 are obtained from (b) and (c) on considering $q = u\, l_\nu$, $\nu = 0, 1$. Equality (g) is also an immediate consequence of the Dirichlet principle.

1 D. Definitions on an Arbitrary Surface. Now let W be an open Riemann surface with more than one boundary component. Choose disjoint subboundaries γ and γ'.

Since $\gamma \cup \gamma'$ is a subboundary of W, we can speak of an exhaustion $V \to W$ towards $\gamma \cup \gamma'$. For each V, the subboundaries $\gamma(V)$ and $\gamma'(V)$ in the sense of III.4 A are disjoint, provided V is sufficiently large. Let $l_{\nu V}, \mu_{\nu V}, q_{\nu V}$ stand for l_ν, μ_ν, q_ν, considered on V with respect to $\gamma(V)$ and $\gamma'(V)$.

By the same argument as in III.2 A and 2 B, we can show the existence of $l_\nu = \lim_{V\to W} l_{\nu V}$ and, if $l_\nu < \infty$, the existence of q_ν with $\lim_{V\to W} D_V[q_{\nu V} - q_\nu] = 0$. The quantities

$$\mu_\nu = \mu_\nu(\gamma, \gamma') = e^{l_\nu}$$

satisfy $1 < \mu_\nu \leq \infty$ and will be called *the moduli* μ_0 *and* μ_1 of W with respect to γ and γ'. The functions

$$q_\nu = q_\nu(z; \gamma, \gamma'),$$

determined up to additive constants, will be referred to as the *modulus functions* of W with respect to γ and γ'.

These functions are seen to coincide with those in 1 A if W satisfies the assumption stated there (cf. III.2 D).

An interchange of γ and γ' gives

$$q_\nu(z;\gamma',\gamma) = -q_\nu(z;\gamma,\gamma') + \text{const},$$

$$\mu_\nu(\gamma',\gamma) = \mu_\nu(\gamma,\gamma').$$

In particular, if $\beta=\gamma\cup\gamma'$, then $\mu_0=\mu_1$ and $q_0=q_1$. We may use the simple notation $\mu(\gamma,\gamma')$, $q(z;\gamma,\gamma')$. Observe that the assumption $\beta=\gamma\cup\gamma'$, $\gamma\cap\gamma'=\emptyset$ means that γ cannot be arbitrary, for both γ and γ' are necessarily isolated subboundaries.

Modulus functions in the case $\mu_\nu=\infty$ will be introduced in 1 G. Harmonic measures will be discussed in 4 A.

Remark. In the case of plane regions modulus functions were first used in a potential-theoretic approach to conformal mappings onto circular and radial slit annuli (see Remark in III.2 B and Chapter VIII). On Riemann surfaces, STREBEL [2] considered q_0 using the Dirichlet principle (cf. 1 C.(g)), JURCHESCU [1] discussed q_1, and recently MARDEN-RODIN [1] introduced more general functions. These functions are, however, u_0 and u_1 (1 B) rather than modulus functions in our sense. Due to the normalization condition 1 A.(1), which is the counterpart of condition III.1 D.(1) for capacity functions, the modulus functions survive even if $\mu_\nu=\infty$ (Theorem 1 G).

1 E. Properties of Modulus Functions. Assume that $\mu_\nu<\infty$ and consider the modulus functions q_ν, $\nu=0,1$. They are real-valued and harmonic on W and have the following properties (for the meaning of (b) see III.2 C.(c)):

(a) $\int_c * dq_\nu = 2\pi$ *for every cycle c which separates γ from γ' and is oriented so that γ lies to the right.*

(b) $L_\nu q_\nu = q_\nu$ *on* $\beta-(\gamma\cup\gamma')$.

(c) $\int_c * dq_\nu = 0$ *for every dividing cycle c which does not separate components of* $\gamma\cup\gamma'$ (cf. III.2 C.(d)).

These properties are insufficient to characterize the functions q_ν in general. However, *if γ and γ' are isolated and $\mu_\nu<\infty$, then properties* (a) *and* (b) *and the following one characterize the q_ν up to an additive constant:*

$$H(q_\nu-k)=q_\nu-k \quad on\ \gamma, \qquad H(q_\nu-k')=q_\nu-k' \quad on\ \gamma'$$

for some constants k, k'. We also have $\log\mu_\nu = k-k'$ (cf. III.2 D).

Next let $KD_{\gamma'\gamma}(W)$ be the family of functions $q \in HD(W)$ (see II.1 F) satisfying (a) and (c) above, and let $KD^0_{\gamma'\gamma}(W)$ consist of functions q in $KD_{\gamma'\gamma}(W)$ with $L_0 q = q$ on $\beta - \gamma \cup \gamma'$. Then

$$q_1 \in KD_{\gamma'\gamma}(W), \quad q_0 \in KD^0_{\gamma'\gamma}(W), \tag{2}$$

$$D_W[q_\nu] = 2\pi\, l_\nu, \tag{3}$$

$$D_W[q - q_1] = D_W[q] - D_W[q_1] \tag{4}$$

for every $q \in KD_{\gamma'\gamma}(W)$,

$$D_W[q - q_0] = D_W[q] - D_W[q_0] \tag{5}$$

for every $q \in KD^0_{\gamma'\gamma}(W)$.

We have the following extremal property:

THEOREM. *If $\mu_\nu < \infty$, then q_0 and q_1 are the unique functions (up to additive constants) in $KD^0_{\gamma'\gamma}(W)$ and $KD_{\gamma'\gamma}(W)$ minimizing the Dirichlet integral. The minimum values are*

$$\min_{KD^0_{\gamma'\gamma}(W)} D_W[q] = 2\pi \log \mu_0, \qquad \min_{KD_{\gamma'\gamma}(W)} D_W[q] = 2\pi \log \mu_1.$$

The proof is similar to that of Theorem III.3 A. For q_1 the result is contained in JURCHESCU [1].

Reasoning parallel to that in III.3 C gives $\mu_1 \le \mu_0$. If $\mu_0 < \infty$, then equality holds if and only if $q_1 = q_0 + \text{const}$.

1 F. Construction by Exhaustion. Results analogous to those in III.4 A–4 F can easily be seen to hold in the present case.

For later reference, we add here the following, which for $\nu = 1$ is a special case of the counterpart of Theorem III.4 B.

THEOREM. *Let γ and γ' be disjoint subboundaries of W. Exhaust W by V towards γ and let $\mu_{\nu V}$ and $q_{\nu V}$ stand for μ_ν and q_ν on V with respect to $\gamma(V)$ and γ'. For $\nu = 0, 1$*

$$\mu_\nu = \lim_{V \to W} \mu_{\nu V}$$

and, if $\mu_V < \infty$,

$$\lim_{V \to W} D_V[q_{\nu V} - q_\nu] = 0.$$

Proof. Only the case $\nu = 0$ needs justification. Apply identity (5) to V:

$$D_V[q - q_{0V}] = D_V[q] - D_V[q_{0V}] \tag{6}$$

for every $q \in KD^0_{\gamma'\gamma(V)}(V)$. For $V \subset V'$, we thus have

$$D_V[q_{0V'} - q_{0V}] \le 2\pi(l_{0V'} - l_{0V}) \tag{7}$$

and, if $\mu_0 < \infty$,

$$D_V[q_0 - q_{0V}] \le 2\pi(l_0 - l_{0V}). \tag{8}$$

By (7), l_{0V} increases with V. If $\lim_{V\to W} l_{0V}=\infty$, (8) shows that $l_0=\infty$. If $\lim_{V\to W} l_{0V}<\infty$, we obtain from (7) a function $\tilde{q}$ such that

$$\lim_{V\to W} D_V[\tilde{q}-q_{0V}]=0.$$

Since $\tilde{q}\in KD^0_{\gamma'\gamma}(W)$, we have by (6)

$$D_V[\tilde{q}-q_{0V}]=D_V[\tilde{q}]-2\pi\, l_{0V}.$$

On letting $V\to W$, we see that

$$D_W[\tilde{q}]=2\pi \lim_{V\to W} l_{0V}<\infty.$$

Thus $l_0<\infty$ by an analogue of Theorem III.3 B. On substituting $q=q_0$ in (6) and letting $V\to W$, we conclude that

$$D_W[q_0-\tilde{q}]=D_W[q_0]-D_W[\tilde{q}].$$

By Theorem 1 E, $q_0=\tilde{q}+\text{const}$ and, consequently, $2\pi\, l_0=D_W[q_0]=D_W[\tilde{q}]=2\pi \lim_{V\to W} l_{0V}$.

1 G. Modulus Functions in the Case $\mu_\nu=\infty$. We have the following counterpart of Theorem III. 2 E:

THEOREM. *Let $\mu_\nu(\gamma,\gamma')=\infty$. Consider an exhaustion $V_n\to W$ towards $\gamma\cup\gamma'$. Denote by $q_{\nu n}$ the modulus function q_ν on V_n with respect to $\gamma(V_n)$ and $\gamma'(V_n)$. If the additive constants are suitably chosen, then $\{q_{\nu n}\}_{n=1}^{\infty}$ contains a subsequence which converges uniformly on every compact set.*

The proof, quite different from that of Theorem III. 2 E, will be given in 1 I. The limiting function is not necessarily unique, even if the additive constants are disregarded. But if no ambiguity arises, any limiting function will be denoted by $q_\nu=q_\nu(z;\gamma,\gamma')$ and we shall speak of *modulus functions q_0 and q_1*.

1 H. Modulus Functions and Principal and Capacity Functions. If W is obtained from a Riemann surface $\tilde{W}$ by deleting two distinct points $\zeta,\zeta'\in\tilde{W}$, then W has boundary components γ and γ' realized in a natural manner as ζ and ζ'. As in III. 5 B, one can establish the following relation between the principal function $p_\nu(z;\zeta,\zeta')$ on $\tilde{W}$ and a modulus function on W:

$$p_\nu(z;\zeta,\zeta')=q_\nu(z;\gamma,\gamma')+\text{const}.$$

In this case, $\mu_\nu(\gamma,\gamma')=\infty$ by Theorem 3 C below.

Next suppose $\tilde{W}$ is an open Riemann surface and choose a subboundary γ. Then a point $\zeta\in\tilde{W}$ can be regarded as a boundary component γ' of the surface $W=\tilde{W}-\{\zeta\}$. The modulus $\mu_\nu(\gamma,\gamma')$ is always infinite. It is plausible that the following relation should hold between capacity functions on $\tilde{W}$ and modulus functions on W:

$$p_{\nu\gamma}(z;\zeta)=q_\nu(z;\gamma,\gamma')+\text{const}.$$

We shall verify this in 3 E under the assumption that $c_{\nu\gamma}>0$ on W. If $c_{\nu\gamma}=0$, however, the question is open except in the following special cases. First, $\tilde{W}$ is obtained from a surface by deleting a single point which is regarded as γ (the proof is analogous to that in III.5 B). Second, for $\nu=1$, $\tilde{W}$ is a plane region and γ is a boundary component (the proof will be given in VIII.2 J).

Relations between modulus functions and principal and capacity functions will be discussed further in 3 D, 3 E, and Chapter V.

1 I. Proof of Theorem 1 G. The following complete proof was communicated to the authors by M. Ohtsuka.

Take a point $\zeta_0\in W\cap V_1$ and normalize the functions by $q_{\nu n}(\zeta_0)=0$. By virtue of the diagonal process, it suffices to show that, for every compact set K with $\zeta_0\in K$, there exists a subsequence of functions with the above normalization which converges uniformly on K.

Exhaust V_n by V_n^m $(m\to\infty)$ towards $\beta(V_n)-(\gamma(V_n)\cup\gamma'(V_n))$. The function $q_{\nu n}$ is approximated by the $q_{\nu n}^m$, which are the functions $q_\nu(z;\gamma(V_n),\gamma'(V_n))$ on V_n^m (use analogues of Theorems III.4 B, 4 F). Take $m=m(n)$ such that $|q_{\nu n}-q_{\nu n}^{m(n)}|<1/n$ on K. Then, as is easily seen, it is sufficient to prove that the sequence of functions $q_\nu^n=q_{\nu n}^{m(n)}$, $n=1,2,\dots$, contains a subsequence which converges uniformly on K.

On the regular subregion $V_n^{m(n)}$, we have

$$\int |{*dq_\nu^n}|\le 2\pi \tag{9}$$

along every level line $c=\{z\,|\,q_\nu^n(z)=\tau\ (=\text{const})\}$ which does not pass through zeros of the differential $dq_\nu^n+i*dq_\nu^n$. For the proof, consider the open set $\{z\,|\,q_\nu^n(z)<\tau\}$. It is bounded by $\gamma'(V_n)$, c, and a part δ of the border of $\overline{V}_n^{m(n)}$. We obtain

$$\int_c |{*dq_\nu^n}|\le \int_{c+\delta} |{*dq_\nu^n}|=\int_{c+\delta} *dq_\nu^n=\int_{-\gamma'(V_n)} *dq_\nu^n=2\pi.$$

About each point of K, consider a parametric disk N and a disk $N'\subset N$ with the same center and with radius one-half that of N. Select a finite number of disks $N_1',\dots,N_k'$ which cover K. We choose ζ_0 for the center of N_1'.

In N_1, the functions

$$\Phi_\nu^n=q_\nu^n+i\,q_\nu^{n*},\qquad n=1,2,\dots,$$

are single-valued. We normalize by $q_\nu^{n*}(\zeta_0)=0$, that is,

$$\Phi_\nu^n(\zeta_0)=0. \tag{10}$$

Consider the image region $\Phi_\nu^n(N_1)$ in the w-plane. For each n, relation (9) holds except for a finite number of τ's. We avoid these, consider two distinct values τ_1 and τ_2, and set $l_j=\{w\,|\,\mathrm{Re}\,w=\tau_j\}, j=1,2$. By (9), the linear measure of $l_j \cap \Phi_\nu^n(N_1)$ does not exceed 2π. As a consequence, we can take points w_{1n} and w_{2n} such that

$$w_{jn}\in l_j-\Phi_\nu^n(N_1), \qquad |\mathrm{Im}\,w_{jn}|\leq 2\pi, \qquad j=1,2.$$

Clearly there are numbers $A<\infty$ and $B>0$ independent of n and such that

$$|w_{1n}|\leq A, \qquad |w_{2n}|\leq A, \qquad |w_{1n}-w_{2n}|\geq B. \tag{11}$$

The family of functions

$$\frac{\Phi_\nu^n(z)-w_{1n}}{w_{2n}-w_{1n}}$$

is normal in N_1, since each member omits the values 0,1, and ∞; condition (10) excludes the possibility that the limiting function is identically infinite. By condition (11), the same is true of the sequence $\{\Phi_\nu^n\}$. Thus we obtain a subsequence $\{q_\nu^{i_n}\}$ which converges uniformly on each compact set in N_1.

We may assume that $N_1'\cap N_2'\neq\emptyset$. The values of $q_\nu^{i_n}$ are bounded at any point $\zeta_1\in N_1'\cap N_2'$. Thus we can carry out the above argument on N_2 by substituting ζ_1 for ζ_0. On repeating this until we have covered $N_1\cup\cdots\cup N_k$, we finally obtain a subsequence of $\{q_\nu^n\}$ which converges uniformly on K.

2. Expression of Modulus in Terms of Extremal Length

2 A. Extremal Length and Extremal Metric. We shall use terminology of Appendix I. On a Riemann surface W, let Γ be a family of locally rectifiable chains. Consider on W a non-negative and lower semicontinuous metric $\rho(z)\,|dz|$; we shall briefly refer to it as a metric ρ. Set

$$L(\Gamma,\rho)=\inf_{c\in\Gamma}\int_c \rho\,|dz|, \qquad A(\rho)=\iint_W \rho^2\,dx\,dy.$$

With the understanding that $0/0=\infty/\infty=0$, we introduce

$$\lambda(\Gamma)=\sup\frac{L(\Gamma,\rho)^2}{A(\rho)},$$

where the supremum is taken over all such metrics ρ. The quantity $\lambda(\Gamma)$ is called the *extremal length* of the family Γ. Its basic properties are found in Appendix I.

Denote by $\mathbf{P}(\Gamma)$ the family of metrics ρ with $L(\Gamma, \rho) \geq 1$ and $A(\rho) < \infty$. It is easily seen that if $\mathbf{P}(\Gamma) = \emptyset$, then $\lambda(\Gamma) = 0$, and otherwise

$$\frac{1}{\lambda(\Gamma)} = \inf_{\rho \in \mathbf{P}(\Gamma)} A(\rho).$$

If there exists a ρ in $\mathbf{P}(\Gamma)$ with $\lambda^{-1} = A(\rho)$, that is, one minimizing $A(\rho)$ in $\mathbf{P}(\Gamma)$, it is called an *extremal metric* with respect to the family Γ. If it exists, it is uniquely determined.

Let $\bar{\mathbf{P}}(\Gamma)$ be the L^2-completion of $\mathbf{P}(\Gamma)$, that is, the family of Borel measurable ρ which have a finite $A(\rho)$ and permit an approximation $A(\rho_n - \rho) \to 0$ by $\rho_n \in \mathbf{P}(\Gamma)$. An element ρ_0 of $\bar{\mathbf{P}}(\Gamma)$ with $A(\rho_0) = \lambda^{-1}$ is called a *generalized extremal metric* with respect to the family Γ. It was introduced by Strebel [3]. Clearly an extremal metric (if it exists) is a generalized extremal metric. *A generalized extremal metric ρ_0 always exists and, if $\lambda(\Gamma) > 0$, it satisfies*

$$A(\rho - \rho_0) \leq A(\rho) - A(\rho_0) \quad \textit{for every } \rho \in \bar{\mathbf{P}}(\Gamma), \tag{1}$$

so that it is unique. The proof of this proposition is given in Appendix I.C and I.D.

Following Ohtsuka [1] (see also Fuglede [1]), we shall say that a property is possessed by *almost all curves in Γ* if there exists a subfamily $\Gamma' \subset \Gamma$ with $\lambda(\Gamma') = \infty$ such that all curves in $\Gamma - \Gamma'$ have the property.

The above concepts are related to each other (Suita [4]; see Appendix I.E): *a Borel measurable ρ with $A(\rho) < \infty$ belongs to $\bar{\mathbf{P}}(\Gamma)$ if and only if $\int_c \rho\,|dz| \geq 1$ for almost all c in Γ.*

2 B. Families to be Considered. Let W be an open Riemann surface with more than one ideal boundary component. Take two disjoint subboundaries γ and γ', and consider the families Γ, Γ^*, $\tilde{\Gamma}$, and $\tilde{\Gamma}^*$ on W defined as follows (for terminology see Appendix I).

Γ consists of all rectifiable cycles of the type $c = \sum_i c_i$ in W which separate γ from γ'. Here the formal sum $\sum_i c_i$ is finite and each c_i is a rectifiable closed curve. Such a cycle is said to separate γ from γ' if it is homologous to ∂U for some neighborhood U of γ with $\beta_U \cap \gamma' = \emptyset$. For the definition of extremal length, the integral $\int_c \rho\,|dz|$ is understood as $\sum_i \int_{c_i} \rho\,|dz|$.

Γ^* is the family conjugate to Γ. It consists of all locally rectifiable curves in W joining γ and γ'; a curve $z(t)$ $(a < t < b)$ is called locally rectifiable if every subarc $z(t)$ $(a_1 \leq t \leq b_1)$ with $a < a_1 < b_1 < b$ is rectifiable (Appendix I.A). It is said to *join* subboundaries γ and γ' if, for all neighborhoods U and U' of γ and γ', there exists a positive number η such that

$$\{z(t) \mid a < t < a + \eta\} \subset U' \quad \text{and} \quad \{z(t) \mid b - \eta < t < b\} \subset U.$$

$\tilde{\Gamma}$ is the union of Γ and the family of chains $c=\sum_i c_i$ with the following properties: (a) c is the formal sum of at most a countable number of locally rectifiable curves in W joining subboundaries disjoint from $\gamma \cup \gamma'$, (b) c meets every curve in Γ^*, and (c) c is disjoint from some neighborhood of $\gamma \cup \gamma'$. The integral $\int_c \rho\,|dz|$ is again defined as $\sum_i \int_{c_i} \rho\,|dz|$. We note that each $c \in \tilde{\Gamma}$ may be considered as the restriction to W of a rectifiable cycle separating γ and γ' which is defined on $W \cup (\beta-(\gamma \cup \gamma'))$ in the sense of the Kerékjártó-Stoïlow compactification.

$\tilde{\Gamma}^*$ is conjugate to $\tilde{\Gamma}$. It consists of all curves in Γ^* and all chains which are at most countable sums of locally rectifiable curves in W joining subboundaries disjoint from $\gamma \cup \gamma'$, and which meet every cycle in Γ.

2 C. Expression of Modulus. For the above W, γ and γ', consider the moduli $\mu_v=\mu_v(\gamma,\gamma')$ and, if $\mu_v<\infty$, the modulus functions $q_v=q_v(z;\gamma,\gamma')$. The case $\infty=\infty$ is included in the following statement:

THEOREM. *The moduli satisfy the relations*

$$\log \mu_1 = \frac{2\pi}{\lambda(\Gamma)} = 2\pi\,\lambda(\tilde{\Gamma}^*),$$

$$\log \mu_0 = \frac{2\pi}{\lambda(\tilde{\Gamma})} = 2\pi\,\lambda(\Gamma^*).$$

If the moduli are finite, then the following are the generalized extremal metrics of the families indicated:

$$\Gamma:\ \frac{1}{2\pi}\,|dq_1+i*dq_1|,$$

$$\tilde{\Gamma}:\ \frac{1}{2\pi}\,|dq_0+i*dq_0|,$$

$$\Gamma^*:\ \frac{1}{\log \mu_0}\,|dq_0+i*dq_0|,$$

$$\tilde{\Gamma}^*:\ \frac{1}{\log \mu_1}\,|dq_1+i*dq_1|.$$

The proof will be given in 2 E – 2 M.

We cannot replace Γ by the subfamily consisting merely of curves instead of cycles. A counterexample will be given in 2 N. The replacement is permissible, however, if W is a plane region and γ, γ' are boundary components (VIII. 1 C).

From the proof, it will be seen that $(2\pi)^{-1}|dq_1+i*dq_1|$ is not only the generalized extremal metric, but also the extremal metric for Γ. That this is not necessarily true in the other cases is easily verified for a plane region W (cf. VII.2, 4).

Remark. The above theorem is a special case of the main theorem of MARDEN-RODIN [1]. The result for Γ^* was obtained by STREBEL [2]. Historically, the results for Γ and Γ^* under a certain restriction were given almost simultaneously with the introduction of extremal length by AHLFORS-BEURLING [1] (cf. Appendix I.J). The basic idea can be traced back to GRÖTZSCH [1]. The family $\tilde{\Gamma}^*$ was first considered by ANDREIAN-CAZACU [1]. Although $\tilde{\Gamma}$ and $\tilde{\Gamma}^*$ are rather intricate they are often useful for obtaining practical estimates (see the last paragraph of XI.5 D).

2 D. Extremal Properties of q_0 and q_1. The second half of the above theorem combined with the two propositions in 2 A.(1) gives extremal properties of q_0 and q_1. For example, the following is an addition to Theorem 1 E:

If $\mu_1<\infty$, then q_1 is the function, unique up to an additive constant, which minimizes $D_W[q]$ in the family of harmonic functions q such that $\int_c|dq+i*dq|\geq 2\pi$ for almost all $c\in\Gamma$.

The corresponding result for q_0 is quite different from those obtained thus far:

THEOREM. *If $\mu_0<\infty$, then $(\log\mu_0)^{-1}q_0$ is the function, unique up to an additive constant, which minimizes $D_W[q]$ in the family of harmonic functions q such that $\int_c|dq+i*dq|\geq 1$ for almost all $c\in\Gamma^*$.*

2 E. Γ and $\tilde{\Gamma}$ on Compact Bordered Surfaces. The proof of Theorem 2 C will occupy us through 2 M. In 2 E – 2 H, we shall give it for the interior W of a compact bordered surface $\overline{W}$.

Consider first the family Γ and the metric $\rho_1|dz|=|dq_1+i*dq_1|$. In view of 1 C.(a) and 1 A.(1),

$$\begin{gathered}\iint_W \rho_1^2\,dx\,dy=D_W[q_1]=2\pi\, l_1=2\pi\log\mu_1,\\ \int_c\rho_1|dz|\geq\left|\int_c *dq_1\right|=2\pi\end{gathered}\tag{2}$$

for all $c\in\Gamma$. Therefore,

$$\frac{2\pi}{\lambda(\Gamma)}\leq l_1.$$

On the other hand, we normalize by $q_1=0$ on γ', and consider the level lines $c_\tau=\{z\,|\,q_1(z)=\tau\}$, $0<\tau<l_1$. There are only a finite number of values of τ for which c_τ either passes through zeros of dq_1+i*dq_1 or meets

$\beta-(\gamma\cup\gamma')$. These values excluded, $c_\tau\in\Gamma$ and $\int_{c_\tau}\rho_1|dz|\le 2\pi$, as seen in 1 I. For any ρ, we then have

$$L(\Gamma,\rho)^2\le\Big(\int_{c_\tau}\rho|dz|\Big)^2\le\int_{c_\tau}\rho_1|dz|\cdot\int_{c_\tau}\frac{\rho^2}{\rho_1}|dz|\le 2\pi\int_{c_\tau}\frac{\rho^2}{\rho_1}|dz|.$$

We integrate with respect to τ over the interval $(0,l_1)$. At every point of every c_τ, the above exceptional values of τ excluded, $q_1+i\,q_1^*$ is a local parameter, and, therefore,

$$l_1\,L(\Gamma,\rho)^2\le 2\pi\int_0^{l_1}d\tau\int_{c_\tau}\frac{\rho^2}{\rho_1}|dz|=2\pi\iint_W\rho^2\,dx\,dy$$

for all ρ. It follows that $l_1\le 2\pi\,\lambda(\Gamma)^{-1}$, and, therefore,

$$\frac{2\pi}{\lambda(\Gamma)}=l_1=\log\mu_1.$$

In view of (2), we conclude that $\rho_1/2\pi$ is the extremal metric.

We consider next the family $\tilde{\Gamma}$. Since we are dealing with a compact bordered Riemann surface, it is not difficult to see that, for every $c\in\tilde{\Gamma}$, there exists a $c'\in\tilde{\Gamma}$ consisting of a finite number of analytic arcs in $\overline{W}$ with $\int_c *dq_0=\int_{c'}*dq_0$. On adding to c' appropriate subarcs of $\beta-(\gamma\cup\gamma')$, we obtain a cycle homologous to γ' and such that $\int_{c'}*dq_0=2\pi$. Thus $\rho_0\,|dz|=|dq_0+i*dq_0|$ satisfies

$$\int_c\rho_0\,|dz|\ge 2\pi$$

for all $c\in\tilde{\Gamma}$. Since $\iint_W\rho_0^2\,dx\,dy=2\pi\,l_0$, we obtain

$$\frac{2\pi}{\lambda(\tilde{\Gamma})}\le l_0=\log\mu_0.$$

As in the above argument for Γ, consideration of level lines $q_0=$const, which in this case belong to $\tilde{\Gamma}$, leads to

$$\frac{2\pi}{\lambda(\tilde{\Gamma})}\ge l_0$$

and the fact that $\rho_0/2\pi$ is the extremal metric.

2 F. Γ^* on Compact Bordered Surfaces. We continue considering the interior W of a compact bordered surface $\overline{W}$. For the above ρ_0,

$$\iint_W\rho_0^2\,dx\,dy=2\pi\,l_0,\qquad \int_c\rho_0\,|dz|\ge\int_c|dq_0|=l_0 \tag{3}$$

for every $c\in\Gamma^*$, so that

$$2\pi\,\lambda(\Gamma^*)\ge l_0.$$

To obtain the inequality in the opposite direction, consider a curve along which $*dq_0=0$, and extend it from a point on γ' as far as possible; if it encounters a zero of dq_0+i*dq_0, continue extending each branch. We call the resulting set a "trajectory". From the set of all trajectories we exclude those which pass through zeros of dq_0+i*dq_0. These are finite in number, and the remaining trajectories do not branch. Since $\int dq_0=0$ along each component of $\beta-(\gamma\cup\gamma')$, a zero of dq_0+i*dq_0 lies on it. Consequently, the remaining trajectories do not meet $\beta-(\gamma\cup\gamma')$. They are single curves in Γ^*, along which $\int\rho_0\,|dz|=\int dq_0=l_0$.

Fix a point $z_0\in\gamma'$. Since $*dq_0<0$ along γ' and $\int_{\gamma'}*dq_0=-2\pi$, we can determine a route which traces γ' exactly once beginning and ending at z_0. By means of this, we parametrize γ' in such a way that for each τ with $0\leq\tau<2\pi$, the point $z(\tau)$ is determined by $\int_{z_0}^{z(\tau)}*dq_0=-\tau$. The correspondence $\tau\to z(\tau)$ is one-to-one.

For each τ, consider the trajectory c_τ which starts from $z(\tau)$. Except for a finite number of values of τ, the trajectories c_τ are those which remain after the above exclusion. For these,

$$L(\Gamma^*,\rho)^2\leq\Big(\int_{c_\tau}\rho\,|dz|\Big)^2\leq\int_{c_\tau}\rho_0\,|dz|\cdot\int_{c_\tau}\frac{\rho^2}{\rho_0}|dz|=l_0\int_{c_\tau}\frac{\rho^2}{\rho_0}|dz|.$$

On integrating with respect to τ from 0 to 2π, we obtain

$$2\pi\,L(\Gamma^*,\rho)^2\leq l_0\iint_W\rho^2\,dx\,dy$$

for every ρ and, therefore, $2\pi\,\lambda(\Gamma^*)\leq l_0$. We infer that

$$2\pi\,\lambda(\Gamma^*)=l_0=\log\mu_0.$$

We see from (3) that ρ_0/l_0 is the extremal metric.

2 G. $\tilde{\Gamma}^*$ on Compact Bordered Surfaces. Again let W be the interior of a compact bordered surface. Normalize q_1 by

$$q_1=0 \quad \text{on } \gamma',$$

so that $q_1=l_1$ on γ. Consider $\rho_1\,|dz|=|dq_1+i*dq_1|$. We shall first show that

$$\int_c\rho_1\,|dz|\geq l_1=\log\mu_1 \tag{4}$$

for every $c=\sum_j c_j\in\tilde{\Gamma}^*$.

Let $\gamma(1),\ldots,\gamma(k)$ be the components of $\beta-(\gamma\cup\gamma')$, and denote by $l(1),\ldots,l(k)$ the constant values of q_1 on them. Assume that $l(1)\leq\cdots\leq l(k)$. On setting $l(0)=0$ and $l(k+1)=l_1$, we have $\sum_{i=0}^{k}(l(i+1)-l(i))=l_1$.

Consider $m_j = \max_{c_j} q_1$ and $m'_j = \min_{c_j} q_1$. By a simple topological consideration, we obtain

$$\bigcup_j [m'_j, m_j] = \bigcup_{i=0}^{k} [l(i), l(i+1)], \tag{5}$$

where the brackets indicate closed intervals. Accordingly,

$$\begin{aligned}\int_c \rho_1 |dz| &\geq \sum_j \int_{c_j} |dq_1| \geq \sum_j (m_j - m'_j) \\ &\geq \sum_i (l(i+1) - l(i)) = l_1,\end{aligned}$$

and (4) is established.

By virtue of

$$L(\tilde{\Gamma}^*, \rho_1) \geq l_1, \tag{6}$$

and

$$A(\rho_1) = D_W[q_1] = 2\pi\, l_1, \tag{7}$$

we have

$$\lambda(\tilde{\Gamma}^*) \geq \frac{l_1}{2\pi}.$$

Once the equality is proved, the fact that ρ_1/l_1 is the extremal metric can be deduced immediately from (6) and (7).

2 H. The Opposite Inequality. It remains to show that

$$\lambda(\tilde{\Gamma}^*) \leq \frac{l_1}{2\pi}.$$

We shall obtain a one-parameter family $\{c_\tau | 0 \leq \tau < 2\pi\} \subset \tilde{\Gamma}^*$ such that $*dq_1 = 0$ along each c_τ.

Fix a point $z_0 \in \gamma'$. We can parametrize γ' as in 2 F: $z = z(t)$, $\int_{z_0}^{z(\tau)} *dq_1 = -\tau$, $0 \leq \tau < 2\pi$. In the sequel, we shall formulate rules for decomposing the interval $0 \leq \tau < 2\pi$ into sets C and S.

Consider the "trajectory" determined by the condition $*dq_1 = 0$, in analogy with 2 F. For each τ, take the trajectory starting from $z(\tau)$, and extend it until it meets either a zero of the differential $dq_1 + i*dq_1$ or a point of $\beta - \gamma'$. In the former case, put τ in C. If the continuation ends at γ, put τ in S and denote the resulting curve by c_τ.

Suppose the continuation ends at a point ζ_τ on a component $\gamma(i)$ of $\beta - (\gamma \cup \gamma')$. If $dq_1 + i*dq_1 = 0$ at ζ_τ, put τ in C. Otherwise, follow $\gamma(i)$ from ζ_τ so that W lies to the left, starting in the direction of positive $*dq_1$. Let ζ'_τ be the first point for which the integral $\int_{\zeta_\tau}^{\zeta'_\tau} *dq_1$ taken along this path vanishes. If $dq_1 + i*dq_1$ vanishes at ζ'_τ, put τ in C. Otherwise a single curve along which $*dq_1 = 0$ starts from ζ'_τ, and we continue it as far as possible. If it encounters a zero of $dq_1 + i*dq_1$, then we put τ in C;

if it arrives at γ, we put τ in S and denote by c_τ the union of the curves from $z(\tau)$ to ζ_τ and from ζ'_τ to γ; if it arrives at a point on $\beta-(\gamma\cup\gamma')$, then we repeat the same procedure.

In this manner, we obtain the decomposition $[0,2\pi)=C\cup S$, $C\cap S=\emptyset$. To each $\tau\in S$ there corresponds a c_τ along which $*dq_1=0$; c_τ ends at a point on γ, so that $c_\tau\in\tilde{\Gamma}^*$. We see further that c_τ consists of a finite number of (connected) curves: $c_\tau=\bigcup_j c_j$. From the construction it is clear that

$$\int_{c_\tau} dq_1 = l_1. \tag{8}$$

Since no point on c_τ is a zero of dq_1+i*dq_1, we can use $q_1+i\,q_1^*$ as a local parameter, in terms of which ρ is given. For an arbitrary metric ρ,

$$\Big(\int_{c_\tau}\rho\,|dz|\Big)^2=\Big(\int_{c_\tau}\rho\,dq_1\Big)^2\le l_1\int_{c_\tau}\rho^2\,dq_1.$$

We integrate with respect to $\tau\in S$. Observe that C consists of only a finite number of values of τ and that any two curves c_τ for different τ's in S are disjoint. Thus

$$\begin{aligned}2\pi\,L(\tilde{\Gamma}^*,\rho)^2 &\le 2\pi\Big(\inf_\tau\int_{c_\tau}\rho\,|dz|\Big)^2\\ &\le l_1\int_{\tau\in S}d\tau\int_{c_\tau}\rho^2\,dq_1\le l_1\,A(\rho)\end{aligned}$$

and, consequently, $\lambda(\tilde{\Gamma}^*)\le l_1/2\pi$.

2 I. Γ and $\tilde{\Gamma}$ in the General Case. Now let W be an arbitrary surface and exhaust it by regular subregions Ω. Denote by Γ_Ω the family Γ defined on Ω with respect to $\gamma(\Omega)$ and $\gamma'(\Omega)$. Since

$$\lambda(\Gamma)\le\lambda(\Gamma_\Omega)=\frac{2\pi}{\log\mu_{1\Omega}},\qquad \lim_{\Omega\to W}\mu_{1\Omega}=\mu_1,$$

we see that $\lambda(\Gamma)\le 2\pi/\log\mu_1$. If $\mu_1<\infty$, consider $\rho_1|dz|=|dq_1+i*dq_1|$. In the same manner as in 2 E, we obtain $2\pi/\log\mu_1\le\lambda(\Gamma)$, hence $\lambda(\Gamma)=2\pi/\log\mu_1$. Again $\rho_1/2\pi$ is the extremal metric.

To prove that

$$\lambda(\tilde{\Gamma})\le\frac{2\pi}{\log\mu_0},$$

we take an exhaustion $V_n\to W$ towards $\gamma\cup\gamma'$. It suffices to show that $\lambda(\tilde{\Gamma}_n)\le 2\pi/\log\mu_{0n}$, since $\lambda(\tilde{\Gamma})\le\lambda(\tilde{\Gamma}_n)$ is trivially true and $\mu_0=\lim\mu_{0n}$ is known. Here $\tilde{\Gamma}_n$ is the family $\tilde{\Gamma}$ defined on V_n with respect to $\gamma(V_n)$ and $\gamma'(V_n)$, and μ_{0n} is the corresponding modulus.

Denote by q_{0n} the modulus function on V_n, and set $\rho_n|dz| = |dq_{0n}+i*dq_{0n}|$. We shall first show that, for almost all values of τ,

$$\int_{c_\tau} \rho_n |dz| \leq 2\pi \tag{9}$$

for the level line c_τ defined by $q_{0n}=\tau$. To this end, exhaust V_n by Ω_n^m $(m\to\infty)$ towards $\beta(V_n)-(\gamma(V_n)\cup\gamma'(V_n))$, and let q_{0n}^m and ρ_n^m correspond to Ω_n^m. The quantities ρ_n and ρ_n^m are different from zero on $c_\tau^m = c_\tau\cap\Omega_n^m$, $m=1,2,\ldots$, for almost all τ. The region $\{z\,|\,q_{0n}^m(z)<\tau\}$ is bounded by $\gamma'(V_n)$, c_τ, and a part of $\beta(\Omega_n^m)-(\gamma(V_n)\cup\gamma'(V_n))$. From this observation we conclude that

$$\int_{c_\tau^m} \rho_n^m |dz| \leq 2\pi,$$

as in 1 I. On letting $m\to\infty$, we obtain (9).

For an arbitrary metric ρ, we have successively

$$L(\tilde{\Gamma},\rho)^2 \leq \Big(\int_{c_\tau}\rho|dz|\Big)^2 \leq \int_{c_\tau}\rho_1|dz| \int_{c_\tau}\frac{\rho^2}{\rho_1}|dz| \leq 2\pi\int_{c_\tau}\frac{\rho^2}{\rho_1}|dz|,$$

$$l_0\, L(\tilde{\Gamma},\rho)^2 \leq 2\pi\, A(\rho),$$

and

$$l_0 \leq \frac{2\pi}{\lambda(\tilde{\Gamma})}.$$

2 J. Completion of the Proof for $\tilde{\Gamma}$. Under the assumption $\mu_0<\infty$, we shall show that

$$\lambda(\tilde{\Gamma}) \geq \frac{2\pi}{\log\mu_0}$$

and that $\rho_0|dz| = (2\pi)^{-1}|dq_0+i*dq_0|$ is the generalized extremal metric.

We use the exhaustion $V_n\to W$ towards $\gamma\cup\gamma'$ as well as $\Omega_n^m\to V_n$ $(m\to\infty)$ towards $\beta(V_n)-(\gamma(V_n)\cup\gamma'(V_n))$, and retain the above notation. For every $c\in\tilde{\Gamma}_n$, $c\cap\Omega_n^m$ belongs to $\tilde{\Gamma}_n^m$, the family $\tilde{\Gamma}$ on Ω_n^m. Therefore, $\int_c \rho_n^m|dz| \geq 2\pi$ (cf. 2 E). Moreover, $A(\rho_n-\rho_n^m)\to 0$ as $m\to\infty$. As a consequence, $(2\pi)^{-1}\rho_n|dz|$ belongs to $\bar{\mathbf{P}}(\tilde{\Gamma}_n)$, and there exists a subfamily $\tilde{\Gamma}_n'\subset\tilde{\Gamma}_n$ with

$$\lambda(\tilde{\Gamma}_n-\tilde{\Gamma}_n')=\infty$$

and

$$\int_c \rho_n|dz| \geq 2\pi \tag{10}$$

for all $c\in\tilde{\Gamma}_n'$.

Let $\tilde{\Gamma}'$ be the family of chains $c \in \tilde{\Gamma}$ which belong to $\tilde{\Gamma}_n'$ for every sufficiently large n; the lower bound of n depends on c. Since $\tilde{\Gamma} - \tilde{\Gamma}' \subset \bigcup_n (\tilde{\Gamma}_n - \tilde{\Gamma}_n')$ and $\lambda(\tilde{\Gamma}_n') = \lambda(\tilde{\Gamma}_n)$, it follows easily that

$$\lambda(\tilde{\Gamma}_n - \tilde{\Gamma}_n') = \infty.$$

Extend ρ_n to W by defining it to be zero on $W - V_n$, and set $\rho_0 \, |dz| = |dq_0 + i * dq_0|$. Because of the property $A(\rho_n - \rho_0) \to 0$ and inequality (10), we obtain $(2\pi)^{-1} \rho_0 \, |dz| \in \overline{\mathbf{P}}(\tilde{\Gamma}')$. Thus there exist a subfamily $\tilde{\Gamma}'' \subset \tilde{\Gamma}'$ with

$$\lambda(\tilde{\Gamma}'') = \lambda(\tilde{\Gamma}')$$

and

$$\int_c \rho \, |dz| \geq 2\pi \tag{11}$$

for every $c \in \tilde{\Gamma}''$. We conclude that

$$\lambda(\tilde{\Gamma}) \geq \frac{(2\pi)^2}{A(\rho_0)} = \frac{2\pi}{\log \mu_0}.$$

Furthermore, (11) shows that $\rho_0 / 2\pi$ belongs to $\overline{\mathbf{P}}(\tilde{\Gamma})$ and is, therefore, the generalized extremal metric.

2 K. Γ^* in the General Case. Take an exhaustion $\Omega_n \to W$ by regular subregions Ω_n satisfying

$$\mu_0 = \lim_{n \to \infty} \mu_{0n}$$

(cf. 1 F and III. 4 C). Here μ_{0n} is the modulus of Ω_n with respect to $\gamma(\Omega_n)$ and $\gamma'(\Omega_n)$. Similarly, denote the modulus function by q_{0n}. Write

$$\rho_n \, |dz| = (\log \mu_{0n})^{-1} \, |dq_{0n} + i * dq_{0n}|$$

and, if $\mu_0 < \infty$,

$$\rho_0 \, |dz| = (\log \mu_0)^{-1} \, |dq_0 + i * dq_0| .$$

Extend the former to W by defining it to be zero on $W - \Omega_n$. We clearly have

$$\int_c \rho_n \, |dz| \geq 1 \tag{12}$$

for every $c \in \Gamma^*$. Thus $\lambda(\Gamma^*) \geq 1/A(\rho_n) = (\log \mu_{0n})/2\pi$ and, consequently,

$$\lambda(\Gamma^*) \geq \frac{\log \mu_0}{2\pi}.$$

From (12) and the fact that $A(\rho_n - \rho_0) \to 0$ if $\mu_0 < \infty$, we see that ρ_0 belongs to $\overline{\mathbf{P}}(\Gamma^*)$. As soon as we verify that $\lambda(\Gamma^*) = (\log \mu_0)/2\pi$, we shall be able to conclude that ρ_0 is the generalized extremal metric.

To prove the opposite inequality

$$\lambda(\Gamma^*) \le \frac{\log \mu_0}{2\pi},$$

it suffices to show that

$$\lambda(\Gamma^*) \le \lim_{n\to\infty} \lambda(\Gamma_n^*), \tag{13}$$

where Γ_n^* stands for the family Γ^* on Ω_n with respect to $\gamma(\Omega_n)$ and $\gamma'(\Omega_n)$.

Consider the family Γ^{**} of curves $z(t)$ $(a<t<b)$ for which there are sequences $t_n' \to a$ and $t_n \to b$ with $z(t_n') \to \gamma'$ and $z(t_n) \to \gamma$. Clearly $\Gamma^* \subset \Gamma^{**}$. But it is easy to verify that $\lambda(\Gamma^{**}-\Gamma^*)=\infty$, so that $\lambda(\Gamma^{**})=\lambda(\Gamma^*)$ (cf. Appendix I. H.). Accordingly, the left-hand side of (13) may be replaced by $\lambda(\Gamma^{**})$.

The following proof of (13) is due to Strebel [2]. Since $\lambda(\Gamma_n^*)$ increases with n, the limiting value for $n\to\infty$ is defined. We may assume that

$$\lim_{n\to\infty} \lambda(\Gamma_n^*) < \infty,$$

since otherwise relation (13) is trivial.

For any $M<\lambda(\Gamma^{**})$, take a metric ρ' with $L(\Gamma^{**},\rho')^2/A(\rho')>M$. The quantity $\Lambda_n = L(\Gamma_n^*, \rho')$ increases with n. Since

$$\Lambda_n^2 \le \lambda(\Gamma_n^*)\, A(\rho') \le \lim \lambda(\Gamma_n^*)\, A(\rho') < \infty,$$

we observe that $\Lambda = \lim_{n\to\infty} \Lambda_n$ exists.

Given $\varepsilon>0$, we shall find a curve $c\in\Gamma^{**}$ with

$$\int_c \rho' |dz| < \Lambda + 4\varepsilon. \tag{14}$$

Then we shall have $M<L(\Gamma^{**},\rho')^2/A(\rho')<(\Lambda+4\varepsilon)^2/A(\rho')$, so that $M\le \Lambda^2/A(\rho')$ and

$$\lim_{n\to\infty} \lambda(\Gamma_n^*) \ge \lim_{n\to\infty} \frac{\Lambda_n^2}{A(\rho')} = \frac{\Lambda^2}{A(\rho')} \ge M,$$

which implies (13) with $\lambda(\Gamma^{**})$ on the left-hand side.

2 L. Completion of the Proof for Γ^*. In order to find a curve $c\in\Gamma^{**}$ satisfying (14), we take $c_n \in \Gamma_n^*$ such that

$$\int_{c_n} \rho' |dz| < \Lambda_n + \varepsilon,$$

$n=1,2,\ldots$. Given $k\ge 1$ and $n>k$ let z_{nk} be a point on $c_n \cap \gamma(\Omega_k)$. By choosing a suitable subsequence, we may assume the existence of $\lim_{n\to\infty} z_{nk} = z_k \in \gamma(\Omega_k)$ for every k.

Consider a small parametric disk punctured at z_k. We know the extremal length vanishes for the family of closed curves encircling z_k in the disk. Therefore, there exists a closed curve c_k' about z_k such that

$$\int_{c_k'} \rho'|dz| < \frac{\varepsilon}{2^k}.$$

For $z_{nk}' \in c_n \cap \gamma'(\Omega_k)$ and $z_k' = \lim_{n\to\infty} z_{nk}'$ we obtain in the same manner a closed curve c_k'' about z_k' with the property

$$\int_{c_k''} \rho'|dz| < \frac{\varepsilon}{2^k}.$$

Given k, the curves c_n, $n = n_0(k), n_0(k)+1, \ldots$ meet c_k' and c_k''. We take a sequence $\{n_j\}$ such that $n_1 < n_2 < \cdots$ and $n_0(j) \le n_j$. Then c_{n_j} meets each c_k' and c_k'', $k = 1, \ldots, j$. Let $\Gamma_{(k)}$ be the family of arcs of c_{n_j} joining c_k' and c_k''. We have

$$\sum_{k=1}^{j} L(\Gamma_{(k)}, \rho') \le \int_{c_{n_j}} \rho'|dz| < \Lambda_{n_j} + \varepsilon < \Lambda + \varepsilon$$

and, therefore,

$$\sum_{k=1}^{\infty} L(\Gamma_{(k)}, \rho') \le \Lambda + \varepsilon.$$

We are ready to construct a curve c satisfying (14). Take a $c_k^* \in \Gamma_{(k)}$ with

$$\int_{c_k^*} \rho'|dz| < L(\Gamma_{(k)}, \rho') + \frac{\varepsilon}{2^k}.$$

First consider c_1^* and choose a subarc of c_2^* joining c_1' and c_2' as well as one joining c_1'' and c_2''. Connect these with c_1^* by using subarcs of c_1' and c_1''. The resulting curve joins c_2' and c_2''. By using c_3^* extend the curve similarly to one joining c_3' and c_3''. Continue this process and, in the end, obtain a $c \in \Gamma^{**}$. It satisfies

$$\begin{aligned}\int_c \rho'|dz| &\le \sum_{k=1}^{\infty} \int_{c_k^*} \rho'|dz| + \sum_{k=1}^{\infty} \int_{c_k'} \rho'|dz| + \sum_{k=1}^{\infty} \int_{c_k''} \rho'|dz| \\ &\le \sum_{k=1}^{\infty} L(\Gamma_{(k)}, \rho') + 3\varepsilon \sum_{k=1}^{\infty} \frac{1}{2^k} < \Lambda + 4\varepsilon,\end{aligned}$$

and (14) is valid.

2 M. $\tilde{\Gamma}^*$ in the General Case. Again exhaust W by a sequence of regular subregions Ω_n. Since $\lambda(\tilde{\Gamma}^*) \ge \lambda(\tilde{\Gamma}_n^*) = (\log \mu_{1n})/2\pi$, we obtain

$$\lambda(\tilde{\Gamma}^*) \ge \frac{\log \mu_1}{2\pi}.$$

The opposite inequality is obtained by verifying

$$\lambda(\tilde{\Gamma}^*) \leq \lim_{n\to\infty} \lambda(\tilde{\Gamma}_n^*). \tag{15}$$

Though technically not quite simple, the proof of (15) is essentially the same as that of inequality (13): We introduce the family $\tilde{\Gamma}^{**}$ corresponding to Γ^{**} in 2 K, and construct a curve $c \in \tilde{\Gamma}^{**}$ satisfying the counterpart of (14). In contrast with the construction in 2 L, the set $c_n \cap \partial\Omega_n$ $(n>k)$ for $c_n \in \tilde{\Gamma}^{**}$ is now not necessarily contained in $\gamma(\Omega_k) \cup \gamma'(\Omega_k)$. However, the diagonal process enables us to obtain the desired c. For details, the interested reader is referred to MARDEN-RODIN [1, pp. 252–256].

If $\mu_1 < \infty$, the metric

$$\rho_1 |dz| = \frac{|dq_1 + i * dq_1|}{\log \mu_1}$$

satisfies $A(\rho_1) = 1/\lambda(\tilde{\Gamma}^*)$. It belongs to $\bar{\mathbf{P}}(\tilde{\Gamma}^*)$, since $A(\rho_{1n} - \rho_1) \to 0$, where $\rho_{1n}|dz| = (\log \mu_{1n})^{-1} |dq_{1n} + i * dq_{1n}|$ and, as was seen in 2 G, $\int_c \rho_{1n}|dz| \geq 1$. Hence $\rho_1 |dz|$ is the generalized extremal metric.

The proof of Theorem 2 C is herewith complete.

2 N. A Counterexample. We stated in 2 C that *the family Γ in Theorem 2 C cannot be replaced by the family Γ_0 of single closed curves in Γ.* We shall now prove this by giving an example in which

$$\lambda(\Gamma) = 0, \qquad \lambda(\Gamma_0) > 0.$$

Let a_n be real numbers with $0 < a_n < 1$, $n = 1, 2, \ldots$. Set $\Delta = \{z \mid 1 < |z| < \infty\} - \bigcup_{n=1}^{\infty} I_n$, where $I_n = \{z \mid n < \operatorname{Re} z < n + a_n, \operatorname{Im} z = 0\}$. Take two replicas of Δ and connect them along $\bigcup I_n$ crosswise. The resulting Riemann surface W has a boundary component γ over $z = \infty$ and a subboundary γ' which is the union of the circles $|z| = 1$ on both sheets. It is not difficult to see that $\lambda(\Gamma) = 0$.

In order to show that $\lambda(\Gamma_0) > 0$, let $\Delta_1 = \Delta - I$, where $I = \{z \mid -\infty < \operatorname{Re} z \leq -1, \operatorname{Im} z = 0\}$. Consider Δ_1 in the upper sheet, which is a subregion of W. For any $c \in \Gamma_0$, the part $c \cap \Delta_1$ contains a subarc joining I with one of the segments I_n. There may be several such arcs, but we always take the one with the smallest n. Denote this n by $n(c)$ and consider $\Gamma_k = \{c \mid c \in \Gamma_0, n(c) = k\}$, $k = 1, 2, \ldots$. Since $\Gamma_0 = \bigcup_{k=1}^{\infty} \Gamma_k$, we have $\lambda(\Gamma_0)^{-1} \leq \sum_{k=1}^{\infty} \lambda(\Gamma_k)^{-1}$ (Appendix I.G). Let Γ_k' be the family of plane curves in Δ_1 joining I and I_k. Then $\lambda(\Gamma_k')$ can be made arbitrarily large by taking a_n sufficiently small (Appendix I.K). For a choice of a_n with $\lambda(\Gamma_k') \geq 2^k$, we have $\lambda(\Gamma_k) \geq 2^k$ and, consequently, $\lambda(\Gamma_0)^{-1} \leq \sum 2^{-k} < \infty$.

3. Capacity and Modulus

3 A. Vanishing of Capacity. Let γ be a subboundary of an open Riemann surface W. Consider its capacities $c_{\nu\gamma}$, $\nu=0,1$, with respect to the reference point $\zeta\in W$. Let U be a connected neighborhood of γ such that $\zeta\notin\bar{U}$, and denote by μ_ν the modulus of U with respect to γ and $\alpha=\partial U$.

THEOREM. $c_{\nu\gamma}=0$ *if and only if* $\mu_\nu=\infty$.

This result and the corollaries in 3 B are due to JURCHESCU [1], CONSTANTINESCU [2], and MARDEN-RODIN [1].

Proof. Exhaust W towards γ by subregions V with $\alpha\cup\{\zeta\}\subset V$. Write k_V and p_V for $k_{\nu\gamma(V)}(\zeta)$ and $p_{\nu\gamma(V)}(z;\zeta)$ on V, and l_V and q_V for $l_\nu(\gamma(V),\alpha)$ and $q_\nu(z;\gamma(V),\alpha)$ on $V\cap U$. We know that $k_{\nu\gamma}=\lim_{V\to W}k_V$ by definition and $l_\nu=\lim_{V\to W}l_V$ by Theorem 1 F.

The properties of the operators L_0 and L_1 (I.2 B and 3 C) imply the equality

$$\int_{\alpha+\gamma(V)} q_V * dp_V = \int_{\alpha+\gamma(V)} p_V * dq_V,$$

so that

$$2\pi\, l_V = 2\pi\, k_V + \int_\alpha p_V * dq_V. \tag{1}$$

Take a sequence $\{V_n\}_{n=1}^\infty$ for which $\lim p_{V_n}$ and $\lim q_{V_n}$ exist (Theorems III.2 E and IV.1 G). The reflection guarantees the existence of $\lim * dq_{V_n}$ on α. Thus the integral in (1) has a finite limit and we arrive at the conclusion.

Remark. In case α is connected we may apply, with VIRTANEN [1], the mean value theorem to (1); see the proof of Theorem 3 C below.

3 B. Corollaries. We have the following consequences of Theorem 3 A:

(a) *For a subboundary* γ, *the property* $c_{\nu\gamma}=0$ *is independent of the reference point* ζ, *and is equivalent to* $\mu_\nu=\infty$, *which in turn is also independent of the neighborhood* U *of* γ.

(b) *If* γ *is an isolated subboundary, then the following four properties are equivalent:*

$$c_{0\gamma}=0, \quad c_{1\gamma}=0, \quad \mu_0=\infty, \quad \mu_1=\infty.$$

For the proof of (b), we only have to take a U with $\beta_U=\gamma$, on which $\mu_0=\mu_1$.

3 C. Further Results on Vanishing Capacity. Let γ and γ' be subboundaries with $\gamma\cap\gamma'=\emptyset$. We consider $c_{\nu\gamma}$ and $c_{\nu\gamma'}$ for W with respect to a point $\zeta \in W$, and also $\mu_\nu(\gamma, \gamma')$ for W. The following theorem is a generalization of Theorem 3 A:

THEOREM. $\mu_\nu(\gamma, \gamma')=\infty$ *if and only if at least one of* $c_{\nu\gamma}$, $c_{\nu\gamma'}$ *vanishes.*

Proof. Let U and U' be neighborhoods of γ and γ' such that $\zeta \notin \bar{U}$, $\zeta \notin \bar{U}'$, $\bar{U} \cap \bar{U}' = \emptyset$ and $\alpha = \partial U$ and $\alpha' = \partial U'$ are connected. Consider the moduli $\mu_\nu(\gamma, \alpha)$ and $\mu_\nu(\gamma', \alpha')$ of U and U'.

Suppose both $c_{\nu\gamma}$ and $c_{\nu\gamma'}$ are positive. Then by the preceding theorem, $\mu_\nu(\gamma, \alpha)$ and $\mu_\nu(\gamma', \alpha')$ are finite. Exhaust W towards $\gamma \cup \gamma'$ by V with $\alpha \cup \alpha' \subset V$, and denote by μ_V and μ'_V the moduli $\mu_\nu(\gamma(V), \alpha)$ and $\mu_\nu(\gamma'(V), \alpha')$ of $U \cap V$ and $U' \cap V$. By Theorem 1 F, these converge to $\mu_\nu(\gamma, \alpha)$ and $\mu_\nu(\gamma', \alpha')$ as $V \to W$. We normalize the functions $q_V = q_\nu(z; \gamma(V), \alpha)$ and $q'_V = q_\nu(z; \gamma'(V), \alpha')$ by

$$q_V = 0 \quad \text{on } \alpha, \qquad q'_V = 0 \quad \text{on } \alpha',$$

and let $\tilde{q}_V$ stand for the function $q_\nu(z; \gamma(V), \gamma'(V))$ on V. Then

$$\int_{\gamma(V)+\alpha} \tilde{q}_V * dq_V = \int_{\gamma(V)+\alpha} q_V * d\tilde{q}_V$$

as before. On denoting by a_V the constant value of $\tilde{q}_V$ on $\gamma(V)$, we obtain $2\pi\, a_V + \int_\alpha \tilde{q}_V * dq_V = 2\pi\, l_V$. Since $*dq_V < 0$ along α, and α is connected, the mean value theorem gives the existence of a point $z_V \in \alpha$ such that the integral equals $-2\pi\, \tilde{q}_V(z_V)$. We obtain

$$a_V - \tilde{q}_V(z_V) = l_V.$$

Similarly, we have a point $z'_V \in \alpha'$ with

$$a'_V - \tilde{q}_V(z'_V) = l'_V,$$

where a'_V is the constant value of $\tilde{q}_V$ on $\gamma'(V)$. Since $a_V - a'_V = l_\nu(\gamma(V), \gamma'(V))$,

$$l_\nu(\gamma(V), \gamma'(V)) - \tilde{q}_V(z_V) + \tilde{q}_V(z'_V) = l_V - l'_V.$$

On letting $V \to W$, we conclude that $l_\nu(\gamma(V), \gamma'(V)) < \infty$, whence $\mu_\nu(\gamma, \gamma') < \infty$.

Conversely, if $\mu_0(\gamma, \gamma') < \infty$, then the restrictions to $\bar{U}$ and $\bar{U}'$ of the function $q_0(z; \gamma, \gamma')$ belong to $KD^0_{\alpha\gamma}(\bar{U})$ and $KD^0_{\alpha'\gamma'}(\bar{U}')$ (see 1 C). Since these families are not void, we conclude by an analogue of Theorem III. 3 B that $\mu_0(\gamma, \alpha)$ and $\mu_0(\gamma', \alpha')$ are finite, so that $c_{0\gamma}$ and $c_{0\gamma'}$ are positive by Theorem 3 A. Similarly, $\mu_1(\gamma, \gamma') < \infty$ implies that both $c_{1\gamma}$ and $c_{1\gamma'}$ are positive.

3 D. Capacity Functions and Modulus Functions. We have already remarked that a capacity function can be regarded as a special modulus function (see 1 H). We shall now show that, conversely, under certain circumstances a modulus function can be regarded as a restriction of a capacity function.

Suppose a subboundary γ of W has positive capacity $c_{\nu\gamma}$, and consider the function $p_{\nu\gamma}(z;\zeta)$, $\zeta\in W$. Let U be a neighborhood of γ with $\zeta\notin\bar{U}$.

THEOREM. *If $c_{\nu\gamma}>0$, and $p_{\nu\gamma}=\text{const}$ on $\alpha=\partial U$, then the restriction of $p_{\nu\gamma}$ to U is a modulus function $q_\nu(z;\gamma,\alpha)$ of U.*

We have not succeeded in furnishing a proof in the case $c_{\nu\gamma}=0$.

Proof. Since the restriction of $p_{0\gamma}$ to U evidently belongs to $KD^0_{\alpha\gamma}(U)$, it suffices to show that

$$D_U[p_{0\gamma}]=2\pi\, l_0(\gamma,\alpha). \tag{2}$$

Exhaust W towards γ by V with $\alpha\subset V$. From equality (1) we see that

$$l_0(\gamma,\alpha)=k_{0\gamma}-a, \tag{3}$$

where a is the constant value of $p_{0\gamma}$ on α.

On the other hand, let U' be a neighborhood of the subboundary $\beta-\beta_U$ such that $\bar{U}\cap\bar{U}'=\emptyset$. By the definition of the quantity $\int_\beta p_{0\gamma}*dp_{0\gamma}$, we see that

$$\begin{aligned}2\pi\, k_{0\gamma}&=\int_\beta p_{0\gamma}*dp_{0\gamma}\\&=D_U[p_{0\gamma}]-\int_\alpha p_{0\gamma}*dp_{0\gamma}+D_{U'}[p_{0\gamma}]-\int_{\alpha'} p_{0\gamma}*dp_{0\gamma},\end{aligned}$$

where $\alpha'=\partial U'$. In view of a property of the operator L_0,

$$D_{U'}[p_{0\gamma}]-\int_{\alpha'} p_{0\gamma}*dp_{0\gamma}=0$$

and, therefore,

$$2\pi\, k_{0\gamma}=D_U[p_{0\gamma}]+2\pi\, a.$$

On combining this with (3) we obtain (2).

The proof in the case $\nu=1$ is completely analogous.

3 E. Further Relations between $p_{\nu\gamma}$ and q_ν. As was stated in 1 H, the above result can be generalized to the case where α degenerates to a single point.

Suppose γ is a subboundary of an open Riemann surface $\tilde{W}$ with $c_{\nu\gamma}>0$. Take a point ζ and consider $p_{\nu\gamma}=p_{\nu\gamma}(z,\zeta)$. The surface $W=\tilde{W}-\{\zeta\}$ has subboundaries γ and $\gamma'=\{\zeta\}$. The function $q_\nu=q_\nu(z;\gamma,\gamma')$ is not known to be unique, even if additive constants are disregarded.

THEOREM. *The restriction to W of the capacity function $p_{\nu\gamma}$ coincides with some modulus function q_ν.*

In the case $c_{\nu\gamma}=0$, the problem is open.

Proof. If $\varepsilon>0$ is sufficiently small, then $N_{\nu\varepsilon}=\{z \mid p_{\nu\gamma}(z)<\log\varepsilon\}$ is a relatively compact subregion of $\tilde{W}$ containing ζ. The region $W_\varepsilon=\tilde{W}-\bar{N}_{\nu\varepsilon}$ exhausts W towards γ' as $\varepsilon\to 0$. The restriction to W_ε of $p_{\nu\gamma}$ is, by Theorem 3 D, the modulus function $q_\nu^\varepsilon=q_\nu(z;\gamma,\partial N_{\nu\varepsilon})$ of W_ε. Exhaust W_ε towards γ by $V_\varepsilon\subset W_\varepsilon$. By Theorem 1 F, the function $q_\nu(z;\gamma(V_\varepsilon),\partial N_{\nu\varepsilon})$ converges to q_ν^ε. Consequently, on choosing ε and V_ε suitably, we obtain an exhaustion $V\to W$ towards $\gamma\cup\gamma'$ such that

$$p_{\nu\gamma}=\lim_{V\to W} q_\nu(z;\gamma(V),\gamma'(V))$$

on W.

3 F. Capacity in Terms of Modulus. Let γ be a subboundary of an open Riemann surface W. Consider the capacity function $p_{\nu\gamma}$ with pole at $\zeta\in W$. If $c_{\nu\gamma}=0$, we take an arbitrary $p_{\nu\gamma}$. Let $N_{\nu\varepsilon}=\{z \mid p_{\nu\gamma}(z)<\log\varepsilon\}$. There exists an $\varepsilon_0>0$ such that $N_{\nu\varepsilon}$ is relatively compact and $W-\bar{N}_{\nu\varepsilon}$ is a connected end for every ε with $0<\varepsilon<\varepsilon_0$.

We compare the capacity $c_{\nu\gamma}$ with respect to the point ζ and the modulus $\mu_\nu(\gamma,\partial N_{\nu\varepsilon})$ of $W-\bar{N}_\varepsilon$. If we exhaust W towards γ by V and use formula 3 A.(1), we obtain in the limit $l=k-\log\varepsilon$, which is valid even in the case $c_{\nu\gamma}=0$, $\mu_\nu=\infty$.

THEOREM. *There exists an ε_0 such that, for every ε with $0<\varepsilon<\varepsilon_0$, the capacity and the modulus are related by*

$$\log\frac{1}{c_{\nu\gamma}}=\log\mu_\nu(\gamma,\partial N_{\nu\varepsilon})+\log\varepsilon. \tag{4}$$

3 G. Further Expressions of Capacity in Terms of Modulus. Above we defined $N_{\nu\varepsilon}$ by means of the capacity function. We shall now make use of a region independent of this function.

For the local parameter t about ζ, in terms of which the capacity $c_{\nu\gamma}$ is considered, write $N_\varepsilon=\{t \mid |t|<\varepsilon\}$. Take an ε_0 so small that N_{ε_0} is a relatively compact simply connected planar region on which the analytic function $f(t)=\exp(p_{\nu\gamma}+i\,p_{\nu\gamma}^*)$ is univalent; then N_ε is relatively compact and $W-\bar{N}_\varepsilon$ is connected. Since $f(0)=0$ and $|f'(0)|=1$, we have by Koebe's distortion theorem (see, e.g., NEHARI [3, p. 217])

$$\varepsilon_1(\varepsilon)\le|f(t)|\le\varepsilon_2(\varepsilon)$$

on $|t|=\varepsilon$, $0<\varepsilon<\varepsilon_0$. Here

$$\varepsilon_1(\varepsilon)=\frac{\varepsilon\,\varepsilon_0^2}{(\varepsilon_0+\varepsilon)^2},\qquad \varepsilon_2(\varepsilon)=\frac{\varepsilon\,\varepsilon_0^2}{(\varepsilon_0-\varepsilon)^2},$$

and $N_{\nu\varepsilon_1} \subset N_\varepsilon \subset N_{\nu\varepsilon_2}$. If ε_0 is small, then $N_{\nu\varepsilon_1}$ and $N_{\nu\varepsilon_2}$ satisfy the assumption for (4). Since $\mu_\nu(\gamma, \partial N_{\nu\varepsilon_1}) \le \mu_\nu(\gamma, \partial N_\varepsilon) \le \mu_\nu(\gamma, \partial N_{\nu\varepsilon_2})$, we obtain:

THEOREM. *There exists an ε_0 such that, for every ε with $0<\varepsilon<\varepsilon_0$, the capacity satisfies the following inequalities in terms of the modulus of $W-\bar{N}_\varepsilon$:*

$$\log \mu_\nu(\gamma, \partial N_\varepsilon) + \log \varepsilon_1(\varepsilon) \le \log \frac{1}{c_{\nu\gamma}} \le \log \mu_\nu(\gamma, \partial N_\varepsilon) + \log \varepsilon_2(\varepsilon). \tag{5}$$

In particular,

$$\log \frac{1}{c_{\nu\gamma}} = \lim_{\varepsilon \to 0} \left(\log \mu_\nu(\gamma, \partial N_\varepsilon) + \log \varepsilon \right). \tag{6}$$

By making use of extremal length, we easily see that the quantity in parentheses increases as ε decreases.

Since the modulus is determined by extremal length, (6) tells us that *the capacity can be expressed in terms of extremal length.*

Remark 1. Relation (6) was first obtained by TEICHMÜLLER [1] for simply connected plane regions.

Remark 2. If W is a plane region and γ is a boundary component, then ε_0 in the theorem is arbitrary as long as $N_{\varepsilon_0} \subset W$. Indeed, $f(t)$ is univalent throughout W (VII.1 B), so that Koebe's distortion theorem can be applied to such an N_{ε_0}.

4. Harmonic Measure

4 A. Harmonic Measure $u_{\nu\gamma}$. In contrast with modulus functions, harmonic measures will be considered only on an end, not on the entire surface.

Let γ be a subboundary of an open Riemann surface. Let U be a connected neighborhood of γ and set $\alpha = \partial U$. If the modulus $\mu_\nu(\gamma, \alpha)$ of U is finite, then the modulus function $q_\nu = q_\nu(z; \gamma, \alpha)$ on U is obtained by an exhaustion of U towards γ (Theorem 1 F); in particular, it is harmonic on $U \cup \alpha$ and constant on α. We use the function with $q_\nu = 0$ on α, and let

$$u_{\nu\gamma} = u_\nu(z; \gamma, \alpha) = \frac{q_\nu(z; \gamma, \alpha)}{\log \mu_\nu(\gamma, \alpha)}.$$

In the case $\mu_\nu = \infty$, we take $u_{\nu\gamma} \equiv 0$. It will be called the *harmonic measure* $u_{\nu\gamma}$ of γ with respect to U. It satisfies

$$D_U[u_{\nu\gamma}] = \frac{2\pi}{\log \mu_\nu(\gamma, \alpha)}. \tag{1}$$

Clearly

$$u_{\nu\gamma} \equiv 0 \text{ if and only if } c_{\nu\gamma} = 0. \tag{2}$$

If γ is realized as a compact border, then $u_{\nu\gamma}$ coincides with the function introduced in 1 B.

Exhaust W towards γ by V with $\alpha \subset V$. If we denote by $u_{\nu\gamma(V)}$ the harmonic measure $u_\nu(z; \gamma(V), \alpha)$ on $V \cap U$, then

$$u_{\nu\gamma(V)} \to u_{\nu\gamma} \quad \text{as } V \to W, \tag{3}$$

$$\lim_{V \to W} D_{U \cap V}[u_{\nu\gamma(V)} - u_{\nu\gamma}] = 0. \tag{4}$$

It is not difficult to verify (cf. III. 3 F): *The harmonic measure* $\mu_{\nu\gamma}$ *is characterized by the extremal property*

$$u_{\nu\gamma}(z) = \max_{u \in \mathscr{H}_\nu(\bar{U})} u(z), \tag{5}$$

where $\mathscr{H}_\nu(\bar{U})$ *is the family of harmonic functions* u *on* $\bar{U}$ *such that* $u = 0$ *on* α, $0 \leq u \leq 1$ *on* U, *and* $L_\nu u = u$ *on* $\beta_U - \gamma$.

From this we infer that, if $u_{\nu\gamma} \neq 0$, then

$$\inf_U u_{\nu\gamma} = 0, \qquad \sup_U u_{\nu\gamma} = 1. \tag{6}$$

4 B. Harmonic Measures for Isolated γ. If γ is isolated, then $u_{\nu\gamma}$ is characterized by the conditions (cf. III. 2 D)

$$u_{\nu\gamma} \in \mathscr{H}_\nu(\bar{U}), \qquad H(u_{\nu\gamma} - 1) = u_{\nu\gamma} - 1 \qquad \text{on } \gamma. \tag{7}$$

In particular, if $\gamma = \beta_U$, then $u_{0\gamma}$ and $u_{1\gamma}$ coincide, and we shall denote the harmonic measure simply by u_γ. It is easily seen that

$$u_\gamma = 1 - H1,$$

where the operator H acts from α into U. Thus u_γ coincides with the function discussed in I. 5 D.

4 C. Harmonic Measure u_γ for Arbitrary γ. Given an arbitrary subboundary γ of an open Riemann surface W, consider a connected neighborhood U of γ. The harmonic measure u_γ was extended to this case by Constantinescu [2].

Exhaust W towards γ by V with $\alpha \subset V$. The entire ideal boundary of $V \cap U$ consists of $\alpha = \partial U$, $\gamma(V)$, and $\beta_U \cap \beta_V$. Here β_V is the subboundary of W in the sense of I. 1 B, where V is regarded as an end of W. Let $u_{\gamma(V)}$ be the harmonic function on $(V \cap U) \cup \alpha \cup \gamma(V)$ satisfying the following conditions: $u_{\gamma(V)} = 0$ on α, $u_{\gamma(V)} = 1$ on $\gamma(V)$, and $H u_{\gamma(V)} = u_{\gamma(V)}$ on $\beta_U \cap \beta_V$. It is readily seen that $u_{\gamma(V)}$ decreases as V increases and we obtain the function

$$u_\gamma = \lim_{V \to W} u_{\gamma(V)},$$

which we shall call the *harmonic measure* u_γ of γ with respect to U.

It is characterized by the extremal property

$$u_\gamma(z) = \max_{u \in \mathscr{H}(\bar{U})} u(z), \tag{8}$$

where $\mathscr{H}(\bar{U})$ is the family of harmonic functions u on $\bar{U}$ such that $u=0$ on α, $0 \leq u \leq 1$ on U, and $Hu = u$ on $\beta_U - \gamma$.

If $u_\gamma \neq 0$, then

$$\inf_U u_\gamma = 0, \qquad \sup_U u_\gamma = 1. \tag{9}$$

In particular, if γ is isolated, then u_γ is characterized by

$$u_\gamma \in \mathscr{H}(\bar{U}), \qquad H(u_\gamma - 1) = u_\gamma - 1 \quad \text{on } \gamma. \tag{10}$$

If in addition $\gamma = \beta_U$, then the present u_γ coincides with the function u_γ in 4 B.

4 D. Vanishing of the Harmonic Measure u_γ. We claim:

THEOREM. *If $c_{0\gamma} = 0$, then $u_\gamma = 0$.*

In fact, from the definition and a property of the operator H, we at once have $u_{\gamma(V)} \leq u_{0\gamma(V)}$. Therefore, $u_\gamma \leq u_{0\gamma}$ and the theorem follows.

Remark 1. The validity of $u_\gamma = 0$ will be seen to be independent of U and therefore a property of γ (Corollary X. 1 J).

Remark 2. If γ is isolated, then $u_\gamma = 0$ if and only if $c_{0\gamma} = 0$. In general, however, the converse of the above theorem is not true as will be seen from an example in XI. 5 D.

Remark 3. In contrast with $u_{v\gamma}$, we do not have

$$\lim_{V \to W} D_{U \cap V}[u_{\gamma(V)} - u_\gamma] = 0.$$

The case $u_\gamma = 0$ and $c_{0\gamma} > 0$ of Remark 2 is a counterexample, since

$$u_\gamma = 0, \qquad \underline{\lim}_{V \to W} D_{U \cap V}[u_{\gamma(V)}] > 0.$$

In fact, if Γ_V^* is the family of curves joining α and $\gamma(V)$ in $U \cap V$ and if $\Gamma_{(V)}^*$ is the family of curves joining $\alpha \cup (\beta_U \cap \beta_V)$ and $\gamma(V)$ in $U \cap V$, then (cf. the first half of 2 K)

$$D_{U \cap V}[u_{\gamma(V)}] \geq \frac{1}{\lambda(\Gamma_{(V)}^*)} \geq \frac{1}{\lambda(\Gamma_V^*)},$$

so that

$$\underline{\lim}_{V \to W} D_{U \cap V}[u_{\gamma(V)}] \geq \frac{2\pi}{\log \mu_0(\gamma, \alpha)} > 0.$$

4 E. Another Approach. CONSTANTINESCU [2] constructed harmonic measures $u_{\nu\gamma}$ and u_γ by the method in III.4 G. We shall show that the functions so formed coincide with those in 4 A – 4 C above.

Recall the definition of a subboundary γ in I.1 B as a correspondence $\Omega \to U_\Omega$. Given a neighborhood U of γ, consider only those Ω for which $U_\Omega \subset U$, and set $\gamma_\Omega = \beta_{U_\Omega}$. It satisfies $\gamma \subset \gamma_\Omega \subset \beta_U$ and is an isolated subboundary, so that the functions $u_{\nu\gamma_\Omega}$ and u_{γ_Ω} are characterized by (7) and (10) in a simpler fashion than are those for arbitrary subboundaries.

THEOREM. *The harmonic measures $u_{\nu\gamma}$ and u_γ have the properties*

$$\lim_{\Omega \to W} D_U[u_{\nu\gamma_\Omega} - u_{\nu\gamma}] = 0,$$

$$u_{\nu\gamma_\Omega} \to u_{\nu\gamma} \quad \text{as } \Omega \to W,$$

$$u_{\gamma_\Omega} \to u_\gamma \quad \text{as } \Omega \to W.$$

Note that $\lim D_U[u_{\gamma_\Omega} - u_\gamma] = 0$ again fails to hold.

Proof. The proof for $u_{\nu\gamma}$ is similar to that of Theorem III.4 G and is left to the reader (cf. also AHLFORS-SARIO [1, p. 147, (C 3)]).

The fact that u_{γ_Ω} decreases as Ω increases is immediate from extremal property (8). For the same reason, $u_\gamma \le \lim_{\Omega \to W} u_{\gamma_\Omega}$. On the other hand, $U - \bar{U}_\Omega$ exhausts U towards γ as $\Omega \to W$, and its relative boundary is $\gamma(\Omega)$. Hence $u_\gamma = \lim u_{\gamma(\Omega)}$. It is easily seen that $u_{\gamma_\Omega} \le u_{\gamma(\Omega)}$ on $U - U_\Omega$, and we conclude that $\lim_{\Omega \to W} u_{\gamma_\Omega} \le u_\gamma$.

Chapter V

Relations between Fundamental Functions

Having thus far discussed properties of principal functions, capacity functions, and modulus functions, we shall, in the present chapter, derive some relations between these functions.

1. Principal Functions and Capacity Functions

1 A. Kernels. For the differentials φ_m and ψ_m (II.3 B), we shall write

$$\varphi_m = \pi\, L_m(z,\zeta)\, dz, \qquad \psi_m = \pi\, K_m(z,\zeta)\, dz.$$

In the case $m=0$, the subindex will often be omitted:

$$L(z,\zeta) = L_0(z,\zeta), \qquad K(z,\zeta) = K_0(z,\zeta).$$

For the differentials $\tilde{\varphi}_m$ and $\tilde{\psi}_m$, we shall similarly use $\tilde{L}_m(z,\zeta)$, $\tilde{K}_m(z,\zeta)$, and

$$\tilde{L}(z,\zeta) = \tilde{L}_0(z,\zeta), \qquad \tilde{K}(z,\zeta) = \tilde{K}_0(z,\zeta).$$

We recall that these kernels depend on the choice of the parameter about ζ. Finally, for the differentials $\varphi(c)$ and $\psi(c)$ (II.3 D), we set

$$\varphi(c) = \pi\, L(z,c)\, dz, \qquad \psi(c) = \pi\, K(z,c)\, dz.$$

The notation $\tilde{L}(z,c)$ and $\tilde{K}(z,c)$ has an analogous meaning.

On an open Riemann surface W, consider the capacity functions $p_\beta(z;\zeta)$ with respect to the ideal boundary β, and $p_{1\gamma}(z;\zeta)$ with respect to an ideal boundary *component* γ. We shall not be concerned with capacity functions for an arbitrary subboundary.

THEOREM. *The kernels can be expressed in terms of the capacity function p_β, including the cases $c_\beta=0$ and $c_{1\gamma}=0$:*

(a) *If $z \neq \zeta$, then*

$$L_m(z,\zeta) = \frac{2}{\pi}\,\frac{1}{(m+1)!}\,\frac{\partial^{m+2} p_\beta(\zeta;z)}{\partial z\,\partial\zeta^{m+1}},$$

$$K_m(z,\zeta) = \frac{2}{\pi}\,\frac{1}{(m+1)!}\,\frac{\partial^{m+2} p_\beta(\zeta;z)}{\partial z\,\partial\bar{\zeta}^{m+1}}.$$

(b) *If* $z \notin c$, *then*

$$L(z,c) = \frac{1}{\pi}\frac{\partial}{\partial z}\int_c (d_\zeta\, p_\beta(\zeta;z) + i * d_\zeta\, p_\beta(\zeta;z)),$$

$$K(z,c) = \frac{1}{\pi}\frac{\partial}{\partial z}\int_c (d_\zeta\, p_\beta(\zeta;z) - i * d_\zeta\, p_\beta(\zeta;z)).$$

Similar relations hold for $\tilde{L}_m(z,\zeta)$, $\tilde{K}_m(z,\zeta)$, $\tilde{L}(z,c)$, *and* $\tilde{K}(z,c)$ *if* $p_\beta(z;\zeta)$ *is replaced by* $p_{1\gamma}(z;\zeta)$ *with respect to an arbitrary ideal boundary component* γ.

If $c_\beta > 0$, then identities (a) for $m=0$ may be written in terms of the Green's function:

$$L(z,\zeta) = -\frac{2}{\pi}\frac{\partial^2 g(z;\zeta)}{\partial z\,\partial \zeta}, \qquad K(z,\zeta) = -\frac{2}{\pi}\frac{\partial^2 g(z;\zeta)}{\partial z\,\partial \bar{\zeta}}.$$

These relations are due to SCHIFFER [3] (see also WIRTINGER [1]).

Since

$$d_\zeta\, p = \frac{\partial p}{\partial \zeta}\, d\zeta + \frac{\partial p}{\partial \bar{\zeta}}\, d\bar{\zeta}, \qquad * d_\zeta\, p = -i\left(\frac{\partial p}{\partial \zeta}\, d\zeta - \frac{\partial p}{\partial \bar{\zeta}}\, d\bar{\zeta}\right),$$

we can modify (b) to the following form:

$$L(z,c) = \frac{2}{\pi}\frac{\partial}{\partial z}\int_c \frac{\partial p_\beta(\zeta;z)}{\partial \zeta}\, d\zeta, \qquad K(z,c) = \frac{2}{\pi}\frac{\partial}{\partial z}\int_c \frac{\partial p_\beta(\zeta;z)}{\partial \bar{\zeta}}\, d\bar{\zeta}.$$

Similar relations hold between $\tilde{L}(z,c)$, $\tilde{K}(z,c)$, and $p_{1\gamma}(z;\zeta)$. These identities are contained in GARABEDIAN-SCHIFFER [1].

The proof of the theorem will be given in 1 B – 1 E.

1 B. Proof of (a). First consider the interior W of a compact bordered surface $\overline{W}$. Denote the functions p_1, p_1' of II.3 B by p_{1m}, p_{1m}'. We shall show that the following relations hold for $\zeta_1 \neq \zeta_2$:

$$\frac{2}{m!}\left(\frac{\partial^{m+1} p_\beta(z;\zeta_2)}{\partial \bar{z}^{m+1}}\right)_{z=\zeta_1} = a(\zeta_1) - (p_{1m}(\zeta_2;\zeta_1) - i\, p_{1m}'(\zeta_2;\zeta_1)), \tag{1}$$

$$\frac{2}{m!}\left(\frac{\partial^{m+1} p_\beta(z;\zeta_2)}{\partial z^{m+1}}\right)_{z=\zeta_1} = b(\zeta_1) - (p_{1m}(\zeta_2;\zeta_1) + i\, p_{1m}'(\zeta_2;\zeta_1)), \tag{2}$$

where $a(\zeta_1)$ and $b(\zeta_1)$ are the constant values of $p_{1m}(z;\zeta_1) - i\, p_{1m}'(z;\zeta_1)$ and $p_{1m}(z;\zeta_1) + i\, p_{1m}'(z;\zeta_1)$ on the border of $\overline{W}$.

For the proof, we consider the reproducing meromorphic differentials

$$\varphi=\frac{\partial}{\partial z}(p_{1m}+i\,p'_{1m})\,dz, \qquad \psi=\frac{\partial}{\partial z}(p_{1m}-i\,p'_{1m})\,dz$$

with $p_{1m}=p_{1m}(z;\zeta_1)$ and $p'_{1m}=p'_{1m}(z;\zeta_1)$, and the differential

$$\omega=dp_\beta+i*dp_\beta$$

with $p_\beta=p_\beta(z;\zeta_2)$. Let N_1 and N_2 be neighborhoods of ζ_1 and ζ_2 with $\bar{N}_1\cap\bar{N}_2=\emptyset$. It is readily seen that

$$-i\int_{W-N_1-N_2} d(p_{1m}-i\,p'_{1m})\wedge\omega=2i\int_{W-N_1-N_2} dp_\beta\wedge\bar{\varphi}, \tag{3}$$

$$-i\int_{W-N_1-N_2} d(p_{1m}+i\,p'_{1m})\wedge\omega=2i\int_{W-N_1-N_2} dp_\beta\wedge\bar{\psi}. \tag{4}$$

Indeed, by virtue of $*\omega=-i\omega$ and the corresponding properties of φ and ψ, the expressions are equal to the inner products (ω,φ) and (ω,ψ), respectively, over $W-N_1-N_2$ (cf. II.3 A).

Concerning the left-hand side of (3), we have

$$-i\int_{W-N_1-N_2} d(p_{1m}-i\,p'_{1m})\wedge\omega=-i\int_{\partial W-\partial N_1-\partial N_2}(p_{1m}-i\,p'_{1m})\,\omega,$$

where

$$\int_{\partial W}(p_{1m}-i\,p'_{1m})\,\omega=a(\zeta_1)\int_{\partial W}\omega=i\,a(\zeta_1)\int_{\partial W}*dp_\beta=2\pi\,i\,a(\zeta_1).$$

If t is a local parameter about ζ_1 such that $p_{1m}-i\,p'_{1m}\sim\bar{t}^{-m-1}$ and $\omega=(c_0+c_1\,t+\cdots)\,dt$, then

$$-\int_{\partial N_1}(p_{1m}-i\,p'_{1m})\,\omega\sim-\int_{\partial N_1}(c_0+c_1\,t+\cdots)\,\bar{t}^{-m-1}\,dt=0,$$

with N_1 the disk $|t|<r$, and $r\to 0$ (cf. II.3 C). If t is a local parameter about ζ_2 such that $\omega\sim t^{-1}\,dt$, then

$$-\int_{\partial N_2}(p_{1m}-i\,p'_{1m})\,\omega\sim-\int_{\partial N_2}(p_{1m}-i\,p'_{1m})\,t^{-1}\,dt=-2\pi\,i\big(p_{1m}(\zeta_2)-i\,p'_{1m}(\zeta_2)\big)$$

as N_2 shrinks to the point ζ_2. As a consequence, the left-hand side of (3) is equal to

$$2\pi\,a(\zeta_1)-2\pi\big(p_{1m}(\zeta_2)-i\,p'_{1m}(\zeta_2)\big). \tag{5}$$

As for the left-hand side of (4), the same reasoning applies except that

$$-\int_{\partial N_1}(p_{1m}+i\,p'_{1m})\,\omega\sim-\int_{\partial N_1}(c_0+c_1\,t+\cdots)\,t^{-m-1}\,dt=-2\pi\,i\,c_m$$

$$=\frac{-4\pi\,i}{m!}\left(\frac{\partial^{m+1}p_\beta}{\partial z^{m+1}}\right)_{z=\zeta_1}.$$

Thus the left-hand side of (4) becomes

$$2\pi\, b(\zeta_1) - \frac{4\pi}{m!}\left(\frac{\partial^{m+1} p_\beta}{\partial z^{m+1}}\right)_{z=\zeta_1} - 2\pi\left(p_{1m}(\zeta_2) + i\, p'_{1m}(\zeta_2)\right). \tag{6}$$

1 C. Proof of (1) and (2). We proceed to the right-hand sides of (3) and (4). We have

$$2i \int_{W-N_1-N_2} dp_\beta \wedge \overline{\varphi} = 2i \int_{\partial W - \partial N_1 - \partial N_2} p_\beta\, \overline{\varphi}.$$

Since $p_\beta = \text{const}$ on β and $\int_{\partial W} \overline{\varphi} = 0$, we obtain $\int_{\partial W} p_\beta \overline{\varphi} = 0$. Let t be a parameter about ζ_1 such that $\varphi \sim -(m+1)\, t^{-m-2}\, dt$. Denote by p_β^* any branch of the locally defined conjugate of p_β on N_1 and set $P_\beta = p_\beta + i\, p_\beta^*$. Then

$$- \int_{\partial N_1} p_\beta \overline{\varphi} \sim (m+1) \int_{\partial N_1} p_\beta\, \bar{t}^{-m-2}\, \overline{dt}$$

$$= \frac{m+1}{2} \int_{\partial N_1} (p_\beta - i\, p_\beta^*)\, \bar{t}^{-m-2}\, \overline{dt} + \frac{m+1}{2} \int_{\partial N_1} (p_\beta + i\, p_\beta^*)\, \bar{t}^{-m-2}\, \overline{dt}.$$

If ∂N_1 corresponds to $|t| = r$, then $\bar{t} = r^2 t^{-1}$ on ∂N_1 and $\overline{dt} = -r^2 t^{-2}\, dt$ along ∂N_1, so that the last integral above becomes

$$\int_{|t|=r} (p_\beta + i\, p_\beta^*)\, t^{2m+4}\, dt.$$

Since the integrand is regular, the integral vanishes, and we obtain

$$- \int_{\partial N_1} p_\beta \overline{\varphi} \sim \frac{m+1}{2} \int_{\partial N_1} \overline{\sum_0^\infty \frac{1}{n!} P_\beta^{(n)}(0) \cdot \frac{dt}{t^{m+2}}}$$

$$= \frac{m+1}{2} \cdot \overline{\frac{2\pi i}{(m+1)!} P_\beta^{(m+1)}(0)}$$

$$= -\frac{\pi i}{m!} \overline{\left(\frac{d^{m+1}}{dt^{m+1}} (p_\beta + i\, p_\beta^*)\right)_{t=0}}$$

$$= -\frac{2\pi i}{m!} \left(\frac{\partial^{m+1} p_\beta}{\partial \bar{z}^{m+1}}\right)_{z=\zeta_1}.$$

At the point ζ_2, we have

$$- \int_{\partial N_2} p_\beta \overline{\varphi} \sim - \int_{\partial N_2} (\log|t|)\, \overline{\varphi} = 0$$

as N_2 shrinks to ζ_2. Thus the right-hand side of (3) is equal to

$$\frac{4\pi}{m!} \left(\frac{\partial^{m+1} p_\beta}{\partial \bar{z}^{m+1}}\right)_{z=\zeta_1}. \tag{7}$$

If ψ is used instead of φ, we obtain

$$\int_{\partial N_1} p_\beta \bar{\psi} \to 0, \qquad \int_{\partial N_2} p_\beta \bar{\psi} \to 0,$$

and the right-hand side of (4) vanishes. This and (6) imply (2). Similarly, (5) and (7) give (1).

1 D. General Case. Now let W be arbitrary. We exhaust it by canonical subregions Ω with $\zeta_1, \zeta_2 \in \Omega$. Principal functions on Ω converge to those on W. The same is true of the capacity functions provided $c_\beta > 0$. If $c_\beta = 0$, we take an arbitrary capacity function p_β, and regions Ω such that the $p_{\beta(\Omega)}$ converge to p_β; this can be done by virtue of Theorems III.4 B, 4 F (cf. Problem 5). The convergence is uniform on every compact set, so that the left-hand sides of (1) and (2) for Ω converge to those for W. As a consequence, $a(\zeta_1)$ and $b(\zeta_1)$ for Ω converge to finite numbers, for which we retain the same notation. We conclude that (1) and (2) hold for W.

The right-hand sides and, consequently, the left-hand sides also, are continuously differentiable in ζ_2 and, therefore,

$$\frac{2}{m!} \frac{\partial}{\partial \zeta_2} \left(\frac{\partial^{m+1} p_\beta(z; \zeta_2)}{\partial \bar{z}^{m+1}} \right)_{z=\zeta_1} = -\frac{\partial}{\partial \zeta_2} (p_{1m}(\zeta_2; \zeta_1) - i\, p'_{1m}(\zeta_2; \zeta_1))$$
$$= \pi (m+1)\, K_m(\zeta_2, \zeta_1),$$

$$\frac{2}{m!} \frac{\partial}{\partial \zeta_2} \left(\frac{\partial^{m+1} p_\beta(z; \zeta_2)}{\partial z^{m+1}} \right)_{z=\zeta_1} = -\frac{\partial}{\partial \zeta_2} (p_{1m}(\zeta_2; \zeta_1) + i\, p'_{1m}(\zeta_2; \zeta_1))$$
$$= \pi (m+1)\, L_m(\zeta_2, \zeta_1).$$

On replacing ζ_1 and ζ_2 by ζ and z, we obtain (a).

1 E. Proof of (b). The idea of the proof of (b) is similar to that of (a). We retain the notation p_{1c}, p'_{1c} of II.3 D. Let ζ be the pole of p_β and consider the interior W of a compact bordered surface $\overline{W}$ with $c \subset W$, $\zeta \in W$, $\zeta \notin c$. For W, we have, corresponding to (1) and (2),

$$-\int_c \bar{\omega} = a(c) - (p_{1c}(\zeta) - i\, p'_{1c}(\zeta)), \tag{8}$$

$$-\int_c \omega = b(c) - (p_{1c}(\zeta) + i\, p'_{1c}(\zeta)), \tag{9}$$

where $a(c)$ and $b(c)$ are the constant values of the functions $p_{1c} - i\, p'_{1c}$ and $p_{1c} + i\, p'_{1c}$ on the border. Since the members of (8) are complex conjugates of those of (9), it suffices to prove the latter. Let N_1 and N_2

be neighborhoods of c and ζ, respectively, such that $\bar{N}_1 \cap \bar{N}_2 = \emptyset$. Regarding $\int_{\partial N_2}$, the result is the same as in 1 B. As for $\int_{\partial N_1}$, we obtain

$$-\int_{\partial N_1} (p_{1c} + i\, p'_{1c}) = 2\pi i \int_c \omega$$

(cf. II.3 E), and

$$-\int_{\partial N_1} p_\beta \bar{\psi} \to 0$$

as N_1 shrinks to c. Eq. (9) follows.

The remaining argument parallels that in 1 D. The proof of Theorem 1 A is herewith complete.

1 F. Expressions for Principal Functions. Eqs. (1) and (8) give relations between principal functions and capacity functions. We have shown that they hold on an arbitrary surface. On taking real parts, we obtain (cf. II.2 G):

THEOREM. *The principal functions for* L_1 *can be expressed in terms of the capacity function* p_β *as follows:*

(a) *If* $z \neq \zeta$, *then* p_{1m} *for* **I** *is*

$$p_{1m}(\zeta; z) = -\frac{1}{m!} \frac{\partial^{m+1} p_\beta(z; \zeta)}{\partial x^{m+1}} + \text{const}.$$

(b) *If* z_1, z_2, *and* ζ *are all different, then* p_1 *for* **I** *is*

$$p_1(\zeta; z_1, z_2) = p_\beta(z_1; \zeta) - p_\beta(z_2; \zeta) + \text{const}.$$

The corresponding expressions of principal functions for **Q** *are obtained by replacing* p_β *by* $p_{1\gamma}$ *with respect to a boundary component* γ.

The authors have not found relations between the principal functions p_0 and capacity functions $p_{0\gamma}$.

However, *on a plane region* W, the principal function p_{0m} for **Q** with $m=0$ has a single-valued conjugate which satisfies

$$p^*_{00}(\zeta; z) = \frac{\partial p_{1\gamma}(z; \zeta)}{\partial y} + \text{const},$$

where γ is a boundary component.

In fact, as we shall show in VI. 1 B, $p^*_{00} = p'_{10}$. Let t be the local parameter about the point z with respect to which the function $p'_{10}(\zeta; z)$ is defined. It has the singularity $\operatorname{Im} t^{-1} = \operatorname{Re}(it)^{-1}$ and is, therefore, the function $p_{10}(\zeta; z)$ with the local parameter it about z. Consequently, the present identity is obtained if the first relation for $m=0$ in the theorem is considered with the parameter *it*.

1 G. Harmonic Period Reproducer. We write relation (8) as follows:

$$\int_c dp_\beta - i * dp_\beta = p_{1c}(\zeta) - i\, p'_{1c}(\zeta) + \text{const}.$$

Differentiation with respect to ζ gives

$$d_\zeta \int_c dp_\beta - i * dp_\beta = dp_{1c} - i\, dp'_{1c},$$

and we have

$$d_\zeta \int_c dp_\beta = dp_{1c}, \qquad * d_\zeta \int_c * dp_\beta = * dp'_{1c}.$$

On exchanging ζ and z, we obtain the following relation (Ahlfors [2]) with respect to the harmonic period reproducer $\omega(c) = \frac{1}{2}(dp_{1c} + * dp'_{1c})$ (see II.3 F):

Theorem. *The harmonic period reproducer $\omega(c)$ has the following expression in terms of the capacity function p_β:*

$$\omega(c) = \tfrac{1}{2}\Big(d_z \int_c d_\zeta\, p_\beta(\zeta; z) + * d_z \int_c * d_\zeta\, p_\beta(\zeta; z)\Big).$$

1 H. Period Reproducers for Dividing Cycles. Given a dividing cycle d, consider the differential $\sigma(d)$ introduced in II.2 A. We know from II.3 D that it is $-i\,\tau(d)/2\pi = i\,\mathrm{Re}\,\psi(d)/\pi$, where we assume that d consists of a finite number of disjoint analytic Jordan curves.

If d is homologous to zero, then $\sigma(d)=0$. In fact, $\|\sigma(d)\|^2 = \int_d * \sigma(d) = 0$ by Theorem II.2 B.

If d is not homologous to zero, then it divides the ideal boundary β into two disjoint isolated subboundaries γ and γ'. Let γ be to the left of d, and consider the modulus $\mu = \mu(\gamma, \gamma')$ and the modulus function $q = q(z; \gamma, \gamma')$ on W; in the present case we need not distinguish between $\nu=0$ and $\nu=1$.

Theorem. *The period reproducer for dividing cycles is a multiple of the differential of the modulus function:*

$$\sigma(d) = (\log \mu)^{-1}\, dq.$$

Note that the right-hand side vanishes identically if $\mu=\infty$. In the case $\mu<\infty$, the additive constant in q vanishes. The function $(\log\mu)^{-1} q$ is the harmonic measure u_γ if W is a compact bordered surface (IV.1 B), but we have not yet considered it in the general case (cf. IV.4 A).

Proof of the Theorem. Let U and U' be neighborhoods of γ and γ' such that $\partial U = -\partial U' = d$. Recall that the differential $\sigma(d)$ was defined in II.2 A as dp_1, where p_1 is a principal function with jump 1 across d. The function

$$u = \begin{cases} p_1 & \text{on } U, \\ p_1 + 1 & \text{on } U' \end{cases}$$

is harmonic throughout W. It is sufficient to show that

$$u + \text{const} = (\log \mu)^{-1} q. \tag{10}$$

To this end, exhaust W by canonical subregions Ω with $d \subset \Omega$. For functions on Ω, (10) is easily verified. On letting $\Omega \to W$, we obtain it in the general case.

We conclude from the above consideration and Theorem II.2 B:

COROLLARY. *$\sigma(d) = 0$ if and only if $\mu(\gamma, \gamma') = \infty$. The validity of $\mu(\gamma, \gamma') = \infty$ implies $\int_d \omega = 0$ for every $\omega \in \Gamma_h$.*

1 I. Relations between Principal Functions p_1 for I and Q. Consider the principal functions $p_1 = p_{1m}(z; \zeta)$ and $p_1 = p_1(z; \zeta, \zeta')$ for both **I** and **Q**. By a property of the operators L_1 for **I** and **Q**, we have

$$\int_{W-N} dp_1 \wedge *\bar{\omega} = -\int_{\partial N} p_1 * \bar{\omega}$$

for every $*\omega \in \Gamma_{hse}(W)$ (see II.2 B), where N is a neighborhood of ζ or $\{\zeta, \zeta'\}$. Consequently,

$$\int_{W-N} d\big(p_1(\mathbf{I}) - p_1(\mathbf{Q})\big) \wedge *\bar{\omega} = 0.$$

On letting $N \to \{\zeta\}$ or $N \to \{\zeta, \zeta'\}$, we obtain $\big(d(p_1(\mathbf{I}) - p_1(\mathbf{Q})), \omega\big) = 0$, that is, $d\big(p_1(\mathbf{I}) - p_1(\mathbf{Q})\big)$ is orthogonal to the space Γ_{hse}^*.

The same argument implies $(\sigma(c) - \tilde{\sigma}(c)) \perp \Gamma_{hse}^*$. If c is a dividing cycle d, we know that $\tilde{\sigma}(d) = 0$ (II.2 B) and, therefore, $\sigma(d) \perp \Gamma_{hse}^*$. Denote by $[\sigma(d)]$ the closed subspace of Γ_h spanned (with real coefficients) by the $\sigma(d)$ for dividing cycles d. If $[\sigma(d)]$ were a proper subspace of $(\Gamma_{hse}^*)^\perp$, there would exist an $\omega \in (\Gamma_{hse}^*)^\perp$ such that $\omega \neq 0$ and $\omega \perp \sigma(d)$ for every d; this is a contradiction, for $\omega \perp \sigma(d)$ implies $\int_d * \omega = 0$, that is, $\omega \in \Gamma_{hse}^*$. Consequently, $[\sigma(d)] = (\Gamma_{hse}^*)^\perp$, and we have proved:

THEOREM. *The principal functions for* **I** *and* **Q** *are related to the period reproducers by*

$$d\big(p_1(\mathbf{I}) - p_1(\mathbf{Q})\big) \in [\sigma(d)]. \tag{11}$$

1 J. Further Relations between Principal Functions p_1 for I and Q. Suppose W is the interior of a compact bordered Riemann surface $\overline{W}$. Let $\beta=\gamma_1\cup\cdots\cup\gamma_n$ be the decomposition into components and write u_k for the harmonic measure u_{γ_k} on W (IV. 1 B), $k=1,\ldots,n$. The differentials $du_1,\ldots,du_{n-1}$ are readily seen to be linearly independent and to span the space $[\sigma(d)]$. Thus there are constants c_k such that

$$d(p_1(\mathbf{I})-p_1(\mathbf{Q}))=\sum_{k=1}^{n-1} c_k\,du_k. \tag{12}$$

In order to determine the coefficients, we introduce the matrix $\mathbf{A}$ whose (j,k)-element for $j,k=1,\ldots,n$ is

$$a_{jk}=\int_{\gamma_j} *du_k=D[u_j,u_k].$$

It is symmetric and positive semidefinite. The matrix $\mathbf{A}_i$ obtained from $\mathbf{A}$ by eliminating the i-th row and column is positive definite. We shall show this only for $i=n$. We start with $\sum x_j x_k a_{jk}=D[\sum_{j=1}^{n-1} x_j u_j]\geq 0$. If equality holds, then $\sum x_j u_j$ is a constant, which is zero because the left-hand side vanishes on γ_n. Thus x_j, the value of $\sum x_j u_j$ on γ_j, is zero for $j=1,\ldots,n-1$. As a consequence, $\mathbf{A}_i$ has an inverse. In terms of this, we claim:

THEOREM. *On a compact bordered surface, the principal functions for* **I** *and* **Q** *are related to the harmonic measures by the identity:*

$$dp_1(\mathbf{I})-dp_1(\mathbf{Q})=\Big(\int_{\gamma_1},\ldots,\int_{\gamma_{n-1}}\Big)*dp_1(\mathbf{I})\,\mathbf{A}_n^{-1}\begin{pmatrix} du_1\\ \vdots\\ du_{n-1}\end{pmatrix}. \tag{13}$$

Indeed, the coefficients c_k are determined by the following simultaneous equations (SCHOTTKY [2], KOEBE [8]), which are derived by taking the periods of the conjugate of (12) along the γ_j, $j=1,\ldots,n-1$:

$$\int_{\gamma_j} *dp_1(\mathbf{I})=\sum_{k=1}^{n-1} c_k\,a_{kj}.$$

1 K. Operators L_1 for I and Q. We continue considering the interior W of a compact bordered surface. Let $(\mathbf{I})\,L_1$ and $(\mathbf{Q})\,L_1$ be the operators L_1 for **I** and **Q** acting from $\alpha=\gamma_n$ into W.

For every $f\in C^\omega(\alpha)$, there are constants $c_1,\ldots,c_{n-1}$ such that the conjugate of

$$(\mathbf{I})\,L_1 f+\sum_{k=1}^{n-1} c_k u_k \tag{14}$$

has no period along $\gamma_1, \ldots, \gamma_{n-1}$. In fact, by solving the simultaneous equations $-\int_{\gamma_j} * d(\mathbf{I}) L_1 f = \sum_{k=1}^{n-1} c_k a_{kj}$, $j=1, \ldots, n-1$, in $c_1, \ldots, c_{n-1}$, we obtain

$$(c_1, \ldots, c_{n-1}) = -\Big(\int_{\gamma_1}, \ldots, \int_{\gamma_{n-1}}\Big) * d(\mathbf{I}) L_1 f \,\mathbf{A}_n^{-1},$$

and conclude that (14) is $(\mathbf{Q}) L_1 f$. Consequently:

THEOREM. *On a compact bordered surface, the function $L_1 f$ for $\mathbf{Q}$ is a linear combination of that for $\mathbf{I}$ and the harmonic measures:*

$$(\mathbf{Q}) L_1 f = (\mathbf{I}) L_1 f - \Big(\int_{\gamma_1}, \ldots, \int_{\gamma_{n-1}}\Big) * d(\mathbf{I}) L_1 f \,\mathbf{A}_n^{-1} \begin{pmatrix} u_1 \\ \vdots \\ u_{n-1} \end{pmatrix}. \tag{15}$$

This is what we referred to in I.3 A.

1 L. Reciprocity Relations. At this point we state in passing:

The kernels are interrelated by the following identities:

(a) *If $\zeta_1 \neq \zeta_2$, then*

$$\frac{1}{(n+1)!}\left(\frac{d^n L_m(z,\zeta_1)}{dz^n}\right)_{z=\zeta_2} = \frac{1}{(m+1)!}\left(\frac{d^m L_n(z,\zeta_2)}{dz^m}\right)_{z=\zeta_1},$$

$$\frac{1}{(n+1)!}\left(\frac{d^n K_m(z,\zeta_1)}{dz^n}\right)_{z=\zeta_2} = \frac{1}{(m+1)!}\left(\overline{\frac{d^m K_n(z,\zeta_2)}{dz^m}}\right)_{z=\zeta_1},$$

and similarly for $\tilde{L}_m(z,\zeta)$ and $\tilde{K}_m(z,\zeta)$.

(b) *If $\zeta \notin c$, then*

$$\int_c L_m(z,\zeta)\, dz = \frac{1}{(m+1)!}\left(\frac{d^m L(z,c)}{dz^m}\right)_{z=\zeta},$$

$$\int_c K_m(z,\zeta)\, dz = \frac{1}{(m+1)!}\left(\overline{\frac{d^m K(z,c)}{dz^m}}\right)_{z=\zeta},$$

and similarly for $\tilde{L}$ and $\tilde{K}$.

(c) *If $c_1 \cap c_2 = \emptyset$, then*

$$\int_{c_2} \varphi(c_1) = \int_{c_1} \varphi(c_2), \qquad \int_{c_2} \psi(c_1) = \int_{c_1} \overline{\psi(c_2)},$$

and similarly for $\tilde{\varphi}(c)$ and $\tilde{\psi}(c)$.

On a closed surface W, these are the classical *reciprocity relations.* In the present general situation, the proof is similar to that in 1 B – 1 E (see also AHLFORS-SARIO [1, pp. 302 ff.]).

The relations contain as special cases the following identities for $z \neq \zeta$ and $z \notin c$:

$$L(z,\zeta)=L(\zeta,z), \qquad K(z,\zeta)=\overline{K(\zeta,z)},$$

$$L_m(z,\zeta)=\frac{1}{(m+1)!}\frac{d^m L(z,\zeta)}{d\zeta^m}, \qquad K_m(z,\zeta)=\frac{1}{(m+1)!}\frac{d^m K(z,\zeta)}{d\bar{\zeta}^m},$$

$$L(z,c)=\int_c L(z,\zeta)\,d\zeta, \qquad K(z,c)=\int_c K(z,\zeta)\,d\bar{\zeta},$$

and similarly for $\tilde{L}$ *and* $\tilde{K}$.

From (a) – (c) above, we obtain the following relations for the principal functions p_0, p_1 for **I** and **Q**, and their combinations $p_2=\frac{1}{2}(p_0+p_1)$, $p_3=\frac{1}{2}(p_0-p_1)$:

COROLLARY. *The principal functions* $p_{\nu m}, p_\nu$ *with* $\nu=0,1,2,3$ *are interrelated as follows:*

(a) *If* $\zeta_1 \neq \zeta_2$, *then*

$$\frac{1}{n!}\left(\frac{\partial^{n+1} p_{\nu m}(z;\zeta_1)}{\partial x^{n+1}}\right)_{z=\zeta_2}=\frac{1}{m!}\left(\frac{\partial^{m+1} p_{\nu n}(z;\zeta_2)}{\partial x^{m+1}}\right)_{z=\zeta_1}.$$

(b) *If* ζ, ζ_1, ζ_2 *are all different, then*

$$p_{\nu m}(\zeta_2;\zeta)-p_{\nu m}(\zeta_1;\zeta)=\frac{1}{m!}\left(\frac{\partial^{m+1} p_\nu(z;\zeta_1,\zeta_2)}{\partial x^{m+1}}\right)_{z=\zeta}.$$

(c) *If* $\zeta_1, \zeta_2, \zeta_1', \zeta_2'$ *are all different, then*

$$p_\nu(\zeta_2;\zeta_1',\zeta_2')-p_\nu(\zeta_1;\zeta_1',\zeta_2')=p_\nu(\zeta_2';\zeta_1,\zeta_2)-p_\nu(\zeta_1';\zeta_1,\zeta_2).$$

(c′) *If* $\zeta \neq \zeta'$, *then*

$$p_\nu(z;\zeta,\zeta')=-p_\nu(z;\zeta',\zeta)+\text{const}.$$

(c″) *If* $\zeta_1, \zeta_2, \zeta_3$ *are all different, then*

$$p_\nu(z;\zeta_1,\zeta_2)=p_\nu(z;\zeta_1,\zeta_3)-p_\nu(z;\zeta_2,\zeta_3)+\text{const}.$$

The proofs are left to the reader.

2. Capacity and Modulus Functions

2 A. Symmetry. The well-known symmetry of the Green's function,

$$g(z,\zeta)=g(\zeta,z)$$

for $z\neq\zeta$, generalizes as follows:

THEOREM. *If $c_{\nu\gamma}>0$, then the capacity function $p_{\nu\gamma}$ has the symmetry property*

$$k_{\nu\gamma}(\zeta)-p_{\nu\gamma}(z;\zeta)=k_{\nu\gamma}(z)-p_{\nu\gamma}(\zeta;z). \tag{1}$$

Proof. Use of an exhaustion towards γ permits us to give the proof only on a bordered surface with compact border γ. We take $\zeta_1\neq\zeta_2$ and show that $k_{1\gamma}(\zeta_1)-p_{1\gamma}(\zeta_2;\zeta_1)=k_{1\gamma}(\zeta_2)-p_{1\gamma}(\zeta_1;\zeta_2)$; the proof for $p_{0\gamma}$ is similar. Let N_1 and N_2 be neighborhoods of ζ_1 and ζ_2 with $\bar{N}_1\cap\bar{N}_2=\emptyset$. For short, write p^k instead of $p_{1\gamma}(z;\zeta_k)$, $k=1,2$. We have $L_1 p^k=p^k$ on $W-N_1-N_2$, where L_1 acts from $\gamma\cup\partial N_1\cup\partial N_2$ into $W-N_1-N_2$. By the definition of L_1 for **Q** (I.3 C), we obtain

$$D_{W-N_1-N_2}[p^1,p^2]=\int_{\gamma-\partial N_1-\partial N_2} p^1*dp^2=2\pi\,k_{1\gamma}(\zeta_1)-\int_{\partial N_1+\partial N_2} p^1*dp^2,$$

$$D_{W-N_1-N_2}[p^2,p^1]=\int_{\gamma-\partial N_1-\partial N_2} p^2*dp^1=2\pi\,k_{1\gamma}(\zeta_2)-\int_{\partial N_1+\partial N_2} p^2*dp^1.$$

Hence

$$\int_{\partial N_1+\partial N_2} p^1*dp^2-p^2*dp^1=2\pi\,k_{1\gamma}(\zeta_1)-2\pi\,k_{1\gamma}(\zeta_2).$$

If N_1 and N_2 shrink to ζ_1 and ζ_2, the left-hand side converges to

$$2\pi\big(p^1(\zeta_2)-p^2(\zeta_1)\big)=2\pi\big(p_{1\gamma}(\zeta_2;\zeta_1)-p_{1\gamma}(\zeta_1;\zeta_2)\big).$$

2 B. Another Symmetry of $p_{1\gamma}$. Suppose γ and γ' are distinct boundary *components* which are realized as contours of W. Then the capacity functions $p_{1\gamma}(z;\zeta)$ and $p_{1\gamma'}(z;\zeta)$ are harmonic and constant on γ and γ' respectively (cf. I.4 E). Let us denote by $p_{1\gamma}(\gamma')$ the constant value of $p_{1\gamma}$ on γ', and by $p_{1\gamma'}(\gamma)$ that of $p_{1\gamma'}$ on γ. We assert:

THEOREM. *For contours γ, γ', the capacity function has the symmetry property*

$$p_{1\gamma}(\gamma')=p_{1\gamma'}(\gamma). \tag{2}$$

For the proof, let N be a neighborhood of ζ. By a property of the operator L_1, we see that

$$\int_{\gamma+\gamma'} (p_{1\gamma}*dp_{1\gamma'}-p_{1\gamma'}*dp_{1\gamma})$$

is equal to the integral along ∂N, which tends to zero as N shrinks to ζ. On the other hand, this integral is equal to $2\pi\big(p_{1\gamma}(\gamma')-p_{1\gamma'}(\gamma)\big)$.

2 C. Analogous Identity for the Modulus Function. If distinct boundary components γ_1, γ_1', γ_2, and γ_2' of W are realized as contours, then we obtain similarly

$$q_1(\gamma_2;\gamma_1,\gamma_1')-q_1(\gamma_2';\gamma_1,\gamma_1')=q_1(\gamma_1;\gamma_2,\gamma_2')-q_1(\gamma_1';\gamma_2,\gamma_2'). \qquad (3)$$

For the sake of comparison, we recall the following relation from IV.1 D:

$$q_\nu(z;\gamma,\gamma')+q_\nu(z;\gamma',\gamma)=\text{const}, \qquad (4)$$

where $\nu=0, 1$.

2 D. Some Relations between Capacity and Modulus Functions. We recall that under certain conditions the capacity function $p_{\nu\gamma}(z;\zeta)$ can be regarded as a modulus function $q_\nu(z;\gamma,\gamma')$, $\nu=0,1$ (IV.1 H and 3 E). We shall give further relations in the case $\nu=1$.

For disjoint subboundaries γ and γ' of an arbitrary surface W, the modulus function and the capacity functions for L_1 are interrelated by

$$q_1(z;\gamma,\gamma')=p_{1\gamma}(z;\zeta)-p_{1\gamma'}(z;\zeta) \qquad (5)$$

with respect to an arbitrary $\zeta\in W$.

Take an exhaustion $V\to W$ towards $\gamma\cup\gamma'$, and first obtain (5) on V, with q_1 normalized to vanish at ζ. On taking the limit, we have the relation on W. The cases $c_{1\gamma}=0$, $c_{1\gamma'}=0$, and $\mu_1(\gamma,\gamma')=\infty$ are not excluded.

If γ and γ' are realized as contours, then the capacity function $p_{1\gamma}$ satisfies the relation

$$l_1(\gamma,\gamma')=k_{1\gamma}(\zeta)+k_{1\gamma'}(\zeta)-p_{1\gamma}(\gamma';\zeta)-p_{1\gamma'}(\gamma;\zeta). \qquad (6)$$

This is deduced from (5) by considering the values of the functions on γ and γ'.

Under the same assumption, the modulus function q_1 has the property

$$q_1(z;\gamma,\gamma')=\tfrac{1}{2}(k_{1\gamma'}(z)-k_{1\gamma}(z))+\text{const}. \qquad (7)$$

For the proof, let N be a neighborhood of ζ. Then $\int_{\gamma+\gamma'}(p_{1\gamma}*dq_1-q_1*dp_{1\gamma})$ is equal to the integral along ∂N and converges to $-2\pi\, q_1(\zeta)$ as N shrinks to ζ. Therefore,

$$k_{1\gamma}(\zeta)-p_{1\gamma}(\gamma')-q_1(\gamma;\gamma,\gamma')=-q_1(\zeta;\gamma,\gamma').$$

Similarly,

$$k_{1\gamma'}(\zeta)-p_{1\gamma'}(\gamma)-q_1(\gamma';\gamma',\gamma)=-q_1(\zeta;\gamma',\gamma).$$

We subtract and use (2) and (4). The difference $q_1(\gamma;\gamma,\gamma')-q_1(\gamma';\gamma',\gamma)$ is a constant independent of ζ. On replacing ζ by z, we obtain (7).

2 E. Harmonic Measure. Let γ with $c_{1\gamma}>0$ be an isolated subboundary of an arbitrary W and consider the harmonic measure u_γ on a neighborhood U of γ with $\beta_U=\gamma$. Let g be the Green's function on U and set $\alpha=\partial U$.

If γ is isolated and $c_{1\gamma}>0$, then the harmonic measure u_γ can be expressed in terms of the Green's function:

$$1-u_\gamma(z)=-\frac{1}{2\pi}\int_\alpha * d_\zeta\, g(\zeta, z). \tag{8}$$

To see this, note that g is harmonic and equal to zero on α (cf. Corollary III. 2 D and Theorem I. 5 E). Let N be a neighborhood of the pole z of g. By a property of the operator H,

$$\int_\alpha ((1-u_\gamma)*dg-g*d(1-u_\gamma))$$

is equal to the integral along ∂N, which converges to $-2\pi(1-u_\gamma(z))$ as N shrinks to ζ. The above integral is $\int_\alpha *dg$, since $u_\gamma=g=0$ on α.

2 F. The Case of a Compact Bordered Surface. Let W be the interior of a compact bordered surface. As in 1 J, we denote by $\gamma_1, \dots, \gamma_n$ the components of the border β, and by $u_1, \dots, u_n$ the harmonic measures. In terms of the matrix $\mathbf{A}_n$ considered in 1 J, we have:

For the contours $\gamma=\gamma_n$ of W, the capacity functions and the harmonic measures are interrelated by

$$p_{1\gamma}=p_\beta-\Big(\int_{\gamma_1}, \dots, \int_{\gamma_{n-1}}\Big)*dp_\beta\, \mathbf{A}_n^{-1}\begin{pmatrix} u_1 \\ \vdots \\ u_{n-1}\end{pmatrix}+\text{const}. \tag{9}$$

The modulus function $q_1=q_1(z; \gamma_n, \gamma_{n-1})$ can be expressed in terms of harmonic measures:

$$q_1=(0, \dots, 0, -2\pi)\, \mathbf{A}_n^{-1}\begin{pmatrix} u_1 \\ \vdots \\ u_{n-1}\end{pmatrix}+\text{const}. \tag{10}$$

The proofs are similar to that in 1 J, and are left to the reader.

From 2 E and 2 F, we conclude that, on a compact bordered surface, $p_{1\gamma}$ and q_1 can be expressed in terms of p_β or g.

Remark. For further identities, valid for plane regions, we refer, for example, to GARABEDIAN-SCHIFFER [1], and SCHIFFER [6]. Several relations are also known (loc. cit.) involving the *Neumann function*, which we have not considered. The results in 2 F, like those in 1 J, are due to SCHOTTKY [2] and KOEBE [8].

PART TWO

Geometric Theory

Chapter VI

Mappings Related to Principal Functions

On plane regions, the principal functions p_0 and p_1 for **Q** have single-valued conjugates outside of a neighborhood of the singularity. Furthermore, for certain singularities, the analytic function $p_\nu + i\,p_\nu^*$ or $\exp(p_\nu + i\,p_\nu^*)$, $\nu = 0, 1$, is univalent. These topics are discussed in 1 and 2. We also review in 2 the extremal property obtained in II.2 C, 2 G to be used in 3 to show that $p_\nu + i\,p_\nu^*$ with singularity $\operatorname{Re}(z-\zeta)^{-1}$ maps the region onto a parallel slit plane. Geometric properties of analytic functions obtained from $\frac{1}{2}(p_0 \pm p_1)$ are investigated in 4 and 5. Finally, in 6, we study the conformal mapping by $\exp(p_\nu + i\,p_\nu^*)$ for the singularity $\log|(z-\zeta)/(z-\zeta')|$ of p_ν.

1. Remarks on the Case of Plane Regions

1 A. Topology. In Chapters VI – IX, we shall be exclusively concerned with plane regions W, subregions of the Riemann sphere **S**.

The complement CE of a set E will mean the complement with respect to **S**, that is, $\mathbf{S} - E$ (not $W - E$).

The topology will always be that of **S**, not the relative topology of W. For example, the boundary operator ∂ is taken in the topology of **S**, so that the relative boundary of an end U of W is $(\partial U) \cap W$, not ∂U.

On **S**, closedness and compactness are synonymous. In the finite plane $|z| < \infty$, a compact set will be referred to as a closed bounded set.

The boundary ∂W of a region W will be denoted by β and identified with the ideal boundary.

1 B. Semiexactness. On a plane region, every cycle divides. As a consequence, exactness and semiexactness of differentials are identical properties.

We also have:

LEMMA. *If U is an end of a plane region W, with connected relative boundary α, then every $u \in KD(\bar{U} \cap W)$ has a single-valued conjugate u^*.*

Furthermore,

$$u = L_1 u \quad \textit{for } \mathbf{Q} \qquad \textit{implies} \qquad u^* = L_0 u^*;$$

$$u = L_0 u \qquad \textit{implies} \qquad u^* = L_1 u^* \quad \textit{for } \mathbf{Q}.$$

This is easily seen by a limiting process; for if U is the interior of a bordered surface, then the conclusion is immediate from the definitions of L_0 and L_1 in I.2 A and I.3 A.

In Chapters VI–IX, we shall not be concerned with L_1 for **I**; even if we do not specify it, L_1 will always be for **Q**.

Consider a singularity (s, E, N) where s is real-valued, s^* single-valued, and ∂N connected. Then by the above lemma, the principal function p_ν and the function p'_ν introduced in II.3 A satisfy

$$p'_1 = p^*_0, \qquad p'_0 = p^*_1.$$

As a consequence, the results in II.3 are essentially identical with those in II.2. We shall return later to this matter in more detail for special singularities (e.g., 2 D, 4 A, 5 A).

1 C. Functions and Differentials. On a plane region W, we shall always fix the local parameter t about $\zeta \in W$ as follows:

$$t(z) = \begin{cases} z - \zeta & \text{if } \zeta \neq \infty, \\ \dfrac{1}{z} & \text{if } \zeta = \infty. \end{cases}$$

As a consequence, we can identify a meromorphic function F with a meromorphic differential $\omega = a(t)\,dt$ by the rule

$$F(z) = \begin{cases} a(z - \zeta) & \text{about } \zeta \neq \infty, \\ -\dfrac{a(1/z)}{z^2} & \text{about } \zeta = \infty. \end{cases}$$

The inner product and norm are defined by

$$(\omega_1, \omega_2) = i \iint_W F_1 \bar{F}_2 \, dz \wedge \overline{dz} = 2 \iint_W F_1 \bar{F}_2 \, dx\, dy,$$

$$\|\omega\|^2 = 2 \iint_W |F|^2 \, dx\, dy.$$

If $\infty \in W$, then the integral is taken over $W - \{\infty\}$.

In the case $\infty \in W$ and $\|\omega\| < \infty$, F has a zero at ∞ of order greater than 1.

The function corresponding to the differential $\omega = dF$ of an analytic function F on W in the above rule is equal to F' in $W - \{\infty\}$, and we have $\|dF\|^2 = 2\mathscr{D}[F]$, where

$$\mathscr{D}[F] = \iint_W |F'|^2\, dx\, dy = D_W[\operatorname{Re} F]$$

$$= \frac{1}{2} \iint_W \left(\left|\frac{\partial F}{\partial x}\right|^2 + \left|\frac{\partial F}{\partial y}\right|^2 \right) dx\, dy = \frac{1}{2} D_W[F]$$

is the *Dirichlet integral* of an analytic function F. If $\infty \in W$, then the integral is taken over $W - \{\infty\}$.

In particular, $\|dF\| < \infty$ implies that F is regular about ∞.

1 D. Boundary Components. In Chapters VI – IX, we are not interested in a general subboundary, but only in an ideal boundary component γ of a plane region W.

We recall that, by definition, γ is a mapping $\Omega \to U_\Omega$, with U_Ω connected. Since $W \subset \mathbf{S}$, γ is realized as the set $E_\gamma = \bigcap_\Omega \bar{U}_\Omega$, which is a component of $\beta = \partial W$ (cf. I.1 B).

It is seen without difficulty that

(a) $E_\gamma = E_{\gamma'}$ if and only if $\gamma = \gamma'$;

(b) every component of β coincides with E_γ for some γ.

In short, an ideal boundary component can be identified with a component of $\beta = \partial W$. For this reason, we shall omit the term "ideal" and simply call γ a *boundary component* of W. We shall also write γ for E_γ.

Given a boundary component γ, there exists a unique component X of $CW = \mathbf{S} - W$ such that $\partial X = \gamma$. It will be denoted by X_γ.

If a region W does not contain ∞, we mean by the *outer boundary* of W the boundary component γ such that $\infty \in X_\gamma$. Similarly, if $0 \notin W$, the component γ with $0 \in X_\gamma$ will be called the *inner boundary*.

Suppose f is a topological mapping of W onto W'. If $\Omega \to U_\Omega$ is a boundary component γ of W, then the mapping $\Omega' \to f(U_{f^{-1}(\Omega')})$ determines a boundary component $f(\gamma)$ of W', the *image* of γ under f.

This notion is visualized in terms of the set E_γ as follows:

LEMMA. *$\gamma' = f(\gamma)$ if and only if, for any sequence $\{z_n\}$ of points of W with accumulation points only on γ, every accumulation point of the sequence $\{f(z_n)\}$ is on γ', and conversely.*

1 E. Slit Planes. In our study the following types of regions will appear as image regions of conformal mappings:

DEFINITION. *A vertical (horizontal) slit plane is a plane region W with $\infty \in W$ such that every component of CW is either a point or a line segment* *parallel to the imaginary (real, resp.) axis. A circular (radial) slit plane is a plane region W with $0, \infty \in W$ such that every component of CW is either a point or a circular arc on $|z| = \text{const}$ (a line segment on $\arg z = \text{const}$).*

The following properties are trivial consequences of the definition:

LEMMA 1. *If E is a closed bounded set every component of which is either a point or a line segment parallel to the imaginary (real) axis, then CE is a vertical (horizontal, resp.) slit plane.*

Proof. Only connectedness must be checked. It follows from a theorem of KERÉKJÁRTÓ [1, p. 49, line 3]: a closed bounded set does not divide the plane if no component of it does.

LEMMA 2. *Let W be a horizontal slit plane. For any z_1 and z_2 in W with $\operatorname{Im} z_1 = \operatorname{Im} z_2$ and any $\varepsilon > 0$, the points z_1 and z_2 can be joined by a curve in $W \cap R$, where $R = \{z \mid \operatorname{Re} z_1 < \operatorname{Re} z < \operatorname{Re} z_2, |\operatorname{Im}(z - z_1)| < \varepsilon\}$. A similar property holds for any vertical slit plane.*

We do not claim that the length of the curve is close to $|z_1 - z_2|$ (cf. IX.4 A).

Proof. Take an $\varepsilon_0 \leq \frac{1}{2}\varepsilon$ so small that the disks $|z - z_1| < \varepsilon_0 \sqrt{2}$ and $|z - z_2| < \varepsilon_0 \sqrt{2}$ are contained in W. Every boundary component γ satisfies $\gamma = X_\gamma$. From the observation in 1 D, we conclude that for every γ there exists an analytic Jordan curve α_γ with the following properties: $\alpha_\gamma \subset W$, the Jordan region D_γ encircled by α_γ contains X_γ, the distance of every point of α_γ to γ is less than ε_0, and $z_1, z_2 \notin \overline{D_\gamma}$. Let $\overline{R_0} = \{z \mid \operatorname{Re} z_1 \leq \operatorname{Re} z \leq \operatorname{Re} z_2, |\operatorname{Im}(z - z_1)| \leq \varepsilon_0\}$, and consider the compact set $CW \cap \overline{R_0}$. From the family $\{D_\gamma \mid \gamma \subset CW \cap \overline{R_0}\}$, we select a finite number of members $D_{\gamma_k} = D_k$, $k = 1, \ldots, n$, with $CW \cap \overline{R_0} \subset \bigcup_{k=1}^n D_k$. The curves $\alpha_{\gamma_k} = \alpha_k$ are contained in $W \cap R$. The set $\delta = \{z \mid \operatorname{Re} z_1 \leq \operatorname{Re} z \leq \operatorname{Re} z_2, \operatorname{Im} z = \operatorname{Im} z_1\} - \bigcup_{k=1}^n D_k$ consists of a finite number of line segments in $W \cap R$ and contains z_1 and z_2. The union of δ and $\alpha_1, \ldots, \alpha_n$ contains a curve in $W \cap R$ joining z_1 and z_2.

Circular slit planes and radial slit planes have properties similar to those in Lemma 1 and Lemma 2. In particular, if E is a totally disconnected closed bounded set, then

(a) CE is connected and is a vertical and horizontal slit plane and, if $0 \notin E$, also a circular and radial slit plane;

(b) for any z_0 and neighborhood N of z_0, there exists a Jordan curve in $N - E$ encircling z_0 and contained in a prescribed arbitrarily narrow circular annulus encircling z_0 in N.

2. Univalent Functions Related to Principal Functions

2 A. Functions P_0 and P_1. Given a point ζ in a plane region W, consider the principal functions $p_\nu = p_{\nu m}(z;\zeta)$ for $m=0$ and $\nu=0,1$ (see II.2 C). Recall that p_1 is always for **Q**. The parameter t about ζ is fixed as in 1 C.

The conjugate p_ν^* of p_ν is single-valued (Lemma 1 B). We normalize it in such a way that the function

$$P_\nu(z;\zeta) = p_\nu(z;\zeta) + i\,p_\nu^*(z;\zeta)$$

is uniquely determined by

$$\lim_{t\to 0}\left(P_\nu(t;\zeta) - \frac{1}{t}\right) = 0 \qquad \text{about } \zeta.$$

Given two points ζ and ζ' in a plane region W, consider the principal functions $p_\nu = p_\nu(z;\zeta,\zeta')$ introduced in II.2 G. The functions

$$P_\nu(z;\zeta,\zeta') = \exp\big(p_\nu(z;\zeta,\zeta') + i\,p_\nu^*(z;\zeta,\zeta')\big)$$

are single-valued, and we normalize them by

$$\lim_{t\to 0}\frac{1}{t}P_\nu(t;\zeta,\zeta') = 1 \qquad \text{about } \zeta.$$

We shall show in 2 B that $P_\nu(z;\zeta)$ and $P_\nu(z;\zeta,\zeta')$ are univalent.

Given a point ζ of a plane region W, we let $\mathscr{V}_\zeta = \mathscr{V}_\zeta(W)$ be the family of univalent meromorphic functions F on W with the following expansion about ζ:

$$F(z) = \begin{cases} \dfrac{1}{z-\zeta} + a(z-\zeta) + \cdots & \text{if } \zeta \neq \infty, \\[2ex] z + \dfrac{a}{z} + \cdots & \text{if } \zeta = \infty. \end{cases}$$

Given points ζ, ζ' of a plane region W, we denote by $\mathscr{V}_{\zeta\zeta'} = \mathscr{V}_{\zeta\zeta'}(W)$ the family of univalent meromorphic functions F on W with a pole at ζ' and such that

$$F(\zeta) = 1 - F'(\zeta) = 0$$

at $\zeta \neq \infty$. If $\zeta = \infty$, this condition is replaced by

$$F(z) = \frac{1}{z} + \frac{c}{z^2} + \cdots \qquad \text{about } \infty.$$

The univalency assured, we obtain the following characterization of P_v as a simple consequence of the definition:

$P_v(z;\zeta)$ is the unique function in $\mathscr{V}_\zeta$ satisfying

$$L_v(\operatorname{Re} P_v)=\operatorname{Re} P_v \quad on\ \beta. \tag{1}$$

$P_v(z;\zeta,\zeta')$ is the unique function in $\mathscr{V}_{\zeta\zeta'}$ satisfying

$$L_v(\log|P_v|)=\log|P_v| \quad on\ \beta. \tag{2}$$

In view of Lemma 1 B, we obtain $P_1(z;\zeta)=i(p_0'+i\,p_0'^*)$ and $P_0(z;\zeta)=i(p_1'+i\,p_1'^*)$, so that we may replace (1) by

$$L_0(\operatorname{Im} P_1)=\operatorname{Im} P_1, \qquad L_1(\operatorname{Im} P_0)=\operatorname{Im} P_0. \tag{3}$$

2 B. Univalency. If W is exhausted by regular subregions Ω containing ζ and if the function on Ω is denoted by $P_{v\Omega}$, then by Theorem II.1 H

$$\lim_{\Omega\to W}(P_{v\Omega}-P_v)=0 \tag{4}$$

uniformly on every compact set in W. Thus it is sufficient to show that $P_{v\Omega}$ is univalent. In other words, we may prove the univalency of P_v under the assumption that W is a regular region.

The function $P_1(z;\zeta)$ is regular on $\overline{W}-\{\zeta\}$ with a simple pole at ζ, and $\operatorname{Re} P_1$ is constant on each component of the border $\beta=\partial W$. The argument principle shows that, if $w\notin P_1(\beta)$, then

$$n(w)-1=\frac{1}{2\pi}\int_\beta d\arg(P_1-w),$$

where $n(w)$ is the number of w-points in W, counted with multiplicities. The integral vanishes since $P_1(\beta)$ is the union of a finite number of line segments parallel to the imaginary axis. Accordingly, $n(w)=1$ for every $w\notin P_1(\beta)$.

Suppose now that z_1 and z_2 in W are such that $P_1(z_1)=P_1(z_2)=w$. If $w\notin P_1(\beta)$, we have $z_1=z_2$ from the above result. If $w\in P_1(\beta)$, we can still conclude that $z_1=z_2$. In fact, if $z_1\neq z_2$, we consider disjoint neighborhoods N_k of z_k, $k=1,2$. Since $P_1(\beta)$ is the union of a finite number of line segments, it is possible to find $z_k'\in N_k$ such that $P_1(z_1')=P_1(z_2')=w'\notin P_1(\beta)$. Then $n(w')\geq 2$, a contradiction. We conclude that P_1 is univalent.

The same reasoning shows that $P_0(z;\zeta)$, $P_1(z;\zeta,\zeta')$, and $P_0(z;\zeta,\zeta')$ are univalent.

We saw that the image of a regular region under $P_1(z;\zeta)$ is a vertical slit plane. On the other hand, it is clear that this property implies (1) for $v=1$, and thus characterizes the function P_1. Similar conclusions hold for $P_0(z;\zeta)$, $P_1(z;\zeta,\zeta')$, and $P_0(z;\zeta,\zeta')$.

THEOREM. *If W is a regular region, then $P_1(z;\zeta)$ (or $P_0(z;\zeta)$) is the unique function in $\mathscr{V}_\zeta$ mapping W conformally onto a vertical (or horizontal, resp.) slit plane; $P_1(z;\zeta,\zeta')$ (or $P_0(z;\zeta,\zeta')$) is the unique function in $\mathscr{V}_{\zeta\zeta'}$ mapping W conformally onto a circular (or radial, resp.) slit plane.*

For an arbitrary W, it will be shown that the image regions are still these slit planes (3 A and 6 A). The proof will be based on an extremal property given below. However, it is to be noted that the above mapping property alone is in general insufficient to characterize P_v (see 3 C).

2 C. Extremal Property. Consider Theorem II. 1 D for the singularity $s=\operatorname{Re}(z-\zeta)^{-1}$. For every $F\in\mathscr{V}_\zeta$ on an arbitrary plane region W, the function $p=\operatorname{Re}F$ belongs to KD_s. As we observed in II. 2 C, $C_s[p]=-2\pi\operatorname{Re}a[F]$, where $a[F]$ is the coefficient a in the expansion

$$F(z)=\begin{cases}\dfrac{1}{z-\zeta}+a(z-\zeta)+\cdots & \text{about } \zeta\neq\infty,\\[2ex] z+\dfrac{a}{z}+\cdots & \text{about } \zeta=\infty.\end{cases}$$

Furthermore, the quantity $-\int_\beta p*dp$ is the area (i.e., the 2-dimensional Lebesgue measure) of the complement of the image region $F(W)$. Thus it is always non-negative. Accordingly, by Theorem II. 1 D (see also II. 2 C),

$$\begin{aligned}2\pi\operatorname{Re}a[P_1]&=2\pi\operatorname{Re}a[P_1]+\int_\beta p_1*dp_1\\&\leq 2\pi\operatorname{Re}a[F]+\int_\beta p*dp\leq 2\pi\operatorname{Re}a[F]\end{aligned}$$

for every F and, if $\operatorname{Re}a[F]=\operatorname{Re}a[P_1]$, then $F=P_1$. Similarly,

$$\begin{aligned}2\pi\operatorname{Re}a[P_0]&=2\pi\operatorname{Re}a[P_0]-\int_\beta p_0*dp_0\\&\geq 2\pi\operatorname{Re}a[F]-\int_\beta p*dp\geq 2\pi\operatorname{Re}a[F].\end{aligned}$$

Analogous inequalities can be obtained for $P_v(z;\zeta,\zeta')$. Here we shall denote by $b[F]$ the coefficient b in the expansion

$$F(z)=\begin{cases}\dfrac{b}{z-\zeta'}+\cdots & \text{about } \zeta'\neq\infty,\\[2ex] b\,z+\cdots & \text{about } \zeta'=\infty,\end{cases}$$

of $F\in\mathscr{V}_{\zeta\zeta'}$. The quantity $-\int_\beta p*dp$ for $p=\log|F|$ is now the *logarithmic area* of $CF(W)$, that is, $\iint_{CF(W)}|w|^{-2}\,du\,dv$. Clearly it vanishes if and only if the area of $CF(W)$ vanishes.

In summary we have:

THEOREM. *$P_1(z;\zeta)$ (or $P_0(z;\zeta)$) is the unique function minimizing (or maximizing, resp.)* $\operatorname{Re} a[F]$ *in $\mathscr{V}_\zeta$, and $P_1(z;\zeta,\zeta')$ (or $P_0(z;\zeta,\zeta')$) is the unique function minimizing (or maximizing, resp.) $|b[F]|$ in $\mathscr{V}_{\zeta\zeta'}$. The complement of the image region of each function has vanishing area.*

Remark. A potential-theoretic approach resulting in Theorem 2 B was used by SCHOTTKY [1, 2], CECIONI [1], and KOEBE [4, 8] (cf. Remark II.1 C). By means of the Dirichlet principle, HILBERT [1], COURANT [1], and KOEBE [1, 5] obtained the mapping of an arbitrary plane region onto a horizontal slit plane. They also established the uniqueness of the mapping function as a consequence of a certain extremal property (different from the one above), as well as relation (4). On the other hand, GRÖTZSCH [8] and DE POSSEL [3, 4] proved that, in the above theorem, the extremal function (the existence of which is directly verified by the compactness of the family $\mathscr{V}_\zeta$) maps the region onto a horizontal or vertical slit plane (Theorem 3 A). They also verified the uniqueness of the extremal function. The corresponding results for mappings onto circular and radial slit planes were obtained by GRÖTZSCH [2, 3, 5] and RENGEL [1, 2].

2 D. Span. The K-span $S_m(\zeta)$ for $m=0$ (II.2 E) will be denoted by $S(\zeta)$. It is given by

$$S(\zeta)=\tfrac{1}{2}\operatorname{Re}(a[P_0]-a[P_1])=\tfrac{1}{2}(\max_{\mathscr{V}_\zeta}\operatorname{Re} a[F]-\min_{\mathscr{V}_\zeta}\operatorname{Re} a[F]). \tag{5}$$

We shall see that $a[P_0]-a[P_1]$ is actually real (Corollary 3 E).

The K-span coincides with the modified K-span $\Sigma(\zeta)=\frac{1}{2}(S(\zeta)+S'(\zeta))$ introduced in II.2 F. Indeed, $S=\frac{1}{2}\operatorname{Re}(a_{(0)}-a_{(1)})$, $dp_\nu+i*dp_\nu=(-t^{-2}+a_{(\nu)}+\cdots)\,dt$ for $\nu=0,1$, and $S'=\frac{1}{2}\operatorname{Re}(a'_{(0)}-a'_{(1)})$, $dp'_\nu+i*dp'_\nu=(-(i\,t)^{-2}+a'_{(\nu)}+\cdots)\,d(i\,t)$, where $p'_1=p_0^*$ and $p'_0=p_1^*$ hold (1 B), so that $a'_{(0)}=-a_{(1)}$ and $a'_{(1)}=-a_{(0)}$.

We shall see in 5 B that $S(\zeta)$ coincides with $\tilde{M}(\zeta)$ of II.3 C.

Similarly, the quantity $S(\zeta,\zeta')$ of II.2 I can be written

$$S(\zeta,\zeta')=\frac{1}{2}\log\left|\frac{b[P_0]}{b[P_1]}\right|=\frac{1}{2}\log\frac{\max_{\mathscr{V}_{\zeta\zeta'}}|b[F]|}{\min_{\mathscr{V}_{\zeta\zeta'}}|b[F]|}. \tag{6}$$

On an arbitrary region W, the family $\mathscr{V}_\zeta$ always contains either the function $(z-\zeta)^{-1}$ if $\zeta\neq\infty$ or the function z if $\zeta=\infty$. In case there is no other function, we see by (5) that the span vanishes. Conversely, if the span vanishes, the uniqueness assertion of Theorem 2 C shows that $\mathscr{V}_\zeta$ consists of a single function.

The function $(z-\zeta)^{-1}$ or z is a linear transformation, and is the unique such member of $\mathscr{V}_\zeta$. On the other hand, for every univalent function Φ on W, there exists a unique linear transformation Λ such that $\Lambda\circ\Phi\in\mathscr{V}_\zeta$. Therefore, the span vanishes if and only if all functions Φ reduce to linear transformations. It is to be noted that this characterization of the vanishing of the span does not refer to the point ζ.

We apply Theorem II.2 E to the present case. The span $S(\zeta)$ vanishes if and only if there is no function u in $KD(W)$ such that $u(\zeta)=0$ and $\partial u(\zeta)/\partial x=1$. This is equivalent to the condition that every $u\in KD$ satisfies $\partial u/\partial x=0$. The above reasoning shows that the choice of the local parameter about ζ is immaterial (cf. II.2 F), so that $S(\zeta)=0$ if and only if $\partial u/\partial y=0$ for every $u\in KD$; moreover, the vanishing of the span is equivalent to $\partial u/\partial x=\partial u/\partial y=0$ for every $u\in KD$ at every point $\zeta\in W$. Hence KD consists only of constants. Since W is a plane region, every $u\in KD$ has a single-valued conjugate. That KD consists only of constants is equivalent to the same property for the family $AD(W)$ of analytic functions F with finite Dirichlet integrals $\mathscr{D}[F]=\iint_W |F'|^2\,dx\,dy$.

A similar conclusion is obtained for $S(\zeta,\zeta')$.

In summary, we have the following result due to AHLFORS-BEURLING [1] (that (d) implies (c) was proved in SARIO [1]):

THEOREM. *The vanishing of* $S(\zeta)$ *is independent of* ζ. *It is equivalent to the vanishing of* $S(\zeta,\zeta')$, *which is also independent of* ζ *and* ζ'. *It is further equivalent to any one of the following:*

(a) $P_1(z;\zeta)=P_0(z;\zeta)$.

(a′) $P_1(z;\zeta,\zeta')=P_0(z;\zeta,\zeta')$.

(b) $\mathscr{V}_\zeta$ *consists of a single element.*

(b′) $\mathscr{V}_{\zeta\zeta'}$ *consists of a single element.*

(c) *All univalent analytic functions on* W *are linear transformations.*

(d) $AD(W)$ *consists only of constants.*

Another important property will be added by Corollary 4 B.

2 E. Examples. For illustration, consider the case $W=\{z\,|\,|z|<r\}$, $r<\infty$. By Theorem 2 B, the functions $P_1(z;\zeta)$ and $P_0(z;\zeta)$ are characterized as functions in $\mathscr{V}_\zeta$ which map W onto the entire plane save a line segment parallel to the imaginary and real axis, respectively. In view of this property, we obtain explicitly

$$P_1(z;\zeta)=\frac{1}{z-\zeta}-\frac{r^2}{r^2-|\zeta|^2}\cdot\frac{z-\zeta}{r^2-\bar\zeta z}=\frac{1}{z-\zeta}-\left(\frac{r}{r^2-|\zeta|^2}\right)^2(z-\zeta)+\cdots,$$

$$P_0(z;\zeta)=\frac{1}{z-\zeta}+\frac{r^2}{r^2-|\zeta|^2}\cdot\frac{z-\zeta}{r^2-\bar\zeta z}=\frac{1}{z-\zeta}+\left(\frac{r}{r^2-|\zeta|^2}\right)^2(z-\zeta)+\cdots.$$

The center and the length of the slits are

$$\frac{\bar{\zeta}}{r^2-|\zeta|^2}, \quad \frac{4r}{r^2-|\zeta|^2},$$

and the span is

$$S(\zeta)=\left(\frac{r}{r^2-|\zeta|^2}\right)^2.$$

Therefore, $\sqrt{S(\zeta)}\,|d\zeta|$ is the Poincaré metric.

The functions $P_1(z;\zeta,\zeta')$ and $P_0(z;\zeta,\zeta')$ are more involved, and we shall only consider the special case

$$\zeta=0, \quad \zeta'=h>0.$$

As above, we obtain

$$P_1(z;\zeta,\zeta')=\frac{h\,z(h\,z-r^2)}{r^2(z-h)}, \quad P_0(z;\zeta,\zeta')=\frac{h\,r^2\,z}{(z-h)(h\,z-r^2)},$$

$$S(\zeta,\zeta')=\log\frac{r^2}{r^2-h^2}.$$

Next, if $W=\{z\,|\,|z|<\infty\}$, then all univalent functions on W are linear transformations. Accordingly,

$$P_1(z;\zeta)=P_0(z;\zeta)=\frac{1}{z-\zeta},$$

$$P_1(z;\zeta,\zeta')=P_0(z;\zeta,\zeta')=\frac{z-\zeta}{z-\zeta'}(\zeta-\zeta'),$$

$$S(\zeta)=S(\zeta,\zeta')=0.$$

3. Parallel Slit Plane

3 A. Conformal Mapping by $P_\nu(z;\zeta)$. We are now in a position to prove:

THEOREM. *An arbitrary plane region W is mapped by $P_1(z;\zeta)$ (or $P_0(z;\zeta)$) conformally onto a vertical (or horizontal, resp.) slit plane. The total area of the slits vanishes.*

The last property was already established in 2 C.

Before giving the proof, we digress slightly and discuss a method, introduced by OIKAWA [3], which will be useful in Chapters VII – IX.

First a comment on terminology. Thus far in our book we have simply used the term "neighborhood" for "boundary neighborhood" as defined in I.1 B. Since imbeddings will now be considered, it will hereafter be necessary to distinguish between these two terms.

Let R be the image region of W under P_1. The function satisfies $L_1(\operatorname{Re} P_1) = \operatorname{Re} P_1$ on ∂W, in the sense of II.1 C. This means that, for a boundary neighborhood U of ∂W, the equation

$$\int_U d(\operatorname{Re} P_1) \wedge \omega = \int_\alpha (\operatorname{Re} P_1)\, \omega$$

with $\alpha = (\partial U) \cap W$ is satisfied by all $\omega \in \Gamma_c^1(\bar{U})$ semiexact relative to α (I.3 C). If we change the variable z to $w = P_1(z)$, then the relation between the transformed integrals shows that $L_1(\operatorname{Re} w) = \operatorname{Re} w$ on $P_1(U)$ or, as one may say, on ∂R, in the sense of II.1 C.

Similarly, for $R = P_0(W)$, we see that $L_0(\operatorname{Re} w) = \operatorname{Re} w$ on ∂R.

3 B. Proof of Theorem 3 A. It is now clear that the proof of Theorem 3 A can be reduced to establishing the following statement:

THEOREM. *Let W be a region in the z-plane containing the point ∞. If*

$$L_1(\operatorname{Re} z) = \operatorname{Re} z \quad \text{on } \partial W, \tag{1}$$

then W is a vertical slit plane. If

$$L_0(\operatorname{Re} z) = \operatorname{Re} z \quad \text{on } \partial W, \tag{2}$$

then W is a horizontal slit plane. In both cases, the total area of the slits vanishes.

In view of Lemma 1 B, conditions (1) and (2) may be replaced by

$$L_0(\operatorname{Im} z) = \operatorname{Im} z \quad \text{on } \partial W, \tag{1'}$$

$$L_1(\operatorname{Im} z) = \operatorname{Im} z \quad \text{on } \partial W. \tag{2'}$$

Proof of the Theorem. The basic idea of the proof is due to DE POSSEL [3]. If (1) is satisfied, then z is the function $P_1(z; \zeta)$ for W with respect to $\zeta = \infty$ (see 2 A). We apply Theorem 2 C. Since the coefficient a for the identity function vanishes, we obtain

$$\operatorname{Re} a[F] \geq 0 \tag{3}$$

for all $F \in \mathscr{V}_\infty$. Suppose there exists a boundary component γ of W such that X_γ is neither a point nor a line segment parallel to the imaginary axis. Let $\tilde{W}$ be the complement of X_γ. Write $\tilde{P}_1$ for the function $P_1(z; \zeta)$ of this simply connected region $\tilde{W}$ with $\zeta = \infty$. The region $\tilde{W}$ is not necessarily regular. But the Riemann mapping theorem permits us to apply Theorem 2 B to obtain $\tilde{P}_1(z) \not\equiv z$. Then by Theorem 2 C applied to $\tilde{W}$, the coefficient $\tilde{a} = a[\tilde{P}_1]$ satisfies $\mathrm{Re}\, \tilde{a} < 0$. Since the restriction to W of $\tilde{P}_1$ evidently belongs to the family $\mathscr{V}_\infty(W)$, this inequality contradicts (3). Thus every component of the complement of W is either a point or a line segment parallel to the imaginary axis.

Since $P_1(z) \equiv z$, the area of CW vanishes by Theorem 2 C.

The proof for a W with (2) is analogous.

3 C. Extremal Slit Plane. In contrast with the case of Theorem 2 B, the function $P_\nu(z; \zeta)$ is not in general characterized as the function in $\mathscr{V}_\zeta$ which maps the region onto a vertical or horizontal slit plane. In other words, an arbitrary vertical (or horizontal) slit plane W does not necessarily satisfy (1) (or (2), resp.). It is easy to construct a slit plane with positive total area of the slits (cf. IX.4 E). Even the requirement that the area vanish is insufficient to ensure that W satisfy (1) or (2); a counterexample will be given in (a) of Theorem IX.4 G.

Accordingly, we shall say that a vertical (or horizontal) slit plane W which does have property (1) (or (2), resp.) is *extremal*. This concept was introduced by KOEBE [7], who used the term "minimal." We shall discuss the matter in Chapter IX.

3 D. Function $P_{(\theta)}$. Let W be an arbitrary plane region. Given $\zeta \in W$ and a real number θ, consider the function

$$P_{(\theta)}(z; \zeta) = e^{i\theta} \left(P_0(z; \zeta) \cos \theta - i\, P_1(z; \zeta) \sin \theta \right). \tag{4}$$

Evidently,

$$P_{(\theta)} = P_{(\theta')} \quad \text{if} \quad \theta \equiv \theta' \bmod \pi,$$

$$P_{(0)} = P_0, \qquad P_{\left(\frac{\pi}{2}\right)} = P_1.$$

Moreover, for any θ_0 we have

$$P_{(\theta)} = e^{i(\theta - \theta_0)} \left(P_{(\theta_0)} \cos(\theta - \theta_0) - i\, P_{\left(\theta_0 + \frac{\pi}{2}\right)} \sin(\theta - \theta_0) \right). \tag{5}$$

Suppose W is a regular region, and let γ be a component of ∂W. Consider the line segments $P_0(\gamma) = \{w_0 + t \mid 0 \le t \le t_0\}$ and $P_1(\gamma) = \{w_1 + i\tau \mid 0 \le \tau \le \tau_0\}$, where t and τ are real parameters. The image of γ under $e^{-i\theta} P_{(\theta)}$ is $\{(w_0 \cos \theta - i\, w_1 \sin \theta) + (t \cos \theta + \tau \sin \theta) \mid 0 \le t \le t_0, 0 \le \tau \le \tau_0\}$, which is a horizontal line segment since the variable $t \cos \theta + \tau \sin \theta$ is real. Accordingly, $P_{(\theta)}(\gamma)$ is a line segment with inclination θ.

For an arbitrary W, we reason as in 2 B, 2 C, 3 A, and 3 B to obtain the following results:

(a) $P_{(\theta)}$ *is univalent. It is characterized as the function in* $\mathscr{V}_\zeta$ *on* W *such that*

$$L_0(\mathrm{Re}(e^{-i\theta} P_{(\theta)})) = \mathrm{Re}(e^{-i\theta} P_{(\theta)}) \quad on\ \partial W.$$

(b) *The image region* $P_{(\theta)}(W)$ *is a parallel slit plane with inclination* θ *of the slits.*

(c) $P_{(\theta)}$ *is the unique function maximizing* $\mathrm{Re}(e^{-2i\theta} a[F])$ *in* $\mathscr{V}_\zeta$.

These results are due to KOEBE [9], GRÖTZSCH [8], and DE POSSEL [3].

3 E. A Property of the Span. Substituting $\theta=0$ and $\theta=\pi/2$ in (5), and replacing θ_0 by θ we obtain $P_0=e^{-i\theta}(P_{(\theta)}\cos\theta+i\,P_{(\theta+\pi/2)}\sin\theta)$ and $P_1=e^{-i\theta}(i\,P_{(\theta)}\sin\theta+P_{(\theta+\pi/2)}\cos\theta)$. Therefore,

$$P_0+P_1=P_{(\theta)}+P_{(\theta+\frac{\pi}{2})}, \tag{6}$$

$$e^{2i\theta}(P_0-P_1)=P_{(\theta)}-P_{(\theta+\frac{\pi}{2})}, \tag{7}$$

from which it follows that

$$P_{(\theta)}=\tfrac{1}{2}(P_0+P_1)+\tfrac{1}{2}e^{2i\theta}(P_0-P_1). \tag{8}$$

We write a_0, a_1, and $a_{(\theta)}$ for the coefficients $a[P_0]$, $a[P_1]$, and $a[P_{(\theta)}]$. From (8), we obtain $e^{-2i\theta}a_{(\theta)}=\frac{1}{2}e^{-2i\theta}(a_0+a_1)+\frac{1}{2}(a_0-a_1)$; from the extremal property (c) of $P_{(\theta)}$ above, we deduce $\mathrm{Re}(e^{-2i\theta}a[F])\le \mathrm{Re}(e^{-2i\theta}a_{(\theta)})$ for every $F\in\mathscr{V}_\zeta$. As a consequence,

$$\mathrm{Re}(e^{-2i\theta}(a[F]-\tfrac{1}{2}(a_0+a_1)))\le \mathrm{Re}\,\tfrac{1}{2}(a_0-a_1),$$

where the right-hand side is equal to the span $S(\zeta)$. We fix F, let θ vary, and conclude that

$$|a[F]-\tfrac{1}{2}(a_0+a_1)|\le S(\zeta).$$

Equality holds for $F=P_{(\theta)}$ and for this function only, as is seen from the uniqueness assertion for the extremal property of $P_{(\theta)}$. In summary, we obtain:

THEOREM. *The set* $\Delta=\{a[F]\,|\,F\in\mathscr{V}_\zeta\}$ *is contained in the (closed) disk with center* $\frac{1}{2}(a[P_0]+a[P_1])$ *and radius* $S(\zeta)$. *Only the points* $a[P_{(\theta)}]$ *with* $0\le\theta<\pi$ *lie on the circumference.*

The result is due to GRÖTZSCH [8], who showed further that Δ coincides with the above disk (see GRÖTZSCH [2, 6] and JENKINS [1, p. 93]).

COROLLARY. *$a[P_0]-a[P_1]$ is real, and*

$$S(\zeta)=\tfrac{1}{2}(a[P_0]-a[P_1]).$$

In fact, a_0 and a_1 contained in Δ satisfy $|a_0-a_1|\leq 2S$ and $\mathrm{Re}(a_0-a_1)=2S$. Hence a_0-a_1 must be real.

The result will also be derived in 5 B.

3 F. Boundary Components with Varying θ. Let us fix a boundary component γ and a point ζ of W. The set $P_{(\theta)}(\gamma)$ is either a point or a line segment. As θ varies, we encounter the following surprising behavior of $P_{(\theta)}(\gamma)$ (SARIO [14]):

THEOREM. *Only three cases can occur:*

(a) *$P_{(\theta)}(\gamma)$ is a point for every θ.*

(b) *There exists a θ_0 such that $P_{(\theta_0)}(\gamma)$ is a point, and $P_{(\theta)}(\gamma)$ is a line segment for every $\theta \not\equiv \theta_0$ mod π.*

(c) *$P_{(\theta)}(\gamma)$ is a line segment for every θ.*

Proof. Suppose neither (a) nor (c) occurs. We shall show that (b) is then true. There exists a θ_0 such that $P_{(\theta_0)}(\gamma)$ is a point, say w_0. If $P_{(\theta_0+\pi/2)}(\gamma)$ is also a single point w_0', we have a contradiction. Indeed, for any sequence $\{z_n\}$ of points in W with accumulation points only on γ, the sequences $\{P_{(\theta_0)}(z_n)\}$ and $\{P_{(\theta_0+\pi/2)}(z_n)\}$ converge to w_0 and w_0', respectively. Then by (5),

$$\lim_{n\to\infty} P_{(\theta)}(z_n)=e^{i(\theta-\theta_0)}(w_0\cos(\theta-\theta_0)-i\,w_0'\sin(\theta-\theta_0)).$$

This shows that $P_{(\theta)}(\gamma)$ is a single point for every θ (Lemma 1 D), in violation of the assumption that (a) does not occur. Thus $P_{(\theta+\pi/2)}(\gamma)$ is a line segment, say $\{(1-t)\,w_0'+t\,w_0''\,|\,0\leq t\leq 1\}$ with $w_0'\neq w_0''$. On considering $\{z_n\}$ as above, we conclude that $P_{(\theta)}(\gamma)$ is the set of points

$$e^{i(\theta-\theta_0)}(w_0\cos(\theta-\theta_0)-i((1-t)\,w_0'+t\,w_0'')\sin(\theta-\theta_0)),$$

$0\leq t\leq 1$. This is a line segment for every $\theta\not\equiv\theta_0$ mod π.

Remark 1. An example of a γ with property (b) is the following: $\gamma=\{z\,|\,\mathrm{Re}\,z=0,\,|\mathrm{Im}\,z|\leq 1\}$ of $W=\{z\,||z|\leq\infty\}-\gamma-\bigcup_{n=1}^{\infty}(\sigma_n\cup\sigma_{-n})$, where $\sigma_n=\{z\,|\,\mathrm{Re}\,z=1/n,\,|\mathrm{Im}\,z|\leq 1\}$. In fact, $P_1(z;\infty)\equiv z$, and $P_0(z;\infty)$ maps γ to a single point. This can be seen by applying to the upper half-plane the Carathéodory theory of the boundary correspondence under conformal mappings of simply connected regions (see, e.g., CARATHÉODORY [1]).

Remark 2. It is not known to the authors whether or not the above classification depends on ζ.

4. Function $\frac{1}{2}(P_0+P_1)$

4 A. Functions $\frac{1}{2}(P_0+P_1)$ and $\tilde{L}$. On an arbitrary plane region W, the combination $\frac{1}{2}(P_0+P_1)$ of the functions $P_\nu(z;\zeta)$ for $\zeta\in W$, $\nu=0,1$, coincides with $p_2+i\,p_2^*$, where $p_2=\frac{1}{2}(p_0+p_1)$ is the function for $m=0$ and $\mathbf{Q}$ discussed in II.2 D, 2 E:

$$\tfrac{1}{2}(P_0(z;\zeta)+P_1(z;\zeta))=p_2(z;\zeta)+i\,p_2^*(z;\zeta). \tag{1}$$

On the other hand, we have $p_1'=p_0^*$ (1 B). This reveals a relationship between the above function and the differential $\tilde{\varphi}_m=-(\partial(p_1+i\,p_1')/\partial z)\,dz$ for $m=0$ introduced in II.3 B:

$$-\frac{\partial}{\partial z}(p_1+i\,p_1')=-\frac{\partial}{\partial z}(p_1+i\,p_0^*)=-\frac{1}{2}\frac{\partial}{\partial z}(P_0+P_1).$$

On putting $\tilde{\varphi}_0=\pi\,\tilde{L}(z,\zeta)\,dz$ as in V.1 A, and identifying the differential with a function as in 1 C, we obtain the following relation between the adjoint kernel $\tilde{L}$ and the mapping functions P_0 and P_1 (Schiffer [3]):

$$\tilde{L}(z,\zeta)=-\frac{1}{2\pi}(P_0'(z;\zeta)+P_1'(z;\zeta)), \tag{2}$$

where the prime indicates differentiation with respect to $z\neq\zeta$.

Theorem II.3 C now takes the following form: *$\tilde{L}$ is the unique function analytic on $W-\{\zeta\}$ and satisfying*

$$p.\,v._{\zeta}\iint_W F'(z)\,\overline{\tilde{L}(z,\zeta)}\,dx\,dy=0 \tag{3}$$

for every $F\in AD(W)$, having a pole at ζ with the principal part $1/\pi(z-\zeta)^2$ if $\zeta\neq\infty$ and z^2/π if $\zeta=\infty$, and possessing a single-valued indefinite integral with a finite Dirichlet integral outside of a neighborhood of ζ. In the case $\infty\in W$, the integral (3) is to be taken over $W-\{\infty\}$.

Note that, because of (1), relation (3) can also be derived from Theorem II.2 D.

The identity

$$\tilde{L}(z,\zeta)=\frac{2}{\pi}\frac{\partial^2 p_{1\gamma}(\zeta;z)}{\partial z\,\partial\zeta}$$

of Theorem V.1 A is also useful. Here $p_{1\gamma}$ is the capacity function with respect to a boundary component γ.

For example, if $W=\{z\,\|z|<r\}$, $r\leq\infty$, we obtain from 2 E:

$$\frac{1}{2}(P_0+P_1)=\frac{1}{z-\zeta}, \tag{4}$$

$$\tilde{L}(z,\zeta)=\frac{1}{\pi(z-\zeta)^2}. \tag{5}$$

4 B. Extremal Property and the Span. Identity (4) shows that $\frac{1}{2}(P_0+P_1)$ is univalent if W is a disk. We shall see that this continues to be true for an arbitrary plane region W, and therefore

$$\tfrac{1}{2}(P_0+P_1)\in\mathscr{V}_\zeta.$$

We postpone the proof of this important fact until 4 C and consider Theorem II.2 E here. In the family $KD_{\zeta m}$, $m=0$, the function $p_2=\frac{1}{2}\operatorname{Re}(P_0+P_1)$ maximizes $-\int_\beta p*dp$. Every $F\in\mathscr{V}_\zeta$ has its real part $p=\operatorname{Re}F$ in $KD_{\zeta m}$, with $-\int_\beta p*dp$ equal to the area of the complement of the image region $F(W)$. We conclude (Grunsky [1], Schiffer [2], Sario [6]):

Theorem. *The function $\frac{1}{2}(P_0(z;\zeta)+P_1(z;\zeta))$ is the unique function in the family $\mathscr{V}_\zeta$ which maximizes the area of $CF(W)$. The maximum value is $\pi S(\zeta)$.*

This result for $W=\{z\,|\,|z|<r\}$ is known as the *Bieberbach-Gronwall area theorem.*

From the theorem we obtain another characterization of the span,

$$S(\zeta)=\frac{1}{\pi}\max_{F\in\mathscr{V}_\zeta} m\big(CF(W)\big),$$

where m stands for the area (i.e., the 2-dimensional Lebesgue measure). In particular, the span vanishes if and only if $m\big(CF(W)\big)=0$ for every $F\in\mathscr{V}_\zeta$. We have the following consequence (Ahlfors-Beurling [1]; cf. also Sario [1, 6]):

Corollary. $S(\zeta)=0$ *if and only if all univalent functions on W have image regions whose complements have vanishing area.*

4 C. Proof of Univalency. It suffices to show that P_0+P_1 is univalent for a regular region W. Decompose the border ∂W into contours $\gamma_1,\ldots,\gamma_k$. On each γ_j,

$$\operatorname{Re}\big(e^{i\theta}(P_0+P_1)\big)-e^{i\theta}P_{(-\theta)}=e^{-i\theta}(\operatorname{Re}P_1-i\operatorname{Im}P_0)$$

is constant. The image of γ_j under $e^{i\theta}P_{(-\theta)}$ is a horizontal line segment. By a well-known result on the boundary correspondence of univalent functions (Carathéodory [1]), $\operatorname{Re}(e^{i\theta}P_{(-\theta)})$ attains every value between its minimum and maximum exactly twice on γ_j. Consequently, the same is true of $\operatorname{Re}\big(e^{i\theta}(P_0+P_1)\big)$ for every θ. A closed curve is, by definition, (strictly) *convex* if it meets every straight line at most twice. Any line in the w-plane can be represented by the equation $\operatorname{Re}(e^{i\theta}w)=\text{const}$. This shows that the image of γ_j under the mapping $w=P_0+P_1$ is a convex curve.

Given an arbitrary w, let n and n' be the numbers of w-points of P_0+P_1 in W and on ∂W, respectively, with due count of multiplicity. By the argument principle, $(2\pi)^{-1}\int_{\partial W} d\arg(P_0+P_1-w)=n+\frac{1}{2}n'-1$. On the other hand, if w is encircled by m and lies on m' image curves of $\gamma_1, \ldots, \gamma_k$, then the left-hand side is equal to $-m-\frac{1}{2}m'$. Hence

$$m+n+\tfrac{1}{2}(m'+n')=1.$$

By definition, $n' \geq m'$. Furthermore, $n'=0$ if $m'=0$. Thus only the following three cases remain: $n=1$, $n'=m=m'=0$; $m=1$, $n=n'=m'=0$; $n'=m'=1$, $n=m=0$. We conclude that P_0+P_1 is univalent.

Since the possibility $n' \geq 2$ is excluded, the derivative of P_0+P_1 never vanishes on ∂W, so that the image curves of $\gamma_1, \ldots, \gamma_k$ are analytic. Thus we also obtain the following result due to GRUNSKY [1] and SCHIFFER [2]:

THEOREM. *If W is a regular region, then its image under $\frac{1}{2}(P_0+P_1)$ is bounded by analytic closed convex curves.*

The image region is bounded by a circle if W is simply connected (cf. (4)). We shall see in 5 E that, if the connectivity is greater than one, the image region is never bounded only by circles.

4 D. Image Region of Arbitrary W. We shall now discuss the shape of the image of an arbitrary plane region W under $P_2=\frac{1}{2}(P_0+P_1)$.

Let γ be a boundary component of W and write Ξ_γ for $X_{P_2(\gamma)}$, the component of the complement of $P_2(W)$ whose boundary coincides with $P_2(\gamma)$. If we denote the end points of $P_{(\theta)}(\gamma)$ by w_θ and w'_θ (they may coincide), then $P_{(\theta)}(\gamma)=\{(1-t)\,w_\theta+t\,w'_\theta \,|\, 0\leq t\leq 1\}$. Consider the following rectangle with inclination θ:

$$R_{\gamma\theta}=\{\tfrac{1}{2}(w+w')\,|\,w\in P_{(\theta)}(\gamma),\ w'\in P_{(\theta+\frac{\pi}{2})}(\gamma)\}.$$

It may degenerate to a line segment or a point.

The shape of the image region is as follows (OIKAWA-SUITA [1]; cf. SARIO [14]):

THEOREM. *The set Ξ_γ is a convex set. More precisely, one of the following three cases occurs:*

(a) *Ξ_γ is a single point and coincides with $R_{\gamma\theta}$ for all θ.*

(b) *Ξ_γ is a line segment and coincides with $R_{\gamma\theta}$ for some θ. Moreover, for every $\theta' \not\equiv \theta \bmod \pi$, Ξ_γ is a diagonal of the nondegenerate rectangle $R_{\gamma\theta'}$.*

(c) *Ξ_γ is a convex set with interior points. For every θ, $R_{\gamma\theta}$ is the smallest rectangle with inclination θ which contains Ξ_γ.*

Remark 1. The proof will show that this classification coincides with that given in Theorem 3 F.

Remark 2. In contrast with the case of regular regions, $P_2(\gamma)$ is not in general an analytic curve. In 4 F we shall give an example of case (c) such that Ξ_γ coincides with the rectangle $R_{\gamma\theta}$ for some θ.

For the proof of the theorem, we need an auxiliary result on the mapping $P_{(\theta)}$:

LEMMA. *Let w_0 be a boundary point of $P_{(\theta)}(W)$. If the boundary component containing w_0 is not a single point, assume further that w_0 is an end point of the component. Then for any $\varepsilon > 0$, there exists a neighborhood N of w_0 such that $N \subset \{w \,|\, |w-w_0| < \varepsilon\}$ and $N \cap P_{(\theta)}(W)$ is connected.*

Proof of the Lemma. Without loss of generality, we may consider the case where $\theta = 0$ and w_0 is the left end point of the segment. It is possible to find a $w_0' \in P_{(0)}(W)$ such that $\operatorname{Re} w_0 - \varepsilon/2 < \operatorname{Re} w_0' < \operatorname{Re} w_0$ and $\operatorname{Im} w_0' = \operatorname{Im} w_0$. Take an $\varepsilon_0 > 0$ with $0 < \varepsilon_0 < \varepsilon/2$ and $\{w \,|\, |w-w_0'| < \varepsilon_0\} \subset P_{(0)}(W) \cap \{w \,|\, |w-w_0| < \varepsilon\}$. Then

$$N = \{w \,|\, |w-w_0'| < \varepsilon_0\} \cup \{w \,|\, |w-w_0| < \varepsilon, \operatorname{Re} w_0' < \operatorname{Re} w, |\operatorname{Im}(w-w_0)| < \varepsilon_0\}$$

is the desired neighborhood of w_0 since, by Lemma 2 in 1 E, every point of $N \cap P_{(0)}(W)$ can be joined to the segment

$$\{w \,|\, \operatorname{Re} w = \operatorname{Re} w_0', |\operatorname{Im}(w-w_0')| < \varepsilon_0\}$$

by a curve in $N \cap P_{(0)}(W)$.

4 E. Proof of Theorem 4 D. We shall prove (c) for a γ with property 3 F.(c). The proofs of (a) and (b) are similar.

First we shall establish the latter half of (c). It suffices to prove that $\Xi_\gamma \subset R_{\gamma\theta}$ and $\sigma \cap P_2(\gamma) \neq \emptyset$ for every side σ (including end points) of the rectangle $R_{\gamma\theta}$. This implies further that $R_{\gamma\theta}$ contains interior points whenever the convexity of Ξ_γ is guaranteed.

We shall give the proof for $\theta = 0$. In view of 3 E.(6), this does not mean loss of generality. For any $w \in P_2(\gamma)$, there exists a sequence of points $z_n \in W$ with accumulation points only on γ and such that $w = \lim P_2(z_n)$. We can find a subsequence for which $\lim P_0(z_{j_n}) = w_0$ and $\lim P_1(z_{j_n}) = w_1$ exist. Since $w = (w_0 + w_1)/2$, $w_0 \in P_0(\gamma)$, and $w_1 \in P_1(\gamma)$, we conclude that $w \in R_{\gamma 0}$ and therefore $\Xi_\gamma \subset R_{\gamma 0}$. Furthermore, since (c) of Theorem 3 F is assumed, $R_{\gamma 0}$ is a non-degenerate rectangle. Let σ be its left vertical side, say, and let w_0 be the left end point of $P_{(0)}(\gamma) = P_0(\gamma)$. Then there exists a sequence of points $z_n' \in W$ with accumulation points only on γ and such that $\lim P_0(z_n') = w_0$. Take a subsequence such that $\lim P_{(\pi/2)}(z_{j_n}') = w' \in P_{(\pi/2)}(\gamma) = P_1(\gamma)$ exists. Then $\lim P_2(z_{j_n}') = (w_0 + w')/2 \in P_2(\gamma)$ and, on the other hand, $(w_0 + w')/2 \in \sigma$. We infer that $\sigma \cap P_2(\gamma) \neq \emptyset$.

It remains to prove that Ξ_γ is convex. If not, then there exists a straight line l and points $w, w', w^* \in l$ with the following properties: Ξ_γ is in one of the closed half-planes determined by l; w and w' are distinct and belong to Ξ_γ; and w^* is not in Ξ_γ but belongs to the line segment determined by w and w'. This is a special case of Carathéodory's theorem (cf. EGGLESTON [1], p. 35).

Let θ be the inclination of l. By 3 E.(6), we may again assume that $\theta=0$ and that Ξ_γ is in the upper half-plane determined by l. Take sequences $z_n, z'_n \in W$ which have accumulation points only on γ and are such that

$$\lim P_2(z_n)=w, \qquad \lim P_2(z'_n)=w'.$$

It is easily verified that

$$\lim P_1(z_n)=\lim P_1(z'_n)=w_{\frac{\pi}{2}},$$

where $w_{\pi/2}$ is the lower end point of the segment $P_1(\gamma)$. Hence

$$\lim P_0(z_n)=w_0, \qquad \lim P_0(z'_n)=w'_0$$

exist and are contained in $P_0(\gamma)$. We have $(w_0+w_{\pi/2})/2=w$, $(w'_0+w_{\pi/2})/2=w'$, and therefore $w_0 \neq w'_0$. Let $w_0^* \in P_0(\gamma)$ be the point defined by $(w_0^*+w_{\pi/2})/2=w^*$ and let l^* be the line passing through w_0^* and perpendicular to $P_0(\gamma)$.

Take $\varepsilon_n \to 0$ such that $|P_1(z_n)-w_{\pi/2}|<\varepsilon_n$ and $|P_1(z'_n)-w_{\pi/2}|<\varepsilon_n$. According to Lemma 4 D, there exists a curve in $P_1(W) \cap \{w \,|\, |w-w_{\pi/2}|<\varepsilon_n\}$ joining $P_1(z_n)$ and $P_1(z'_n)$. Its inverse image c_n under P_1 clusters to γ only as $n \to \infty$. For all sufficiently large n, the points $P_0(z_n)$ and $P_0(z'_n)$ lie on different sides of l^* and, therefore, $P_0(c_n) \cap l^* \neq \emptyset$. Thus we can find $z_n^* \in c_n$ with $P_0(z_n^*) \in l^*$. Accumulation points of z_n^* lie only on γ, so that the points $P_0(z_n^*)$ cluster to $P_0(\gamma)$ only. We conclude that $\lim P_0(z_n^*)=w_0^*$. Clearly $\lim P_1(z_n^*)=w_{\pi/2}$, and therefore $w^*=\lim P_2(z_n^*) \in P_2(\gamma)$, which violates the assumption $w^* \notin \Xi_\gamma$. It follows that Ξ_γ is convex.

4 F. An Example. As anticipated in Remark 2 of 4 D, we shall now construct a region of case (c) having a component γ such that $\Xi_\gamma=R_{\gamma\theta}$ for some θ.

In the w-plane, take the region considered in Remark 1 of 3 F, that is, $W'=\{w \,|\, |w| \leq \infty\}-\gamma'-\bigcup_{n=1}^{\infty}(\sigma_n \cup \sigma_{-n})$, where $\gamma'=\{w \,|\, \operatorname{Re} w=0, |\operatorname{Im} w| \leq 1\}$ and $\sigma_n=\{w \,|\, \operatorname{Re} w=1/n, |\operatorname{Im} w| \leq 1\}$. Write $f(w)$ for the function $P_0(w;\infty)$ of this region, and put $s_n=f(\sigma_n)$, $n=\pm 1, \pm 2, \ldots$. Observe that f is symmetric about the real and imaginary axes (Problem 9), and, therefore, s_n is a line segment on the real axis which is symmetric with s_{-n} about the origin. Moreover, as we saw in Remark 1 of 3 F, $f(\gamma')=\{0\}$. Rotate s_n about the origin by $\pi/2$ and let the resulting line segment be s'_n. We shall infer that the boundary component $\gamma=\{0\}$ of the region

$$W=\{z \,|\, |z| \leq \infty\}-\gamma-\bigcup_{n=1}^{\infty}(s_n \cup s_{-n} \cup s'_n \cup s'_{-n})$$

has the desired properties.

By symmetry, $f^{-1}(s_n')$ is a line segment on the imaginary axis. Thus $f^{-1}(W)$ is a region bounded by a countable number of vertical slits. By a criterion to be given in IX.4 B, it is an extremal vertical slit plane. Consequently, the restriction of f^{-1} to W coincides with the function $P_1(z)=P_1(z;\infty)$ of W, so that $P_1(\gamma)=[-i,i]$, the line segment connecting $-i$ and i. The symmetry of W about the real and the imaginary axes implies $P_0(z)=-i\,P_1(i\,z)$ (Problem 9), so that $P_0(\gamma)=[-1,1]$.

We see that $R_{\gamma 0}$ is the square with vertices $\pm 1/2$ and $\pm i/2$. Using the convexity of Ξ_γ, we can deduce $R_{\gamma 0}=\Xi_\gamma$ by showing that the four points $\pm 1/2 \pm i/2$ belong to $P_2(\gamma)$.

It suffices to consider $1/2+i/2$. Observe that the first quadrant corresponds to itself under both P_0 and P_1. Draw a line segment in $P_0(W)\cap$(1st quad.) with an end point at 1. Its image under P_0^{-1} is a curve in $W\cap$(1st quad.) terminating at $z=0$, and its image under $P_1 \circ P_0^{-1}$ is a curve in $P_1(W)\cap$(1st quad.) clustering to the segment $[0,i]$. On the latter image curve, there exists a sequence of points w_n with $\lim w_n = i$. Thus we obtain $P_1^{-1}(w_n)=z_n \in W$ with the properties $\lim z_n = 0$ and $\lim P_2(z_n)=\lim P_0 \circ P_1^{-1}(w_n)/2+\lim w_n/2=1/2+i/2$. Consequently, $1/2+i/2 \in P_2(\gamma)$.

5. Function $\frac{1}{2}(P_0-P_1)$

5 A. Functions $\frac{1}{2}(P_0-P_1)$ and $\tilde{K}$. On an arbitrary plane region W, the function $\frac{1}{2}(P_0-P_1)$ for $\zeta \in W$ is identical with $p_3+i\,p_3^*$, where $p_3=\frac{1}{2}(p_0-p_1)$ is the function for $m=0$ and $\mathbf{Q}$ discussed in II.2 D, 2 E:

$$\tfrac{1}{2}(P_0(z;\zeta)-P_1(z;\zeta))=p_3(z;\zeta)+i\,p_3^*(z;\zeta). \tag{1}$$

On the other hand, because of the relation $p_1'=p_0^*$ (1 B) satisfied on plane regions, we see that (1) is closely related to the differential $\tilde{\psi}_m$ for $m=0$ introduced in II.3 B. In fact, as in 4 A, we obtain for $\tilde{\psi}_0=\pi\,\tilde{K}(z,\zeta)\,dz$

$$\tilde{K}(z,\zeta)=\frac{1}{2\pi}(P_0'(z;\zeta)-P_1'(z;\zeta)), \tag{2}$$

where the prime means differentiation with respect to $z \neq \zeta$.

Theorem II.3 C now takes the following form: *The function $\tilde{K}$ is the unique function with a single-valued indefinite integral in $AD(W)$ and such that*

$$\iint_W F'(z)\,\overline{\tilde{K}(z,\zeta)}\,dx\,dy=\begin{cases} F'(\zeta) & \text{if } \zeta \neq \infty,\\ -\lim\limits_{z\to\infty} z^2 F'(z) & \text{if } \zeta=\infty \end{cases} \tag{3}$$

for every $F \in AD(W)$. The above integral is taken over $W-\{\infty\}$ if $\infty \in W$.

Relation (3) can also be derived from Theorem II.2 D.

The identity

$$\tilde{K}(z,\zeta) = \frac{2}{\pi} \frac{\partial^2 p_{1\gamma}(\zeta; z)}{\partial z\, \partial \bar{\zeta}}$$

(Theorem V.1 A) is also useful; here $p_{1\gamma}$ is the capacity function with respect to an arbitrary boundary component γ.

For example, if $W = \{z \,|\, |z| < r\}$, $r < \infty$, then (cf. 2 E)

$$\frac{1}{2}(P_0 - P_1) = \frac{r^2}{r^2 - |\zeta|^2} \cdot \frac{z - \zeta}{r^2 - \bar{\zeta} z}, \tag{4}$$

$$\tilde{K}(z,\zeta) = \frac{1}{\pi} \frac{r^2}{(r^2 - \bar{\zeta} z)^2}. \tag{5}$$

Remark. Relation (2) above and Theorem 5 B below are due to GRUNSKY [1], BERGMAN [2], SCHIFFER [3], and SARIO [6]; for simply connected regions these results were first obtained by BERGMAN [1] (cf. also 5 C). LEHTO [1] and GARABEDIAN-SCHIFFER [2], on the other hand, started with $\tilde{K}$ and studied the mapping functions P_0 and P_1; cf. also NEHARI [1].

5 B. Extremal Property. From (2) it follows that

$$\tilde{K}(\zeta,\zeta) = \frac{1}{2\pi}(a[P_0] - a[P_1]).$$

The left-hand side is equal to the quantity $\pi^{-1}\tilde{M}(\zeta)$ introduced in II.3 C. Since this is real-valued, we obtain an alternative proof of Corollary 3 E.

Furthermore,

$$\tilde{M}(\zeta) = S(\zeta).$$

In particular, by Theorem 2 D, $\tilde{M}(\zeta)$ is identically zero whenever it vanishes at a point.

To state the extremal property of the function $\frac{1}{2}(P_0 - P_1)$, we introduce, for a given point ζ of an arbitrary plane region W, the family $\mathscr{A}_\zeta = \mathscr{A}_\zeta(W)$ of analytic functions F on W such that

$$F(\zeta) = 1 - F'(\zeta) = 0$$

if $\zeta \neq \infty$; in the case $\zeta = \infty$ the last condition is replaced by $F(z) = z^{-1} + \cdots$ about ∞.

Theorem II.2 E now reads (cf. 1 C):

THEOREM. *If $S(\zeta) > 0$, then the function $(P_0(z;\zeta) - P_1(z;\zeta))/2S(\zeta)$ is the unique function minimizing the Dirichlet integral $\mathscr{D}[F]$ in $\mathscr{A}_\zeta$. The minimum value is $\pi/S(\zeta)$.*

From this we obtain another expression for the span:

$$S(\zeta)=\frac{\pi}{\min\limits_{F\in\mathscr{A}_\zeta}\mathscr{D}[F]}.$$

Thus $S(\zeta)=0$ if and only if $\mathscr{D}[F]=\infty$ for every $F\in\mathscr{A}_\zeta$; this in turn is equivalent to the property that $AD(W)$ consists only of constants (cf. Theorem 2 D).

5 C. Bergman Metric. In place of $\tilde{M}(\zeta)$ we again consider the Bergman metric $(\pi^{-1}M(\zeta))^{\frac{1}{2}}|d\zeta|$ (II.3 C) and obtain as above

$$M(\zeta)=\frac{\pi}{\min\iint\limits_W |F|^2\,dx\,dy},$$

where the minimum is taken over the family of analytic functions F on W with $F(\zeta)=1$ if $\zeta\neq\infty$ and $\lim_{z\to\infty} z^2F(z)=-1$ if $\zeta=\infty$. It will be seen in X.2 B.(b) that $M(\zeta)=0$ if and only if the capacity $c_\beta(\zeta)$ (III.2 B) vanishes; this property is independent of ζ (cf. II.3 C and IV.3 B).

If W is simply connected, then we have trivially $M(\zeta)=\tilde{M}(\zeta)$, $K(z,\zeta)=\tilde{K}(z,\zeta)$, and $L(z,\zeta)=\tilde{L}(z,\zeta)$.

5 D. Mapping by $\frac{1}{2}(P_0-P_1)$. The following result is due to GARABEDIAN-SCHIFFER [1]:

THEOREM. *If W is a regular region of connectivity $n\geq 2$, the function $\frac{1}{2}(P_0-P_1)$ is at most n-valent, but not univalent. It is regular on $\overline{W}$, and the image of an arbitrary contour γ of W is an analytic convex curve which is the reflection of the image of γ under $\frac{1}{2}(P_0+P_1)$ about a line (depending on γ) parallel to the real axis.*

If W is simply connected, then the function is univalent (cf. 5 A.(4)). The authors do not know if the above results extend to non-regular regions.

Proof. The function $\frac{1}{2}(P_0-P_1)$ is analytic on $\overline{W}$. Along an arbitrary contour γ of W,

$$d(P_0+P_1)-\overline{d(P_0-P_1)}=2(dp_1+i*dp_0)=0.$$

Therefore, on γ,

$$\tfrac{1}{2}(P_0-P_1)=\tfrac{1}{2}\overline{(P_0+P_1)}+\text{const}, \tag{6}$$

where the constant depends on γ. This and Theorem 4 C give the latter half of the above theorem.

We see from this reasoning that if γ is oriented positively with respect to W, then its image γ' under $\frac{1}{2}(P_0-P_1)$ is a Jordan curve oriented positively with respect to its interior. If the function is univalent and if the connectivity n is ≥ 2, then there exists a γ such that γ' is not the outer boundary, the region encircled by it being disjoint from the image region. This is a contradiction, and we conclude that $\frac{1}{2}(P_0-P_1)$ is not univalent.

Finally, for any analytic function f on $\overline{W}$ and any $w\notin f(\partial W)$, the number $n(w)$ of w-points of f in W is $(2\pi)^{-1}\int_{\partial W} d\arg(f-w)$. For $f=\frac{1}{2}(P_0-P_1)$, the integral along each γ is 2π or zero according as w is encircled by γ or not. Therefore, $n(w)\le n$. For an arbitrary w, we conclude exactly as in 2 B that $n(w)\le n$.

5 E. Remark on the Mapping $\frac{1}{2}(P_0+P_1)$. We return to the question raised at the end of 4 C. By a *circle region* we shall mean a plane region containing ∞, every component of whose complement is either a single point or a disk.

THEOREM. *If W is a regular region of connectivity ≥ 2, then its image under $\frac{1}{2}(P_0+P_1)$ can never be a circle region.*

We do not exclude the possibility that some contours of the image region may be circles.

For the proof, we make use of the following classical theorem of KOEBE [2,3] and COURANT [1] (see also HURWITZ-COURANT [1, pp. 531 ff.]): *If W is a regular region, then there exists a unique function in $\mathscr{V}_\zeta(W)$ which maps W onto a circle region.*

Proof of the Theorem. We shall derive a contradiction to the hypothesis that $\frac{1}{2}(P_0+P_1)$ gives a mapping onto a circle region. Because of the theorem of KOEBE and COURANT, we may assume in advance that W is a circle region; then the mapping onto another circle region is a linear transformation. Thus we are to derive a contradiction from

$$\frac{1}{2}(P_0+P_1)=\frac{1}{z-\zeta}.$$

Without loss of generality, we may suppose further that $\zeta=0$ and the outer boundary γ_0 of W is the circle $|z|=r$. Let

$$\gamma_k\colon\ |z-c_k|=r_k,\qquad k=1,\dots,n-1,$$

be the other contours. The contours $\gamma_0,\dots,\gamma_{n-1}$ are oriented positively with respect to W. The hypothesis is now $\frac{1}{2}(P_0+P_1)=z^{-1}$, so that

$$\frac{1}{2}(P_0-P_1)=P_0-\frac{1}{z}.$$

By reproducing property 5 A.(3) and Stokes' formula, we obtain for every function F analytic on $\overline{W}$

$$F'(0)=\frac{1}{\pi}\iint_W F'\overline{\left(P_0-\frac{1}{z}\right)'}\,dx\,dy=\frac{-1}{2\pi i}\int_{\partial W} F\,d\overline{P_0}+\frac{1}{2\pi i}\int_{\partial W} F\,d\overline{\left(\frac{1}{z}\right)}.$$

On using the fact that $dP_0 = \overline{dP_0}$ along ∂W and considering the residue of FP_0', we have

$$F'(0) = F'(0) + \frac{1}{2\pi i} \int_{\partial W} F \, \overline{d\left(\frac{1}{z}\right)},$$

and therefore

$$\int_{\gamma_0} F \frac{\overline{dz}}{\bar{z}^2} = -\sum_k \int_{\gamma_k} F \frac{\overline{dz}}{\bar{z}^2}.$$

In order to transform $\overline{dz}$ to dz, we denote by c_k^* the mirror images of 0 about the circles γ_k, $k = 1, \ldots, n-1$; note that they are all different. We obtain

$$\frac{1}{r^2} \int_{\gamma_0} F \, dz = -\sum_k \frac{r_k^2}{\bar{c}_k^2} \int_{\gamma_k} \frac{F \, dz}{(z - c_k^*)^2}.$$

In particular, if F is analytic throughout $|z| < \infty$, then

$$\sum_k \frac{r_k^2}{\bar{c}_k^2} F'(c_k^*) = 0.$$

This must be satisfied, e.g., by all polynomials F, a contradiction.

6. Circular and Radial Slit Planes

6 A. Conformal Mapping by $P_v(z; \zeta, \zeta')$. We maintain:

THEOREM. *An arbitrary plane region W is mapped by $P_1(z; \zeta, \zeta')$ (or $P_0(z; \zeta, \zeta')$) with $\zeta, \zeta' \in W$ conformally onto a circular (or radial, resp.) slit plane. The total area of the slits vanishes.*

As in the case of Theorem 3 A, the proof is reduced to establishing the theorem below.

6 B. Extremal Slit Plane. In terms of the operators L_1 and L_0 we state:

THEOREM. *Let W be a plane region in the z-plane such that $0, \infty \in W$. If*

$$L_1(\log|z|) = \log|z| \quad \text{on } \partial W, \tag{1}$$

then W is a circular slit plane. If

$$L_0(\log|z|) = \log|z| \quad \text{on } \partial W, \tag{2}$$

then W is a radial slit plane. In both cases the total area of the slits vanishes.

The proof is analogous to that of Theorem 3 B and is omitted.

A circular (or radial) slit plane with property (1) (or (2), resp.) will be called *extremal*.

6 C. Combinations of P_0 and P_1. Let c be a chain with $c=\overrightarrow{\zeta\zeta'}$ and consider the differentials $\tilde{\varphi}(c)=\pi\,\tilde{L}(z,c)\,dz$ and $\tilde{\psi}(c)=\pi\,\tilde{K}(z,c)\,dz$ introduced in II.3 D (cf. V.1 A). As in 4 A and 5 A we obtain the relations

$$\tilde{L}(z,c)=\frac{1}{2\pi}\left(\frac{P_0'}{P_0}+\frac{P_1'}{P_1}\right),\qquad \tilde{K}(z,c)=\frac{1}{2\pi}\left(\frac{P_1'}{P_1}-\frac{P_0'}{P_0}\right),$$

with $P_\nu=P_\nu(z;\zeta,\zeta')$, $\nu=0,1$. Thus we have independence of the choice of c joining ζ' and ζ. Moreover,

$$\iint_W F'(z)\,\overline{\tilde{L}(z,c)}\,dx\,dy=F(\zeta')-F(\zeta),$$

$$\iint_W F'(z)\,\overline{\tilde{K}(z,c)}\,dx\,dy=F(\zeta)-F(\zeta')$$

for all $F\in AD(W)$ (Theorem II.3 E). The following expressions are also obtained involving the capacity function $p_{1\gamma}$ with respect to a boundary component γ (Theorem V.1 A):

$$\tilde{L}(z,c)=\frac{1}{\pi}\frac{\partial}{\partial z}\int_{\zeta'}^{\zeta}(d_u\,p_{1\gamma}(u;z)+i*d_u\,p_{1\gamma}(u;z)),$$

$$\tilde{K}(z,c)=\frac{1}{\pi}\frac{\partial}{\partial z}\int_{\zeta'}^{\zeta}(d_u\,p_{1\gamma}(u;z)-i*d_u\,p_{1\gamma}(u;z)).$$

One can show that $\sqrt{P_0 P_1}$ has single-valued branches which are univalent. The branch with

$$(P_0(z;\zeta,\zeta')\,P_1(z;\zeta,\zeta'))^{\frac{1}{2}}\in\mathscr{V}_{\zeta\zeta'}$$

has properties similar to those in 4 B – 4 F.

If $\mathscr{A}_{\zeta\zeta'}=\mathscr{A}_{\zeta\zeta'}(W)$ is the family of analytic functions F on W with $F(\zeta)=0$ and $F(\zeta')=1$, then there exists a branch such that

$$\frac{1}{2S(\zeta,\zeta')}\log\frac{P_0(z;\zeta,\zeta')}{P_1(z;\zeta,\zeta')}\in\mathscr{A}_{\zeta\zeta'},$$

provided $S(\zeta,\zeta')>0$. It has properties analogous to those in 5 B – 5 D.

The details will be left to the reader; reference is made to GRUNSKY [1] and GARABEDIAN-SCHIFFER [1].

For example, if $W=\{z\,|\,|z|<r\}$, $r<\infty$, $\zeta=0$, and $\zeta'=h>0$, then

$$\tilde{L}(z,c)=\frac{1}{\pi}\left(\frac{1}{z}-\frac{1}{z-h}\right),\qquad \tilde{K}(z,c)=\frac{1}{\pi}\frac{h}{r^2-h\,z},$$

$$(P_0\,P_1)^{\frac{1}{2}}=\frac{-h\,z}{z-h},\qquad \frac{1}{2S}\log\frac{P_0}{P_1}=\frac{\log\dfrac{r^2}{r^2-h\,z}}{\log\dfrac{r^2}{r^2-h^2}}.$$

Chapter VII

Mappings Related to Capacity Functions

On a plane region, capacity functions $p_{0\gamma}$ and $p_{1\gamma}$ with respect to a boundary component γ provide univalent functions $P_{\nu\gamma}=\exp(p_{\nu\gamma}+i\,p_{\nu\gamma}^*)$, $\nu=0, 1$. The function $P_{1\gamma}$ maps the region onto a circular slit disk, whereas $P_{0\gamma}$ maps it onto a radial slit disk with possible radial incisions at the periphery. These topics will be discussed in 1, 2, 3, and 4. Relations between $P_{1\gamma}$ and functions considered in Chapter VI will be derived in 5. Logarithmic potentials are used in 6 to study the boundary behavior of the capacity function p_β with respect to the entire boundary β. A geometric expression of the capacity c_β in terms of the transfinite diameter will also be given.

1. Univalent Functions Related to Capacity Functions

1 A. Functions $P_{0\gamma}$ and $P_{1\gamma}$. Let W be an arbitrary plane region. We shall be concerned exclusively with a *boundary component* γ of W, not an arbitrary subboundary.

Given a point $\zeta\in W$, we consider the capacity functions $p_{\nu\gamma}(z;\zeta)$ always with respect to the local parameter t of VI.1 C:

$$t(z)=\begin{cases} z-\zeta & \text{if } \zeta\neq\infty, \\ \dfrac{1}{z} & \text{if } \zeta=\infty. \end{cases}$$

The functions satisfy $L_\nu\, p_{\nu\gamma}=p_{\nu\gamma}$ on every end which is disjoint from a neighborhood of γ. Accordingly, $\int_c * dp_{\nu\gamma}$ vanishes for a cycle $c\subset W-\{\zeta\}$ which does not separate γ from ζ, whereas it is equal to an integral multiple of 2π for other cycles. It follows that the functions

$$P_{\nu\gamma}(z;\zeta)=e^{p_{\nu\gamma}+i\,p_{\nu\gamma}^*}$$

are single-valued. We shall normalize the conjugate $p_{\nu\gamma}^*$ by

$$\lim_{t\to 0}\frac{P_{\nu\gamma}}{t}=1$$

at ζ.

Let $\mathscr{S}_{\zeta\gamma}=\mathscr{S}_{\zeta\gamma}(W)$ be the family of univalent analytic functions $F\in\mathscr{A}_\zeta$ (VI.5 B) such that $F(\gamma)$ is the outer boundary (VI.1 D) of the image region $F(W)$. We shall show in 1 B that $P_{v\gamma}$ is univalent and, therefore,

$$P_{v\gamma}(z;\zeta)\in\mathscr{S}_{\zeta\gamma}.$$

Recall that the capacity function was defined by using an exhaustion $V\to W$ towards γ with $\zeta\in V$. If the capacity $c_{v\gamma}(\zeta)$ is positive, then the function $P_{v\gamma}$ is uniquely determined as

$$P_{v\gamma}=\lim_{V\to W}P_{v\gamma(V)},$$

where $P_{v\gamma(V)}(z;\zeta)$ is considered on V, and $\gamma(V)$ is assumed to be a boundary *component* of V. Moreover (III.2 B.(2)),

$$\lim_{V\to W}\iint_{V}\left|\frac{P'_{v\gamma(V)}}{P_{v\gamma(V)}}-\frac{P'_{v\gamma}}{P_{v\gamma}}\right|^2 dx\,dy=0.$$

In the case $c_{v\gamma}=0$, we denoted by $P_{v\gamma}$ an arbitrary limiting function $\lim_{n\to\infty}P_{v\gamma(V_n)}$, the uniqueness of which has not been guaranteed. However, $P_{1\gamma}$ will be seen to be unique (2 H).

We remark in passing that Theorem III.2 E for the present case is easily established since $\mathscr{S}_{\zeta\gamma}$, or more precisely $\bigcup\mathscr{S}_{\zeta\gamma(V)}(V)$, is known to be a normal family.

1 B. Univalency. To prove the univalency of $P_{v\gamma}$, we may assume that W is a regular region (cf. Problem 5). Then $P_{1\gamma}$ is analytic on $\overline{W}$ and satisfies $|P_{1\gamma}|=\text{const}$ on each contour of W. Furthermore, $\int_\gamma d\arg P_{1\gamma}=2\pi$ and $\int_{\gamma'} d\arg P_{1\gamma}=0$ if γ' is a contour different from γ. Therefore the image $P_{1\gamma}(\gamma)$ is a curve on $|w|=r_{1\gamma}=1/c_{1\gamma}$ with winding number 1 about the origin, and $P_{1\gamma}(\gamma')$ is a curve on $|w|=\text{const}$ with winding number 0 about the origin. If $w\notin P_{1\gamma}(\partial W)$, then

$$n(w)=\frac{1}{2\pi}\int_{\partial W} d\arg(P_{1\gamma}-w),$$

where $n(w)$ is the number of w-points in W, counted with multiplicities. We have

$$\int_{\gamma'} d\arg(P_{1\gamma}-w)=0,\qquad \int_{\gamma} d\arg(P_{1\gamma}-w)=0 \quad\text{or}\quad 2\pi,$$

according as $|w|>r_{1\gamma}$ or $|w|<r_{1\gamma}$. Therefore, if $w\notin P_{1\gamma}(\partial W)$, then

$$n(w)=\begin{cases}1 & \text{for } |w|<r_{1\gamma},\\ 0 & \text{for } |w|>r_{1\gamma}.\end{cases}$$

Suppose that z_1 and z_2 in W are such that $P_{1\gamma}(z_1)=P_{1\gamma}(z_2)=w$. If $w\notin P_{1\gamma}(\partial W)$, then $z_1=z_2$ as we have just seen. If $w\in P_{1\gamma}(\partial W)$, we can still draw this conclusion. In fact, if $z_1 \neq z_2$, consider disjoint neighborhoods N_k of z_k, $k=1,2$. Since $P_{1\gamma}(\partial W)$ is a finite union of circles or circular arcs about the origin, it is possible to find $z'_k \in N_k$ such that $P_{1\gamma}(z'_1) = P_{1\gamma}(z'_2) = w' \notin P_{1\gamma}(\partial W)$. Then $n(w') \geq 2$, which is impossible, and we infer that $P_{1\gamma}$ is univalent.

The proof of the univalency of $P_{0\gamma}$ is the same, except that $\arg P_{0\gamma} = $ const on each contour of W different from γ.

Next we show that $P_{v\gamma}(\gamma)$ is the outer boundary of $P_{v\gamma}(W)$. We do not assume now that W is a regular region, and we exhaust it towards γ by V with $\zeta \in V$. Since $\gamma(V)$ is a boundary component and $|P_{v\gamma(V)}| \leq \exp k_{v\gamma(V)}$ (Lemma III.2 F), it is clear that $P_{v\gamma(V)}(\gamma(V))$ is the outer boundary of $P_{v\gamma(V)}(V)$.

It is not difficult to see in general that a boundary component γ_0 of a region W_0 with $\infty \notin W_0$ is its outer boundary if and only if there exists a boundary neighborhood U_0 of γ_0 such that the winding number of $\partial U_0 \cap W_0$ about a point $\zeta_0 \in W_0 - \overline{U}_0$ is 1.

Let U be a boundary neighborhood of γ such that $0 \notin \overline{P_{v\gamma}(U)}$. It suffices to show that the winding number $(1/2\pi)\int_\alpha d \arg P_{v\gamma}$ of $P_{v\gamma}(\alpha)$, $\alpha = \partial U$, about the origin is 1. We have seen that $P_{v\gamma(V)}(\gamma(V))$ is the outer boundary of $P_{v\gamma(V)}(V)$, so that

$$\frac{1}{2\pi} \int_\alpha d \arg P_{v\gamma(V)} = 1 .$$

The conclusion is immediately obtained by a limiting process.

1 C. Circular and Radial Slit Disks. The above proof shows that if W is a regular region, then the image region $P_{1\gamma}(W)$ (or $P_{0\gamma}(W)$) is a circular (or radial, resp.) slit disk of radius $1/c_{1\gamma}$ (or $1/c_{0\gamma}$, resp.). Here we are making use of the following concepts:

DEFINITION. *A circular (radial) slit disk of radius* r ($\leq \infty$) *is a plane region which contains* 0, *is contained in the disk* $|z| < r$, *and has outer boundary* $|z| = r$ *and other boundary components either points or circular arcs on* $|z| = $ const *(or line segments on* $\arg z = $ const, *resp.).*

Suppose γ is an isolated boundary component of a plane region W which is not necessarily regular. If γ consists of a single point, say ζ', then by Theorem III.5 B,

$$P_{v\gamma}(z; \zeta) = P_v(z; \zeta, \zeta'), \tag{1}$$

where the right-hand side is considered on $W \cup \{\zeta'\}$. In this case $c_{v\gamma} = 0$, but the capacity function is uniquely determined. If γ consists of more than one point, then by the Riemann mapping theorem we may assume that γ is an analytic Jordan curve, so that the definitions of capacity functions in III.1 C, 1 D may be used. We thus arrive at the following proposition, in which the case $c_{\nu\gamma}=0$ is not excluded:

(a) *If γ is an isolated boundary component of a plane region W, then $P_{\nu\gamma}(z;\zeta)$ is the unique function in $\mathscr{S}_{\zeta\gamma}$ with the properties*

$$\lim_{z\to\gamma}|P_{\nu\gamma}|=\text{const},$$

$$L_\nu(\log|P_{\nu\gamma}|)=\log|P_{\nu\gamma}| \quad \textit{on } \beta-\gamma.$$

The constant on γ is $1/c_{\nu\gamma}$.

Now apply the counterpart of Theorem VI.6 B in analogy with the proof of Theorem VI.6 A, to obtain:

(b) *If γ is an isolated boundary component of a plane region W, then for any $c_{\nu\gamma}$, the function $P_{1\gamma}(z;\zeta)$ (or $P_{0\gamma}(z;\zeta)$) maps W conformally onto a circular (or radial, resp.) slit disk of radius $1/c_{\nu\gamma}$. The total area of the slits vanishes.*

In general, this mapping property is insufficient to characterize the functions. We shall introduce *extremal* slit disks, to be studied in Chapter IX. In the following case, however, the mapping property does characterize the functions:

(c) *If W is a regular region punctured at a finite number of points, then $P_{1\gamma}(z;\zeta)$ (or $P_{0\gamma}(z;\zeta)$) is the unique function in $\mathscr{S}_{\zeta\gamma}$ which maps W conformally onto a circular (or radial, resp.) slit disk.*

In fact, the assumption implies the condition in (a).

The following relation, different from (1), holds for the function $P_\nu(z;\zeta,\zeta')$:

(d) *Suppose $0\in W$, $\infty\notin W$, and the outer boundary γ is isolated and is a circle about the origin. Consider the restriction to W of $P_\nu(z;0,\infty)$ on the double $\hat{W}$ of $\overline{W}$ about γ. Then*

$$P_{\nu\gamma}(z;0)=P_\nu(z;0,\infty).$$

The proof is immediate in view of (a) and the symmetry of P_ν about γ (cf. Problem 9).

For example, if

$$W=\{z\,|\,|z|<r\},\qquad \gamma=\{z\,|\,|z|=r\},\qquad r\le\infty,$$

then P_1 and P_0 coincide and we easily obtain the following from (c): If $r<\infty$, then

$$P_{1\gamma}(z;\zeta)=\frac{(z-\zeta)(r^2-|\zeta|^2)}{r^2-\bar{\zeta}z},\qquad c_{1\gamma}(\zeta)=\frac{r}{r^2-|\zeta|^2},$$

so that $c_{1\gamma}(\zeta)|d\zeta|$ is again the Poincaré metric; if $r=\infty$, then

$$P_{1\gamma}(z;\zeta)=z-\zeta, \qquad c_{1\gamma}(\zeta)=0.$$

Remark. As mentioned in Remark III.2 B, a potential-theoretic approach was used and (c) established by KOEBE [8]. For a regular region, (d) was also given by KOEBE [10] (cf. RENGEL [1, 2]). For arbitrary regions, the study was initiated by GRÖTZSCH (see 2, 3, and 4 below). NEHARI [1] obtained the identity expressing $P_{1\gamma}$ in terms of the kernel function K (V.1 A); see also BERGMAN [3].

2. Extremal Properties of $P_{1\gamma}$ and Conformal Mapping by $P_{1\gamma}$

2 A. Reduced Logarithmic Area. Let γ be a boundary component of a plane region W. Among the functions $F\in\mathcal{S}_{\zeta\gamma}$, the function $P_{1\gamma}$ minimizes $\int_\beta p*dp$, where $p=\log|F|$ (Theorem III.3 A). We consider the geometric meaning of this integral.

In general, let $E\subset\{|z|<\infty\}$ be a measurable set which contains 0 as an interior point. For any $\varepsilon>0$ with $\{z\,|\,|z|\le\varepsilon\}\subset E$, we set $E_\varepsilon=E-\{z\,|\,|z|\le\varepsilon\}$. The quantity

$$2\pi\log\varepsilon+\iint_{E_\varepsilon}\frac{dx\,dy}{|z|^2}, \qquad z=x+iy,$$

which is clearly independent of ε, will be called the *reduced logarithmic area of* E. In particular, if E is a region, then

$$\iint_{E_\varepsilon}\frac{dx\,dy}{|z|^2}=\int_{\partial E}\log|z|\,d\arg z-\int_{|z|=\varepsilon}\log|z|\,d\arg z$$

and, therefore,

$$2\pi\log\varepsilon+\iint_{E_\varepsilon}\frac{dx\,dy}{|z|^2}=\int_{\partial E}\log|z|\,d\arg z, \tag{2}$$

where the last integral is to be understood in the sense of II.1 D.

For any F in the family $\mathcal{S}_{\zeta\gamma}$, the integral $\int_\beta p*dp$ for $p=\log|F|$ is thus equal to the reduced logarithmic area of the image region $F(W)$. We express this by writing

$$\tilde{\mathcal{D}}[F]=\int_{\partial W}\log|F|\,d\arg F.$$

Theorem III.3 A can then be stated as follows:

THEOREM. *The minimum in $\mathcal{S}_{\zeta\gamma}$ of the reduced logarithmic area is*

$$\min_{F\in\mathcal{S}_{\zeta\gamma}}\tilde{\mathcal{D}}[F]=2\pi\log\frac{1}{c_{1\gamma}}.$$

If $c_{1\gamma}>0$, then $P_{1\gamma}$ is the unique minimizing function.

For later reference, we include here the following identity (III. 3 A. (4)) which plays a fundamental role in the proof:

$$\tilde{\mathscr{D}}[F]-\tilde{\mathscr{D}}[P_{1\gamma}]=\iint\left|\frac{F'}{F}-\frac{P'_{1\gamma}}{P_{1\gamma}}\right|^2 dx\,dy \tag{3}$$

for every $F\in\mathscr{S}_{\zeta\gamma}$.

2 B. Maximum Modulus. For an analytic function F on W, set

$$M[F]=\sup_{z\in W}|F(z)|.$$

THEOREM. *The minimum in $\mathscr{S}_{\zeta\gamma}$ of the maximum modulus is*

$$\min_{F\in\mathscr{S}_{\zeta\gamma}} M[F]=\frac{1}{c_{1\gamma}}.$$

If $c_{1\gamma}>0$, then $P_{1\gamma}$ is the unique minimizing function.

For $W=\{z\,|\,|z|<r\}$, $r<\infty$, and $\zeta=0$, the theorem is a part of Schwarz's lemma.

Remark. The above theorem (extremal property) and Theorem 2 E below (mapping property) are due to GRÖTZSCH [3, 5], RENGEL [1, 2], GRUNSKY [1] and SARIO [13]. SCHIFFER [1] derived the latter from the former using his variational method.

Proof. From the definition of reduced logarithmic area, we obtain

$$\tilde{\mathscr{D}}[F]\leq 2\pi\log M[F] \tag{4}$$

for every $F\in\mathscr{S}_{\zeta\gamma}$. If $c_{1\gamma}=0$, then $\tilde{\mathscr{D}}[F]=\infty$ by Theorem 2 A and, consequently, $M[F]=\infty$ for every F. If $c_{1\gamma}>0$, we have $M[P_{1\gamma}]\leq 1/c_{1\gamma}$ by Lemma III. 2 F. Combining this with (4) and using the relation $\tilde{\mathscr{D}}[P_{1\gamma}]=2\pi\log(1/c_{1\gamma})$ (Theorem 2 A), we obtain $M[P_{1\gamma}]=1/c_{1\gamma}$. On substituting this and (4) into (3) we conclude that

$$2\pi\log\frac{M[F]}{M[P_{1\gamma}]}\geq\iint\left|\frac{F'}{F}-\frac{P'_{1\gamma}}{P_{1\gamma}}\right|^2 dx\,dy. \tag{5}$$

Consequently $M[F]\geq M[P_{1\gamma}]$, where equality implies $F=P_{1\gamma}$.

2 C. Weak Boundary Components. By the above theorem, the capacity $c_{1\gamma}$ vanishes if and only if $M[F]=\infty$, that is,

$$F(\gamma)=\{\infty\}$$

for every $F\in\mathscr{S}_{\zeta\gamma}$. For an arbitrary univalent function Φ on W, there exists a linear transformation Λ such that $\Lambda\circ\Phi\in\mathscr{S}_{\zeta\gamma}$. Therefore, $c_{1\gamma}=0$

if and only if $\Phi(\gamma)$ consists of a single point for every Φ. This again shows that the property $c_{1\gamma}(\zeta)=0$ is independent of ζ (cf. IV.3 B). We shall use the following suggestive terminology (SARIO [13, 15]):

DEFINITION. *A boundary component γ of a plane region W is weak if its image under every univalent mapping of W consists of a single point.*

Weak components were first considered by GRÖTZSCH [10], who called them *vollkommen punktförmig;* later results concerning such components are due mainly to CONSTANTINESCU [2], JURCHESCU [1], OIKAWA [1, 2], SARIO [13, 15], and SAVAGE [1].

The above discussion implies (SARIO [13]):

(a) *A boundary component γ is weak if and only if $c_{1\gamma}=0$.*

Suppose that a compact set γ is a boundary component of two plane regions W and W' and that a boundary neighborhood of γ with respect to W is conformally equivalent to one with respect to W'. Then by IV.3 B, γ is a weak boundary component of W if and only if it is a weak boundary component of W'. This result may be stated as follows:

(b) *Weakness is a boundary property.*

We also have trivially:

(c) *If γ is an isolated boundary component, then it is weak if and only if it consists of a single point.*

A number of criteria for weakness of non-isolated boundary components will be given in Chapter XI.

2 D. Dirichlet Integral. We shall establish the following property of our function $P_{1\gamma}$:

THEOREM. *The minimum in $\mathscr{S}_{\zeta\gamma}$ of the Dirichlet integral is*

$$\min_{F\in\mathscr{S}_{\zeta\gamma}} \mathscr{D}[F]=\frac{\pi}{c_{1\gamma}^2}.$$

If $c_{1\gamma}>0$, then $P_{1\gamma}$ is the unique minimizing function.

For $W=\{z\,|\,|z|<r\}$, $r<\infty$, this is a special case of Theorem VI.5 B, because now $(P_0-P_1)/2S=(z-\zeta)(r^2-|\zeta|^2)/(r^2-\bar{\zeta}z)=P_{1\gamma}$ and $\sqrt{S}=r/(r^2-|\zeta|^2)=c_{1\gamma}$. This simplification is not true for an arbitrary W, and we shall give the general relation in 5 D – 5 F.

Remark. A result analogous to the above theorem was first obtained for mappings onto circular slit annuli (cf. VIII.1 D.(3)) by GRÖTZSCH [5]. The present theorem is contained implicitly in AHLFORS-BEURLING [1].

The proof below is due to REICH-WARSCHAWSKI [1]. See also SARIO [10] and STONE [1, 2].

Proof. Suppose $F \in \mathcal{S}_{\zeta\gamma}$ has a finite $\mathcal{D}[F]$. Consider the real number r such that

$$\mathcal{D}[F] = \pi r^2.$$

Let $R_\varepsilon = F(W) - \{w \,|\, |w| \le \varepsilon\}$ for sufficiently small $\varepsilon > 0$. In terms of the logarithmic area $\tilde{m}$ we have

$$\begin{aligned}
\tilde{\mathcal{D}}[F] &= 2\pi \log \varepsilon + \tilde{m}(R_\varepsilon) \\
&= 2\pi \log \varepsilon + \tilde{m}(R_\varepsilon \cap \{w \,|\, |w| < r\}) + \tilde{m}(R_\varepsilon \cap \{w \,|\, |w| > r\}) \\
&\le 2\pi \log \varepsilon + \tilde{m}(R_\varepsilon \cap \{w \,|\, |w| < r\}) + \tilde{m}(\{w \,|\, \varepsilon < |w| < r\} - R_\varepsilon) \\
&= 2\pi \log \varepsilon + \tilde{m}(\{w \,|\, \varepsilon < |w| < r\}) = 2\pi \log r.
\end{aligned}$$

Thus

$$\tilde{\mathcal{D}}[F] \le \pi \log \frac{\mathcal{D}[F]}{\pi} \tag{6}$$

for every $F \in \mathcal{S}_{\zeta\gamma}$ with a finite $\mathcal{D}[F]$. If $\mathcal{D}[F] = \infty$, this relation is trivially true.

Now if $c_{1\gamma} = 0$, then $\tilde{\mathcal{D}}[F] = \infty$ by Theorem 2 A, so that $\mathcal{D}[F] = \infty$ for every $F \in \mathcal{S}_{\zeta\gamma}$.

If $c_{1\gamma} > 0$, we obtain

$$\mathcal{D}[P_{1\gamma}] = \frac{\pi}{c_{1\gamma}^2} \tag{7}$$

as follows: $|P_{1\gamma}| \le 1/c_{1\gamma}$ (Lemma III.2 F) implies that $\mathcal{D}[P_{1\gamma}] \le \pi/c_{1\gamma}^2$; on the other hand, $\mathcal{D}[P_{1\gamma}] \ge \pi \exp(\pi^{-1} \tilde{\mathcal{D}}[P_{1\gamma}])$ by (6), and $\tilde{\mathcal{D}}[P_{1\gamma}] = 2\pi \log(1/c_{1\gamma})$ by Theorem 2 A. On substituting (6) and (7) into (3), we obtain

$$\pi \log \frac{\mathcal{D}[F]}{\mathcal{D}[P_{1\gamma}]} \ge \iint \left| \frac{F'}{F} - \frac{P'_{1\gamma}}{P_{1\gamma}} \right|^2 dx\, dy. \tag{8}$$

We conclude that $\mathcal{D}[F] \ge \mathcal{D}[P_{1\gamma}]$ and that if $\mathcal{D}[F] = \mathcal{D}[P_{1\gamma}]$, then $F = P_{1\gamma}$.

2 E. Conformal Mapping by $P_{1\gamma}$. By any one of the extremal properties in Theorems 2 A, 2 B, and 2 C, we can now determine the shape of the image region with no restriction on γ. We shall make use of Theorem 2 B.

THEOREM. *If γ is an arbitrary boundary component of an arbitrary plane region W, then $P_{1\gamma}(z;\zeta)$ maps W onto a circular slit disk of radius $1/c_{1\gamma}(\zeta)$, $\zeta \in W$. The total area of the slits vanishes.*

The case $c_{1\gamma}=0$ is not excluded.

Proof. Let $R=P_{1\gamma}(W)$ and $\gamma^*=P_{1\gamma}(\gamma)$. By using the argument in VI.6 A and 6 B, we obtain

$$L_1(\log|w|)=\log|w|$$

on $\partial R-\gamma^*$, or more precisely, on every end of R disjoint from a boundary neighborhood of γ^*. Thus, by Theorem VI.6 B, every component of $\partial R-\gamma^*$ is either a point or a circular arc. Furthermore, the area of $\partial R-\gamma^*$ vanishes.

We shall show that γ^* coincides with the circle $|w|=1/c_{1\gamma}$.

If $c_{1\gamma}=0$, then $\gamma^*=\{\infty\}$, as seen in 2 C.

If $c_{1\gamma}>0$, we let $\tilde{R}$ be the complement of X_{γ^*}, that is, of the component of CR whose boundary coincides with γ^* (VI.1 D). It is a simply connected region containing R. Since $c_{1\gamma}>0$, $\partial\tilde{R}=\gamma^*$ is not a single point (2 C), and hence $\tilde{c}$, the capacity $c_{1\gamma^*}(0)$ considered on $\tilde{R}$, is positive. Let $\tilde{P}$ be the function $P_{1\gamma^*}(w;0)$ for $\tilde{R}$, and compare it with the identity function $I(w)=w$. Since $M[I]=\sup_{w\in\tilde{R}}|w|=M[P_{1\gamma}]=1/c_{1\gamma}$, we obtain $1/\tilde{c}\leq 1/c_{1\gamma}$ by Theorem 2 B for $\tilde{R}$. On the other hand, $\tilde{P}\circ P_{1\gamma}\in\mathscr{S}_{\zeta\gamma}$, so that $1/\tilde{c}=M[\tilde{P}\circ P_{1\gamma}]\geq 1/c_{1\gamma}$ by the same theorem for W. Thus $1/\tilde{c}=1/c_{1\gamma}$ and, therefore, $\tilde{P}(w)=w$ again by Theorem 2 B for $\tilde{R}$. Since $\tilde{P}(w)$ maps $\tilde{R}$ conformally onto the disk $|w|<1/\tilde{c}$ (1 C.(b)), we see that $\tilde{R}$ coincides with this disk and, therefore, γ^* coincides with the circle $|w|=1/c_{1\gamma}$.

The property in Theorem 2 E is again insufficient to characterize the function $P_{1\gamma}$. We shall introduce extremal circular slit disks in Chapter IX.

2 F. Distance to the Outer Boundary. Instead of the maximum modulus $M[F]$, we consider

$$m_\gamma[F]=\min_{w\in F(\gamma)}|w|$$

for $F\in\mathscr{S}_{\zeta\gamma}$. The family $\mathscr{S}_{\zeta\gamma}$ is compact and the functional m_γ is lower semicontinuous, that is,

$$m_\gamma[F]\leq\varliminf_{n\to\infty} m_\gamma[F_n]$$

if $F_n\in\mathscr{S}_{\zeta\gamma}$ and $\lim_{n\to\infty}F_n=F$ uniformly on every compact set. As a consequence, there exists a function minimizing $m_\gamma[F]$ in $\mathscr{S}_{\zeta\gamma}$.

If W is the disk $|z|<r<\infty$ and $\zeta=0$, then Koebe's quarter theorem shows that the minimizing functions of $m_\gamma[F]$ are the so-called Koebe functions

$$k_{r\theta}(z)=\frac{r^2 z}{(r+e^{i\theta}z)^2},\qquad 0\leq\theta<2\pi.$$

In the general case,

$$K_\theta = k_{r\theta} \circ P_{1\gamma} \quad \text{with} \quad r = \frac{1}{c_{1\gamma}}$$

are minimizing functions:

THEOREM. *The minimum in $\mathscr{S}_{\zeta\gamma}$ of the distance to the outer boundary is one quarter of the radius of the circular slit disk,*

$$\min_{F \in \mathscr{S}_{\zeta\gamma}} m_\gamma[F] = \frac{1}{4c_{1\gamma}}.$$

If $c_{1\gamma} > 0$, then the only minimizing functions are K_θ, $0 \le \theta < 2\pi$.

The image of W under an extremal function for positive $c_{1\gamma}$ is, therefore, as follows. For $\theta = 0$, the image of γ is the ray $[1/4c_{1\gamma}, \infty)$ and the image of every other boundary component is either a point or an arc on a curve which is the image under $k_{r\theta}$ of $|z| = \text{const} < r = 1/c_{1\gamma}$. For an arbitrary θ, the image region is obtained by rotating the above by $-\theta$ about the origin.

Proof of the Theorem. By 2 C the result is trivial for $c_{1\gamma} = 0$. We assume in the sequel that $c_{1\gamma} > 0$. Let F_0 be a function minimizing $m_\gamma[F]$ in $\mathscr{S}_{\zeta\gamma}$, and set $m_0 = m_\gamma[F_0]$. We know that $K_\theta \in \mathscr{S}_{\zeta\gamma}$ and

$$m_\gamma[K_\theta] = \frac{1}{4c_{1\gamma}}, \tag{9}$$

and therefore $m_0 \le 1/4c_{1\gamma}$. The image $F_0(\gamma)$ must be a ray on $\arg w = \text{const}$, since otherwise there would exist an $F \in \mathscr{S}_{\zeta\gamma}$ with $m_\gamma[F] < m_0$, as is seen by a simple application of Koebe's quarter theorem to the w-plane. Suppose $F_0(\gamma)$ is on the ray $\arg w = \theta$. If $m_0 < 1/4c_{1\gamma}$, then $k_{r\theta}^{-1} \circ F_0 \in \mathscr{S}_{\zeta\gamma}$ satisfies $M[k_{r\theta}^{-1} \circ F_0] = 1/4c_{1\gamma}$ but is still different from $P_{1\gamma}$. This contradicts Theorem 2 B. Hence $m_0 = 1/4c_{1\gamma}$ and $k_{r\theta}^{-1} \circ F_0 = P_{1\gamma}$.

Remark. For a generalization of the Koebe function, consider the coefficients of the following expansion:

$$F(z) = (z - \zeta) + a_2(z - \zeta)^2 + \cdots \in \mathscr{S}_{\zeta\gamma},$$

where for the sake of simplicity one supposes $\zeta \neq \infty$. If W is simply connected, then $\max |a_2|$ is attained only by Koebe functions, and the points a_2 fill a disk (which has radius 2 and is centered at the origin provided W is the unit disk and $\zeta = 0$). For a W of greater connectivity the result was generalized by GRÖTZSCH [6, 9]. Although the a_2 again fill a disk, the function whose a_2 is on the periphery of this disk is not K_θ but a mapping such that the image of γ is a ray and the images of the other boundary components are arcs on confocal parabolas.

2 G. Diameter of a Boundary Component. Our function $P_{1\gamma}$ is also related to a function which has the following extremal property in the family $\mathscr{V}_\zeta = \mathscr{V}_\zeta(W)$ (VI.2 A) (GRÖTZSCH [2] and KOMATU [3]):

THEOREM. *The maximum in $\mathscr{V}_\zeta$ of the diameter of the image of γ is*

$$\max_{F \in \mathscr{V}_\zeta}(\operatorname{diam} F(\gamma)) = 4c_{1\gamma}(\zeta).$$

If $c_{1\gamma} > 0$, the only maximizing functions are

$$\frac{1}{K_\theta} + \text{const}, \qquad 0 \leq \theta < 2\pi,$$

where the constant is chosen so that the function belongs to $\mathscr{V}_\zeta$.

Thus the image of W under a maximizing function for a positive $c_{1\gamma}$ has these properties: the image of γ is a line segment with length $4c_{1\gamma}$ and inclination $-\theta$, and the image of every other boundary component is either a point or an arc on an ellipse whose foci are the end points of the image of γ.

Proof of the Theorem. For an arbitrary $F \in \mathscr{V}_\zeta$, consider points w_1 and w_2 of $F(\gamma)$ such that $|w_1 - w_2| = \operatorname{diam} F(\gamma)$. Then $G(z) = 1/(F(z) - w_1) \in \mathscr{S}_{\zeta\gamma}$ and $m_\gamma[G] = 1/|w_1 - w_2|$. We, therefore, see from Theorem 2 F that $\operatorname{diam} F(\gamma) \leq 4c_{1\gamma}$ for all $F \in \mathscr{V}_\zeta$ and that the equality holds if G is an extremal function K_θ. If $c_{1\gamma} = 0$, the proof is complete. If $c_{1\gamma} > 0$, let $F \in \mathscr{V}_\zeta$ be an extremal function, that is, $\operatorname{diam} F(\gamma) = 4c_{1\gamma}$. Then $G \in \mathscr{S}_{\zeta\gamma}$ and $m_\gamma[G] = 1/\operatorname{diam} F(\gamma) = 1/4c_{1\gamma}$. By Theorem 3 F, we have $G = K_\theta$ for some $0 \leq \theta < 2\pi$. Therefore, $F = 1/K_\theta + \text{const}$, as desired.

Remark. A function minimizing $\operatorname{diam} F(\gamma)$ in $\mathscr{V}_\zeta$ does not necessarily exist. If one does, then the minimum value is $2c_{0\gamma}(\zeta)$, and the extremal function maps W onto the exterior of a disk with radial slits. This result is due to GRÖTZSCH [2]. (See Problem 12.)

2 H. Uniqueness of $P_{1\gamma}$ for $c_{1\gamma} = 0$. At this point, we shall show that the function $P_{1\gamma}(z; \zeta)$ is always unique, including the case where γ has vanishing capacity $c_{1\gamma}$. The proof is reduced to showing the following (OIKAWA-SUITA [2]):

THEOREM. *On a plane region W and for a boundary component γ, the capacity function $p_{1\gamma}(z; \zeta)$ is uniquely determined even if $c_{1\gamma} = 0$.*

Proof. Suppose there are two such functions $p_{1\gamma}$ and $\tilde{p}_{1\gamma}$. The difference

$$u = p_{1\gamma} - \tilde{p}_{1\gamma}$$

is harmonic throughout W, vanishes at ζ, and satisfies

$$L_1 u = u \quad \text{on } \beta - \gamma,$$

where $\beta = \partial W$. We are to prove that $u = \text{const.}$

Let us first show that u is bounded from above. Consider $P_{1\gamma} = \exp(p_{1\gamma} + i\, p_{1\gamma}^*)$ and $\tilde{P}_{1\gamma} = \exp(\tilde{p}_{1\gamma} + i\, \tilde{p}_{1\gamma}^*)$, which map W conformally onto circular slit disks of infinite radius (see the proof of Theorem 2 E). For $n = 1, 2, \ldots$, the region $R_n = P_{1\gamma}(W) \cap \{w \mid |w| < n\}$ is a circular slit disk of radius n. Theorem IX.2 B will show that it is extremal, so that the capacity of the circle $|w| = n$ with respect to R_n and $\zeta = 0$ is $1/n$. By Theorem 2 F, we see that the image of $|w| = n$ under the mapping $\tilde{P}_{1\gamma} \circ P_{1\gamma}^{-1}$ lies outside of the disk $|w| \le n/4$. Take an analytic Jordan curve $\alpha_n \subset W$ such that $P_{1\gamma}(\alpha_n)$ lies in the annulus $n < |w| < n+1$ and separates 0 from ∞ (cf. an analogue of Lemma 2 in VI.1 E). We have $|P_{1\gamma}| < n+1$ and $|\tilde{P}_{1\gamma}| \ge n/4$ on α_n and, therefore,

$$u \le \log \frac{4(n+1)}{n} \quad \text{on } \alpha_n.$$

The curve α_n divides W into two subregions. Let W_n be the one containing ζ. Then $L_1 u = u$ on W_n, where the operator L_1 acts from α_n into W_n. The maximum principle for normal operators (I.1 D) implies that $u \le \log(4(n+1)\, n^{-1})$ on W_n. On letting $n \to \infty$, we conclude that

$$u \le \log 4 \quad \text{on } W,$$

that is, u is bounded from above.

Now let U be a boundary neighborhood of β. The harmonic function u on $\bar{U} \cap W$ is bounded from above and satisfies $L_1 u = u$ on $\beta_U - \gamma$. Since γ is weak, we obtain by the maximum principle proved in X.1 G,

$$u \le \max_{\alpha} u \quad \text{on } U,$$

where $\alpha = (\partial U) \cap W$. Clearly $u \le \max_\alpha u$ on $W - U$. Accordingly, u attains its maximum over W at a point on α. By the standard maximum principle for harmonic functions, we conclude that $u = \text{const.}$

3. Extremal Properties of $P_{0\gamma}$

3 A. Reduced Logarithmic Area. Let W be a plane region and γ a boundary component. Take a point $\zeta \in W$. In order to study the extremal property of $P_{0\gamma}$ given in 1 A, we introduce the family $\mathscr{S}_{\zeta\gamma}^0 = \mathscr{S}_{\zeta\gamma}^0(W)$ of functions $F \in \mathscr{S}_{\zeta\gamma}$ such that $L_0(\log|F|) = \log|F|$ on $\partial W - \gamma$ (in the sense of III.2 C.(c)).

Geometrically, a function $F \in \mathscr{S}_{\zeta\gamma}$ belongs to $\mathscr{S}_{\zeta\gamma}^0$ if and only if it has an image region such that $\partial F(W) - F(\gamma)$ consists of points or radial segments and, for every analytic Jordan curve $\alpha \subset F(W)$ for which the region $\operatorname{Int}\alpha$ enclosed by it is disjoint from $F(\gamma)$, the region $C((\operatorname{Int}\alpha) \cap CF(W))$ is an extremal radial slit plane (VI.6 B). This is seen by an analogue of the reasoning used in the first paragraph of the proof of Theorem 2 E; essential use is made of Theorem VI.6 B.

In terms of this family and the reduced logarithmic area $\tilde{\mathscr{D}}[F]$ of the image region $F(W)$, we now state Theorem III.3 A as follows:

THEOREM. *The minimum in* $\mathscr{S}_{\zeta\gamma}^0$ *of the reduced logarithmic area is*

$$\min_{F \in \mathscr{S}_{\zeta\gamma}^0} \tilde{\mathscr{D}}[F] = 2\pi \log \frac{1}{c_{0\gamma}}.$$

If $c_{0\gamma} > 0$, *then* $P_{0\gamma}$ *is the unique extremal function.*

The proof was based on the following identity (cf. III.3 A.(3)) of which we shall make further use:

$$\tilde{\mathscr{D}}[F] - \tilde{\mathscr{D}}[P_{0\gamma}] = \iint \left| \frac{F'}{F} - \frac{P'_{0\gamma}}{P_{0\gamma}} \right|^2 dx\, dy \tag{1}$$

for every $F \in \mathscr{S}_{\zeta\gamma}^0$.

3 B. Maximum Modulus. As in 2 B we obtain from (1) the relation

$$2\pi \log \frac{M[F]}{M[P_{0\gamma}]} \geq \iint \left| \frac{F'}{F} - \frac{P'_{0\gamma}}{P_{0\gamma}} \right|^2 dx\, dy \tag{2}$$

for every $F \in \mathscr{S}_{\zeta\gamma}^0$. Consequently (OIKAWA [4]):

THEOREM. *The minimum in* $\mathscr{S}_{\zeta\gamma}^0$ *of the maximum modulus is*

$$\min_{F \in \mathscr{S}_{\zeta\gamma}^0} M[F] = \frac{1}{c_{0\gamma}}.$$

If $c_{0\gamma} > 0$, *then* $P_{0\gamma}$ *is the unique extremal function.*

We also obtain a counterpart of Theorem 2 D showing that $P_{0\gamma}$ minimizes $\mathscr{D}[F]$ in $\mathscr{S}_{\zeta\gamma}^0$, and the value of the minimum is $\pi/c_{0\gamma}^2$. The details will be left to the reader.

3 C. Distance to the Outer Boundary. It is not easy to characterize $P_{0\gamma}$ in the family $\mathscr{S}_{\zeta\gamma}$ in terms of a simple geometric extremal property. The quantity

$$m_\gamma[F] = \min_{w \in F(\gamma)} |w|$$

discussed in 2 F was considered by RENGEL [1, 2], who established the following result as well as Theorem 4 B under a certain restriction (cf. also SARIO [12]):

THEOREM. *Let W be an arbitrary plane region, γ a boundary component, and $\zeta \in W$. Then*

$$\sup_{F \in \mathcal{S}_{\zeta\gamma}} m_\gamma[F] = \frac{1}{c_{0\gamma}}.$$

If $c_{0\gamma} > 0$ and there exists an $F \in \mathcal{S}_{\zeta\gamma}$ with $m_\gamma[F] = 1/c_{0\gamma}$, then F is necessarily $P_{0\gamma}$.

If γ is an isolated boundary component consisting of more than one point, then $c_{0\gamma} > 0$ and the extremal function always exists.

The proof, due to OIKAWA [4], will be given in 3 E – 3 G.

An extremal function does not exist in general. This was first pointed out by GRÖTZSCH [7] who dealt with a related extremal problem (cf. Remark 2 G). In 3 H, we shall exhibit a counterexample given by STREBEL [1].

In the case $c_{0\gamma} = 0$, the extremal function F in $\mathcal{S}_{\zeta\gamma}$ with $F(\gamma) = \{\infty\}$ (if it exists), is not necessarily identical with $P_{0\gamma}$. A counterexample will be found in XI. 5 K (Remark 2).

3 D. Strong Boundary Components. Before proving the above theorem, we make the following observation. In the case $c_{0\gamma} > 0$, the theorem shows that $m_\gamma[F] < \infty$, and therefore $F(\gamma) \neq \{\infty\}$ for every $F \in \mathcal{S}_{\zeta\gamma}$. In other words, $F(\gamma)$ consists of more than one point. Every univalent function on W has a linear transform belonging to the family $\mathcal{S}_{\zeta\gamma}$. Thus if $c_{0\gamma} > 0$, the image of γ under an arbitrary univalent function contains more than one point. We introduce (SARIO [13, 15]):

DEFINITION. *A boundary component γ of a plane region W is said to be strong if its image under every univalent mapping of W consists of more than one point.*

The above discussion implies the following proposition (SARIO [13]):

If $c_{0\gamma} > 0$, then γ is strong.

The authors do not know whether the converse is true. It would not be true if there existed a region such that $c_{0\gamma} = 0$ and the supremum in the theorem cannot be replaced by the maximum. The counterexample to be given in 3 H has a positive $c_{0\gamma}$ and does not answer the present question.

If γ is isolated, it is strong if and only if it consists of more than one point, and the condition is equivalent to $c_{0\gamma} > 0$. We conclude in this case that if γ is not strong then it is weak.

In general, there exist components γ which are neither strong nor weak. Case (b) of Theorem VI.3 F is an example (cf. Remark 1 in VI.3 F).

3 E. Proof of Theorem 3 C. We begin with the case in which γ is isolated and consists of more than one point.

In view of the Riemann mapping theorem, we may assume that γ is an analytic Jordan curve. Therefore, W satisfies the requirements in III.1 C–1 F and we can freely use results derived from them. In particular, $c_{0\gamma}>0$ (III.2 D) and $m_\gamma[P_{0\gamma}]=M[P_{0\gamma}]$ (VII.1 C).

Consider an $F\in\mathcal{S}_{\zeta\gamma}$ with $m_\gamma[F]=M[F]$. It is analytic on $W\cup\gamma$, and $p=\log|F|$ is constant on γ. By identity III.1 F.(7), we have

$$\begin{aligned} D_W[p-p_{0\gamma}] &= \int_{\beta-\gamma} p*dp-\int_\gamma p*dp+\int_\gamma p_{0\gamma}*dp_{0\gamma} \\ &= \int_{\beta-\gamma} p*dp-2\pi\log m_\gamma[F]+2\pi k_{0\gamma}. \end{aligned}$$

The quantity $\int_{\beta-\gamma}p*dp$ is non-positive, since it is the negative of the logarithmic area of $CF(W)\cap\{w\,|\,|w|<m_\gamma[F]\}$. We know that $k_{0\gamma}=\log m_\gamma[P_{0\gamma}]$ and, therefore,

$$D_W[p-p_{0\gamma}]\le 2\pi\log\frac{m_\gamma[P_{0\gamma}]}{m_\gamma[F]}.$$

Consequently,

$$m_\gamma[F]\le m_\gamma[P_{0\gamma}], \tag{3}$$

with equality holding only for $F=P_{0\gamma}$.

Next consider an arbitrary element F of $\mathcal{S}_{\zeta\gamma}$. Let $\tilde{R}$ be the region enclosed by $F(\gamma)$, that is, the interior of the bounded component of the complement of $F(\gamma)$. Write $\tilde{P}$ for the function $P_{0\tilde{\gamma}}(w;0)$ on $\tilde{R}$ with respect to $\tilde{\gamma}=\partial\tilde{R}=F(\gamma)$. Since $\tilde{R}$ contains the disk $|w|<m_\gamma[F]$, Schwarz's lemma applied to the disk $|w|<m_\gamma[F]$ shows that

$$m_\gamma[F]\le M[\tilde{P}]. \tag{4}$$

Equality holds if and only if $\tilde{P}(w)=w$, in which case $\tilde{R}$ must be a disk and $m_\gamma[F]=M[F]$. The function $\tilde{P}\circ F$ belongs to $\mathcal{S}_{\zeta\gamma}$ and satisfies $M[\tilde{P}\circ F]=m_\gamma[\tilde{P}\circ F]$, so that (3) is applicable. We combine this with (4):

$$m_\gamma[F]\le M[\tilde{P}]=m_\gamma[\tilde{P}\circ F]\le m_\gamma[P_{0\gamma}],$$

with equalities only for $F=P_{0\gamma}$.

Thus the proof is complete if γ is isolated and consists of more than one point.

3 F. General Case. If γ is arbitrary, we exhaust W towards γ by subregions V with $\zeta\in V$. On each V consider the functional $m_{\gamma(V)}$ on $\mathcal{S}_{\zeta\gamma(V)}(V)$. If F belongs to $\mathcal{S}_{\zeta\gamma}(W)$, then its restriction to V belongs to

$\mathscr{S}_{\zeta\gamma(V)}(V)$. Since $m_{\gamma(V)}[F]\leq 1/c_{0\gamma(V)}$ has been proved and $m_\gamma[F]\leq \underline{\lim}\, m_{\gamma(V)}[F]$ is easily verified, we conclude that $m_\gamma[F]$ does not exceed the reciprocal of $c_{0\gamma}=\lim c_{0\gamma(V)}$. Thus

$$\sup_{F\in\mathscr{S}_{\zeta\gamma}} m_\gamma[F]\leq\frac{1}{c_{0\gamma}}. \tag{5}$$

To prove the opposite inequality, we first note that if γ consists of a single point, then $\sup m_\gamma[F]=\infty$. In fact, by considering a linear transformation which takes γ to ∞, we can easily find a function $F\in\mathscr{S}_{\zeta\gamma}$ with $F(\gamma)=\{\infty\}$. For this F we have $m_\gamma[F]=\infty$, so that the equality in (5) is valid in the degenerate sense $\infty=\infty$.

Now assume that γ consists of more than one point. Denote by X_γ and $X_{\gamma(V)}$ the components of CW and CV whose boundaries are γ and $\gamma(V)$, respectively (VI.1 D), and set $W_V=V\cup(X_{\gamma(V)}-X_\gamma)$. This is a region with $V\subset W\subset W_V$ which has γ as an isolated boundary component. Let P_V be the function $P_{0\gamma}(z;\zeta)$ on W_V. The conclusion in 3 E applies to W_V:

$$m_\gamma[P_V]=\frac{1}{c_V},$$

where c_V is the capacity $c_{0\gamma}(\zeta)$ considered for W_V. The restriction of P_V to W belongs to $\mathscr{S}_{\zeta\gamma}(W)$, and the value of m_γ is not changed. Therefore, the left-hand side is dominated by $\sup m_\gamma[F]$ in $\mathscr{S}_{\zeta\gamma}(W)$. The quantity c_V is dominated by $c_{0\gamma(V)}$ with respect to V; this is apparent from the expression of $c_{0\gamma}$ in terms of extremal length (Theorem IV.3 G). Thus

$$\sup_{F\in\mathscr{S}_{\zeta\gamma}} m_\gamma[F]\geq\frac{1}{c_{0\gamma(V)}}.$$

On letting $V\to W$, we obtain

$$\sup_{F\in\mathscr{S}_{\zeta\gamma}} m_\gamma[F]\geq\frac{1}{c_{0\gamma}}. \tag{6}$$

The first part of the theorem has been proved.

3 G. Completion of the Proof of Theorem 3 C. It remains to establish the second part of the theorem. Suppose that $c_{0\gamma}>0$ and that an extremal function F exists.

We have $M[F]=m_\gamma[F]$, that is,

$$F(\gamma)=\left\{w\,\middle|\,|w|=\frac{1}{c_{0\gamma}}\right\}. \tag{7}$$

In fact, as in 3 E we consider the simply connected region $\tilde{R}$ enclosed by $F(\gamma)$ and map it onto a disk by $\tilde{P}$ with $\tilde{P}\circ F\in\mathscr{S}_{\zeta\gamma}$. If $F(\gamma)$ is not a circle, then $\tilde{P}(w) \not\equiv w$, so that by Schwarz's lemma, $m_\gamma[\tilde{P}\circ F] = M[\tilde{P}\circ F] > m_\gamma[F]$. This contradicts the maximality of F, and we obtain (7).

Consider an exhaustion $V \to W$ towards γ with $\zeta \in V$. On V we apply identity III.1 F.(6) to $p = \log|F|$ and $p^0 = p_{0\gamma(V)} = \log|P_{0\gamma(V)}|$:

$$D_V[p - p_{0\gamma(V)}] = \int_{\beta(V)} p * dp - 2\int_{\gamma(V)} p * dp_{0\gamma(V)} + \int_{\gamma(V)} p_{0\gamma(V)} * dp_{0\gamma(V)}$$
$$= \tilde{\mathscr{D}}_V[F] - 2\int_{\gamma(V)} p * dp_{0\gamma(V)} + 2\pi\, k_{0\gamma(V)}$$
$$\le 2\pi\, k_{0\gamma} - 2\int_{\gamma(V)} p * dp_{0\gamma(V)} + 2\pi\, k_{0\gamma}.$$

Given $\varepsilon > 0$, take V sufficiently close to W so that

$$p > k_{0\gamma} - \varepsilon \quad \text{on } \gamma(V);$$

this is actually possible because of (7). Since $*dp_{0\gamma(V)} > 0$ along $\gamma(V)$, we obtain

$$D_V[p - p_{0\gamma(V)}] \le 4\pi\,\varepsilon.$$

On letting $V \to W$ and then $\varepsilon \to 0$, we conclude that $p = p_{0\gamma}$, that is, $F = P_{0\gamma}$.

The proof of Theorem 3 C is herewith complete.

3 H. A Counterexample. We shall show by an example that the supremum in Theorem 3 C cannot be replaced by the maximum. Consider

$$W = \{z \,|\, |z| < 1\} - \sigma - \bigcup_{n=2}^{\infty} (\sigma_n \cup \sigma_{-n}),$$

where $\sigma = \{z \,|\, \frac{1}{2} \le |z| \le 1, \arg z = 0\}$ and $\sigma_n = \{z \,|\, \frac{1}{2} \le |z| \le 1 - 1/|n|, \arg z = \pi/n\}$, $n = \pm 2, \pm 3, \ldots$. With respect to $\zeta = 0$ and $\gamma = \{z \,|\, |z| = 1\} \cup \sigma$, we claim that

$$c_{0\gamma} = 1 \tag{8}$$

and

$$m_\gamma[F] < 1 \quad \text{for all } F \in \mathscr{S}_{\zeta\gamma}. \tag{9}$$

We shall thus have the desired example.

To prove (8), use the equality $\log(1/c_{0\gamma}) = \lim_{\varepsilon\to 0}(\log\varepsilon + 2\pi\,\lambda(\Gamma_\varepsilon^*))$, which was obtained in IV.2 C and IV.3 G. Here Γ_ε^* is the family of curves joining γ and $|z| = \varepsilon$ in $W - \{z \,|\, |z| \le \varepsilon\}$. Let Γ_1 be the subfamily of curves joining σ and $|z| = \varepsilon$ and set $\Gamma_2 = \Gamma_\varepsilon^* - \Gamma_1$. Then $\lambda(\Gamma_2)^{-1} \le \lambda(\Gamma_\varepsilon^*)^{-1} \le \lambda(\Gamma_1)^{-1} + \lambda(\Gamma_2)^{-1}$. Since every $c \in \Gamma_1$ terminates only at $z = \frac{1}{2}$ or $z = 1$, we have $\lambda(\Gamma_1) = \infty$ (Appendix I.K). Clearly $2\pi\,\lambda(\Gamma_2) = \log(1/\varepsilon)$. Consequently, $2\pi\,\lambda(\Gamma_\varepsilon^*) = \log(1/\varepsilon)$ and, therefore, $\log(1/c_{0\gamma}) = 0$.

To prove (9), we assume the existence of an F with $m_\gamma[F] = 1$ and derive a contradiction. By the second part of Theorem 3 C, the function F is necessarily $P_{0\gamma}$. By Theorem 4 B, the image region $P_{0\gamma}(W)$ is a radial slit disk of radius 1 with no incisions. On the other hand, the identity function I has maximum modulus $M(I)=1$; hence $P_{0\gamma}(z)=z$ by Theorem 3 B. The fact that I belongs to the family $\mathscr{S}^0_{\zeta\gamma}$ is easily verified by observing that a radial slit plane with a finite number of slits is always extremal. Thus $W=P_{0\gamma}(W)$, which is a contradiction.

3 I. Characterization of $P_{0\gamma}$ in $\mathscr{S}_{\zeta\gamma}$. By using the idea of Theorem IV. 2 D, we can characterize $P_{0\gamma}$ in the family $\mathscr{S}_{\zeta\gamma}$ in terms of an extremal property. The result was obtained by Suita [3].

We shall introduce a modification $m^*_\gamma[F]$ of $m_\gamma[F]$. Let γ be a boundary component of a plane region W. Given a point $\zeta\in W$, let N be a simply connected regular region containing ζ and consider the family Γ^* of curves in $W-\bar{N}$ joining γ and ∂N (IV. 2 B). Given $F\in\mathscr{S}_{\zeta\gamma}$ and $c\in\Gamma^*$, set

$$\kappa_c[F]=\overline{\lim_{\substack{z\to\gamma\\ z\in c}}}\,|F(z)|.$$

Consider the supremum of numbers a with

$$\kappa_c[F]\geq a \qquad \text{for almost all } c\in\Gamma^*. \tag{10}$$

We shall call it the *modified distance* $m^*_\gamma[F]$ *to the outer boundary*. Clearly,

$$\kappa_c[F]\geq m^*_\gamma[F]$$

for almost all $c\in\Gamma^*$.

We claim that $m^*_\gamma[F]$ *does not depend on the choice of* N. For the proof, take two neighborhoods N_1 and N_2 and write Γ_1^* and Γ_2^* for the families Γ^* on $W-\bar{N}_1$ and $W-\bar{N}_2$. It is possible to take a third neighborhood N_3, containing $\bar{N}_1\cup\bar{N}_2$, and a quasi-conformal mapping T of $W-\bar{N}_1$ onto $W-\bar{N}_2$ which is the identity mapping on $W-\bar{N}_3$ (for the definition of quasi-conformal mappings, see Appendix I. M). Clearly $\kappa_c[F]=\kappa_{T(c)}[F]$, so that for every a, the families $\{c\,|\,c\in\Gamma_1^*, \kappa_c[F]<a\}$ and $\{c\,|\,c\in\Gamma_2^*, \kappa_c[F]<a\}$ correspond under the mapping $c\to T(c)$. Accordingly, if the extremal length of the latter family is infinite, then so is that of the former, and conversely (cf. Appendix I. M). The validity of (10) is, therefore, independent of N.

Theorem. *The maximum in* $\mathscr{S}_{\zeta\gamma}$ *of the modified distance to the outer boundary is*

$$\max_{F\in\mathscr{S}_{\zeta\gamma}} m^*_\gamma[F]=\frac{1}{c_{0\gamma}}.$$

If $c_{0\gamma}>0$, *then* $P_{0\gamma}$ *is the unique extremal function.*

3 J. Proof. If $c_{0\gamma}=0$, then $m^*_\gamma[F]=\infty$ for every $F\in\mathscr{S}_{\zeta\gamma}$. In fact, $\lambda(\Gamma^*)=\infty$, so that (10) is satisfied by an arbitrary a.

Next suppose $c_{0\gamma}>0$, let $N_{0\varepsilon}=\{z\,|\,|P_{0\gamma}(z;\zeta)|<\varepsilon\}$, and consider the family Γ^* in $W-\bar{N}_{0\varepsilon}$. We know that the metric

$$\rho_0\,|dz|=\frac{|dP_{0\gamma}|}{\log\left(\dfrac{1}{\varepsilon\, c_{0\gamma}}\right)|P_{0\gamma}|}$$

on $W-\bar{N}_{0\varepsilon}$ has the property $\int_c \rho_0\,|dz|\geq 1$ for almost all $c\in\Gamma^*$ (IV.2 C and 3 D). Therefore, $\rho_0\in\bar{\mathbf{P}}(\Gamma^*)$.

For an arbitrary $F\in\mathscr{S}_{\zeta\gamma}$, we shall use the metric

$$\rho\,|dz|=\frac{|dF|}{\log\left(\dfrac{m_\gamma^*[F]}{\varepsilon_F}\right)|F|},$$

where ε_F is the maximum of $|F|$ on $\partial N_{0\varepsilon}$. Define $\tilde{\rho}\,|dz|$ on $W-\bar{N}_{0\varepsilon}$ by $\tilde{\rho}(z)=\rho(z)$ for $\varepsilon_F<|F(z)|<m_\gamma^*[F]$ and $\tilde{\rho}(z)=0$ elsewhere. It is easily verified that $\int_c \tilde{\rho}\,|dz|<1$ implies $\kappa_c[F]<m_\gamma^*[F]$. Accordingly, $\int_c \tilde{\rho}\,|dz|\geq 1$ for almost all $c\in\Gamma^*$, so that $\tilde{\rho}\in\bar{\mathbf{P}}(\Gamma^*)$.

An application of inequality IV.2 A.(1) gives

$$A(\tilde{\rho}-\rho_0)\leq A(\tilde{\rho})-A(\rho_0)\leq\frac{2\pi}{\log\left(\dfrac{m_\gamma^*[F]}{\varepsilon_F}\right)}-\frac{2\pi}{\log\left(\dfrac{1}{\varepsilon\, c_{0\gamma}}\right)}.$$

For a relatively compact open set G contained in $\{z\in W\,|\,\varepsilon_F<|F(z)|<m_\gamma^*[F]\}$, we obtain

$$\iint_G\left(\frac{1}{\log\dfrac{m_\gamma^*[F]}{\varepsilon_F}}\left|\frac{F'}{F}\right|-\frac{1}{\log\dfrac{1}{\varepsilon\, c_{0\gamma}}}\left|\frac{P'_{0\gamma}}{P_{0\gamma}}\right|\right)^2 dx\,dy$$

$$\leq 2\pi\frac{\log\dfrac{1}{c_{0\gamma}\,m_\gamma^*[F]}+\log\dfrac{\varepsilon_F}{\varepsilon}}{\left(\log\dfrac{m_\gamma^*[F]}{\varepsilon_F}\right)\left(\log\dfrac{1}{\varepsilon\, c_{0\gamma}}\right)}.$$

We multiply both sides by $\log(m_\gamma^*[F]/\varepsilon_F)\cdot\log(1/\varepsilon\, c_{0\gamma})$ and let $\varepsilon\to 0$. Since $\lim \varepsilon_F/\varepsilon=1$, it follows that

$$\iint_G\left(\left|\frac{F'}{F}\right|-\left|\frac{P'_{0\gamma}}{P_{0\gamma}}\right|\right)^2 dx\,dy\leq 2\pi\log\frac{1}{c_{0\gamma}\,m_\gamma^*[F]}.$$

Consequently, $m_\gamma^*[F]\leq 1/c_{0\gamma}$, and if equality occurs, then $|F'/F|=|P'_{0\gamma}/P_{0\gamma}|$ on the open set G and therefore $F=P_{0\gamma}$ on W.

To verify that

$$m_\gamma^*[P_{0\gamma}]=\frac{1}{c_{0\gamma}},$$

it suffices to prove that the left-hand side is not less than the right-hand side. To this end, we set $\Gamma_{(\infty)}=\{c\in\Gamma^*|\kappa_c[P_{0\gamma}]<1/c_{0\gamma}\}$ and show that $\lambda(\Gamma_{(\infty)})=\infty$.

Since $|P_{0\gamma}|<1/c_{0\gamma}$ on W, we see that $\Gamma_{(\infty)}\subset\bigcup_{k=1}^{\infty}\Gamma_{(k)}$, where

$$\Gamma_{(k)}=\left\{c\,|\,c\in\Gamma^*,\ \sup_{z\in c}\log|P_{0\gamma}(z)|<\log\left(\frac{1}{c_{0\gamma}}\right)-\frac{1}{k}\right\},$$

$k=1,2,\ldots$. It suffices to prove that $\lambda(\Gamma_{(k)})=\infty$.

Let $\zeta\in V_n\to W$ be an exhaustion towards γ. Write c_n and P_n for $c_{0\gamma(V_n)}(\zeta)$ and $P_{0\gamma(V_n)}(z;\zeta)$ on V_n. We know that $\lim P_n=P_{0\gamma}$ uniformly on $\partial N_{0\varepsilon}$, $\lim \mathscr{D}_{V_n}[P_n'/P_n-P_{0\gamma}'/P_{0\gamma}]=0$, and $\lim c_n=c_{0\gamma}$. It is possible to take n so large that

$$\log|P_n|<\log\varepsilon+\frac{1}{4k}\qquad \text{on } \partial N_{0\varepsilon},$$

$$\log|P_n|>\log\left(\frac{1}{c_{0\gamma}}\right)-\frac{1}{2k}\qquad \text{on } \gamma(V_n).$$

An arbitrary $c\in\Gamma_{(k)}$ may be assumed to be oriented in such a way that the initial point z_c lies on $\partial N_{0\varepsilon}$. Let z'_{nc} be the point where c meets $\gamma(V_n)$ for the first time. For the integral along the subarc between z_c and z'_{nc}, we have

$$\int(d\log|P_n|-d\log|P_{0\gamma}|)=\log\left|\frac{P_n(z'_{nc})}{P_{0\gamma}(z'_{nc})}\right|-\log\left|\frac{P_n(z_c)}{P_{0\gamma}(z_c)}\right|>\frac{1}{4k}.$$

Accordingly, if n is sufficiently large, the metric $|d(\log P_n-\log P_{0\gamma})|=|P_n^{-1}\,dP_n-P_{0\gamma}^{-1}\,dP_{0\gamma}|$ has an integral along every $c\in\Gamma_{(k)}$ greater than $1/4k$.

Define the metric $\rho_n|dz|$ as $|d(\log P_n-\log P_{0\gamma})|$ on $V_n-\overline{N}_{0\varepsilon}$ and as $|P_{0\gamma}|^{-1}|dP_{0\gamma}|$ on $W-V_n$. We have $\lim A(\rho_n)=0$. On the other hand, $\int_c\rho_n|dz|\geq 1/4k$ for every $c\in\Gamma_{(k)}$. Consequently, $\lambda(\Gamma_{(k)})\geq(4k)^{-2}/A(\rho_n)\to\infty$ as $n\to\infty$.

4. Conformal Mapping by $P_{0\gamma}$

4 A. Incised Radial Slit Disks. Let W be an arbitrary plane region, γ a boundary component, and ζ a point of W. In contrast with $P_{1\gamma}$ (Theorem 2 E), the result in 1 C.(b) for $P_{0\gamma}$ cannot be generalized to an arbitrary case. As we saw in 3 H, the image region $P_{0\gamma}(W)$ is not necessarily a radial slit disk and may have incisions.

DEFINITION. *A plane region is called an incised radial slit disk of radius r* ($\leq\infty$) *if it has the following three properties:*

(a) *It contains* 0 *and is contained in the disk* $|z|<r$.

(b) *The outer boundary consists of the circle* $|z|=r$ *and possibly a number of line segments, referred to as incisions, on rays* $\arg z=\text{const}$.

(c) *Every other boundary component is either a point or a line segment on a ray* $\arg z=\text{const}$.

A radial slit disk is a special case of an incised radial slit disk.

THEOREM. *The function* $P_{0\gamma}(z;\zeta)$ *maps W onto an incised radial slit disk of radius* $1/c_{0\gamma}$. *The total angular measure of the incisions and the total area of the slits vanish.*

The case $c_{0\gamma}=0$ is not excluded.

The theorem is due to STREBEL [2] and REICH [2]. The proof will be given in 4 C – 4 G.

One might ask if there always exists a function (perhaps different from $P_{0\gamma}$) in the family $\mathscr{S}_{\zeta\gamma}$ mapping W onto a radial slit disk without incisions. That *the answer is negative* is observed by applying Theorem IX. 3 B as follows: If a region has at most a countable number of boundary components (like the one in 3 H), then the relevant mapping function necessarily coincides with $P_{0\gamma}$.

4 B. Radial Slit Disks. The above theorem is supplemented by the following (cf. RENGEL [1, 2], REICH-WARSCHAWSKI [1]), which contains a generalization of 1 C.(b) (cf. Theorem 3 C):

THEOREM. *If* $P_{0\gamma}$ *maximizes* $m_\gamma[F]$ *in* $\mathscr{S}_{\zeta\gamma}$, *then* $P_{0\gamma}(W)$ *is a radial slit disk of radius* $1/c_{0\gamma}$. *Conversely, if* $P_{0\gamma}(W)$ *has no incisions, then* $P_{0\gamma}$ *maximizes* $m_\gamma[F]$ *in* $\mathscr{S}_{\zeta\gamma}$.

The case $c_{0\gamma}=0$ is not excluded, but then the result is trivial.

Proof. We assume that $c_{0\gamma}>0$ and that Theorem 4 A has been proved. If $P_{0\gamma}$ maximizes m_γ in $\mathscr{S}_{\zeta\gamma}$, then by 3 F.(6), $P_{0\gamma}(\gamma)$ is the circle $|w|=1/c_{0\gamma}$ and has no incisions. Conversely, if there is no incision, then $m_\gamma[P_{0\gamma}]=1/c_{0\gamma}$, so that $m_\gamma[P_{0\gamma}]=\max_{\mathscr{S}_{\zeta\gamma}} m_\gamma[F]$ by Theorem 3 C.

4 C. Reich's Proof of Theorem 4 A. The proof by REICH [2] is applicable only to the case of a positive $c_{0\gamma}$, but it is direct.

For any r with $0<r<1/c_{0\gamma}$, let

$$R_r=P_{0\gamma}(W)\cup\left\{w\,|\,r<|w|<\frac{1}{c_{0\gamma}}\right\}.$$

Assume for a moment that R_r is connected and observe that $\gamma' = \{w \mid |w| = 1/c_{0\gamma}\}$ is an isolated boundary component of the region R_r. Let P_r be the function $P_{0\gamma'}(w; 0)$ on R_r. The main part of Reich's proof is a demonstration of the following auxiliary result:

LEMMA. *If* $c_{0\gamma} > 0$, *then* R_r *is connected and* $P_r(w) \equiv w$ *for every* r.

Remark. The conclusion holds for $c_{0\gamma} = 0$ as well; the proof will be found in IX.3 G.

As soon as the lemma is established, the proof of Theorem 4 A for $c_{0\gamma} > 0$ follows immediately. In fact, we can apply 1 C.(b) to the function P_r, since γ' is an isolated boundary component of R_r. From $P_r(w) \equiv w$ we infer that R_r is a radial slit disk of radius $1/c_{0\gamma}$ such that the total area of the slits vanishes. By Theorem 3 B, $P_{0\gamma}(W)$ is contained in $|w| < 1/c_{0\gamma}$. For any r with $0 < r < 1/c_{0\gamma}$, the sets $P_{0\gamma}(\gamma) \cap \{w \mid |w| < r\}$ and $(\partial F_{0\gamma}(W) - P_{0\gamma}(\gamma)) \cap \{w \mid |w| \leq r\}$ both consist of points and/or radial segments and have vanishing area. We conclude that $P_{0\gamma}(W)$ has the desired properties.

4 D. Connectedness of R_r. The proof of Lemma 4 C will be given in 4 D – 4 F.

If there exists an r such that $0 < r < 1/c_{0\gamma}$ and R_r is not connected, then it is easily seen that $P_{0\gamma}(W)$ is contained in the disk $|w| < r$. This violates Theorem 3 B, so that R_r must be connected.

We shall also give a direct proof (REICH [2]), which has interest in its own right. Eq. (2) to be derived will be used later.

We consider an exhaustion $\zeta \in V_n \to W$ towards γ and use the notation $\gamma_n = \gamma(V_n)$, $P_n = P_{0\gamma(V_n)}(z; \zeta)$ on V_n, and $c_n = c_{0\gamma(V_n)}(\zeta)$ for V_n. On each γ_m, the convergence $\lim_{n\to\infty} P_n = P_{0\gamma}$ is uniform; therefore, for any r' with $r < r' < 1/c_{0\gamma}$, we have

$$|P_n| < r' \quad \text{on } \gamma_m \tag{1}$$

if n $(> m)$ is sufficiently large.

On V_m, apply identity III.1 F.(6) for $p = \log|P_n| = p_n$ and $p^0 = \log|P_m| = p_m$:

$$D_{V_m}[p_n - p_m] = \int_{\partial V_m} p_n * dp_n - 2\int_{\gamma_m} p_n * dp_m + 2\pi \log \frac{1}{c_m}.$$

On the other hand, by III.1 F.(8),

$$D_{V_m}[p_n - p_m] = \int_{\partial V_m} p_n * dp_n - 2\pi \log \frac{1}{c_m}.$$

From these it follows that

$$2\pi \log \frac{1}{c_m} = \int_{\gamma_m} p_n * dp_m. \tag{2}$$

By (1) and the fact that $*dp_m > 0$ along γ_m,

$$2\pi \log \frac{1}{c_m} < 2\pi \log r' < 2\pi \log \frac{1}{c_{0\gamma}}.$$

On letting $m \to \infty$, we obtain $2\pi \log(1/c_{0\gamma}) \leq 2\pi \log r' < 2\pi \log(1/c_{0\gamma})$, a contradiction. We conclude that R_r is connected.

4 E. Identity $P_r(w) \equiv w$. Next we prove the second part of the lemma. To this end, let c_r be the capacity $c_{0\gamma'}(0)$ of $\gamma' = \{w \,|\, |w| = 1/c_{0\gamma}\}$ with respect to the region R_r. The function $I(w) \equiv w$ belongs to the family $\mathscr{S}_{0\gamma'}(R_r)$ and satisfies $m_\gamma[I] = 1/c_{0\gamma}$. Therefore, as soon as we prove that

$$c_{0\gamma} \leq c_r, \tag{3}$$

we can conclude by Theorem 3 C that $P_r(w) \equiv w$. Inequality (3) can be easily verified by means of extremal length (IV.3 G and Appendix I.G.(a)). Here we shall, however, give a direct proof due to Reich [2].

The function $|P_r|$ takes the constant value $1/c_r$ on γ'. For any $\varepsilon > 0$, there exists a $\delta > 0$ such that $r + \delta < 1/c_{0\gamma}$ and $1/c_r - \varepsilon < |P_r(w)|$ in the annulus $1/c_{0\gamma} - \delta < |w| < 1/c_{0\gamma}$. For these ε and δ, there exists an m such that

$$\frac{2\pi \log \dfrac{1}{c_m} - \log\left(\dfrac{1}{c_{0\gamma}} - \dfrac{\delta}{2}\right)}{2\pi \log \dfrac{1}{c_n} - \log\left(\dfrac{1}{c_{0\gamma}} - \dfrac{\delta}{2}\right)} > 1 - \varepsilon \tag{4}$$

for every $n > m$, where c_m and c_n are as in 4 D. Finally, for these δ and m, we can find an m' such that $|P_n - P_{0\gamma}| < \delta/2$ on γ_m for every $n > m'$, where P_n and γ_m are again as in 4 D. At every point $z \in \gamma_m$ with $|P_{0\gamma}(z)| \leq 1/c_{0\gamma} - \delta$, we see that $|P_n(z)| < 1/c_{0\gamma} - \delta/2$. Therefore, on setting

$$E = \left\{ z \,|\, z \in \gamma_m, |P_{0\gamma}(z)| > \frac{1}{c_{0\gamma}} - \delta \right\},$$

we obtain

$$|P_r \circ P_{0\gamma}(z)| > \frac{1}{c_r} - \varepsilon, \qquad z \in E, \tag{5}$$

$$|P_n(z)| \leq \frac{1}{c_{0\gamma}} - \frac{\delta}{2}, \qquad z \in \gamma_m - E. \tag{6}$$

Now set $\theta=\int_E *dp_m$. By (2) and (6),

$$2\pi\log\frac{1}{c_m}=\int_E p_n*dp_m+\int_{\gamma_m-E} p_n*dp_m$$
$$\le\theta\log\frac{1}{c_{0\gamma}}+(2\pi-\theta)\log\left(\frac{1}{c_{0\gamma}}-\frac{\delta}{2}\right).$$

Hence

$$2\pi\log\frac{1}{c_m}-2\pi\log\left(\frac{1}{c_{0\gamma}}-\frac{\delta}{2}\right)\le\theta\left(\log\frac{1}{c_{0\gamma}}-\log\left(\frac{1}{c_{0\gamma}}-\frac{\delta}{2}\right)\right)$$

and, by (4),

$$\theta\ge 2\pi(1-\varepsilon). \tag{7}$$

4 F. Identity $P_r(w)\equiv w$ (continued). We apply identity III.1 F.(6) to $p=\log|P_r\circ P_{0\gamma}|$ and $p^0=\log|P_m|$ on V_m:

$$0\le\tilde{\mathscr{D}}_{V_m}[P_r\circ P_{0\gamma}]-2\int_{\gamma_m}\log|P_r\circ P_{0\gamma}|*dp_m+2\pi\log\frac{1}{c_m}$$
$$\le 2\pi\log\frac{1}{c_r}-2\int_{\gamma_m-E}\log|P_r\circ P_{0\gamma}|*dp_m-2\int_E\log|P_r\circ P_{0\gamma}|*dp_m+2\pi\log\frac{1}{c_m}$$
$$\le 2\pi\log\frac{1}{c_r}-2(2\pi-\theta)\log a-2\theta\log\left(\frac{1}{c_r}-\varepsilon\right)+2\pi\log\frac{1}{c_m},$$

where a is an arbitrary positive number with $\{w\,|\,|w|\le a\}\subset P_r\circ P_{0\gamma}(W)$. Hence

$$2\theta\log\frac{\frac{1}{c_r}-\varepsilon}{a}+4\pi\log a\le 2\pi\log\frac{1}{c_r}+2\pi\log\frac{1}{c_m}.$$

If a is taken so small that $(1/c_r-\varepsilon)/a>1$, then on using (7) we see that the left-hand side dominates

$$4\pi(1-\varepsilon)\log\frac{\frac{1}{c_r}-\varepsilon}{a}+4\pi\log a.$$

On letting $m\to\infty$ and then $\varepsilon\to 0$, we obtain

$$4\pi\log\frac{1}{c_r}\le 2\pi\log\frac{1}{c_r}+2\pi\log\frac{1}{c_{0\gamma}}$$

and, therefore, (3).

The proof of Theorem 4 A for $c_{0\gamma}>0$ is herewith complete.

4 G. Strebel's Proof of Theorem 4 A. The proof due to STREBEL [2] covers the case $c_{0\gamma}=0$.

Since $P_{0\gamma}\in\mathscr{S}^0_{\zeta\gamma}$, $\partial P_{0\gamma}(W)-P_{0\gamma}(\gamma)$ consists of points and radial segments and has vanishing area (3 A). Therefore, the only question that needs attention is the shape of $P_{0\gamma}(\gamma)$.

On the ray $\arg w=\theta$, consider the point $w(\theta)\in P_{0\gamma}(\gamma)$ nearest the origin. We know that

$$|w(\theta)|\leq\frac{1}{c_{0\gamma}},\qquad 0\leq\theta<2\pi, \tag{8}$$

so that the proof will be complete if equality holds for almost all θ.

For $\varepsilon>0$ with $\{w\,|\,|w|\leq\varepsilon\}\subset P_{0\gamma}(W)$, the modulus μ_0 of the region $P_{0\gamma}(W)-\{w\,|\,|w|\leq\varepsilon\}$ with respect to $P_{0\gamma}(\gamma)$ and $|w|=\varepsilon$ is, as was shown in IV. 3 F, equal to $1/(\varepsilon\, c_{0\gamma})$. It will be seen that

$$\log\frac{1}{\varepsilon\, c_{0\gamma}}\leq\frac{2\pi}{\int_0^{2\pi}\frac{d\theta}{l_\varepsilon(\theta)}},$$

where $l_\varepsilon(\theta)=\log(|w(\theta)|/\varepsilon)$. We postpone the proof of this *Strebel inequality* to 4 H, where it will be established in a somewhat more general situation. The right-hand side is dominated by $(2\pi)^{-1}\int_0^{2\pi} l_\varepsilon(\theta)\,d\theta$, which by (8) does not exceed $\log(1/\varepsilon\, c_{0\gamma})$. Accordingly,

$$\log\frac{1}{\varepsilon\, c_{0\gamma}}=\frac{1}{2\pi}\int_0^{2\pi} l_\varepsilon(\theta)\,d\theta.$$

Again by (8), we conclude that $|w(\theta)|=1/c_{0\gamma}$ for almost all θ. The proof of Theorem 4 A is herewith complete.

4 H. Strebel's Inequality. Consider a multiply connected region W in the z-plane with $0,\infty\notin W$. Let γ be its outer boundary and γ' its inner boundary (VI. 1 D). Suppose further that

$$L_0(\log|z|)=\log|z|$$

on $\partial W-(\gamma\cup\gamma')$, or more precisely, on $(\operatorname{Int}\alpha)\cap W$ for every analytic Jordan curve $\alpha\subset W$ with $(\operatorname{Int}\alpha)\cap(\gamma\cup\gamma')=\emptyset$, where $(\operatorname{Int}\alpha)$ stands for the region enclosed by α. Thus $E=\partial W-(\gamma\cup\gamma')$ consists of points and radial segments, and the complement of $E_\alpha=(\operatorname{Int}\alpha)\cap E$ is an extremal radial slit plane for every α.

On the ray $\arg z=\theta$, let $z(\theta)$ be the point of γ nearest the origin and $z'(\theta)$ the point of γ' farthest from the origin. Let $l(\theta)$ be the logarithmic length of the line segment joining these points,

$$l(\theta)=\left|\log\left|\frac{z(\theta)}{z'(\theta)}\right|\right|.$$

STREBEL'S INEQUALITY. *The modulus $\mu_0(\gamma, \gamma')$ of W has the following upper bound:*

$$\log \mu_0 \leq \frac{2\pi}{\int_0^{2\pi} \frac{d\theta}{l(\theta)}}.$$

Proof. Let Γ^* be the family of locally rectifiable curves joining γ and γ' in W. We know that

$$\log \mu_0 = 2\pi \, \lambda(\Gamma^*).$$

First assume that W is of finite connectivity. Let c_θ be the line segment joining $z(\theta)$ and $z'(\theta)$. Except for a finite number of values of θ, c_θ does not meet the slits and, therefore, contains a subarc belonging to Γ^*. We have

$$L(\Gamma^*, \rho)^2 \leq \Big(\int_{c_\theta} \rho \, |dz|\Big)^2 \leq l(\theta) \left| \int_{|z'(\theta)|}^{|z(\theta)|} \rho^2 \, r \, dr \right|,$$

and

$$\int_0^{2\pi} \frac{d\theta}{l(\theta)} \leq \frac{1}{L(\Gamma^*, \rho)^2} \iint_W \rho^2 \, r \, dr \, d\theta.$$

Since ρ is arbitrary,

$$\int_0^{2\pi} \frac{d\theta}{l(\theta)} \leq \frac{1}{\lambda(\Gamma^*)}.$$

Next, if W is arbitrary, we exhaust it by regions V towards $\gamma \cup \gamma'$ with $\gamma(V)$ and $\gamma'(V)$ analytic Jordan curves. We indicate by the subscript V quantities considered on V, and have $l_V(\theta) \leq l(\theta)$. Since, by definition, $\mu_0 = \lim \mu_{0V}$, the proof will be complete if we show that

$$\log \mu_{0V} \leq \frac{2\pi}{\int_0^{2\pi} \frac{d\theta}{l_V(\theta)}}.$$

In other words, it suffices to prove Strebel's inequality for a W whose γ and γ' are isolated boundary components and analytic Jordan curves.

The set $E = \partial W - (\gamma \cup \gamma')$ is compact in this case, and $R = CE$ is an extremal radial slit plane. We exhaust R by regular regions Ω with $0, \infty \in \Omega$, and denote by P_Ω the function $P_0(z; 0, \infty)$ on Ω. Then

$$\lim_{\Omega \to W} P_\Omega(z) = z \tag{9}$$

uniformly on every compact set in R. We may assume that γ, γ' are contained in Ω. Since Strebel's inequality has been verified for the region

$$W_\Omega = P_\Omega(W \cap \Omega),$$

we have for the quantities $l_\Omega(\theta)$ and $\mu_{0\Omega}$ corresponding to W_Ω,

$$\log \mu_{0\Omega} \le \frac{2\pi}{\int_0^{2\pi} \frac{d\theta}{l_\Omega(\theta)}}.$$

Here $\mu_{0\Omega}$ is equal to the modulus of $W \cap \Omega$ with respect to γ and γ'. As $\Omega \to R$, $W \cap \Omega$ exhausts W towards $\partial W - (\gamma \cup \gamma')$, so that

$$\mu_0 = \lim_{\Omega \to R} \mu_{0\Omega}.$$

Hence the proof will be complete if we show that

$$\overline{\lim_{\Omega \to R}}\, l_\Omega(\theta) \le l(\theta) \tag{10}$$

for almost all θ.

To prove (10), we use the quantities $l(\theta) = |\log|z(\theta)/z'(\theta)||$ and $l_\Omega(\theta) = |\log|w_\Omega(\theta)/w'_\Omega(\theta)||$, where $w_\Omega(\theta) \in P_\Omega(\gamma)$ is the point on $\arg w = \theta$ nearest the origin and $w'_\Omega(\theta) \in P_\Omega(\gamma')$ is the point farthest from the origin. The convergence in (9) is uniform on $\gamma \cup \gamma'$. As a consequence, it is not difficult to see that (10) holds for every θ such that neither $z(\theta)$ nor $z'(\theta)$ is a point where the ray $\arg z = \theta$ is tangent to γ or γ'. These θ's are finite in number. This completes the proof of Strebel's inequality.

5. Extremal Functions of the Families $\mathscr{S}_\zeta$ and $\mathscr{A}_\zeta$

5 A. Extremal Problem for $\mathscr{S}_\zeta$. Let W be an arbitrary plane region. For $\zeta \in W$ consider the family $\mathscr{S}_\zeta = \mathscr{S}_\zeta(W) = \{F | F \in \mathscr{A}_\zeta, F \text{ is univalent}\}$, where $\mathscr{A}_\zeta$ was defined in VI.5 B. It is identical with the union $\bigcup \mathscr{S}_{\zeta\gamma}$ taken over all boundary components γ of W.

We introduce the quantity

$$c_1(\zeta) = \sup c_{1\gamma}(\zeta),$$

where the supremum is taken over all boundary components γ.

If W is exhausted by regular subregions Ω with $\zeta \in \Omega$, it is easily verified that

$$\lim_{\Omega \to W} c_{1\Omega} = c_1.$$

Here $c_{1\Omega}$ is the quantity $c_1(\zeta)$ considered for Ω.

THEOREM. *The supremum of the capacities $c_{1\gamma}$ of the boundary components is attained:*

$$c_1(\zeta) = \max_{\gamma \in \partial W} c_{1\gamma}(\zeta).$$

It is related to the minima in $\mathcal{S}_\zeta$ of the maximum modulus and the Dirichlet integral as follows:

$$\frac{1}{c_1(\zeta)} = \min_{F\in\mathcal{S}_\zeta} M[F] = \min_{F\in\mathcal{S}_\zeta} \left(\frac{1}{\pi}\mathcal{D}[F]\right)^{\frac{1}{2}}.$$

If $c_1(\zeta)>0$, the function minimizing $M[F]$ minimizes $\mathcal{D}[F]$ as well, is equal to $P_{1\gamma}(z;\zeta)$ for a γ with $c_{1\gamma}=c_1$, and is the only extremal function for this γ.

The extremal function is not necessarily determined uniquely, since there may exist more than one γ with $c_{1\gamma}=c_1$.

Proof. The family $\mathcal{S}_\zeta$ is compact and the functional $M[F]$ is lower semicontinuous, that is,

$$M[F] \le \varliminf_{n\to\infty} M[F_n]$$

if $F_n\in\mathcal{S}_\zeta$ and $F=\lim F_n$ uniformly on every compact set in W. As a standard argument shows, these facts guarantee the existence of a function F_0 minimizing $M[F]$ in $\mathcal{S}_\zeta$. Let γ^* be the boundary component which is mapped by F_0 to the outer boundary of $F_0(W)$. Then F_0 belongs to $\mathcal{S}_{\zeta\gamma^*}$ and minimizes $M[F]$ in it. Thus $M[F_0]=1/c_{1\gamma^*}$ and $F_0=P_{1\gamma^*}$ (Theorem 2 B). For any γ, we obtain $M[F_0]\le M[P_{1\gamma}]=1/c_{1\gamma}$. Consequently, $1/c_1=1/c_{1\gamma^*}=\min 1/c_{1\gamma}=\min M[F]$.

Regarding $\mathcal{D}[F]$, the proof is similar (use Theorem 2 D), since the functional is again lower semicontinuous.

5 B. Angle Subtended by a Circular Slit. If $c_1>0$, the extremal function $P_{1\gamma}$ of the above theorem maps W onto a circular slit disk of radius $1/c_1$. In addition to the properties discussed in 2, the fact that $c_{1\gamma}$ is maximized gives a further restriction to the image region. The following result is due to REICH-WARSCHAWSKI [1]:

THEOREM. *Every circular slit under a mapping $P_{1\gamma}$ for maximal $c_{1\gamma}$ subtends an angle less than π.*

Proof. Let γ be such that $c_1=c_{1\gamma}>0$. For any other γ' we are to show that $P_{1\gamma}(\gamma')$ subtends an angle less than π. We may assume that $P_{1\gamma}(\gamma')$ is not a single point.

Consider in the w-plane the doubly connected circular slit disk

$$\tilde{R} = \left\{w \,\middle|\, |w| < \frac{1}{c_1}\right\} - P_{1\gamma}(\gamma')$$

bounded by $\tilde{\gamma}=P_{1\gamma}(\gamma)=\{w \mid |w|=1/c_1\}$ and $\tilde{\gamma}'=P_{1\gamma}(\gamma')$. For the capacities $c_{1\tilde{\gamma}}=c_{1\tilde{\gamma}}(0)$ and $c_{1\tilde{\gamma}'}=c_{1\tilde{\gamma}'}(0)$ on $\tilde{R}$, we have $c_1=c_{1\tilde{\gamma}}$. Furthermore, the functions $P_{1\tilde{\gamma}}(w;0)$ and $P_{1\tilde{\gamma}'}(w;0)$ on $\tilde{R}$ satisfy $P_{1\tilde{\gamma}}\circ P_{1\gamma}\in\mathscr{S}_{\zeta\gamma}(W)$ and $M[P_{1\tilde{\gamma}'}\circ P_{1\gamma}]=1/c_{1\tilde{\gamma}'}$. It follows that

$$c_{1\tilde{\gamma}'}\le c_{1\tilde{\gamma}}. \tag{1}$$

On the other hand, we use the identities in V.2 B and 2 D on $\tilde{R}$. In terms of the capacity functions with pole at 0, the modulus function $q_1(w;\tilde{\gamma}',\tilde{\gamma})$ normalized by $q_1=0$ on $\tilde{\gamma}$ can be expressed as

$$q_1(w)=p_{1\tilde{\gamma}'}(w)-p_{1\tilde{\gamma}}(w)+q_1(0)$$

(V.2 D. (5)). On $\tilde{\gamma}'$ this equality reads

$$l_1(\tilde{\gamma}',\tilde{\gamma})=\log\frac{1}{c_{1\tilde{\gamma}'}}-p_{1\tilde{\gamma}}(\tilde{\gamma}')+q_1(0),$$

and on $\tilde{\gamma}$,

$$0=p_{1\tilde{\gamma}'}(\tilde{\gamma})-\log\frac{1}{c_{1\tilde{\gamma}}}+q_1(0).$$

By V.2 B. (2), $p_{1\tilde{\gamma}'}(\tilde{\gamma})=p_{1\tilde{\gamma}}(\tilde{\gamma}')$, so that

$$2q_1(0)=l_1-\log\frac{c_{1\tilde{\gamma}}}{c_{1\tilde{\gamma}'}},$$

which is dominated by l_1 (>0) in view of (1). The function q_1/l_1 is the harmonic measure $u_{\tilde{\gamma}'}$ on $\tilde{R}$ (IV.4 B), and we obtain

$$u_{\tilde{\gamma}'}(0)\le\tfrac{1}{2}.$$

Denote by θ the angle at the origin subtended by the arc $\tilde{\gamma}'$, and let r be the distance from the origin to $\tilde{\gamma}'$. By the mean-value property, we have

$$\frac{1}{2}\ge u_{\tilde{\gamma}'}(0)=\frac{1}{2\pi}\int_{|w|=r}u_{\tilde{\gamma}'}\,d\arg w>\frac{1}{2\pi}\int_{\tilde{\gamma}'}u_{\tilde{\gamma}'}\,d\arg w=\frac{\theta}{2\pi}.$$

5 C. Vanishing of c_1. We know that the property $c_1(\zeta)=0$ is independent of the reference point ζ. From Theorem 5 A we see that it is equivalent to any one of the following:

(a) Every boundary component of W is weak.

(b) $C\Phi(W)$ is totally disconnected for every univalent function Φ on W.

(c) Every univalent analytic function on W is unbounded.

(d) Every univalent analytic function on W has an infinite Dirichlet integral.

5 D. Extremal Problems for $\mathcal{A}_\zeta$. We continue considering an arbitrary plane region W. In view of Theorem 5 A, we introduce the quantities $c_B(\zeta)$ and $c_D(\zeta)$ with $\zeta \in W$ for the family $\mathcal{A}_\zeta$ (VI.5 B) as follows:

$$\frac{1}{c_B} = \min_{F \in \mathcal{A}_\zeta} M[F], \qquad \frac{1}{c_D} = \min_{F \in \mathcal{A}_\zeta} \left(\frac{1}{\pi} \mathcal{D}[F]\right)^{\frac{1}{2}}.$$

The existence of minimizing functions is assured by virtue of the compactness of $\mathcal{A}_\zeta$ and the lower semicontinuity of the functionals.

For an exhaustion of W by regular subregions Ω with $\zeta \in \Omega$, the relations

$$\lim_{\Omega \to W} c_{B\Omega} = c_B, \qquad \lim_{\Omega \to W} c_{D\Omega} = c_D$$

are easily verified.

Since $\mathcal{S}_\zeta \subset \mathcal{A}_\zeta$, we see that

$$c_1 \leq c_B, \qquad c_1 \leq c_D.$$

By Theorem VI.5 B, c_D is closely related to the span and the kernel:

$$c_D(\zeta) = \sqrt{S(\zeta)} = \sqrt{\pi \tilde{K}(\zeta, \zeta)}.$$

Furthermore, if $c_D > 0$, then the function minimizing $\mathcal{D}[F]$ in $\mathcal{A}_\zeta$ is uniquely determined and is equal to

$$\frac{1}{2S(\zeta)} (P_0(z;\zeta) - P_1(z;\zeta)) = \frac{\pi}{S(\zeta)} \int_\zeta^z \tilde{K}(z;\zeta)\, dz.$$

We recall that the property $c_D(\zeta) = 0$ is independent of ζ and is equivalent to the fact that $AD(W)$ consists only of constants (Theorem VI.2 D).

5 E. Extremal Problems for $\mathcal{A}_\zeta$ (continued). Concerning the quantity c_B, Schwarz's lemma gives us the unique function minimizing $M[F]$ if W is simply connected. The problem has not been solved completely for an arbitrary region W. However, CARLESON [1] established the uniqueness of the minimizing function if $c_B > 0$.

For a regular region W (in which case $c_B > 0$), further results have been obtained by AHLFORS [1, 2], GARABEDIAN [1], and NEHARI [2, 3]. For the function minimizing $M[F]$, they obtained a characterization which in particular implies that the function maps W onto an n-sheeted disk of radius $1/c_B$, where n is the connectivity of W; note that this property does not in turn characterize the minimizing function. GARABEDIAN and NEHARI further derived a relationship with Szegö's kernel function (SZEGÖ [1], SCHIFFER [5]). However, we shall not go into a more detailed discussion of these interesting results.

From the definition, we see immediately that $c_B(\zeta)=0$ at every point if and only if the family $AB(W)$ of bounded analytic functions on W consists only of constants. Concerning this we have the following proposition:

The property $c_B(\zeta)=0$ is independent of ζ.

The proof is fairly direct, in contrast with the case of c_D (VI.2 D). For the sake of simplicity, suppose $\zeta \neq \infty$. If $c_B(\zeta')>0$, then there exists an $F \in \mathscr{A}_{\zeta'}$ with a finite $M[F]$. Since $F \neq \text{const}$, it has an expansion $F(z)=F(\zeta)+a(z-\zeta)^k+\cdots$, $a \neq 0$, $k \geq 1$, about ζ. Consequently, $F_1(z)=(F(z)-F(\zeta))\,a^{-1}(z-\zeta)^{1-k}$ belongs to $\mathscr{A}_\zeta$. For $\varepsilon>0$, it is bounded on $W-\{z \mid |z-\zeta| \leq \varepsilon\}$ by definition and, therefore, throughout W. In other words, $M[F_1]<\infty$; a fortiori $c_B(\zeta)>0$.

Remark. On a Riemann surface W, the quantities $c_B(\zeta)$ and $c_D(\zeta)$ can be introduced in the same manner with an additional choice of the local parameter about ζ. The fact that $\mathscr{A}_\zeta \neq \emptyset$ is guaranteed by a result in GUNNING-NARASIMHAN [1]. The validity of $c_B(\zeta)=0$ or $c_D(\zeta)=0$ at every ζ is equivalent to the fact that AB or AD, respectively, consists only of constants. However, VIRTANEN [2] pointed out, by constructing a counterexample, that the vanishing of c_B or c_D at a single point is *not* sufficient for the vanishing throughout W. His proof will be given in X.2 K.

5 F. Relations between Minima. For an arbitrary point ζ of an arbitrary plane region W we shall prove:

THEOREM. *The minima of $c_{1\gamma}(\zeta)$, $\mathscr{D}[F]$, and $M[F]$ satisfy the relations*

$$c_1(\zeta) \leq c_D(\zeta) \leq c_B(\zeta).$$

In particular, the inequality $c_D \leq c_B$ implies:

COROLLARY. *If $AB(W)$ consists only of constants, then so does $AD(W)$.*

This result can also be proved directly as follows: If $AD(W)$ contains a nonconstant function, then $P_0-P_1 \neq \text{const}$ by Theorem VI.2 D, and evidently $P_0-P_1 \in AB(W)$.

Proof of the Theorem. We have already shown that $c_1 \leq c_D$. It suffices to establish $c_D \leq c_B$ under the assumption $c_D>0$. The proof below is due to AHLFORS-BEURLING [1].

Consider the function

$$P_2(z)=\tfrac{1}{2}\left(P_0(z;\zeta)+P_1(z;\zeta)\right)$$

discussed in VI.4 B. The set $E=CP_2(W)$ has area $mE=\pi S(\zeta)=\pi c_D^2>0$. The analytic function

$$F(z)=\iint_E \frac{du\,dv}{P_2(z)-w}, \qquad w=u+iv,$$

on W satisfies $F(\zeta)=0$ and $F'(\zeta)=mE$. Hence $F_1=F/mE$ belongs to $\mathcal{A}_\zeta$. If we prove that

$$\left|\iint_I \frac{du\,dv}{a-w}\right|\leq\sqrt{\pi\, mE} \tag{2}$$

for an arbitrary $a\notin E$, then we can deduce

$$M[F_1]\leq\sqrt{\frac{\pi}{mE}}=\frac{1}{c_D},$$

and therefore $1/c_B\leq 1/c_D$.

For the proof of (2), we may assume without loss of generality that $a=0$:

$$\left|\iint_I \frac{du\,dv}{w}\right|\leq\sqrt{\pi\, mE}, \qquad 0\notin E.$$

By rotating the plane if necessary, we further assume that the integral is real and positive. We use polar coordinates and obtain

$$\iint_E \frac{1}{w}\,du\,dv=\iint_I \frac{1}{e^{i\theta}}\,dr\,d\theta=\iint_E \cos\theta\,dr\,d\theta\leq\iint_{E^+}\cos\theta\,dr\,d\theta,$$

where $E^+=\{w\,|\,w\in E,\ \operatorname{Re} w>0\}$. In terms of the quantities

$$l(\theta)=\int_{\substack{w\in E^+\\ \arg w=\theta}} d|w|, \qquad l(r,\theta)=\int_{\substack{w\in E^+,\ |w|\leq r\\ \arg w=\theta}} d|w|,$$

we have

$$\iint_{E^+}\cos\theta\,dr\,d\theta=\int_{-\pi/2}^{\pi/2} l(\theta)\cos\theta\,d\theta\leq\left(\frac{\pi}{2}\int_{-\pi/2}^{\pi/2} l(\theta)^2\,d\theta\right)^{\frac{1}{2}}$$

and

$$mE\geq\iint_{E^+} r\,dr\,d\theta\geq\int_{-\pi/2}^{\pi/2} d\theta\int l(r,\theta)\,dl(r,\theta)=\tfrac{1}{2}\int_{-\pi/2}^{\pi/2} l(\theta)^2\,d\theta.$$

Therefore,

$$\iint_E \frac{1}{w}\,du\,dv\leq(\pi\, mE)^{\frac{1}{2}}.$$

5 G. Capacity c_β. The capacity $c_\beta(\zeta)$ of the entire boundary $\beta=\partial W$ has the following expression similar to that of $c_1(\zeta)$ in Theorem 5 A. Let $\mathcal{A}_\zeta^*=\mathcal{A}_\zeta^*(W)$ be the family of multiple-valued analytic functions F such that $|F|$ is single-valued, $F(\zeta)=0$, and $F'(\zeta)=1$ for one of the branches. We state (SARIO[10]):

The capacity of the entire boundary and the minimum in $\mathscr{A}_\zeta^*$ of the maximum modulus are reciprocals,

$$\frac{1}{c_\beta(\zeta)}=\min_{F\in\mathscr{A}_\zeta^*} M[F].$$

If $c_\beta(\zeta)>0$, then the only minimizing function is $F_\beta=\exp(p_\beta+i\,p_\beta^*)$, where p_β is the capacity function with pole at ζ.

The proof is easily obtained by using Corollary III.3 F, the details being left to the reader.

From this expression we obtain

$$c_B(\zeta)\le c_\beta(\zeta).$$

We also note that

$$\lim_{\Omega\to W} c_{\beta(\Omega)}=c_\beta$$

for an exhaustion by regular regions Ω with $\zeta\in\Omega$ and that the validity of $c_\beta(\zeta)=0$ is independent of ζ (IV.3 B).

We shall return to c_β in 6, where a relation with the logarithmic capacity of the set CW will be given.

5 H. Mapping Radius. If W is the disk $|z|<r\le\infty$, then

$$c_1(\zeta)=c_D(\zeta)=c_B(\zeta)=c_\beta(\zeta)=\frac{r}{r^2-|\zeta|^2}.$$

The values of c_1 and $c_D=\sqrt{S}$ were obtained in 1 C and VI.2 E. The value of c_B is easily computed using Schwarz's lemma, and $c_1=c_\beta$ is trivial.

As a consequence, *if W is simply connected, then the quantities $c_1(\zeta)$, $c_D(\zeta)$, $c_B(\zeta)$, and $c_\beta(\zeta)$ are equal.* The reciprocal of the quantity is called the *mapping radius* of W with respect to ζ. There is a function in $\mathscr{S}_\zeta(W)$ which maps W conformally onto a disk of this radius. The mapping radius is ∞ if and only if W is the punctured Riemann sphere.

If the connectivity is greater than one, then these four quantities are never all equal. We shall construct in XI.3 G and 3 J examples in which one of the quantities vanishes and another does not. Here we shall establish the following proposition:

If W is a regular region of connectivity ≥ 2, then

$$c_1(\zeta)<c_D(\zeta),\qquad c_B(\zeta)<c_\beta(\zeta)$$

for every $\zeta\in W$.

In fact, if $c_1(\zeta)=c_D(\zeta)$, then $P_{1\gamma}$ with $c_{1\gamma}=c_1$ must be identical with $(P_0-P_1)/2S$ by the uniqueness assertion of Theorem VI.5 B. By Theorem VI.5 D, this function is not univalent, a contradiction. If $c_B(\zeta)=c_\beta(\zeta)$, then $\exp(p_\beta+i\,p_\beta^*)$ must be single-valued by the uniqueness assertion in 5 G. But this is impossible, since $\int_\gamma * dp_\beta=2\pi\, u_\gamma(\zeta)\not\equiv 0 \bmod 2\pi$ (V.2 E) for an arbitrary contour γ of W.

Remark. It is not known whether $c_D(\zeta)<c_B(\zeta)$ for every $\zeta\in W$, if W is a regular region of connectivity ≥ 2.

5 I. Bergman Metric. Let us compare the Bergman metric

$$\sqrt{M(\zeta)}\,|d\zeta|=\sqrt{\pi\, K(\zeta,\zeta)}\,|d\zeta|$$

with $c_D(\zeta)=\sqrt{\pi\,\tilde{K}(\zeta,\zeta)}$ (II.3 C). By the expression in VI.5 C, we clearly have

$$c_D(\zeta)\leq\sqrt{M(\zeta)}.$$

If W is simply connected, then equality holds. If W is a regular region of connectivity ≥ 2, the two quantities never coincide.

We know that the property $M(\zeta)=0$ is independent of ζ (II.3 C).

It will be seen in X.2 B that $M(\zeta)=0$ if and only if $c_\beta(\zeta)=0$. Except for this fact, the relation between the magnitudes of $\sqrt{M}$ and c_β or c_B is unknown.

6. Capacity Function p_β and Logarithmic Potential

6 A. Mass Distribution. The capacity function $p_\beta(z;\zeta)$ of the entire boundary $\beta=\partial W$ can be expressed as a logarithmic potential. This relation permits us to study the boundary behavior of the capacity function and to derive a geometric expression for the capacity c_β in terms of the transfinite diameter.

For preciseness we first agree, following Royden [3], on some terminology in real analysis.

Let R^2 be the 2-dimensional Euclidean space $|z|<\infty$. Denote by $\mathscr{B}=\mathscr{B}(R^2)$ the σ-algebra of Borel sets (=Baire sets) in R^2. By a *mass distribution* μ we mean a Baire measure on the measurable space $(R^2,\mathscr{B})$, that is, a non-negative countably additive set function on $\mathscr{B}$ satisfying $\mu(\emptyset)=0$ and $\mu(K)<\infty$ for every closed bounded set K; note that the value ∞ is admissible and that, for every $E\in\mathscr{B}$, we have $\mu(E)=\sup\mu(K)$, where the K are closed bounded subsets of E.

For a set $E\subset R^2$, we designate by $C'E$ its complement with respect to R^2; the prime is used in order to distinguish it from the complement CE with respect to the Riemann sphere (VI.1 A). By the *support* S_μ of a mass distribution μ, we mean the smallest closed set the complement of which has vanishing μ-value, that is, $S_\mu = \bigcap F$, where F is closed and $\mu(C'F)=0$.

We remark that if E is closed and bounded, then $S_\mu \subset E$ if and only if

$$\int f\, d\mu = 0$$

for every continuous function f on R^2 which vanishes on E.

Given a closed bounded set E, we consider the family

$$\mathbf{M}(E) = \{\mu \mid S_\mu \subset E,\ \mu(E)=1\}.$$

It is compact in the sense of the following theorem:

SELECTION THEOREM. *For any sequence $\mu_n \in \mathbf{M}(E)$, $n=1,2,\ldots$, there exists a $\mu \in \mathbf{M}(E)$ and a subsequence $\{\mu_{i_n}\}$ such that*

$$\lim_{n\to\infty} \int f\, d\mu_{i_n} = \int f\, d\mu$$

for every continuous function f on R^2, and

$$\lim_{n\to\infty} \int h\, d(\mu_{i_n} \times \mu_{i_n}) = \int h\, d(\mu \times \mu)$$

for every continuous function h on R^4.

A proof is found in Appendix II.B.

For the above subsequence $\{\mu_{i_n}\}$, we also have

$$\underline{\lim} \int f\, d\mu_{i_n} \geq \int f\, d\mu$$

and

$$\underline{\lim} \int h\, d(\mu_{i_n} \times \mu_{i_n}) \geq \int h\, d(\mu \times \mu)$$

for all lower semicontinuous functions f and h on R^2 and R^4, respectively.

6 B. Logarithmic Potential and Logarithmic Capacity. Given a mass distribution μ with bounded support S_μ and finite total mass $\mu(R^2)$, the function

$$u(z) = \int \log \frac{1}{|z-\zeta|}\, d\mu(\zeta)$$

is called the *logarithmic potential* with respect to μ; it will also be denoted by $u_\mu(z)$. It is well-defined and superharmonic in $|z|<\infty$ (see, e.g., AHLFORS-SARIO [1, p. 248]). Furthermore, it is harmonic in $C'S_\mu$ and has the property

$$u(z) - \mu(R^2) \log \frac{1}{|z|} \to 0 \quad \text{as } |z| \to \infty.$$

Accordingly,

$$-\infty < u(z) \leq \infty$$

in $|z|<\infty$, and u is bounded from below on every closed bounded set.

The correspondence $\mu \to u_\mu$ is one-to-one in the following sense: If the potentials u_μ and u_ν are identical, then μ and ν coincide. A proof is found in Appendix II.I.

For a μ with bounded S_μ and finite $\mu(R^2)$, the quantity

$$I(\mu)=\int \log \frac{1}{|z-\zeta|}\, d(\mu\times\mu)(z,\zeta)=\int u_\mu(z)\, d\mu(z),$$

which is well-defined and satisfies $-\infty < I(\mu) \leq \infty$, is called the *energy integral* of μ.

Let E be a closed bounded set. We set

$$V_E = \inf_{\mu\in \mathbf{M}(E)} I(\mu)$$

$(-\infty < V_E \leq \infty)$ and call the quantity

$$\operatorname{Cap} E = e^{-V_E}$$

the *logarithmic capacity* of E. It satisfies $0 \leq \operatorname{Cap} E < \infty$.

If E is an arbitrary set in R^2, we introduce the *inner logarithmic capacity* by

$$\operatorname{Cap}^{(i)} E = \sup(\operatorname{Cap} F),$$

where the supremum is taken over all closed bounded subsets F of E. Clearly $\operatorname{Cap}^{(i)} E = \operatorname{Cap} E$ if E is closed and bounded.

6 C. Conductor Potential. Suppose $E \subset R^2$ is closed and bounded. Take a sequence $\mu_n \in \mathbf{M}(E)$ such that $\lim I(\mu_n) = V_E$. By the inequality at the end of 6 A, we can find a $\mu \in \mathbf{M}(E)$ and a subsequence with $I(\mu) \leq \underline{\lim}\, I(\mu_{i_n})$ and, therefore, $I(\mu) = V_E$. For the potential of (any such) μ, the following theorem (FROSTMAN [1]) is of importance:

FROSTMAN'S THEOREM. *If E is a closed bounded set with* $\operatorname{Cap} E > 0$, *then the potential u_μ of a mass distribution $\mu \in \mathbf{M}(E)$ with*

$$I(\mu) = V_E = \min_{\nu\in \mathbf{M}(E)} I(\nu) \tag{1}$$

satisfies the following two conditions:

$$u_\mu \leq V_E \tag{2}$$

in $|z| < \infty$, and

$$u_\mu = V_E \tag{3}$$

on E, except for an F_σ-set of inner logarithmic capacity zero.

In Appendix II.C–I, we give the proof of this theorem as well as that of the *uniqueness of μ* in the following sense: $\mathbf{M}(E)$ contains only one μ with (2) and (3). Thus $\mathbf{M}(E)$ contains only one μ with (1). The uniquely determined μ will be called the *equilibrium distribution* on E, and its potential the *conductor potential* (or equilibrium potential) for E.

In AHLFORS-SARIO [1, pp. 248 ff.], logarithmic capacity and conductor potential are introduced in a different fashion. The equivalence of the definitions is clear from the following proposition:

If E is a closed bounded set, then

$$V_E = \min_{\nu \in \mathbf{M}(E)} \sup_{|z|<\infty} u_\nu(z). \tag{4}$$

Furthermore, if Cap $E>0$, *then the equilibrium distribution is the only one furnishing the minimum.*

In fact, $I(\nu)=\int_E u_\nu \, d\nu \leq \sup u_\nu$ implies that V_E is dominated by the right-hand side of (4). For the case Cap $E=0$, that is, $V_E=\infty$, the proof is complete. If Cap $E>0$, the conductor potential u_μ satisfies $V_E=\sup u_\mu$ by (3), since E actually contains a point where u_μ attains the value V_E (cf. Appendix II.D). Thus (4) is satisfied. Finally, if $\nu \in \mathbf{M}(E)$ is such that $V_E=\sup u_\nu$, then $I(\nu)=\int u_\nu \, d\nu \leq V_E$, that is, $I(\nu)=V_E$. Consequently, ν is the equilibrium distribution.

6 D. Capacity Function p_β. Let W be a plane region with $\infty \in W$. The boundary $\beta=\partial W$ and the complement $E=CW$ are both closed and bounded.

THEOREM. *The capacity $c_\beta(\infty)$ coincides with* Cap E, *which in turn is equal to* Cap β. *If $c_\beta>0$, then the capacity function $p_\beta(z;\infty)$ coincides with the restriction to W of the conductor potential for E, and this is identical with that for β.*

Proof. Exhaust W by regular subregions Ω with $\infty \in \Omega$. Write p_Ω and k_Ω for $p_{\beta(\Omega)}(z;\infty)$ and $k_{\beta(\Omega)}(\infty)=\log(1/c_{\beta(\Omega)}(\infty))$ for Ω. By Green's formula, we obtain

$$p_\Omega(z)=\frac{1}{2\pi}\int_{\beta(\Omega)} \log\frac{1}{|z-\zeta|} * dp_\Omega(\zeta) \quad \text{if } z \in \Omega, \tag{5}$$

$$k_\Omega=\frac{1}{2\pi}\int_{\beta(\Omega)} \log\frac{1}{|z-\zeta|} * dp_\Omega(\zeta) \quad \text{if } z \in C\bar{\Omega}. \tag{6}$$

Denote by $\tilde{\mu}_\Omega$ the mass distribution defined by

$$\tilde{\mu}_\Omega(e)=\frac{1}{2\pi}\int_{e\cap\beta(\Omega)} * dp_\Omega$$

for every Borel set e. Fix an Ω, say Ω_0. Then $\tilde{\mu}_\Omega \in \mathbf{M}(C\Omega_0)$ for every $\Omega \supset \bar{\Omega}_0$. By the Selection Theorem and Theorem III.2 E, we obtain a mass distribution $\tilde{\mu} \in \mathbf{M}(C\Omega_0)$ and a nested sequence $\{\Omega_n\}$ with $\bar{\Omega}_0 \subset \Omega_n \to W$ such that $\lim p_{\Omega_n}=p_\beta$ and $\lim \int f \, d\tilde{\mu}_{\Omega_n}=\int f \, d\tilde{\mu}$ for every continuous function f on $|z|<\infty$.

We first show that

$$\tilde{\mu} \in \mathbf{M}(\beta), \qquad u_{\tilde{\mu}} = p_\beta \tag{7}$$

on W. In fact, if a continuous function f vanishes identically on $\overline{W}$, then $\int f\, d\tilde{\mu}_{\Omega_n} = 0$, so that $\int f\, d\tilde{\mu} = 0$, and $S_{\tilde{\mu}} \subset \overline{W}$. If $f \equiv 0$ on $C\Omega_j$, then $\int f\, d\tilde{\mu}_{\Omega_n} = 0$ for every $n > j$ and, therefore, $\int f\, d\tilde{\mu} = 0$. A fortiori, $S_{\tilde{\mu}} \subset C\Omega_j$, $j = 1, 2, \ldots$, whence $S_{\tilde{\mu}} \subset \overline{W} \cap CW = \beta$ and, consequently, $\tilde{\mu} \in \mathbf{M}(\beta)$. Given $z \in \Omega_n$, the function $\log(1/|z-\zeta|)$ of ζ is continuous on $C\Omega_n$. This function can be extended to a continuous function on $|\zeta| < \infty$, and we see from (5) that

$$u_{\tilde{\mu}}(z) = \lim_n \int \log \frac{1}{|z-\zeta|}\, d\tilde{\mu}_{\Omega_n}(\zeta) = \lim_n p_{\Omega_n}(z) = p_\beta(z).$$

Second, we shall show that

$$\sup_{|z| < \infty} u_{\tilde{\mu}}(z) = k_\beta, \tag{8}$$

where $k_\beta = \log(1/c_\beta(\infty))$ is for W. By Corollary III.3 F, we have $\sup_W u_{\tilde{\mu}} = \sup_W p_\beta = k_\beta$. If $z \in CW$, then the function $\log(1/|z-\zeta|)$ of ζ is lower semicontinuous on $|\zeta| < \infty$, so that, by (6),

$$u_{\tilde{\mu}}(z) \le \underline{\lim} \int \log \frac{1}{|z-\zeta|}\, d\tilde{\mu}_{\Omega_n}(\zeta) = \lim k_{\Omega_n} = k_\beta.$$

Thus (8) is satisfied.

Now let μ be the equilibrium distribution on β. From (8) and (4) we have

$$V_\beta \le k_\beta. \tag{9}$$

For the case Cap $\beta = 0$, that is, $V_\beta = \infty$, the proof of Cap $\beta = c_\beta$ is complete.

If Cap $\beta > 0$, the conductor potential u_μ satisfies $V_\beta - u_\mu \ge 0$, so that $k_\beta < \infty$ and $V_\beta - u_\mu \ge k_\beta - p_\beta$ in view of Theorem III.3 F. By (7) and (9), the difference $u_{\tilde{\mu}} - u_\mu$ is non-negative on W. It is harmonic on W and vanishes at ∞. Consequently, by the maximum principle for harmonic functions, $u_{\tilde{\mu}} = u_\mu$, that is,

$$u_\mu = p_\beta \qquad \text{on } W.$$

Since $k_\beta = \sup_W p_\beta \le \sup_{|z|<\infty} u_\mu = V_\beta$, we conclude, in view of (9), that

$$V_\beta = k_\beta, \qquad \text{i.e., Cap } \beta = c_\beta. \tag{10}$$

Finally, let μ be the equilibrium distribution on E. Since $\tilde{\mu} \in \mathbf{M}(\beta) \subset \mathbf{M}(E)$, the above reasoning applies without change to this μ.

The proof of Theorem 6 D is herewith complete.

Remark. From the proof we observe that the equilibrium distributions on β and on E are identical.

6 E. Conformal Invariance of the Vanishing of Logarithmic Capacity. Given a compact set E contained in a region Ω, let f be a univalent analytic function on Ω, and set $E'=f(E)$. We maintain:

$$\operatorname{Cap} E=0 \quad \text{implies } \operatorname{Cap} E'=0.$$

In short, *the vanishing of the logarithmic capacity is conformally invariant.*

For the proof, observe that E and E' are totally disconnected, and denote by W and W' the regions CE and CE'. We may assume without loss of generality that Ω is a regularly imbedded region and f is analytic on $\bar{\Omega}$. Then $\Omega-E$ and $f(\Omega)-E'$ are boundary neighborhoods of $\beta=\partial W$ and $\beta'=\partial W'$. By IV.3 B, $c_\beta(\infty)=0$ implies $c_{\beta'}(\infty)=0$, so that $\operatorname{Cap} E'=0$.

6 F. Boundary Behavior of p_β. The conductor potential u_μ of a closed bounded set E with $\operatorname{Cap} E>0$ attains its maximum value V_E at every point $z_0\in\partial E$ except for a subset A of inner logarithmic capacity zero. Because of the lower semicontinuity, we have

$$\lim_{z\to z_0} u_\mu(z)=V_E,$$

where $z_0\in\partial E-A$. Here z may be limited to the complement of E. Consequently:

THEOREM. *If W is a plane region such that $\infty\in W$ and $c_\beta>0$, then the capacity function satisfies*

$$\lim_{z\to z_0} p_\beta(z;\infty)=\log\frac{1}{c_\beta(\infty)}, \qquad z\in W,$$

at every $z_0\in\beta$ except for an F_σ-set of inner logarithmic capacity zero.

Observe the contrast with the following boundary behavior of the capacity function $p_{1\gamma}$ of a boundary component γ with $c_{1\gamma}>0$:

$$\lim_{z\to z_0} p_{1\gamma}(z;\infty)=\log\frac{1}{c_{1\gamma}(\infty)}, \qquad z\in W,$$

for every $z_0\in\gamma$ without exception (Theorem 2 E).

On the other hand,

$$\lim_{z\to z_0} p_{0\gamma}(z;\infty)=\log\frac{1}{c_{0\gamma}(\infty)}, \qquad z\in W,$$

is satisfied for $z_0\in\gamma$ with some points excluded (Theorem 4 A). However, for such exceptional points properties analogous to those given in the above theorem and Theorem 6 G below are not known.

6 G. Boundary Behavior of p_β (continued). For the exceptional points z_0 in the above theorem, the following theorem of BOULIGAND [1] is important; originally it was stated for the Green's function:

THEOREM. *If W is a plane region such that $\infty\in W$ and $c_\beta>0$, then*

$$\lim_{z\to z_0} p_\beta(z;\infty)=\log\frac{1}{c_\beta(\infty)}, \qquad z\in W, \tag{11}$$

if and only if z_0 is a regular point for the Dirichlet problem on W.

We remark that the following theorem of KELLOGG [1] is in turn a consequence of the theorems of FROSTMAN and BOULIGAND: The set of irregular boundary points is an F_σ-set of inner logarithmic capacity zero.

Proof of the Theorem. The sufficiency is easily established as a combination of property I.5 E of the operator H (Theorem I.5 E) and the fact that $H(p_\beta-k_\beta)=p_\beta-k_\beta$ on β (Theorem III.2 D).

Conversely, suppose z_0 is a point at which (11) is valid. Then the conductor potential u_μ for $E=CW$ satisfies $u_\mu(z_0)=V_E$. Take $a>0$ so large that the disk $|z|<a$ contains E and consider the region

$$W_0=W\cap\{z\,||z|<a\}.$$

Because of the local nature of regularity, it suffices to show that z_0 is a regular boundary point of W_0. This means the existence of a barrier b, which by definition has the following properties (e.g., AHLFORS-SARIO [1, p.139]):

(a) b is subharmonic on W_0,

(b) $\lim_{z\to z_0} b(z)=0$, $z\in W_0$,

(c) $\overline{\lim}_{z\to z_1} b(z)<0$, $z\in W_0$ for every $z_1\neq z_0$, $z_1\in\partial W_0$.

The following construction of a barrier is due to BRELOT [1]. Let $A=\max_{z\in\partial W_0}|z-z_0|$. For an arbitrary r with $0<r<A$, set $N_r=\{z\,||z-z_0|<r\}$ and $\alpha_r=W_0\cap\{z\,||z-z_0|=r\}$. Take a closed set $e_r\subset\alpha_r$ such that the linear measure of α_r-e_r does not exceed $2\pi r^2/A$. Using the Poisson integral, first construct an auxiliary function h_r, harmonic on N_r with boundary values A on α_r-e_r and 0 on $\partial N_r-(\alpha_r-e_r)$. It satisfies

$$0<h_r(z_0)\le r.$$

6 H. Proof of Theorem 6 G (continued). Consider the continuous function $f(z)=|z-z_0|$ on ∂W_0. Let $u_f(z)$ be the solution of the Dirichlet problem on W_0 for boundary values f. Since the subharmonic function $|z-z_0|$ on W_0 belongs to the Perron family $\mathscr{V}(f)$ (AHLFORS-SARIO [1, p.139]), we see that

$$|z-z_0|\le u_f(z), \qquad z\in W_0.$$

Furthermore, $f\le A$ on ∂W_0, so that

$$u_f(z)\le A, \qquad z\in W_0.$$

We shall show that

$$b(z) = -u_f(z)$$

is a barrier. It is harmonic on W_0 and, at every $z_1 \neq z_0$, $z_1 \in \partial W_0$,

$$\overline{\lim_{z \to z_1}} \, b(z) \leq -|z_1 - z_0| < 0$$

for $z \in W_0$. Therefore, it only remains to prove (b): $\lim_{z \to z_0} b(z) = 0$.

Set $m_r = \min_{z \in e_r} g(z)$, where $g(z) = g(z; \infty) = k_\beta - p_\beta(z; \infty)$ is the Green's function of W. For every $v \in \mathscr{V}(f)$, the function

$$\varphi(z) = v(z) - r - \frac{A\, g(z)}{m_r} - h_r(z)$$

is subharmonic on $W_0 \cap N_r$. If $z \to z_1 \in N_r \cap \partial W_0$, then $\overline{\lim}\, \varphi \leq \overline{\lim}\, v - r \leq |z_1 - z_0| - r < 0$; if $z \to z_1 \in \alpha_r - e_r$, then $\overline{\lim}\, \varphi \leq \overline{\lim}(v - h_r) \leq v(z_1) - A \leq u_f(z_1) - A \leq 0$; if $z \to z_1 \in e_r$, then $\overline{\lim}\, \varphi \leq \overline{\lim}(v - A\, g(z)\, m_r^{-1}) \leq v(z_1) - A\, g(z_1)\, m_r^{-1} \leq v(z_1) - A \leq u_f(z_1) - A \leq 0$. Thus $\varphi \leq 0$ on $N_r \cap W_0$, that is,

$$v(z) \leq r + \frac{A\, g(z)}{m_r} + h_r(z)$$

on $N_r \cap W_0$. Accordingly,

$$-b(z) = u_f(z) \leq r + \frac{A\, g(z)}{m_r} + h_r(z).$$

Now use the assumption that $\lim_{z \to z_0} g(z) = 0$. The inequality $h_r(z_0) \leq r$ gives

$$\underline{\lim_{z \to z_0}}\, b(z) \geq -2r.$$

Since r is arbitrary, we obtain $\underline{\lim}\, b \geq 0$. On the other hand, it is evident from $b \leq 0$ that $\overline{\lim}\, b \leq 0$.

Consequently, b is a barrier and, therefore, z_0 is a regular point.

Remark. A non-positive function satisfying only (a) and (b) of the definition of a barrier is sometimes referred to as a *generalized barrier*. In the above proof, we may replace $-g$ by an arbitrary generalized barrier on W, so that the existence of a generalized barrier implies that of a barrier. In other words, *a boundary point is regular if a generalized barrier exists.*

6 I. Transfinite Diameter. One of the important applications of potential theory is a geometric representation of the capacity c_β.

Let E be a closed bounded set in the plane. For $z_1, \ldots, z_n \in E$, $n \geq 2$, set

$$V(z_1, \ldots, z_n) = \prod_{\substack{i,j=1 \\ (i<j)}}^{n} |z_i - z_j|,$$

$$V_n = \max_{z_1, \ldots, z_n \in E} V(z_1, \ldots, z_n).$$

The quantities

$$\delta_n = V_n^{\frac{2}{n(n-1)}}$$

satisfy $\delta_2 \geq \delta_3 \geq \cdots$. In fact, for $V_{n+1} = V(\zeta_1, \ldots, \zeta_{n+1})$, we have $V_{n+1} = |\zeta_1 - \zeta_2| \cdots |\zeta_1 - \zeta_{n+1}|\, V(\zeta_2, \ldots, \zeta_n) \leq |\zeta_1 - \zeta_2| \cdots |\zeta_1 - \zeta_{n+1}|\, V_n$. Similarly, $V_{n+1} \leq |\zeta_2 - \zeta_1| \cdots |\zeta_2 - \zeta_{n+1}|\, V_n, \ldots, V_{n+1} \leq |\zeta_{n+1} - \zeta_1| \cdots |\zeta_{n+1} - \zeta_n|\, V_n$. On multiplying we obtain $V_{n+1}^{n+1} \leq V_{n+1}^2\, V_n^{n+1}$, hence $\delta_{n+1} \leq \delta_n$.

The quantity

$$\delta(E) = \lim_{n \to \infty} \delta_n$$

is the *transfinite diameter* of E, introduced by Fekete [1] (see also Szegö [2]). They obtained the following result:

Theorem. *The transfinite diameter and the logarithmic capacity are equal,*

$$\delta(E) = \operatorname{Cap} E.$$

Proof. We first show that $\delta(E) \leq \operatorname{Cap} E$. For an arbitrary $\varepsilon > 0$, there exists a relatively compact open set D with $E \subset D$ and $\operatorname{Cap} \bar{D} \leq \operatorname{Cap} E + \varepsilon$; this is clear by virtue of the property of k_β. Take n so large that $\Delta_i = \{z \,|\, |z - z_i| \leq 1/\sqrt{n\pi}\} \subset D$ for every $z_i \in E$, $i = 1, \ldots, n$. Let μ be the mass distribution on D which is the sum of those with density 1 on Δ_i and 0 elsewhere. Then

$$\log \frac{1}{\operatorname{Cap} \bar{D}} \leq \int_{\bar{D} \times \bar{D}} \log \frac{1}{|z - \zeta|}\, d(\mu \times \mu)(z, \zeta)$$
$$\leq \sum_{j=1}^{n} \iint_{\Delta_j} dx\, dy \sum_{i=1} \iint_{\Delta_i} \log \frac{1}{|z - \zeta|}\, d\xi\, d\eta,$$

where $z = x + iy$, $\zeta = \xi + i\eta$.

Since the function $\log(1/|z - \zeta|)$ is superharmonic in ζ, we have

$$\iint_{\Delta_i} \log \frac{1}{|z - \zeta|}\, d\xi\, d\eta \leq \log \frac{1}{|z - z_i|} \iint_{\Delta_i} d\xi\, d\eta = \frac{1}{n} \log \frac{1}{|z - z_i|}.$$

Consequently,

$$\log\frac{1}{\operatorname{Cap} E+\varepsilon}\le\frac{1}{n}\sum_{i,j=1}^{n}\iint_{\Delta_j}\log\frac{1}{|z-z_i|}\,dx\,dy.$$

For $i\neq j$, the superharmonicity implies that

$$\iint_{\Delta_j}\log\frac{1}{|z-z_i|}\,dx\,dy\le\frac{1}{n}\log\frac{1}{|z_j-z_i|},$$

and for $i=j$, direct computation gives

$$\iint_{\Delta_j}\log\frac{1}{|z-z_j|}\,dx\,dy=O\left(\frac{\log n}{n}\right).$$

Therefore,

$$\begin{aligned}\log\frac{1}{\operatorname{Cap} E+\varepsilon}&\le\frac{1}{n}\left(\frac{2}{n}\sum_{i<j}\log\frac{1}{|z_i-z_j|}+O(\log n)\right)\\&=-\frac{2}{n^2}\log V(z_1,\ldots,z_n)+O\left(\frac{\log n}{n}\right).\end{aligned}$$

If we select $z_1,\ldots,z_n\in E$ such that $-\log V(z_1,\ldots,z_n)=(n(n-1)/2)\log(1/\delta_n)$, then

$$\log\frac{1}{\operatorname{Cap} E+\varepsilon}\le\frac{n-1}{n}\log\frac{1}{\delta_n}+O\left(\frac{\log n}{n}\right)$$

and $\delta(E)\le\operatorname{Cap} E$.

6 J. Opposite Inequality. To show that $\delta(E)\ge\operatorname{Cap} E$, we may assume that $\operatorname{Cap} E>0$. Take $\zeta_1,\ldots,\zeta_n\in E$ with $V_n=V(\zeta_1,\ldots,\zeta_n)$. Then $V_n=\max_{z\in E}V(z,\zeta_2,\ldots,\zeta_n)$. Since $V(z,\zeta_2,\ldots,\zeta_n)=|z-\zeta_2|\cdots|z-\zeta_n|\,V(\zeta_2,\ldots,\zeta_n)$, we see that $|\zeta_1-\zeta_2|\cdots|\zeta_1-\zeta_n|=\max_{z\in E}(|z-\zeta_2|\cdots|z-\zeta_n|)$. In general, we similarly have

$$\prod_{\substack{j=1\\j\neq i}}^{n}|\zeta_i-\zeta_j|=\max_{z\in E}\prod_{\substack{j=1\\j\neq i}}^{n}|z-\zeta_j| \tag{12}$$

for $i=1,\ldots,n$. It follows that

$$\begin{aligned}\frac{n(n-1)}{2}\log\frac{1}{\delta_n}&=\sum_{\substack{i,j=1\\i<j}}^{n}\log\frac{1}{|\zeta_i-\zeta_j|}=\frac{1}{2}\sum_{\substack{i,j=1\\i\neq j}}^{n}\log\frac{1}{|\zeta_i-\zeta_j|}\\&=\frac{1}{2}\sum_{i=1}^{n}\sum_{\substack{j=1\\j\neq i}}^{n}\log\frac{1}{|\zeta_i-\zeta_j|}=\frac{1}{2}\sum_{i=1}^{n}\min_{z\in E}\sum_{\substack{j=1\\j\neq i}}^{n}\log\frac{1}{|z-\zeta_j|}\\&=\frac{n-1}{2}\sum_{i=1}^{n}\min_{z\in E}u_i(z),\end{aligned}$$

where $u_i(z)$ is the logarithmic potential of the discrete mass distribution $u_i \in \mathbf{M}(E)$ defined as follows: for every Borel set e, $\mu_i(e)=0$ if $\zeta_j \notin e$ for all $j=1,\dots,n$, $j \neq i$, and $\mu_i(\{\zeta_j\})=(n-1)^{-1}$, $j=1,\dots,n$, $j \neq i$. If μ is the equilibrium distribution on E (Cap $E>0$), then by Fubini's theorem

$$\int_E u_i\, d\mu = \int_E u_\mu\, d\mu_i \leq V_E$$

and, consequently, $\min_{z\in E} u_i(z) \leq V_E$. We conclude that

$$\frac{n(n-1)}{2} \log \frac{1}{\delta_n} \leq \frac{n(n-1)}{2} V_E$$

and $\delta(E) \geq \operatorname{Cap} E$.

6 K. Evans-Selberg Potentials. The following theorem was established by Evans [1] for sets in 3-space and by Selberg [1] in the 2-dimensional case:

Theorem. *If a closed bounded set E has vanishing logarithmic capacity, then there exists a $\mu \in \mathbf{M}(E)$ such that $u_\mu(z)=\infty$ at every $z \in E$.*

Lower semicontinuity implies that

$$\lim_{z\to z_0} u_\mu(z)=\infty, \qquad z\in W,$$

for every $z_0 \in \partial E$.

Proof of the Theorem. We consider the transfinite diameter of E. Let $\zeta_1,\dots,\zeta_n \in E$ be such that $V(\zeta_1,\dots,\zeta_n)=V_n$ and set

$$m_n = \min_{z_1,\dots,z_n \in E} \left(\max_{z\in E} \prod_{i=1}^{n} |z-z_i| \right).$$

Clearly there exist points $\zeta'_1,\dots,\zeta'_n$ with $m_n = \max_{z\in E} \prod_{i=1}^n |z-\zeta'_i|$. From this we infer that

$$\lim_{n\to\infty} m_n^{1/n}=0. \tag{13}$$

In fact, by (12),

$$\prod_{\substack{i=1\\(i\neq j)}}^{n} |\zeta_i-\zeta_j| \geq m_{n-1},$$

and on multiplying for $j=1,\dots,n$ we obtain $V_n^2 \geq m_{n-1}^n$. Therefore, $\delta_n \geq m_{n-1}^{1/(n-1)} \to 0$.

The left-hand side of

$$\frac{1}{n}\sum_{i=1}^{n}\log\frac{1}{|z-\zeta_i'|}\geq\frac{1}{n}\log\frac{1}{m_n},\qquad z\in E,$$

is the logarithmic potential of the discrete mass distribution μ_n defined, in the same manner as in 6 J, by assigning the mass $1/n$ to each $\zeta_1',\ldots,\zeta_n'$. The right-hand side diverges to ∞ as $n\to\infty$. Given an arbitrary integer j, there exists an $n=n(j)$ such that $(1/n)\log(1/m_n)\geq 2^j$. The function

$$v_j(z)=\frac{1}{2^j}\,\frac{1}{n(j)}\sum_{i=1}^{n(j)}\log\frac{1}{|z-\zeta_i'|}$$

is the logarithmic potential of a discrete mass distribution on E with total mass 2^{-j}. It satisfies $v_j(z)\geq 1$ at every $z\in E$. As a consequence,

$$u(z)=\sum_{j=1}^{\infty}v_j(z)$$

is the logarithmic potential of a mass distribution in $\mathbf{M}(E)$ with $u(z)=\infty$ at every $z\in E$.

6 L. Evans-Selberg Potentials and Capacity Functions. Let W be a plane region with $\infty\in W$ and $c_\beta=0$. We recall that the capacity function p_β is not determined uniquely and that the complement $E=CW$ has vanishing Cap E.

THEOREM. *The restriction to W of any* EVANS-SELBERG *potential u_μ coincides with some capacity function $p_\beta(z,\infty)$.*

Proof. Consider an increasing sequence $\{a_n\}$ of real numbers such that $\lim a_n=\infty$ and $\operatorname{grad} u_\mu\neq 0$ on the level lines $u_\mu=a_n$, $n=1,2,\ldots$. The component Ω_n about ∞ of $\{z\,|\,u_\mu(z)<a_n\}$ is a regular subregion exhausting W as $n\to\infty$. The restriction of u_μ to Ω_n coincides with the capacity function $p_{\beta(\Omega_n)}(z;\infty)$ on Ω_n, and the statement follows.

Thus, if $c_\beta=0$, we obtain a capacity function $p_\beta(z;\infty)$ such that

$$\lim_{n\to\infty}p_\beta(z_n;\infty)=\infty \tag{14}$$

for every sequence of points $z_n\in W$ with no accumulation point in W. Again observe the contrast with the capacity functions $p_{1\gamma}$ and $p_{0\gamma}$ for a boundary component γ with $c_{1\gamma}=0$ and $c_{0\gamma}=0$, respectively (Theorems 2 E, 4 A).

For an arbitrary open Riemann surface W with $c_\beta=0$, the existence of p_β satisfying (14) was established by NAKAI [1] by extending Theorem 6 K to a compactification of W; see also SARIO-NOSHIRO [1, pp. 98 ff.].

Chapter VIII

Mappings Related to Modulus Functions

Conformal mappings onto circular and radial slit annuli (possibly with incisions) are obtained by means of modulus functions. Basically this is done as in the case of capacity functions, but for some theorems, additional technical devices are needed. These will be discussed in 1. We shall study the simplest case, that of a doubly connected region, in more detail in 2.

1. Mappings onto Slit Annuli

1 A. Univalent Functions Q_0 and Q_1. Let W be a plane region whose connectivity exceeds one. Given two distinct boundary components γ and γ', consider the modulus functions $q_\nu = q_\nu(z; \gamma, \gamma')$. The functions

$$Q_\nu = Q_\nu(z; \gamma, \gamma') = e^{q_\nu + i q_\nu^*}, \qquad \nu = 0, 1,$$

are single-valued and belong to the family

$$\mathcal{S}_{\gamma'\gamma} = \mathcal{S}_{\gamma'\gamma}(W)$$

of univalent analytic functions F on W such that (a) $F \neq 0$ on W and (b) $F(\gamma)$ and $F(\gamma')$ are the outer and inner boundaries (VI. 1 D) of the image region $F(W)$.

Recall that q_ν has the ambiguity of an additive constant. Similarly, we shall not determine q_ν^* uniquely. If the modulus

$$\mu_\nu = \mu_\nu(\gamma, \gamma') = \mu_\nu(\gamma', \gamma)$$

of W is finite, Q_ν is uniquely determined up to a constant factor; if $\mu_0 = \infty$, Q_0 is not necessarily unique even in this sense. However, for boundary components γ, γ' of a plane region W, the function Q_1 is unique up to a constant factor, as we shall see in 1 I.

In particular, if γ and γ' are isolated boundary components, then (including the case $\mu_\nu = \infty$) $Q_\nu(z; \gamma, \gamma')$ is the unique function in $\mathcal{S}_{\gamma'\gamma}$, up to a constant factor, such that

$$\lim_{z\to\gamma} |Q_\nu| = \text{const} = r_{\nu\gamma},$$

$$\lim_{z\to\gamma'} |Q_\nu| = \text{const} = r'_{\nu\gamma},$$

$$L_\nu(\log|Q_\nu|) = \log|Q_\nu| \quad \text{on } \partial W - (\gamma\cup\gamma').$$

The ratio $r_{\nu\gamma}/r'_{\nu\gamma}$ is equal to the modulus μ_ν.

We recall that $\mu_\nu = \infty$ if and only if at least one of the capacities $c_{\nu\gamma}$ and $c_{\nu\gamma'}$ vanishes (Theorem IV.3 C). For isolated components γ and γ', we have $c_{\nu\gamma} = c_{\nu\gamma'} = 0$ if and only if both γ and γ' are single points, say ζ and ζ'; then $r'_{\nu\gamma} = 0$, $r_{\nu\gamma} = \infty$, and

$$Q_\nu = \text{const}\cdot P_\nu(z;\zeta,\zeta'),$$

where P_ν is the function for $W\cup\{\zeta,\zeta'\}$, introduced in VI.2 A. Similarly, $c_{\nu\gamma} > 0$ and $c_{\nu\gamma'} = 0$ if and only if γ contains more than one point and γ' consists of a single point, say ζ; then $r'_{\nu\gamma} = 0$, $r_{\nu\gamma} < \infty$, and

$$Q_\nu = \text{const}\cdot P_{\nu\gamma}(z;\zeta),$$

where $P_{\nu\gamma}$ is the function for $W\cup\{\zeta\}$ introduced in VII.1 A.

1 B. Circular and Radial Slit Annuli. We introduce:

DEFINITION. *A circular (radial) slit annulus of inner radius* $r'(\geq 0)$ *and outer radius* $r(\leq\infty)$ *is a subregion of the annulus* $r' < |z| < r$ *whose inner boundary is the circle* $|z| = r'$, *outer boundary the circle* $|z| = r$, *and other boundary components are either points or circular arcs on* $|z| = \text{const}$ *(line segments on* $\arg z = \text{const}$, *resp.).*

For short, we shall use the term circular (radial) slit annulus $r' < |z| < r$.

If γ and γ' are isolated boundary components of a plane region W, then the function $Q_1(z;\gamma,\gamma')$ (or $Q_0(z;\gamma,\gamma')$) maps W onto a circular (or radial, resp.) slit annulus $r' < |z| < r$, where r/r' is equal to the modulus $\mu_1(\gamma,\gamma')$ (or $\mu_0(\gamma,\gamma')$, resp.).

In particular, if W is a regular region, then the Q_ν are characterized as the functions in $\mathcal{S}_{\gamma'\gamma}$ with the above mapping property.

In the general case, this mapping property is insufficient to characterize the functions, and it will be necessary to introduce extremal slit annuli.

All results in 1 A and 1 B are proved as in VII.1 A – 1 C. The details are left to the reader.

1 C. Modulus μ_1 and Extremal Length. Among the results in Theorem IV.2 C, we shall discuss the identity

$$\log \mu_1 = \frac{2\pi}{\lambda(\Gamma)}.$$

In the present case, where W is a plane region and γ, γ' are boundary components, it is not necessary to consider all cycles in Γ:

Γ may be restricted to the family of rectifiable closed curves in W which separate γ from γ'.

For the proof, first observe that $\lambda(\Gamma) \geq L(\Gamma, \rho_1)^2 A(\rho_1)^{-1} = 2\pi(\log \mu_1)^{-1}$ if $\rho_1 |dz| = (1/2\pi)|dq_1 + i * dq_1|$. To establish the opposite inequality, exhaust W by regular subregions Ω. The function $Q_{1\Omega} = Q_1(z; \gamma(\Omega), \gamma'(\Omega))$ on Ω normalized by $|Q_{1\Omega}| = 1$ on $\gamma'(\Omega)$ maps Ω conformally onto a circular slit annulus $1 < |z| < \mu_{1\Omega}$, where $\mu_{1\Omega}$ is the modulus of Ω with respect to $\gamma(\Omega)$ and $\gamma'(\Omega)$. The image region has only a finite number of slits. Accordingly, except for a finite number of values τ with $0 < \tau < \log \mu_{1\Omega}$, the level line $c_{\tau\Omega} = \{z \mid q_{1\Omega}(z) = \tau\}$ is a single rectifiable closed curve separating γ from γ'. For $\rho_{1\Omega} = |Q'_{1\Omega}/Q_{1\Omega}|$ and every metric ρ on W, we have as in IV.2 E,

$$L(\Gamma, \rho)^2 \leq \Big(\int_{c_{\tau\Omega}} \rho |dz|\Big)^2 \leq 2\pi \int_{c_{\tau\Omega}} \frac{\rho^2 |dz|}{\rho_{1\Omega}},$$

$$(\log \mu_{1\Omega})\, L(\Gamma, \rho)^2 \leq 2\pi \iint_{\Omega} \rho^2 \, dx\, dy \leq 2\pi A(\rho),$$

$$\lambda(\Gamma) \leq \frac{2\pi}{\log \mu_{1\Omega}}.$$

On letting $\Omega \to W$, we conclude that $\lambda(\Gamma) \leq 2\pi(\log \mu)^{-1}$.

1 D. Extremal Properties of Q_1 and Q_0. The extremal properties are analogous to those of $P_{\gamma\gamma}$ discussed in VII.2 A – 2 D and VII.3 A – 3 H. In what follows, we shall merely present the results and leave the proofs to the reader.

The modulus $\mu_1(\gamma, \gamma')$ is characterized by any one of these properties:

$$2\pi \log \mu_1 = \min_{F \in \mathcal{S}_{\gamma'\gamma}} D[\log |F|], \tag{1}$$

$$\mu_1 = \min_{F \in \mathcal{S}_{\gamma'\gamma}} \frac{M[F]}{m[F]}, \tag{2}$$

where $m[F]=\inf_{z\in W}|F(z)|$. If γ' is not weak, then

$$\pi(\mu_1^2-1)=\min \mathscr{D}[F], \tag{3}$$

where the minimum is taken over all $F\in\mathscr{S}_{\gamma'\gamma}$ with $F(\gamma')=\{w\,|\,|w|=1\}$. If $\mu_1<\infty$, then $Q_1(z;\gamma,\gamma')$ is, up to a constant factor, the unique function minimizing the functionals in the family.

Denote by $\mathscr{S}_{\gamma'\gamma}^0(W)$ the family of functions $F\in\mathscr{S}_{\gamma'\gamma}$ with $L_0(\log|F|)=\log|F|$ on $\partial W-(\gamma\cup\gamma')$. The modulus $\mu_0(\gamma,\gamma')$ is given by

$$2\pi\log\mu_0=\min_{F\in\mathscr{S}_{\gamma'\gamma}^0} D[\log|F|], \tag{4}$$

$$\mu_0=\min_{F\in\mathscr{S}_{\gamma'\gamma}^0}\frac{M[F]}{m[F]}. \tag{5}$$

If $\mu_0<\infty$, then $Q_0(z;\gamma,\gamma')$ is, up to a constant factor, the unique function minimizing the functionals in the family.

Furthermore, for

$$m_\gamma[F]=\min_{w\in F(\gamma)}|w|, \qquad M_{\gamma'}[F]=\max_{w\in F(\gamma')}|w|,$$

we obtain

$$\mu_0=\sup_{F\in\mathscr{S}_{\gamma'\gamma}}\frac{m_\gamma[F]}{M_{\gamma'}[F]}. \tag{6}$$

If $\mu_0<\infty$ and a maximizing function exists, it is necessarily $Q_0(z;\gamma,\gamma')$; if γ and γ' are isolated and each consists of more than one point, then $\mu_0<\infty$ and a maximizing function exists.

1 E. Conformal Mapping by Q_1. Let W be a plane region of connectivity greater than one, and choose two boundary components γ and γ'. Recall that $\mu_1(\gamma,\gamma')=\infty$ if and only if $c_{1\gamma}=0$ and/or $c_{1\gamma'}=0$, that is, at least one of γ and γ' is weak (Theorem IV.3 C).

The first part of the following theorem is due to Grötzsch [3, 5] and Rengel [1, 2], the second part to Oikawa-Suita [2].

Theorem. *If $\mu_1(\gamma,\gamma')<\infty$, then $Q_1(z;\gamma,\gamma')$ maps W conformally onto a circular slit annulus $r'<|z|<r$ with $r/r'=\mu_1$.*

If $\mu_1(\gamma,\gamma')=\infty$, any $Q_1(z;\gamma,\gamma')$ maps W conformally onto a circular slit annulus $r'<|z|<r$. If $c_{1\gamma'}=c_{1\gamma}=0$, then $r'=0$ and $r=\infty$; if $c_{1\gamma'}=0$ and $c_{1\gamma}>0$, then $r'=0$ and $r<\infty$.

In every case the total area of the slits vanishes.

Proof. The entire reasoning is analogous to that of Theorem VII.2 E, except that, in the last case, $Q_1(\gamma)$ is a circle of finite radius.

To prove this, take a positive number a such that $Q_1(\gamma)$ is outside of the disk $|w|\leq a$. For an arbitrary $\eta>0$, there exists an analytic Jordan curve $\alpha_\eta\subset W$ separating γ from γ' such that $Q_1(\alpha_\eta)\subset\{w\,|\,a<|w|<a+\eta\}$. This follows from Lemma 2 in VI.1 E. Here α_η divides W into boundary neighborhoods U and U' of γ and γ'. We are concerned with the third case, so that the moduli of U and U' satisfy $\mu_1(\gamma,\alpha_\eta)<\infty$ and $\mu_1(\gamma',\alpha_\eta)=\infty$.

By definition, Q_1 is obtained by means of an exhaustion $V_n\to W$ towards $\gamma\cup\gamma'$:

$$Q_1=\lim_{n\to\infty} Q_{1n},$$

where $Q_{1n}=Q_1(z;\gamma(V_n),\gamma'(V_n))$ on V_n and the convergence is uniform on every compact set in W. Let r_n and r_n' be the outer and inner radii of the circular slit annulus $Q_{1n}(V_n)$. As a straightforward argument shows, $Q_1(W)$ is contained in the annulus $\lim r_n'<|w|<\underline{\lim}\, r_n$. By the assumption that γ' is weak, $Q_1(\gamma')$ consists of a single point 0; hence $\lim r_n'=0$. Clearly $a<\underline{\lim}\, r_n$.

We claim that

$$r^*=\underline{\lim}\, r_n<\infty.$$

For the proof, take n so large that $\alpha_\eta\subset V_n$ and consider the modulus $\mu_1(\gamma(V_n),\alpha_\eta)$ of the region $U\cap V_n$; clearly $\mu_1(\gamma(V_n),\alpha_\eta)\leq\mu_1(\gamma,\alpha_\eta)$. On the other hand, for n sufficiently large, $Q_{1n}(\alpha_\eta)$ lies in the annulus $a<|w|<a+\eta$. The image region $Q_{1n}(U\cap V_n)$ contains a circular slit annulus $\{w\,|\,a+\eta<|w|<r_n\}\cap Q_{1n}(U\cap V_n)$, whose modulus is $r_n(a+\eta)^{-1}$. This follows from IX.2 A, 2 C. Consequently, $r_n(a+\eta)^{-1}\leq\mu_1(\gamma,\alpha_\eta)<\infty$ and, therefore, $r^*<\infty$.

To continue, let

$$\tilde{R}=Q_1(W)\cap\{w\,|\,a<|w|\};$$

that this is a region is again seen by Lemma 2 in VI.1 E. It is contained in $a<|w|<r^*$ and its modulus $\tilde{\mu}_1$ is not less than $\mu_1(\gamma,\alpha_\eta)$. Thus $r^*(a+\eta)^{-1}\leq\tilde{\mu}_1\leq r^*a^{-1}$. Since η is arbitrary, it follows that $\tilde{\mu}_1=r^*a^{-1}$.

We use 1 D.(2) on $\tilde{R}$. The identity function coincides with a constant multiple of the function Q_1 on $\tilde{R}$ with respect to the boundary components $|w|=a$ and $Q_1(\gamma)$. This is now known to map $\tilde{R}$ onto a circular slit annulus of positive inner radius and finite outer radius. Consequently, $Q_1(\gamma)$ is a circle of finite radius, and the proof of Theorem 1 E is complete.

1 F. Conformal Mapping by Q_0. An *incised radial slit annulus* $r'<|w|<r$ $(0\leq r'<r\leq\infty)$ is defined in the same fashion as an incised disk (VII.4 A).

Let W be a plane region of connectivity greater than one and take two boundary components γ and γ'. We know that $\mu_0(\gamma,\gamma')=\infty$ if and only if at least one of $c_{0\gamma}$ and $c_{0\gamma'}$ vanishes.

THEOREM. *If* $\mu_0(\gamma,\gamma')<\infty$, *then* $Q_0(z;\gamma,\gamma')$ *maps* W *conformally onto an incised radial slit annulus* $r'<|w|<r$ *with* $r/r'=\mu_0$.

If $\mu_0(\gamma,\gamma')=\infty$, *any* $Q_0(z;\gamma,\gamma')$ *maps* W *conformally onto an incised radial slit annulus* $r'<|w|<r$ *as follows: if* $c_{0\gamma'}=c_{0\gamma}=0$, *then* $r'=0$ *and* $r=\infty$; *if* $c_{0\gamma'}=0$ *and* $c_{0\gamma}>0$, *then* $r'=0$ *and* $r<\infty$.

The total area of the slits and the angular measure of the incisions both vanish.

The proof will extend to the end of 1 H.

Remark. The part concerning $\mu_0<\infty$ was established by STREBEL [2]. Under a certain restriction, it had been previously obtained by RENGEL [1, 2] and REICH-WARSCHAWSKI [2]. The case $\mu_0=\infty$ was verbally suggested to the authors by SUITA.

Proof. As before, the case $c_{0\gamma}>0$, $c_{0\gamma'}=0$ will require most of our attention. Strebel's inequality (VII.4 H) will play a fundamental role.

The set $\partial Q_0(W)-(Q_0(\gamma)\cup Q_0(\gamma'))$ consists of points or radial segments, for we have $L_0(\log|w|)=\log|w|$ on it. In the sequel, we shall be concerned only with $Q_0(\gamma)$ and $Q_0(\gamma')$.

Take an analytic Jordan curve α in W separating γ and γ'. Let U and U' be the components of $W-\alpha$ which are boundary neighborhoods of γ and γ', and consider their moduli $\mu_0(\gamma,\alpha)$ and $\mu_0(\gamma',\alpha)$. Set $b'=\min|w|$ and $b=\max|w|$ on $Q_0(\alpha)$; then the closed annulus $0<b'\leq|w|\leq b<\infty$ contains $Q_0(\alpha)$.

The function Q_0 is defined by means of an exhaustion $\alpha\subset V_n\to W$ towards $\gamma\cup\gamma'$:

$$Q_0=\lim_{n\to\infty}Q_{0n},$$

where $Q_{0n}=Q_0(z;\gamma(V_n),\gamma'(V_n))$ on V_n, and the convergence is uniform on every compact set in W. Let r_n and r_n' be the outer and inner radii of the radial slit annulus $Q_{0n}(V_n)$. As before, we see that $Q_0(W)$ is contained in the annulus $r'^*<|w|<r^*$, where $r'^*=\overline{\lim}\,r_n'$ and $r^*=\underline{\lim}\,r_n$.

First consider the case in which $\mu_0(\gamma,\gamma')<\infty$. We have

$$\frac{r_n}{r_{n'}}=\mu_0(\gamma(V_n),\gamma'(V_n))\leq\mu_0$$

by definition, and

$$\log\mu_0\leq\frac{2\pi}{\int_0^{2\pi}\frac{d\theta}{l(\theta)}}\leq\log\frac{r^*}{r'^*}$$

by Strebel's inequality for $Q_0(W)$. Thus $l(\theta)=\log(r^*/r'^*)$ for almost all θ. On combining this with the fact that $Q_0(W)$ is contained in $r'^*<|w|<r^*$, we conclude that $Q_0(W)$ is an incised radial slit annulus $r'^*<|w|<r^*$ with $r^*/r'^*=\mu_0$.

Second, let $c_{0\gamma}=c_{0\gamma'}=0$. Apply Strebel's inequality to $Q_0(U)$:

$$\infty=\log\mu_0(\gamma,\alpha)\le\frac{2\pi}{\int_0^{2\pi}\frac{d\theta}{l(\theta)}}.$$

Thus $l(\theta)=\infty$ for almost all θ, so that $Q_0(\gamma)$ consists of the point ∞ and possibly incisions with vanishing angular measure. By applying the same reasoning to $Q_0(U')$, we draw a similar conclusion with respect to $Q_0(\gamma')$.

1 G. Proof of Theorem 1 F (continued). Suppose finally that $c_{0\gamma}>0$ and $c_{0\gamma'}=0$. On applying Strebel's inequality to $Q(U')$, we obtain

$$\infty=\log\mu_0(\gamma',\alpha)\le\frac{2\pi}{\int_0^{2\pi}\frac{d\theta}{l(\theta)}}\le\log\frac{b}{r'^*},$$

and conclude as before that $Q_0(\gamma')$ consists of the point 0 and possibly incisions with vanishing angular measure. Moreover, $r'^*=0$.

Next, observe that r^* is finite. In fact, a comparison of $Q_0(U\cap V_n)$ with the annulus bounded by the circles $|w|=r_n$ and $|w|=b$ shows that $r_n/b\le\mu_0(\gamma(V_n),\alpha)\le\mu_0(\gamma,\alpha)<\infty$. Furthermore,

$$Q_0(W)\subset\{w\,|\,|w|<r^*\}.$$

To complete the proof of the theorem, we must show that $Q_0(\gamma)$ consists of the circle $|w|=r^*$ and possibly incisions with vanishing angular measure. Let D be the region encircled by the analytic Jordan curve $Q_0(\alpha)$. We consider the region

$$\tilde{R}=Q_0(W)\cup D,$$

whose outer boundary $\tilde{\gamma}$ is precisely $Q_0(\gamma)$. Take $\varepsilon>0$ such that $|w|\le\varepsilon$ is contained in D. On the ray $\arg w=\theta$, let $w(\theta)\in\tilde{\gamma}$ be the point nearest the origin and set $l_\varepsilon(\theta)=\log|w(\theta)|/\varepsilon$. We shall show that $l_\varepsilon(\theta)=\log(r^*/\varepsilon)$ for almost all θ, $0\le\theta<2\pi$.

Let $\tilde{\mu}(\varepsilon)$ be the modulus of the region $\tilde{R}-\{w\,|\,|w|\le\varepsilon\}$ with respect to $\tilde{\gamma}$ and $|w|=\varepsilon$. Strebel's inequality shows that

$$\log\tilde{\mu}(\varepsilon)\le\frac{2\pi}{\int_0^{2\pi}\frac{d\theta}{l_\varepsilon(\theta)}}\le\log\frac{r^*}{\varepsilon}.$$

The regions $\tilde{R}_n=Q_0(V_n)\cup D$ exhaust $\tilde{R}$ towards $\tilde{\gamma}$ as $V_n\to W$. Denote by $\tilde{\mu}_n(\varepsilon)$ the modulus of $\tilde{R}_n-\{w\,|\,|w|\le\varepsilon\}$ with respect to $\tilde{\gamma}_n=Q_0(\gamma(V_n))$ and $|w|=\varepsilon$. Evidently,

$$\log\tilde{\mu}_n(\varepsilon)\le\log\tilde{\mu}(\varepsilon).$$

In order to estimate the left-hand side of the above inequality, we need the following auxiliary result of Suita [2] to be proved in 1 H:

Lemma. *Denote by* $\tilde{P}$ *and* $\tilde{P}_n$ *the capacity functions* $P_{0\tilde{\gamma}}(w;0)$ *on* $\tilde{R}$ *and* $P_{0\tilde{\gamma}_n}(w;0)$ *on* $\tilde{R}_n$. *Take an* $\varepsilon_1>\varepsilon$ *with* $\{w\,|\,|w|\le\varepsilon_1\}\subset D$. *For every sufficiently large n, there exists a quasi-conformal mapping* $T_n(w)$ *of* $\tilde{R}_n$ *with the following properties:*

(a) $T_n(w)=w$ *on* $|w|\le\varepsilon_1$,

(b) $T_n(w)=\tilde{P}_n\circ\tilde{P}^{-1}(w)$ *on* $\tilde{R}_n-D$,

(c) T_n *is continuously differentiable on* $D-\{w\,|\,|w|\le\varepsilon_1\}$,

(d) $\lim_{n\to\infty}K_n=1$, *where* K_n *is the maximal dilatation of* T_n.

For the definition of quasi-conformal mapping and maximal dilatation, see Appendix I. M.

To continue the proof of Theorem 1 F, observe that the image $T_n(\tilde{R}_n)$ is a radial slit disk of radius r_n. The radial slits are $\partial\tilde{P}_n(\tilde{R}_n)-\tilde{P}_n(\tilde{\gamma}_n)$, so that $T_n(\tilde{R}_n)$ is an extremal radial slit disk of radius r_n (VII. 1 C). The modulus of $T_n(\tilde{R}_n)-\{w\,|\,|w|\le\varepsilon\}$ is, therefore, r_n/ε (IV. 3 F). By expressing the modulus in terms of extremal length and by considering the change of extremal length under quasi-conformal mappings (Appendix I. M), we obtain

$$\frac{1}{K_n}\log\frac{r_n}{\varepsilon}\le\log\tilde{\mu}_n(\varepsilon).$$

We combine the above three inequalities and let $n\to\infty$ to obtain

$$\frac{2\pi}{\int_0^{2\pi}\frac{d\theta}{l_\varepsilon(\theta)}}=\log\frac{r^*}{\varepsilon}.$$

Since $l_\varepsilon(\theta)\le\log(r^*/\varepsilon)$, we infer that $l_\varepsilon(\theta)=\log(r^*/\varepsilon)$ for almost all θ. It remains only to establish the lemma.

1 H. Proof of Lemma 1 G. On the doubly connected region $\Delta=D-\{w\,|\,|w|\le\varepsilon_1\}$, consider the harmonic measure $u(w)$, which is 0 on $|w|=\varepsilon_1$ and 1 on ∂D. The function

$$T_n(w)=\begin{cases}\tilde{P}_n(\tilde{P}^{-1}(w)) & \text{for } w\in\tilde{R}_n-D,\\ w+u(w)(\tilde{P}_n(\tilde{P}^{-1}(w))-w) & \text{for } w\in\Delta,\\ w & \text{for } |w|\le\varepsilon_1,\end{cases}$$

is continuous on $\tilde{R}_n$ and univalent on $\tilde{R}_n-\Delta$. Note that

$$\lim_{n\to\infty}\tilde{P}_n(\tilde{P}^{-1}(w))=w$$

uniformly on every compact set in $\tilde{R}$. It is easily seen that, if n is sufficiently large, the Jacobian of T_n does not vanish on Δ and T_n maps $\partial\Delta$ topologically onto two disjoint Jordan curves. This shows that T_n is univalent throughout the region $\tilde{R}_n$. By a simple calculation, we see that the maximal dilatation K_n of T_n converges to 1 as $n \to \infty$.

The proof of the lemma and thus of Theorem 1 F is herewith complete.

1 I. Uniqueness of Q_1 in the Case $\mu_1 = \infty$. The function $Q_1(z; \gamma, \gamma')$ is always unique up to a constant factor, even if the modulus $\mu_1(\gamma, \gamma')$ is infinite. The proof reduces to showing the following (OIKAWA-SUITA [2]):

THEOREM. *Even if $\mu_1(\gamma, \gamma') = \infty$, the modulus function $q_1(z; \gamma, \gamma')$ is uniquely determined up to an additive constant, provided W is a plane region and both γ and γ' are boundary components.*

To prove this in analogy with Theorem VII.2 H, we need a result corresponding to Theorem VII.2 F (generalization of Koebe's quarter theorem). It will be given in 2 I. We shall then prove the above theorem in 2 J.

2. Doubly Connected Regions

2 A. Modulus. Let W be a doubly connected plane region, with boundary components γ and γ'. The quantities $\mu_1(\gamma, \gamma')$, $\mu_1(\gamma', \gamma)$, $\mu_0(\gamma, \gamma')$, and $\mu_0(\gamma', \gamma)$ are equal. We shall call their common value the *modulus of W* and denote it by $\mu(W)$.

Evidently, $\mu(W) = \infty$ if and only if at least one of γ, γ' degenerates to a single point.

The functions $Q_1(z; \gamma, \gamma')$ and $Q_0(z; \gamma, \gamma')$ are identical and map W onto a concentric annulus. Conversely, such a mapping function must be $Q_\nu(z; \gamma, \gamma')$ or $Q_\nu(z; \gamma', \gamma)$, $\nu = 0, 1$. If the image annulus is $r' < |w| < r$, then

$$\mu(W) = \frac{r}{r'}.$$

Consequently, two doubly connected regions with finite moduli are conformally equivalent if and only if their moduli are equal. Moreover:

LEMMA. *Let W be a doubly connected region with a finite modulus $\mu(W)$, and W' a doubly connected subregion which separates the components of ∂W. Then*

$$\mu(W') \leq \mu(W),$$

with equality if and only if $W = W'$.

Proof. The inequality is evident from extremal property 1 D.(2) or alternatively from the extremal length expression in 1 C. Suppose next that $\mu(W)=\mu(W')$ and $W'<W$, where the inequality sign stands for strict inclusion. Since $\mathcal{S}_{\gamma'\gamma}(W)\subset\mathcal{S}_{\gamma'(W')\,\gamma(W')}(W')$, the function Q_1 for W must coincide on W' with the one for W'; this is a consequence of extremal property 1 D.(2). Therefore, if $Q_1(W)=\{w\,|\,r'<|w|<r\}$ and $Q_1(W')=\{w\,|\,r_1'<|w|<r_1\}$, then $r/r'=r_1/r_1'$. But this contradicts $Q_1(W')<Q_1(W)$, and we must have $W=W'$.

2 B. Golusin's Inequality. In addition to the general properties discussed thus far, the following result is well known (Golusin [1]):

Let W be a doubly connected region bounded by analytic Jordan curves. Given an $F\in\mathcal{S}_{\gamma'\gamma}$ analytic and univalent on $\overline{W}$, denote by I_F and I_F' the areas of the regions enclosed by the curves $F(\gamma)$ and $F(\gamma')$. Then

$$\mu(W)^2\leq\frac{I_F}{I_F'},$$

with equality only for $F=\text{const}\cdot Q_1$.

Proof. Without loss of generality, we may assume that W is the annulus $1<|z|<\mu$. In terms of the coefficients of the Laurent expansion $\sum_{-\infty}^{\infty}a_n z^n$ of $F(z)=u+i\,v$, we have

$$I_F=\Big|\int_\gamma u\,dv\Big|=\Big|\pi\sum_{-\infty}^{\infty}n\,|a_n|^2\,\mu^{2n}\Big|$$

and similarly

$$I_F'=\Big|\pi\sum_{-\infty}^{\infty}n\,|a_n|^2\Big|.$$

Therefore,

$$\frac{I_F}{I_F'}=\left|\frac{\sum n\,|a_n|^2\,\mu^{2n}}{\sum n\,|a_n|^2}\right|.$$

On the other hand,

$$\frac{\sum n\,|a_n|^2\,\mu^{2n}}{\sum n\,|a_n|^2}-\mu^2=\frac{\sum\limits_{n\neq 0,\,1} n\,|a_n|^2(\mu^{2(n-1)}-1)\,\mu^2}{\sum\limits_{n\neq 0} n\,|a_n|^2}\geq 0,$$

with equality if and only if $a_n=0$, $n\neq 0, 1$. Thus,

$$\frac{I_F}{I_F'}=\frac{\sum n\,|a_n|^2\,\mu^{2n}}{\sum n\,|a_n|^2}\geq\mu^2,$$

where equality occurs if and only if $a_n=0$, $n\neq 0, 1$.

2 C. Extremal Region of Grötzsch. In contrast with extremal property 1 D.(6), let us consider, as in Theorem VII.2 F, the problem of minimizing $m_\gamma[F]/M_{\gamma'}[F]$ in $\mathscr{S}_{\gamma'\gamma}(W)$.

We begin with a special case and shall deal with the general case in 2 G. Consider the family $\mathscr{S}_1(W)$ of functions F in $\mathscr{S}_{\gamma'\gamma}(W)$ such that $M_{\gamma'}[F]=m_{\gamma'}[F]=1$, that is, $F(\gamma')$ coincides with the unit circle. In analogy with Koebe's quarter theorem, it is natural to consider the region

$$D_h=\{z\,|\,|z|>1\}-[h,\infty), \qquad 1<h<\infty,$$

where $[h,\infty)$ is the interval $\{z\,|\,h\le \operatorname{Re} z<\infty,\ \operatorname{Im} z=0\}$. This is called the *extremal region of* Grötzsch [1] (cf. Teichmüller [1], Ahlfors [4]).

Theorem. *If a doubly connected region W has a finite modulus $\mu(W)$, then a function minimizes $m_\gamma[F]$ in $\mathscr{S}_1(W)$ if and only if it maps W conformally onto the extremal region D_h of* Grötzsch *with $\mu(W)=\mu(D_h)$ or onto a rotation of D_h about the origin.*

We shall give the proof in 2 F. First we discuss, in 2 D – 2 E, general properties of the extremal region of Grötzsch. In particular, we shall show that for a given W, it is possible to determine an h with $\mu(D_h)=\mu(W)$.

2 D. Properties of D_h. We claim that the function

$$\Phi(h)=\mu(D_h), \qquad 1<h<\infty,$$

satisfies the following (partly overlapping) conditions:

(a) *Φ is continuous and strictly increasing;*

(b) *$h<\Phi(h)<4h$ and*

$$\frac{4h}{1+\dfrac{1}{2(h-1)}}<\Phi(h);$$

(c) *$\Phi(h)/h$ increases with h and*

$$\lim_{h\to\infty}\frac{\Phi(h)}{h}=4;$$

(d) $\lim_{h\to 1}\Phi(h)=1$ *and*

$$\lim_{h\to 1}\log\Phi(h)\log\frac{1}{h-1}=\frac{\pi^2}{2};$$

(e) $\log\Phi(h)=\pi^2(4\log\Phi(h'))^{-1}$, *where* $h'=h(h^2-1)^{-\frac{1}{2}}$.

Remark. An explicit expression for Φ may be found in AHLFORS [4; p. 47]; cf. also 2 G. (a).

Proof. (a) The function is strictly increasing since $h<h'$ implies $D_h<D_{h'}$ and, therefore, $\Phi(h)<\Phi(h')$ by Lemma 2 A. The continuity is a consequence of the fact that the function takes every value >1. To see this, it suffices to prove that, for any $\mu>1$, there exists an h such that D_h is conformally equivalent to the annulus $1<|w|<\mu$. By the Schwarz-Christoffel theorem (e.g., NEHARI [3]), it is possible to map the region $\{w|1<|w|<\mu,\ \mathrm{Im}\, w>0\}$ onto the region $\{z||z|>1,\ \mathrm{Im}\, z>0\}$ in such a way that the points $w=-\mu, -1, 1$ correspond to $z=\infty, -1, 1$, respectively. The image of $w=\mu$ is on the real axis and we denote it by h. The reflection about the real axis then gives the desired mapping.

2 E. Properties of D_h (continued). (c) It is convenient to prove this property prior to (b). Set $W_\gamma=\{z||z|<\infty\}-\gamma$, where $\gamma=\{z|1\le \mathrm{Re}\, z<\infty,\ \mathrm{Im}\, z=0\}$. By Koebe's quarter theorem, $c_{1\gamma}(0)=\frac{1}{4}$. For the region $W_\varepsilon= W_\gamma-\{w||w|\le\varepsilon\}$, we have by Theorem IV. 3 G, $\log\varepsilon+\log\mu(W_\varepsilon)\uparrow\log 4$ as $\varepsilon\downarrow 0$ and, therefore,

$$\log\mu(W_{1/h})-\log h\uparrow\log 4 \qquad \text{as } h\uparrow\infty.$$

Since $W_{1/h}$ and D_h are conformally equivalent, $\mu(W_{1/h})$ is equal to $\Phi(h)$, so that

$$\log\Phi(h)-\log h\uparrow\log 4 \qquad \text{as } h\uparrow\infty.$$

Thus we obtain (c).

(b) The inequality $\Phi(h)<4h$ is contained in (c). Since $\{z|1<|z|<h\}<D_h$, it is easy to obtain the inequality $h<\Phi(h)$. Next, we map W_γ conformally onto the unit disk $|w|<1$ by means of the function $w=w(z)$ defined by

$$z=\frac{4w}{(1+w)^2}, \qquad w=\frac{1}{z}\left(-z+2-2(1-z)^{\frac{1}{2}}\right).$$

The image of $|z|<1/h$ is contained in the disk $|w|<h\left(2-1/h-2(1-1/h)^{\frac{1}{2}}\right)$ and, therefore,

$$\Phi(h)=\mu(W_{1/h})>\frac{1}{h\left(\left(2-\frac{1}{h}\right)-2\left(1-\frac{1}{h}\right)^{\frac{1}{2}}\right)}>\frac{4h}{1+\frac{1}{2(h-1)}}.$$

(d) The relations are obtained directly from (e) and (c).

(e) Let $[a,b]$ be the interval $\{z|a\le\mathrm{Re}\, z\le b,\ \mathrm{Im}\, z=0\}$. Concerning the changes of moduli or extremal length under reflections, we shall use the identities in Appendix I.I. By reflecting D_h about $|z|=1$, we obtain

$$2\log\mu(D_h)=\log\mu(\hat{D}_h),$$

where $\hat{D}_h = C([0, 1/h] \cup [h, \infty])$. If $\hat{\Gamma}^*$ and Γ^* are the families of curves joining $[0, 1/h]$ and $[h, \infty]$ in $\hat{D}_h$ and in $\operatorname{Im} z > 0$, respectively, then

$$\log \mu(\hat{D}_h) = 2\pi\, \lambda(\hat{\Gamma}^*) = \pi\, \lambda(\Gamma^*).$$

On the other hand, if $D'_h = C([-\infty, 0] \cup [1/h, h])$ and $\hat{\Gamma}$ is the family of closed curves in D'_h separating $[-\infty, 0]$ from $[1/h, h]$, then

$$\log \mu(D'_h) = \frac{2\pi}{\lambda(\hat{\Gamma})} = \frac{\pi}{\lambda(\Gamma^*)}.$$

Clearly D'_h and $C([0, 1/h'], [h', \infty]) = \hat{D}_{h'}$, where $h' = h(h^2-1)^{-\frac{1}{2}}$, are conformally equivalent, and we have

$$2 \log \mu(D_{h'}) = \log \mu(D'_h).$$

On combining these four relations, we obtain the equality $\log \mu(D_h) = \pi^2/4 \log \mu(D_{h'})$.

2 F. Proof of Theorem 2 C. The following proof is patterned after that of Koebe's quarter theorem (Jenkins [1, p. 26ff.]). By 2 D.(a), there exists an h with $\Phi(h) = \mu = \mu(W)$. We may assume that

$$W = \{z \mid 1 < |z| < \mu\}.$$

Let f be the conformal mapping of W onto D_h such that $f(\mu) = h$, and let $g = f^{-1}$. By reflecting about the circle $|z| = \mu$, we see that f is single-valued and analytic on $1 < |z| < \mu^2$. The function g is two-valued and analytic on $\{z \mid 1 < |z| < \infty\} - \{h\}$, with single-valued $|g'/g|$. We are going to derive a contradiction by assuming the existence of an $F \in \mathscr{S}_1(W)$ such that

$$F \neq e^{i\alpha} f, \quad m_\gamma[F] \le h,$$

where α is a real constant.

By the assumption, there exists a point $e^{i\theta} h$ which does not belong to $F(W)$. We may assume $\theta = 0$, that is,

$$h \notin F(W).$$

Let Γ be the family of circles $|z| = r$, $1 < r < \mu$. Denote by c_r and C_r the images of $|z| = r$ under f and F, and consider the families Γ_f and Γ_F of these images. Clearly,

$$\lambda(\Gamma) = \lambda(\Gamma_f) = \lambda(\Gamma_F) = \frac{2\pi}{\log \mu}.$$

On the w-plane, consider the metric

$$\rho(w) = \begin{cases} |g'(w)/g(w)| & \text{on } D_h \cap F(W), \\ 0 & \text{elsewhere}. \end{cases}$$

We have

$$\int_{C_r} \rho\,|dw| \geq 2\pi, \qquad \iint_{F(W)} \rho^2\,du\,dv \leq 2\pi \log\mu.$$

Furthermore, the assumption $F \neq e^{i\alpha} f$ shows that there exists an r with $c_r \neq C_{r'}$ for every r', and, consequently, an r_0 such that

$$\int_{C_{r_0}} \rho\,|dw| > 2\pi.$$

It is then possible to find $\eta > 0$ and $\delta > 0$ such that

$$\int_{C_r} \rho\,|dw| \geq 2\pi + \delta$$

for all r with $r_0 - \eta < r < r_0 + \eta$. We denote by A the union of the C_r, $r_0 - \eta < r < r_0 + \eta$, and modify ρ on A by introducing another density ρ_1 as follows:

$$\rho_1 = \begin{cases} \dfrac{2\pi\,\rho}{2\pi + \delta} & \text{on } A, \\ \rho & \text{elsewhere}. \end{cases}$$

For the family Γ_F, we obtain

$$\lambda(\Gamma_F) \geq \frac{(2\pi)^2}{\iint_{F(W)} \rho_1^2\,du\,dv} = \frac{(2\pi)^2}{\iint_{F(W)} \rho^2\,du\,dv - \dfrac{\delta}{2\pi + \delta} \iint_A \rho^2\,du\,dv}$$
$$> \frac{(2\pi)^2}{\iint_{F(W)} \rho^2\,du\,dv} \geq \frac{2\pi}{\log\mu},$$

in violation of the previous identity.

2 G. Extremal Region of Teichmüller. We return to the family $\mathscr{S}_{\gamma'\gamma}(W)$ for a doubly connected region W.

Given h with $0 < h < \infty$, we call

$$\tilde{D}_h = C([-1, 0] \cup [h, \infty))$$

an *extremal region of* Teichmüller [1] (cf. Ahlfors [4]).

Theorem. *If a doubly connected region W has a finite modulus $\mu(W)$, then a function minimizes $m_\gamma[F]/M_{\gamma'}[F]$ in $\mathscr{S}_{\gamma'\gamma}(W)$ if and only if a constant multiple of it maps W conformally onto the extremal region $\tilde{D}_h$ of* Teichmüller *with $\mu(W) = \mu(\tilde{D}_h)$.*

The proof will be given in 2 H.

The function
$$\Psi(h)=\mu(\tilde{D}_h), \qquad 0<h<\infty,$$
has the following properties:

(a) $\Psi(h)=\Phi\big(1+2h(1+(1+h^{-1})^{\frac{1}{2}})\big)=\Phi\big((1+h)^{\frac{1}{2}}\big)^2$,

(b) $\lim_{h\to\infty}\Psi(h)/h=16$,

(c) $\Psi(h)<16h+8$,

(d) $\lim_{h\to 0}\log\Psi(h)\log(1/h)=\pi^2$,

(e) $\Psi(1)=e^{\pi}$.

For the proof, we use the notation adopted in 2 E. The region $\tilde{D}_h$ is conformally equivalent to $C([0,1]\cup[1+h,\infty])$ and to
$$C\big([0,(1+h)^{-\frac{1}{2}}]\cup[(1+h)^{\frac{1}{2}},\infty]\big)=\hat{D}_{\sqrt{1+h}}.$$

For this reason, $\mu(\tilde{D}_h)=\mu(\hat{D}_{\sqrt{1+h}})=\mu(D_{\sqrt{1+h}})^2$ and $\Psi(h)=\Phi\big((1+h)^{\frac{1}{2}}\big)^2$, which is the second part of (a). This in turn gives (b) and (d):
$$\lim_{h\to\infty}\frac{\Psi(h)}{h}=\lim_{h\to\infty}\left(\frac{\Phi\big((1+h)^{\frac{1}{2}}\big)}{(1+h)^{\frac{1}{2}}}\right)^2=16,$$
$$\lim_{h\to 0}\log\Psi(h)\log\frac{1}{h}=2\lim_{h\to 0}\log\Phi\big((1+h)^{\frac{1}{2}}\big)\log\frac{1}{h}$$
$$=2\lim_{k\to 1}\log\Phi(k)\log\frac{1}{k^2-1}=\pi^2.$$

Next, we map the complement of $[-1,0]$ onto $|w|>1$ by $z=(w-1)^2/4w$, that is, $w=1+2z\big(1+(1+z^{-1})^{\frac{1}{2}}\big)$. Then $\tilde{D}_h$ in the z-plane is mapped onto $D_{h'}$ with $h'=1+2h\big(1+(1+h^{-1})^{\frac{1}{2}}\big)$, and we have the first part of (a). As a consequence, we obtain (c):
$$\Psi(h)<4\left(1+2h\left(1+\left(1+\frac{1}{h}\right)^{\frac{1}{2}}\right)\right)=4+8h+8h\left(1+\frac{1}{h}\right)^{\frac{1}{2}}<16h+8.$$

To prove (e), note that the upper half of $\tilde{D}_1$ can be mapped conformally onto a square of side π in such a way that the points $-1, 0, 1$, and ∞ correspond to the vertices $i\pi$, 0, π, and $\pi(1+i)$. The square in turn is mapped onto the upper half of the annulus $1<|w|<e^{\pi}$ by the exponential function. On reflecting about the real axis, we see that $\tilde{D}_1$ and $1<|w|<e^{\pi}$ are conformally equivalent.

Remark. In Beurling-Ahlfors [1, p. 131], the behavior of Ψ was described in more detail than is given by (d). Komatu [1, pp. 39 ff.] and Ahlfors [4, p. 46] gave an explicit expression for Ψ.

2 H. Proof of Theorem 2 G. We normalize F by $M_{\gamma'}[F]=1$ and prove that

$$m_\gamma[F]\geq h$$

with $\mu(W)=\Psi(h)$ and that equality holds if and only if $e^{i\alpha}F$ maps W onto $\tilde{D}_h$ for a suitable real α. We write m for $m_\gamma[F]$.

For the region $F(W)$ in the w-plane, let $X_{F(\gamma)}$ and $X_{F(\gamma')}$ be the components of $CF(W)$ corresponding to $F(\gamma)$ and $F(\gamma')$ (cf. VI.1 D). By rotating, if necessary, we obtain

$$m\in F(\gamma).$$

Note that $\infty\in X_{F(\gamma)}\subset\{w\,|\,|w|\geq m\}$ and $0\in X_{F(\gamma')}\subset\{w\,|\,|w|\leq 1\}$.

Map the w-plane by $\tilde{w}=1/w$ and let $\tilde{W}$, $\tilde{X}$, and $\tilde{X}'$ be the images of $F(W)$, $X_{F(\gamma)}$, and $X_{F(\gamma')}$, respectively. Observe that $0\in\tilde{X}\subset\{\tilde{w}\,|\,|\tilde{w}|\leq 1/m\}$, $1/m\in\tilde{X}$, and $\infty\in\tilde{X}'\subset\{\tilde{w}\,|\,|\tilde{w}|\geq 1\}$. Then map the simply connected region $C\tilde{X}'$ onto the unit disk in the ζ-plane by $\zeta=H(\tilde{w})$ with $H(0)=0$ and $H'(0)=c>0$. Koebe's quarter theorem applied to the function $c\,H^{-1}(\zeta)$ gives

$$\tfrac{1}{4}\leq c,$$

where equality holds if and only if $\tilde{X}'$ is a ray $\{\tilde{w}\,|\,|\tilde{w}|\geq 1, \arg\tilde{w}=\text{const}\}$.

We set $b=\max_{\zeta\in H(\tilde{X})}|\zeta|$ and obtain by Koebe's distortion theorem applied to $c\,H^{-1}(\zeta)$,

$$\frac{c}{m}\leq\frac{b}{(1-b)^2},$$

with equality if and only if $\tilde{X}'=\{\tilde{w}\,|\,|\tilde{w}|\geq 1, \arg\tilde{w}=\pi\}$. Consequently,

$$m\geq\frac{(1-b)^2}{4b},$$

where equality occurs if and only if $\tilde{X}'=\{\tilde{w}\,|\,|\tilde{w}|\geq 1, \arg\tilde{w}=\pi\}$, in which case

$$H^{-1}(\zeta)=\frac{4\zeta}{(1-\zeta)^2}.$$

Next, we apply Theorem 2 C to the $(1/\zeta)$-plane:

$$\frac{1}{b}\geq h',$$

where $\Phi(h')=\mu(W)$ and equality is valid if and only if $H(\tilde{X})$ is a line segment with 0 at an end point. Therefore,

$$m\geq\frac{(h'-1)^2}{4h'},$$

and we have equality if and only if $\tilde{X}'=[-\infty,-1]$, $H^{-1}(\zeta)=4\zeta/(1-\zeta)^2$, and $H(\tilde{X})$ is the above line segment; it is easily seen that $\tilde{X}=[0,1/m]$.

We set $(h'-1)^2/4h'=h$. From $h'>1$, we see that

$$h'=1+2h(1+(1+h^{-1})^{\frac{1}{2}}),$$

whence $\Phi(h')=\Psi(h)$. It follows that $m\geq h$, with equality if and only if $\tilde{X}=[0, 1/h]$ and $\tilde{X}'=[-\infty, -1]$, in which case $F(W)=\tilde{D}_h$.

2 I. Generalization of Theorems 2 C and 2 G. Now let W be an arbitrary plane region of connectivity greater than one. Given two boundary components γ and γ', consider the modulus $\mu_1(\gamma, \gamma')$ of W. We have this extension of Theorems 2 C and 2 G:

THEOREM. *Given an arbitrary plane region, the* GRÖTZSCH *and* TEICHMÜLLER *functions Φ and Ψ provide the following lower bounds:*

(a) *If $c_{1\gamma'}>0$, then*

$$m_\gamma[F]\geq\Phi^{-1}(\mu_1)$$

for every $F\in\mathscr{S}_{\gamma'\gamma}$ such that $F(\gamma')=\{w\,|\,|w|=1\}$.

(b) *For arbitrary γ and γ',*

$$\frac{m_\gamma[F]}{M_{\gamma'}[F]}\geq\Psi^{-1}(\mu_1)$$

in $\mathscr{S}_{\gamma'\gamma}$.

These statements reduce to Theorems 2 C and 2 G in the same fashion as Theorem VII. 2 F reduced to Koebe's quarter theorem. In this way we can also obtain, under the assumption $\mu_1<\infty$, functions satisfying the above equalities. The proof is left to the reader.

2 J. Proof of Theorem 1 I. Suppose there are two modulus functions, say q_1 and $\tilde{q}_1$. We shall prove that

$$u=q_1-\tilde{q}_1$$

is constant. To this end, take an analytic Jordan curve $\alpha\subset W$ separating γ and γ'. We shall show that either $\max_W u$ or $\min_W u$ is taken at a point on α; then clearly $u=\text{const}$.

Let U and U' be the components of $W-\alpha$ which are boundary neighborhoods of γ and γ'. It suffices to prove that

$$|u|\leq\max_\alpha|u| \tag{1}$$

on U and U'.

First assume $c_{1\gamma'}>0$ and $c_{1\gamma}=0$, and normalize q_1 and $\tilde{q}_1$ so that $Q_1=\exp(q_1+i\,q_1^*)$ and $\tilde{Q}_1=\exp(\tilde{q}_1+i\,\tilde{q}_1^*)$ map γ' onto the unit circle $|w|=1$ (cf. Theorem 1 E). Then $\lim u(z_n)=0$ for every sequence of points $z_n\in U'$ with accumulation points only on γ'. Moreover, $L_1\,u=u$ on $\beta_{U'}-\gamma'$, and (1) follows on U'. On the other hand, γ is weak and $L_1\,u=u$ on $\beta_U-\gamma$.

As in VII.2 H, the maximum principle in X.1 G implies that (1) satisfied on U, provided u is bounded on U.

In order to verify the boundedness, take n so large that $Q_1(\alpha)$ contained in $|w|<n$. By IX.2 B and 2 C, $R_n = Q_1(W) \cap \{w \,|\, |w|<n\}$ is extremal circular slit annulus $1<|w|<n$; therefore, its modulus is equ to n. On applying Theorem 2 I to $\tilde{Q}_1 \circ Q_1^{-1}$ on R_n, we see that the ima of $|w|=n$ under this mapping lies outside of the disk $|w| \leq \Phi^{-1}(n)$. L $\alpha_n \subset U$ be a curve such that $Q_1(\alpha_n)$ is contained in $n<|w|<n+1$ a separates $Q_1(\gamma)$ from $|w|=1$. We obtain

$$u \leq \log\big((n+1)(\Phi^{-1}(n))^{-1}\big)$$

on α_n. Since $(n+1)(\Phi^{-1}(n))^{-1} \to 4$ as $n \to \infty$, we conclude that

$$u \leq \max\big(\max_{\alpha} u, \log 4\big),$$

that is, u is bounded from above on U. On interchanging Q_1 and $\tilde{Q}_1$, see that u is bounded from below on U.

Next, assume $c_{1\gamma'} = c_{1\gamma} = 0$, in which case γ' and γ are both weak. T essential part of the proof is the application of Theorem X.1 G, whi again reduces the problem to verifying the boundedness of u on U and U' The interchange of Q_1 and $\tilde{Q}_1$ shows that it suffices to prove only t boundedness from above of u on U and U'.

Take a, b, $\tilde{a}$, and $\tilde{b}$ such that

$$a \leq |Q_1| \leq b, \quad \tilde{a} \leq |\tilde{Q}_1| \leq \tilde{b}$$

on α. The region $Q_1(W) \cap \{w \,|\, b<|w|<n\}$ with $n>b$ is an extrem circular slit annulus. By Theorem 2 I, the image of $|w|=n$ und $\tilde{Q}_1 \circ Q_1^{-1}$ lies outside of the disk $|w| \leq \tilde{a}\,\Psi^{-1}(n/b)$. Using the fact th $\lim_{h \to \infty} h/\Psi^{-1}(h) = 16$, we conclude as before that

$$u \leq \max\left(\max_{\alpha} u, \log \frac{16b}{\tilde{a}}\right) \quad \text{on } U.$$

On U', we consider $\tilde{Q}_1(W) \cap \{w \,|\, 1/n<|w|<\tilde{a}\}$ and obtain analogous the same inequality. Consequently, u is bounded on both U and U'.

The proof of Theorem 1 I is herewith complete.

Chapter IX

Extremal Slit Regions

In Chapters VI – VIII, we singled out those slit regions which can be image regions under conformal mappings P_ν or Q_ν. For a more detailed study of such regions, called extremal, we shall make use of normal operators and extremal length.

1. Extremal Slit Plane

1 A. Definition. A plane region W with $\infty \in W$ is said to be an *extremal vertical slit plane* if

$$P_1(z; \infty) \equiv z \tag{1}$$

on W. Similarly, an *extremal horizontal slit plane* is defined by

$$P_0(z; \infty) \equiv z. \tag{2}$$

A plane region W with $0, \infty \in W$ is called an *extremal circular slit plane* if

$$P_1(z; 0, \infty) \equiv z \tag{3}$$

on W. Analogously, an *extremal radial slit plane* is one with

$$P_0(z; 0, \infty) \equiv z. \tag{4}$$

These regions are vertical, horizontal, circular, and radial slit planes in the sense of VI.1 E, as is immediately seen by Theorems VI.3 A and 6 A. For a region of finite connectivity, the extremality imposes no restriction, whereas for a general region it does: the area of the complementary set must vanish (Theorems VI.3 A and 6 A). This condition, however, is not sufficient (Theorem 4 G.(a) below).

Remark. Extremal slit regions were introduced by KOEBE [7], who called them minimal. His definition differs from the one above and will be given in Theorem 1 B. Extremal regions were later studied by GRÖTZSCH [4, 8], DE POSSEL [4], CARTWRIGHT [1], and recently by STREBEL [2, 3], RENGGLI [1], JENKINS [1], SAKAI [1], REICH-WARSCHAWSKI [1], SUITA [2, 3], OIKAWA [3], MARDEN-RODIN [1], and others.

1 B. Characterization by Normal Operators. Using the characterization of $P_1(z;\infty)$ by VI.2 A.(1), we obtain: A region W in the extended $(z=x+iy)$-plane with $\infty\in W$ is an extremal vertical slit plane if and only if

$$L_1 x = x \quad \text{on } \partial W; \tag{5}$$

more precisely, (5) holds on a boundary neighborhood U of $\beta=\partial W$ with $\infty\notin\bar{U}$. The operator L_1 is for **Q**. Similarly, extremal horizontal, circular, and radial slit planes are characterized by

$$L_0 x = x \quad \text{on } \partial W, \tag{6}$$

$$L_1(\log|z|) = \log|z| \quad \text{on } \partial W, \tag{7}$$

$$L_0(\log|z|) = \log|z| \quad \text{on } \partial W. \tag{8}$$

By VI.1 B, conditions (5) and (6) are equivalent, respectively, to

$$L_0 y = y \quad \text{on } \partial W, \tag{5'}$$

$$L_1 y = y \quad \text{on } \partial W. \tag{6'}$$

We return to the definition of the operator L_1 for **Q** (I.3 C). If $\alpha=(\partial U)\cap W$, then (5) holds if and only if

$$\int_U dx\wedge\omega = \int_\alpha x\,\omega$$

for every C^1-differential ω on $U\cup\alpha$ which is semiexact relative to α. If α consists of a single contour (such a U always exists), then ω is necessarily exact, $\omega=dh$, and we obtain

$$\iint_U \frac{\partial h}{\partial y}\,dx\,dy = \int_\alpha x\,dh = -\int_\alpha h\,dx. \tag{9}$$

We extend the function h of class C^2 on $U\cup\alpha$ to W so that $h\in C^2(W)$ and $h=0$ on $W-U'$, where U' is an end of W with $W\cap\bar{U}\subset U'$. Then

$$\iint_W \frac{\partial h}{\partial y}\,dx\,dy = 0.$$

Conversely, it is easily verified that this implies (9).

Similar identities are derived from (6), (7), and (8). The above reasoning, due to Oikawa [3], has thus lead us to the following properties, adopted as definitions by Koebe [7]:

Theorem. *A region W with $\infty\in W$ in the extended $(z=x+iy)$-plane is an extremal vertical slit plane if and only if*

$$\iint_W \frac{\partial h}{\partial y}\,dx\,dy = 0 \tag{10}$$

for every $h \in C^2(W)$ *which vanishes identically on a neighborhood of* ∞. *Similarly, an extremal horizontal slit plane is a region* W *with*

$$\iint_W \frac{\partial h}{\partial x}\, dx\, dy = 0. \tag{11}$$

A region W *with* $0, \infty \in W$ *in the extended* $(z = r\, e^{i\theta})$*-plane is an extremal circular slit plane if and only if*

$$\iint_W \frac{\partial h}{\partial \theta} \frac{dr\, d\theta}{r} = 0 \tag{12}$$

for every $h \in C^2(W)$ *which vanishes identically on some neighborhoods of* 0 *and* ∞. *An extremal radial slit plane is characterized similarly by*

$$\iint_W \frac{\partial h}{\partial r}\, dr\, d\theta = 0. \tag{13}$$

1 C. Elementary Properties. We shall present a number of basic properties of extremal slit planes. Whereas it may not be easy to derive them directly from our definitions in 1 A, they are almost trivial consequences of Theorem 1 B. The proofs are left to the reader.

For short, we shall call a closed bounded set an *extremal set of vertical, horizontal, circular, or radial slits* if its complement is a corresponding extremal slit plane.

(a) *If* E *is an extremal set of vertical, horizontal, circular, or radial slits, then so is any closed subset of* E.

(b) *If* $E_1, \ldots, E_n$ *are disjoint extremal sets of vertical, horizontal, circular, or radial slits, then so is their union.*

(c) *If* E *is an extremal set of vertical (or horizontal) slits, then so is its image under an affine transformation* $x + i\,y \to a\,x + i\,b\,y + c$, *with real* b *(*a*, respectively) and* $\operatorname{Re} a\,\bar{b} \neq 0$.

(d) *If* E *is an extremal set of circular or radial slits, then so is its image under each of the transformations* $z \to c\,z$ *and* $z \to c\,\bar{z}, c \neq 0$. *If a set* E *contained in a sector* $\theta_0 < \arg z < \theta_1$ *is an extremal set of circular slits, then so is its image under the transformation* $r\,e^{i\theta} \to r\,e^{ia\theta}$, *where* a *is a non-zero real number with* $|a|(\theta_1 - \theta_0) < 2\pi$.

(e) *Rotation about the origin by* $\pi/2$ *transforms an extremal set of horizontal slits into one of vertical slits, and vice versa.*

(f) *Let* $E \subset \{z \,|\, |\operatorname{Im} z| < \pi\}$ *and let* E' *be its image under the transformation* $z \to e^z$. *Then* E' *is an extremal set of circular or radial slits if and only if* E *is an extremal set of vertical or horizontal slits, respectively.*

1 D. Characterization by Extremal Length. Let W be a plane region containing ∞. For a rectangle

$$R=\{z\,|\,a<\operatorname{Re} z<a', b<\operatorname{Im} z<b'\}, \qquad CW\subset R,$$

we shall consider the family Γ_R (or Γ_R^*) of locally rectifiable curves joining horizontal (or vertical, resp.) sides of R in $R\cap W$.

If W is a plane region containing 0 and ∞, we take an annulus

$$A=\{z\,|\,a<|z|<b\}, \qquad CW\subset A,$$

and recall that $\mu(A)=b/a$ (VIII.2 A). We shall be concerned with the family Γ_A of rectifiable closed curves in $A\cap W$ separating the components of ∂A and the family Γ_A^* of locally rectifiable curves joining $|z|=a$ and $|z|=b$ in $A\cap W$.

THEOREM. *A plane region W with $\infty\in W$ is an extremal vertical slit plane if there exists an R with*

$$\lambda(\Gamma_R)=\frac{b'-b}{a'-a}. \tag{14}$$

Conversely, if W is an extremal slit plane, then (14) *holds for every R. An extremal horizontal slit plane is characterized similarly by*

$$\lambda(\Gamma_R^*)=\frac{a'-a}{b'-b}. \tag{15}$$

A plane region W with $0, \infty\in W$ is an extremal circular slit plane if there exists an A with

$$\frac{2\pi}{\lambda(\Gamma_A)}=\log\mu(A). \tag{16}$$

Conversely, if W is such a region, then (16) *is satisfied for every A. Similarly, W is an extremal radial slit plane if and only if*

$$2\pi\,\lambda(\Gamma_A^*)=\log\mu(A). \tag{17}$$

We shall give the proof in $2F$ by making use of an extremal circular slit annulus.

Remark. The theorem, in different terminology, is due to GRÖTZSCH [4]. The above form was presented by JENKINS [1], who assumed, as the sufficient condition, the validity of (14) or (15) for *all* R. The replacing of "all" by "some" is due to SUITA [2].

1 E. Vanishing of the Span. From Theorem VI. 2 D we immediately obtain the following proposition:

A plane region W with $\infty \in W$ has vanishing span if and only if it is simultaneously an extremal vertical and horizontal slit plane or, equivalently (provided $0, \infty \in W$), an extremal circular and radial slit plane.

Accordingly, W has vanishing span if and only if

$$\lambda(\Gamma_R) = \frac{1}{\lambda(\Gamma_R^*)} = \frac{b'-b}{a'-a}.$$

This criterion is due to AHLFORS-BEURLING [1]. It is equivalent to the simultaneous validity of

$$\lambda(\Gamma_A) = \frac{2\pi}{\log \mu(A)} \quad \text{and} \quad \lambda(\Gamma_A^*) = \frac{\log \mu(A)}{2\pi}.$$

2. Extremal Circular Slit Disk and Annulus

2 A. Definition. A circular slit disk W with outer boundary γ is said to be *extremal* if

$$P_{1\gamma}(z; 0) = z$$

on W.

The radius is necessarily equal to $1/c_{1\gamma}(0)$. If the radius is finite, this equality is also sufficient:

THEOREM. *Let W be a region with*

$$0 \in W \subset \{z \mid |z| < r\}, \qquad r < \infty,$$

and let γ be its outer boundary. Then W is an extremal circular slit disk of radius r if and only if $r = 1/c_{1\gamma}(0)$.

Proof. If $r = 1/c_{1\gamma}(0)$, apply Theorem VII. 2 B to the identity function $I(z) = z$. It satisfies $M[I] \leq r = 1/c_{1\gamma}(0)$, so that $I(z) \equiv P_{1\gamma}(z; 0)$.

An *extremal circular slit annulus*, defined similarly by

$$Q_1(z; \gamma, \gamma') \equiv \text{const} \cdot z,$$

is characterized in analogy with the above theorem.

In the sequel, we shall be concerned only with disks; annuli can be treated in a similar manner.

2 B. A Characterization. The following result is due to SUITA [2] and OIKAWA-SUITA [2]:

THEOREM. *A plane region W with*

$$0 \in W \subset \{z \mid |z| < r\}, \qquad r \leq \infty,$$

is an extremal circular slit disk of radius r if and only if every compact subset of

$$E = CW \cap \{z \mid |z| < r\}$$

is an extremal set of circular slits.

If the outer boundary of W coincides with the circle $|z| = r$ and if it is an isolated boundary component of W, then the result is immediately obtained by combining criterion 1 B.(7) with VII. 1 C.(a).

There is, of course, an analogous criterion for extremal circular slit annuli.

Proof. Let γ be the outer boundary of W. The necessity is easily deduced from 1 B.(7), Theorem VII. 2 E, and the fact that $L_1(\log|z|) = \log|z|$ on $\partial W - \gamma$.

Conversely, suppose every compact subset of E is an extremal set of circular slits. Then it is not difficult to see that γ is the circle $|z| = r$. We leave this part of the proof to the reader.

If $r = \infty$, we repeat the proof of Theorem VII. 2 H; the hypothesis implies $L_1(\log|z|) = \log|z|$ on $\partial W - \gamma$, from which we conclude that $P_{1\gamma}(z; 0) \equiv z$. Hence W is an extremal circular slit disk of infinite radius.

If $r < \infty$, we use Theorem 2 A. It suffices to find an exhaustion $0 \in \Omega_n \to W$ by regular subregions Ω_n such that

$$\lim_{n \to \infty} \frac{1}{c_n} \geq r, \tag{1}$$

where $c_n = c_{1\gamma(\Omega_n)}(0)$, considered for Ω_n.

As mentioned previously, γ coincides with the circle $|z| = r$. Using the definition of a boundary component in I. 1 B and VI. 1 D, it is possible to find, for every $n = 1, 2, \ldots$, an analytic Jordan curve $\gamma_n \subset W$ which is contained in the annulus $r - 1/n < |z| < r$, separates γ from 0, and is such that the closure of the region $(\operatorname{Int} \gamma_n)$ enclosed by γ_n is contained in $(\operatorname{Int} \gamma_{n+1})$. By assumption,

$$E_n = \overline{E \cap (\operatorname{Int} \gamma_n)}$$

is an extremal set of circular slits, so that, as stated immediately after the theorem,

$$W_n = \{z \mid |z| < r\} - E_n$$

is an extremal circular slit disk of radius r. Exhaust W_n by $0 \in V_m^n \to W_n (m \to \infty)$ towards $\partial W_n - \gamma$. We may assume that $\gamma_n \subset V_m^n$. The functions $P_m^n = P_{1\gamma}(z; 0)$ for V_m^n converge to the identity function z uniformly on γ_n as $m \to \infty$. Thus there exists an $m = m(n)$ such that $P_m^n(\gamma_n)$ is contained in $r - 1/n < |w| < r$. Take $m = m(n)$ so large that the regions

$$\Omega_n = V_{m(n)}^n \cap (\operatorname{Int} \gamma_n), \qquad n = 1, 2, \ldots,$$

satisfy $\bar{\Omega}_n \subset \Omega_{n+1}$ and $W = \bigcup_{n=1}^{\infty} \Omega_n$. This is the desired exhaustion. In fact, the image $R_n = P_{m(n)}^n(\Omega_n)$ contains $R_n \cap \{w \mid |w| < r - 1/n\}$. The latter is a circular slit disk of radius $r - 1/n$ with only a finite number of slits. Consequently, $r - 1/n \leq 1/c_n$, and we have (1).

2 C. Relations between Circular Slit Planes, Disks, and Annuli. Because of the results in 1 C and 2 B, relations between circular slit planes, disks, and annuli are fairly simple. We shall give two typical examples.

(a) *Let W be an extremal circular slit plane, or an extremal circular slit disk of radius $r \leq \infty$, and let r_1 and r_2 be such that $0 < r_1 < r_2 < \infty$ or $0 < r_1 < r_2 < r$, respectively. Then*

$$W_0 = W \cap \{z \mid r_1 < |z| < r_2\}$$

is an extremal circular slit annulus.

The connectedness of W_0 is verified by Lemma 2 in VI.1 E.

(b) *Let W be a region with $0 \in W \subset \{z \mid |z| < r\}, r \leq \infty$. Given an r_1 with $0 < r_1 < r$ and*

$$\{z \mid |z| = r_1\} \subset W, \tag{2}$$

suppose that $W \cap \{z \mid |z| < r_1\}$ is an extremal circular slit disk of radius r_1 and $W \cap \{z \mid r_1 < |z| < r\}$ is an extremal circular slit annulus $r_1 < |z| < r$. Then W is an extremal circular slit disk of radius r.

Assumption (2) may be omitted (2 E).

2 D. Characterization by Extremal Length. Let W be a plane region with

$$0 \in W \subset \{z \mid |z| < r\}, \qquad r < \infty,$$

and let γ be its outer boundary. Set $N_\varepsilon = \{z \mid |z| < \varepsilon\}$. For every $\varepsilon > 0$ with $\bar{N}_\varepsilon \subset W$, denote by Γ_ε and $\tilde{\Gamma}_\varepsilon^*$ the families Γ and $\tilde{\Gamma}^*$ in IV.2 B for the region $W-\overline{N}_\varepsilon$ and its subboundaries γ and ∂N_ε; the family Γ_ε may be limited as in VIII.1 C.

THEOREM. *W is an extremal circular slit disk of radius $r<\infty$ if and only if one of the following conditions is satisfied for some (or equivalently every) $\varepsilon>0$ with $\overline{N}_\varepsilon\subset W$:*

(a) $2\pi/\lambda(\Gamma_\varepsilon)=\log(r/\varepsilon)$,

(b) $2\pi\,\lambda(\tilde{\Gamma}_\varepsilon^*)=\log(r/\varepsilon)$,

(c) $\int_c |z|^{-1}|dz|\geq\log(r/\varepsilon)$ *for almost all* $c\in\tilde{\Gamma}_\varepsilon^*$.

The case $r=\infty$ is excluded.

Condition (a) and its analogue for annuli are explicitly stated in REICH-WARSCHAWSKI [1] and SAKAI [1]. Conditions (b), (c), and their counterparts for annuli are due to MARDEN-RODIN [1].

Proof. By 2 C.(b), W is extremal if and only if $W-\overline{N}_\varepsilon$ is extremal for some (or equivalently every) ε with $\overline{N}_\varepsilon\subset W$. Thus (a) and (b) are direct consequences of Theorem 2 A for an annulus and Theorem IV.2 C.

The necessity of (c) is a simple consequence of Theorem IV.2 C. Conversely, if (c) holds, then $\rho|dz|=(\log(r/\varepsilon))^{-1}|z|^{-1}|dz|$ belongs to the family $\overline{\mathbf{P}}(\tilde{\Gamma}_\varepsilon^*)$ (IV.2 A). Since $A(\rho)\leq 2\pi/\log(r/\varepsilon)$, we have $2\pi\,\lambda(\tilde{\Gamma}_\varepsilon^*)\geq\log(r/\varepsilon)$. Evidently, $2\pi\,\lambda(\tilde{\Gamma}_\varepsilon^*)\leq\log(r/\varepsilon)$, and (b) follows.

2 E. Redundancy. That assumption (2) can be dispensed with is seen as follows. In the case $r<\infty$, divide W by the circle $|z|=r_1$ into a disk and an annulus and consider the corresponding families Γ_ε' and Γ. Then $2\pi/\lambda(\Gamma_\varepsilon)\geq 2\pi/\lambda(\Gamma_\varepsilon')+2\pi/\lambda(\Gamma)=\log(r_1/\varepsilon)+\log(r/r_1)=\log(r/\varepsilon)$. In the case $r=\infty$, the region $W\cap\{z\,|\,|z|<r'\}$ is extremal for every $r'<\infty$ as above, and, therefore, W is extremal by Theorem 2 B.

2 F. Proof of Theorem 1 D. The proof of 1 D.(16) is immediate. By 2 C, W is an extremal circular slit plane if and only if $W\cap A$ is an extremal circular slit annulus. The latter property is equivalent to 1 D.(16) because of the analogue for annuli of (a) in Theorem 2 D.

Property 1 D.(17) is reduced to 3 C below and to (b) of Theorem 3 D in the same fashion.

Next suppose W is an extremal vertical slit plane. To derive 1 D.(14) for all R, it suffices to show that

$$\lambda(\Gamma_R)\leq\frac{b'-b}{a'-a},$$

since the inequality in the opposite direction is trivial. We may assume that $b'-b=2\pi$, since homothetic stretching changes neither $\lambda(\Gamma_R)$ nor the extremality (1 C.(c)). The image E' of $E = CW$ under $w = \text{const} \cdot e^z$ is an extremal set of circular slits (1 C.(f)) contained in the annulus $1 < |w| < e^{a'-a}$, provided the constant is suitably chosen. Therefore, $W' = \{w \mid 1 < |w| < e^{a'-a}\} - E'$ is an extremal circular slit annulus. Let Γ_R' be the image of Γ_R and let Γ be the family of curves in W' separating $|w| = 1$ from $|w| = e^{a'-a}$. By the extremality of W', Theorem 2 D applies and gives $\lambda(\Gamma) = (b'-b)/(a'-a)$. Since every element of Γ contains an element of Γ_R' as a subarc, we see that $\lambda(\Gamma_R') \leq \lambda(\Gamma)$. Hence we obtain 1 D.(14).

Conversely, suppose 1 D.(14) holds for an R. By the same reasoning as above, we may assume that $b'-b=\pi$ and prove that the image E' of E under the mapping $w = \text{const} \cdot e^z$ is an extremal set of circular slits. If we suitably choose the constant of the mapping, the set E' will be in $\{w \mid 1 < |w| < e^{a'-a}, 0 < \arg w < \pi\}$. Let E'' be the reflection of E' about the real axis and let $\hat{E} = E' \cup E''$. It suffices to show that $\hat{E}$ is an extremal set of circular slits (1 C.(a)). If Γ_R' is the image of Γ_R and if Γ is the family of closed curves in $\Delta = \{w \mid 1 < |w| < e^{a'-a}\} - \hat{E}$ separating $|w| = 1$ from $|w| = e^{a'-a}$, then

$$\lambda(\Gamma) = 2\lambda(\Gamma_R')$$

(see Appendix I.I). Since $\lambda(\Gamma_R') = \pi/(a'-a)$ by assumption 1 D.(14), we obtain

$$\frac{2\pi}{\lambda(\Gamma)} = a'-a = \log e^{a'-a}.$$

This shows, by Theorem 2 D, that Δ is extremal, that is, $\hat{E}$ is an extremal set of circular slits. We conclude that W is an extremal vertical slit plane.

To prove the corresponding result for horizontal slit planes, it is preferable not to use a mapping into radial slit annuli. By means of a mapping into circular slit annuli by $w = \text{const} \cdot e^{iz}$, the proof becomes completely analogous.

3. Extremal Radial Slit Disk and Annulus

3 A. Definition. By an *extremal radial slit disk of radius* $r < \infty$, we mean an incised radial slit disk of radius r (VII.4 A) on which $P_{0\gamma}(z; 0) \equiv z$, with γ the outer boundary.

To be exact, we should say an extremal "incised" radial slit disk, but this is tacitly understood.

Here we concern ourselves only with the case of finite radius, for otherwise $P_{0\gamma}$ may not be uniquely defined. The case of infinite radius will be considered in 3 G – 3 K.

An *extremal radial slit annulus* of inner radius $r' > 0$ and outer radius $r < \infty$ is defined similarly; incisions may occur. We shall call it an extremal radial slit annulus $r' < |z| < r$.

3 B. A Characterization. If W is an extremal radial slit disk of radius $r<\infty$, and if γ is its outer boundary, then they have the following properties (cf. Theorem VII.4 A and Proposition VII.4 C):

(a) $1/c_{0\gamma}(0)=r$.

(b) For any r_1 with $0<r_1<r$, the set $CW\cap\{z\,|\,|z|\le r_1\}$ is an extremal set of radial slits.

(c) Any compact subset of $\partial W-\gamma$ is an extremal set of radial slits.

(d) γ consists of the circle $|z|=r$ and possibly a number of segments on rays $\arg z=$const whose total angular measure is zero.

In general, however, none of the conditions alone is sufficient for W to be an extremal radial slit disk. In fact, the disk $|z|<r$ with one incision and no slit satisfies (b)–(d) but is not extremal. As for (a), a counterexample is the disk with one incision and one hole.

Sufficient conditions are obtained by taking combinations as follows:

THEOREM. *Let W be a region such that $0\in W\subset\{z\,|\,|z|<r\}$, $r<\infty$, and let γ be its outer boundary. W is an extremal radial slit disk of radius r if* (a) *and either* (b) *or* (c) *are satisfied.*

In particular, if γ coincides with the circle $|z|=r$, then any one of (a), (b), (c) *is sufficient for W to be an extremal radial slit disk (without incisions) of radius r.*

Condition (d) is then automatically fulfilled.

An analogous result holds for slit annuli.

Proof. If either (b) or (c) is satisfied, the identity function $I(z)=z$ belongs to the family $\mathcal{S}^0_{0\gamma}(W)$ of Theorem VII.3 B. Condition (a) implies $M[I]\le r=1/c_{0\gamma}$, so that $P_{0\gamma}(z;0)\equiv z$ by that theorem.

In particular, if $\gamma=\{z\,|\,|z|=r\}$, we have $m_\gamma[I]=r=1/c_{0\gamma}$ from (a). By Theorem VII.3 C, we obtain $P_{0\gamma}=I$. From (b) or (c), we see that $I\in\mathcal{S}^0_{0\gamma}(W)$ and, therefore, $1/c_{0\gamma}\le M[I]$ by Theorem VII.3 B. On the other hand, $M[I]=m_\gamma[I]\le 1/c_{0\gamma}$ by Theorem VII.3 C. Thus $m_\gamma[I]=1/c_{0\gamma}$, and $P_{0\gamma}(z;0)\equiv z$ as above.

3 C. Relations between Extremal Radial Slit Planes, Disks, and Annuli. As a consequence of Theorem 3 B, we obtain simple relations between radial slit planes, disks, and annuli. For example:

(a) *Let W be an extremal radial slit plane, or an extremal radial slit disk of radius $r<\infty$. Let r_1 and r_2 be such that $0<r_1<r_2<\infty$ or $0<r_1<r_2<r$, respectively. If $W_0=W\cap\{z\,|\,r_1<|z|<r_2\}$ has no incisions, then it is an extremal radial slit annulus.*

Lemma 3 D shows that, if W_0 has incisions, it may not be extremal.

(b) *Let W be a region with $0 \in W$, $\infty \notin W$, and with the circle $|z| = r < \infty$ as its outer boundary. For an r_1 with $0 < r_1 < r$ and $\{z \mid |z| = r_1\} \subset W$, suppose that $W \cap \{z \mid |z| < r_1\}$ is an extremal radial slit disk of radius r_1 and $W \cap \{z \mid r_1 < |z| < r\}$ is an extremal radial slit annulus $r_1 < |z| < r$. Then W is an extremal radial slit disk (without incisions) of radius r.*

Regarding the case in which W has incisions, see also Lemma 3 D.

As an application, we obtain a connection between the result in 1 E and a property of a region which is simultaneously an extremal circular and radial slit disk. Let W be a region with $0 \in W \subset \{z \mid |z| < r\}$ such that the outer boundary γ coincides with the circle $|z| = r < \infty$. Consider all functions $F \in \mathscr{S}_{0\gamma}(W)$ such that the image $F(\gamma)$ is a circle with center at the origin, that is, $M[F] = m_\gamma[F]$. If the radius does not depend on F, we shall say that W is a *rigid disk* (SARIO [13]). By Theorems VII.2 B and VII.4 B, W is a rigid disk if and only if $c_{0\gamma}(0) = c_{1\gamma}(0)$. As is easily seen by considering the identity function $I(z) = z$, this condition is equivalent to the requirement that W be at once an extremal circular and radial slit disk of radius r. Accordingly, we obtain from 1 E:

W is a rigid disk of radius r if and only if every compact subset E of $CW \cap \{z \mid |z| < r\}$ has the property that the region CE has vanishing span.

3 D. Characterization by Extremal Length. Let W be a region with

$$0 \in W \subset \{z \mid |z| < r\}, \qquad r < \infty,$$

and let γ be its outer boundary. Set $N_\varepsilon = \{z \mid |z| < \varepsilon\}$ with $\overline{N}_\varepsilon \subset W$, and denote by $\mu_0(\varepsilon)$ the modulus $\mu_0(\gamma, \partial N_\varepsilon)$ of the region $W_\varepsilon = W - \overline{N}_\varepsilon$.

We begin with the following generalization of the special cases (a) and (b) above:

LEMMA. *W is an extremal radial slit disk of radius r if and only if W_ε is an extremal radial slit annulus $\varepsilon < |z| < r$ for some (or equivalently every) $\varepsilon > 0$ with $\overline{N}_\varepsilon \subset W$.*

Proof. Suppose first that W_ε is an extremal slit annulus for some ε. Then $\log(1/c_{0\gamma}(0)) = \lim_{\varepsilon \to 0}(\log \varepsilon + \log \mu_0(\varepsilon)) \geq \log \varepsilon + \log \mu_0(\varepsilon) = \log r$ by IV.3 G and an analogue of 3 B.(a) for annuli. On the other hand, W clearly satisfies 3 B.(c) and, therefore, the identity function $I(z) = z$ belongs to $\mathscr{S}^0_{0\gamma}(W)$. Since $M[I] \leq r \leq 1/c_{0\gamma}(0)$, we conclude that $P_{0\gamma}(z; 0) \equiv z$ by Theorem VII.3 B, so that W is extremal.

Conversely, we suppose that W is an extremal slit disk and show that W_ε is extremal for every ε. We apply an analogue of Theorem 3 B for annuli. Since 3 B.(c) is evidently satisfied, it suffices to prove 3 B.(a), that is, $\mu_0(\varepsilon) = r/\varepsilon$. Let $\overline{N}_\varepsilon \subset V_n \to W$ be an exhaustion towards γ and set $\gamma_n = \gamma(V_n)$. Since W is extremal, we have $\lim r_n = r$ and $\lim P_n(z) = z$, where

$r_n = 1/c_{0\gamma_n}(0)$ and $P_n = P_{0\gamma_n}(z;0)$, both considered for V_n. Let $\mu_{0n}(\varepsilon)$ be the modulus $\mu_0(\gamma_n, \partial N_\varepsilon)$ of $V_{\varepsilon n} = V_n - \bar{N}_\varepsilon$. By applying (5) and (6) in VIII.1 D, we see that

$$\frac{r_n}{\max\limits_{\partial N_\varepsilon} |P_n|} \le \mu_{0n}(\varepsilon) \le \frac{r_n}{\min\limits_{\partial N_\varepsilon} |P_n|}.$$

As $n \to \infty$, $P_n(z)$ converges to z uniformly on ∂N_ε. Thus $\lim \mu_{0n}(\varepsilon) = r/\varepsilon$. On the other hand, $V_{\varepsilon n}$ exhausts W_ε towards γ. By Theorem IV.1 F, we obtain $\lim \mu_{0n}(\varepsilon) = \mu_0(\varepsilon)$. Consequently, $\mu_0(\varepsilon) = r/\varepsilon$, and we have proved the lemma.

Denote by Γ_ε^* and $\tilde{\Gamma}_\varepsilon$ the families of curves Γ^* and $\tilde{\Gamma}$ in IV.2 B with respect to the region W_ε and its subboundaries γ and ∂N_ε. We have

$$\log \mu_0(\varepsilon) = 2\pi \lambda(\Gamma_\varepsilon^*) = \frac{2\pi}{\lambda(\tilde{\Gamma}_\varepsilon)}.$$

On applying Theorem 3 B for annuli we obtain:

THEOREM. *Under the assumption of either* 3 B.(b) *or* 3 B.(c), *W is an extremal slit disk of radius r if and only if one of the following is satisfied for some (or equivalently every) $\varepsilon > 0$ with $\bar{N}_\varepsilon \subset W$:*

(a) $2\pi \lambda(\Gamma_\varepsilon^*) = \log(r/\varepsilon)$.

(b) $2\pi/\lambda(\tilde{\Gamma}_\varepsilon) = \log(r/\varepsilon)$.

In the case $\gamma = \{z \mid |z| = r\}$, assumptions 3 B.(b) *and* 3 B.(c) *are redundant.*

Property (a) for an isolated boundary component $\gamma = \{z \mid |z| = r\}$ is explicitly stated in REICH-WARSCHAWSKI [1] and SAKAI [1]. Property (b) is due to MARDEN-RODIN [1] (cf. 3 E).

3 E. Logarithmic Length of Curves. We retain the previous notation. The following result is due, in essence, to MARDEN-RODIN [1]:

THEOREM. *W is an extremal radial slit disk of radius r if and only if one of the following conditions is satisfied for some (or equivalently every) $\varepsilon > 0$ with $\bar{N}_\varepsilon \subset W$:*

(a) $\int_c |z|^{-1} |dz| \ge \log(r/\varepsilon)$ *for almost all $c \in \Gamma_\varepsilon^*$, and* $\int_c |z|^{-1} |dz| \ge 2\pi$ *for almost all $c \in \tilde{\Gamma}_\varepsilon$.*

(b) 3 B.(b) *or* 3 B.(c), *and* $\int_c |z|^{-1} |dz| \ge \log(r/\varepsilon)$ *for almost all $c \in \Gamma_\varepsilon^*$.*

(c) $\mu_0(\varepsilon) \le r/\varepsilon$ *and* $\int_c |z|^{-1} |dz| \ge \log(r/\varepsilon)$ *for almost all $c \in \Gamma_\varepsilon^*$.*

(d) $\mu_0(\varepsilon) \ge r/\varepsilon$ *and* $\int_c |z|^{-1} |dz| \ge 2\pi$ *for almost all $c \in \tilde{\Gamma}_\varepsilon$.*

In the case $\gamma = \{z \mid |z| = r\}$, it is sufficient that

$$\int_c \frac{|dz|}{|z|} \ge 2\pi$$

for almost all $c \in \tilde{\Gamma}_\varepsilon$.

Proof. The necessity is a simple consequence of Lemma 3 D and Theorem IV.2 C. We shall show that any one of (a)–(d) for an $\varepsilon>0$ implies that W_ε is extremal.

Let $\rho|dz|=|z|^{-1}|dz|$ and consider the families $\bar{\mathbf{P}}(\Gamma_\varepsilon^*)$ and $\bar{\mathbf{P}}(\tilde{\Gamma}_\varepsilon)$ discussed in IV.2 A. From (a) we see that $(\log(r/\varepsilon))^{-1}\rho|dz|$ belongs to the former and $(2\pi)^{-1}\rho|dz|$ to the latter. Since $A(\rho)\leq 2\pi\log(r/\varepsilon)$, we obtain

$$\begin{aligned}\log\frac{r}{\varepsilon}&\leq\frac{2\pi\left(\log\frac{r}{\varepsilon}\right)^2}{A(\rho)}\\&\leq 2\pi\lambda(\Gamma_\varepsilon^*)=\log\mu_0(\varepsilon)=\frac{2\pi}{\lambda(\tilde{\Gamma}_\varepsilon)}\\&\leq\frac{2\pi A(\rho)}{(2\pi)^2}\leq\log\frac{r}{\varepsilon}.\end{aligned}$$

Thus $A(\rho)=2\pi\log(r/\varepsilon)$ and, therefore, $(\log(r/\varepsilon))^{-1}\rho|dz|$ is the generalized extremal metric for the family Γ_ε^*. Thus by Theorem IV.2 C, the identity function I coincides with $Q_0=Q_0(z;\gamma,\partial N_\varepsilon)$ on W_ε.

Condition (c) implies

$$\log(r/\varepsilon)\geq\log\mu_0(\varepsilon)=2\pi\lambda(\Gamma_\varepsilon^*)\geq 2\pi(\log(r/\varepsilon))^2/A(\rho)\geq\log(r/\varepsilon),$$

so that $I=Q_0$ as above.

From (d) we deduce similarly that $(2\pi)^{-1}\rho|dz|$ is the generalized extremal metric for $\tilde{\Gamma}_\varepsilon$, and, therefore, $I=Q_0$.

The latter half of (b) gives $\mu_0(\varepsilon)\geq r/\varepsilon$ as above. Since the inequality in the opposite direction is a consequence of Strebel's inequality (VII.4 H), we obtain $\mu_0(\varepsilon)=r/\varepsilon$. In view of Theorem 3 B for annuli, this equality implies that W_ε is extremal.

Finally, if $\gamma=\{z\,||z|=r\}$, then $\mu_0(\varepsilon)\geq r/\varepsilon$ is trivially satisfied, and this assumption may be omitted from (d).

Remark. A slightly different characterization was obtained by SUITA [6] as follows: Under the assumption $\gamma=\{z\,||z|=r\}$, W is extremal if and only if $|\int_c d\arg z|\geq 2\pi$ for almost all $c\in\tilde{\Gamma}_\varepsilon$. Note that $\int_c|dz|/|z|\geq|\int_c d\arg z|$.

3 F. Curves Terminating at Incisions or Periphery. Using previous notation, we introduce two subfamilies Γ_ε^p and Γ_ε^i of Γ_ε^* as follows: Γ_ε^p (or Γ_ε^i) consists of rectifiable curves in W_ε which have their initial points on ∂N_ε and terminal points on $\gamma\cap\{z\,||z|=r\}$ (or $\gamma\cap\{z\,||z|<r\}$, resp.). Intuitively, these are the families of curves connecting ∂N_ε and the periphery $|z|=r$ or the incisions, respectively.

Note that $\Gamma_\varepsilon^* - \Gamma_\varepsilon^p - \Gamma_\varepsilon^i$ consists only of curves lacking at least one end point. Accordingly, $\lambda(\Gamma_\varepsilon^* - \Gamma_\varepsilon^p - \Gamma_\varepsilon^i) = \infty$, so that

$$\lambda(\Gamma_\varepsilon^p \cup \Gamma_\varepsilon^i) = \lambda(\Gamma_\varepsilon^*)$$

(cf. Appendix I.H).

THEOREM. *W is an extremal radial slit disk of radius r if and only if one of the following conditions is satisfied for some (or equivalently every) $\varepsilon > 0$ with $\bar{N}_\varepsilon \subset W$:*

(a) $\mu_0(\varepsilon) \geq r/\varepsilon$ *and* $2\pi\lambda(\Gamma_\varepsilon^p) = \log(r/\varepsilon)$,

(b) $\mu_0(\varepsilon) \leq r/\varepsilon$ *and* $\lambda(\Gamma_\varepsilon^i) = \infty$,

(c) 3 B.(b) *or* 3 B.(c), *and* $\lambda(\Gamma_\varepsilon^i) = \infty$.

Property (a) is due to STREBEL [3]; (b) and (c) were given by SUITA [3]. We have already had occasion to use this theorem in VII. 3 H.

Proof. To prove the sufficiency, we first observe that (a) implies $\log \mu_0(\varepsilon) = 2\pi\lambda(\Gamma_\varepsilon^*) \leq 2\pi\lambda(\Gamma_\varepsilon^p) = \log(r/\varepsilon)$ and, consequently, $\mu_0(\varepsilon) = r/\varepsilon$. In view of Theorem 3 B for annuli, we only have to show that, for every r_1 with $0 < r_1 < r$, the set $E = CW \cap \{z \mid |z| \leq r_1\}$ is an extremal set of radial slits. To this end, we denote by Γ^* the family of curves joining ∂N_ε and $|z| = r$ in $\{z \mid \varepsilon < |z| < r\} - E$. We have $\log(r/\varepsilon) \leq 2\pi\lambda(\Gamma^*) \leq 2\pi\lambda(\Gamma_\varepsilon^p) = \log(r/\varepsilon)$, whence $2\pi\lambda(\Gamma^*) = \log(r/\varepsilon)$. Consequently, E has the desired property. (b) and (c) reduce to statements (c) and (b) in Theorem 3 E, since from $\lambda(\Gamma_\varepsilon^i) = \infty$ we conclude that $\int_c |z|^{-1}|dz| \geq \log(r/\varepsilon)$ for almost all $c \in \Gamma_\varepsilon^*$.

Conversely, if W is an extremal radial slit disk of radius r, then (a) reduces to (b) since $\lambda(\Gamma_\varepsilon^p)^{-1} \leq \lambda(\Gamma_\varepsilon^*)^{-1} \leq \lambda(\Gamma_\varepsilon^p)^{-1} + \lambda(\Gamma_\varepsilon^i)^{-1}$. The first halves of (b) and (c) are trivial. Finally, $\lambda(\Gamma_\varepsilon^i) = \infty$ is immediate from Theorem VII. 3 I, where it is shown that $\sup_{z \in c} |z| = r$ for almost all $c \in \Gamma_\varepsilon^*$.

3 G. Infinite Radius. Strictly speaking, we cannot use the term *extremal radial slit disk of infinite radius*, since we do not yet know whether $P_{0\gamma}$ is unique if $c_{0\gamma} = 0$. But we define it for the present as an incised radial slit disk of infinite radius on which

$$P_{0\gamma}(z; 0) \equiv z$$

for *one of the functions* $P_{0\gamma}$. Here γ is the outer boundary of the region.

Such a region satisfies $c_{0\gamma}(0) = 0$, that is, 3 B.(a) for $r = \infty$. It has properties 3 B.(c) and 3 B.(d) for $r = \infty$ (Theorem VII. 4 A). Condition 3 B.(b) is also met:

LEMMA. *Under the assumption* $c_{0\gamma}(0) = 0$, *an incised radial slit disk of radius* $r = \infty$ *has property* 3 B.(b) *if and only if* 3 B.(c) *holds.*

We shall give the proof in 3 K.

The following analogue of Theorem 3 B is due to SUITA [3]:

THEOREM. *If an incised radial slit disk W of radius* $r=\infty$ *has property* 3 B.(a) *and either* 3 B.(b) *or* 3 B.(c), *then it is extremal.*

The proof will be furnished in 3 H – 3 J. Condition 3 B.(d) is then automatically satisfied. If there is no incision, 3 B.(a) is trivially valid, so that this assumption may be omitted.

EXAMPLE. Consider the region

$$W=C\left(\gamma\cup\bigcup(\sigma_n\cup\sigma_n')\right),$$

where

$$\gamma=\{z\,|\,\mathrm{Re}\,z\le -1,\ \mathrm{Im}\,z=0\}\cup\{z\,|\,1\le \mathrm{Re}\,z,\ \mathrm{Im}\,z=0\},$$

$$\sigma_n=\{z\,|\,1\le|z|\le|n|,\ \arg z=\pi/n\},$$

$$\sigma_n'=\{z\,|\,1\le|z|\le|n|,\ \arg z=\pi(1+1/n)\},$$

$n=\pm 4, \pm 5, \ldots$. It is an extremal incised radial slit disk of infinite radius. In fact, condition 3 B.(c) is evidently satisfied, and $c_{0\gamma}=0$ by the fact that $\lambda(\Gamma_\varepsilon^i)=\infty$.

3 H. Proof of Theorem 3 G. Let γ be the outer boundary of W. For an arbitrary R with $0<R<\infty$, we construct a subregion W_R of W with the following properties:

(a) $0\in W_R\subset W$.

(b) The outer boundary γ_R of W_R is such that $\gamma_R\cap\{z\,||z|<R\}\subset\gamma$.

(c) The reduced logarithmic area of W_R does not exceed $2\pi\log R+\eta$, where η is an arbitrarily given positive number.

(d) If $\eta_1>\eta_2>\cdots\to 0$ and $R_1<R_2<\cdots\to\infty$ with $R_n\ge\max\{|z|\,|z\in\gamma_{R_{n-1}}\}$, then $W_{R_{n-1}}\subset W_{R_n}$ and $W=\bigcup_1^\infty W_{R_n}$.

We postpone the construction until 3 J, and continue the proof of the theorem. Denote by p_R the capacity function $p_{0\gamma_R}(z;0)$ on W_R and set $p(z)=\log|z|$. We shall show that

$$D_{W_R}[p_R-p]\le\eta. \tag{1}$$

The proof will then be completed as follows.

Exhaust W_R towards γ_R by $0\in V\to W_R$. Then $p_V=p_{0\gamma(V)}(z;0)$ satisfies $\lim D_V[p_V-p_R]=0$. We construct $V_n\subset W_{R_n}$ recursively so that $\{V_n\}$ exhausts W towards γ, $D_{V_n}[p_{V_n}-p_{R_n}]<\eta_n$, and $D_{V_n}[p-p_{R_n}]<\eta_n$. This gives an exhaustion $0\in V_n\to W$ towards γ which satisfies $\lim D_{V_n}[p_{V_n}-p]=0$. Consequently, $p=\log|z|$ is one of the capacity functions $p_{0\gamma}(z;0)$ of W and, therefore, W is an extremal radial slit disk of infinite radius.

3 I. Proof of Theorem 3 G (continued). In order to prove (1), set $r_R=1/c_{0\gamma_R}(0)$ on W_R. By Theorem III.3 A and property 3 H.(c) of W_R, we have

$$2\pi \log r_R \leq \int_{\partial W_R} p * dp \leq 2\pi \log R + \eta. \tag{2}$$

Let $N_{R\varepsilon} = \{z \mid p_R(z) < \log \varepsilon\}$, and denote by Γ^* the family of curves in $W_R - N_{R\varepsilon}$ joining γ_R and $\partial N_{R\varepsilon}$. By IV.2 C and IV.3 D, the metric

$$\rho_R |dz| = \frac{|dp_R + i * dp_R|}{\log \dfrac{r_R}{\varepsilon}}$$

is the generalized extremal metric for the family Γ^*, so that by IV.2 A.(1),

$$A(\rho - \rho_R) \leq A(\rho) - \frac{2\pi}{\log \dfrac{r_R}{\varepsilon}} \tag{3}$$

for every $\rho \in \bar{\mathbf{P}}(\Gamma^*)$.

Let ε' be the number with $\varepsilon_1(\varepsilon') = \varepsilon$ (IV.3 G). It satisfies $\lim_{\varepsilon \to 0} (\varepsilon'/\varepsilon) = 1$ and $N_{R\varepsilon} \subset N_{\varepsilon'} = \{z \mid |z| < \varepsilon'\}$. Then

$$\rho |dz| = \frac{|dp + i * dp|}{\log \dfrac{R}{\varepsilon'}}$$

belongs to $\bar{\mathbf{P}}(\Gamma^*)$. For the proof, observe that the family $\Gamma_{\varepsilon'}^i$ (3 F) for the present W trivially has $\lambda(\Gamma_{\varepsilon'}^i) = \infty$. Consequently, the family of curves in $W_R - N_{R\varepsilon}$ joining $\partial N_{R\varepsilon}$ and $\gamma \cap \gamma_R$ has infinite extremal length. Thus there is a sequence of $\rho_n |dz|$ with $\lim A(\rho_n) = 0$ such that $\int_c \rho_n |dz| \geq 1$ for every $c \in \Gamma^*$ joining $\partial N_{R\varepsilon}$ and $\gamma \cap \gamma_R$. For $\rho_n' = \max(\rho_n, \rho)$, we have $A(\rho - \rho_n') \leq A(\rho_n) \to 0$. If $c \in \Gamma^*$ joins $\partial N_{R\varepsilon}$ and $\gamma \cap \gamma_R$, then $\int_c \rho_n' |dz| \geq \int_c \rho_n |dz| \geq 1$; if c joins $\partial N_{R\varepsilon}$ and $\gamma_R - \gamma$, then $\int_c \rho_n' |dz| \geq \int_c \rho |dz| \geq 1$. Hence $\rho_n' \in \mathbf{P}(\Gamma^*)$ and, consequently, $\rho \in \bar{\mathbf{P}}(\Gamma^*)$.

Apply (3) to this ρ. In terms of the Dirichlet integral over $W_R - N_{R\varepsilon}$, the inequality takes the form

$$D\left[\frac{p}{\log \dfrac{R}{\varepsilon'}} - \frac{p_R}{\log \dfrac{r_R}{\varepsilon}}\right] \leq \frac{D[p]}{\left[\log \dfrac{R}{\varepsilon'}\right]^2} - \frac{2\pi}{\log \dfrac{r_R}{\varepsilon}}. \tag{4}$$

Let ε'' be defined by $\varepsilon_2(\varepsilon'') = \varepsilon$ (IV.3 G). It satisfies $\lim_{\varepsilon \to 0} (\varepsilon''/\varepsilon) = 1$ and $N_{\varepsilon''} \subset N_{R\varepsilon}$. From (2) it follows that

$$D[p] \leq 2\pi \log \frac{R}{\varepsilon''} + \eta,$$

where the Dirichlet integral is taken over $W_R - N_{R\varepsilon}$. On multiplying both sides of (4) by $(\log(R/\varepsilon'))(\log(r_R/\varepsilon))$ and letting $\varepsilon \to 0$, we obtain (1).

3 J. Construction of W_R. To complete the proof, we show how to construct the region W_R with properties 3 H.(a)–(d).

To begin, we remark that our region has property 3 B.(d) with $r=\infty$. In fact, if we use the symbol $l_\varepsilon(\theta)$ in exactly the same way as in VII.4 G, we obtain

$$\frac{2\pi}{\int_0^{2\pi} \frac{d\theta}{l_\varepsilon(\theta)}} \geq 2\pi\,\lambda(\Gamma_\varepsilon^*) \geq \log\frac{1}{\varepsilon_2(\varepsilon)\,c_{0\gamma}} = \infty$$

from assumption 3 B.(b) or 3 B.(c), so that $l_\varepsilon(\theta)=\infty$ for almost every θ. Thus γ consists of $\{\infty\}$ and possibly a number of radial segments (incisions) with total angular measure zero.

Now remove from the circle $|z|=R$ all points on the incisions of W. We obtain a countable number of disjoint open arcs A_i with $\{z\,||z|=R\}-\gamma=\bigcup_{i=1}^\infty A_i$. Let I_i be the set of arguments of points in A_i, that is, $A_i=\{z\,||z|=R, \arg z\in I_i\}$. The open interval I_i can be expressed as the union $\bigcup_{j=1}^\infty I_{ij}$ of a countable number of closed intervals which have only end points in common. Let $A_{ij}=\{z\,||z|=R, \arg z\in I_{ij}\}$. It is possible to pick the I_{ij} so that the slits $E=\partial W-\gamma$ do not pass through the end points of A_{ij}, since $E\cap\{z\,||z|=R\}$ is totally disconnected. Then $E\cap A_{ij}$ is compact and, therefore, the supremum l_{ij} of the logarithmic length of the slits meeting A_{ij} is finite.

For a given $\eta>0$, the set of points in I_{ij} whose image points in A_{ij} are on the slits with logarithmic length $\geq\eta/8\pi$ is compact, and this set has vanishing linear measure since the area of E vanishes by 3 B.(a). We cover the set by a finite number of disjoint open intervals $I_{ijk}\subset I_{ij}$, $k=0,\dots,k(i,j)$, such that the sum of the lengths is less than $\eta/l_{ij}\,2^{i+j+k(i,j)+3}$. We may assume again that the end points of the arcs A_{ijk} corresponding to I_{ijk} are not on slits. The remaining part $I_{ij}-\bigcup_k I_{ijk}$ consists of a finite number of closed intervals or points, whose interiors (if non-void) are denoted by I'_{ijk}. Let A'_{ijk} be the arc corresponding to I'_{ijk}.

Denote by E_{ijk} the union of slits meeting A_{ijk}, and by E'_{ijk} the corresponding union with respect to A'_{ijk}. The assumption that the end points of A_{ijk} and A'_{ijk} do not meet E implies that E_{ijk} and E'_{ijk} are both compact. Consider the regions

$$S_{ijk}=\{z\,|\log R-2l_{ij}<\log|z|<\log R+2l_{ij},\ \arg z\in I_{ijk}\},$$

$$S'_{ijk}=\left\{z\,|\log R-\frac{\eta}{8\pi}<\log|z|<\log R+\frac{\eta}{8\pi},\ \arg z\in I'_{ijk}\right\}.$$

We have

$$E_{ijk}\subset S_{ijk},\qquad E'_{ijk}\subset S'_{ijk},$$

and the logarithmic areas of S_{ijk} and S'_{ijk} are less than $\eta/2^{i+j+k(i,j)+1}$ and $|I'_{ijk}|\,\eta/4\pi$, respectively, where $|I'_{ijk}|$ is the length of the interval I'_{ijk}.

A slit σ of E_{ijk} is a boundary component of W. It can be encircled by a Jordan curve in $W \cap S_{ijk}$ (cf. Lemma 2 of VI.1 F). If we take all slits $\sigma \subset E_{ijk}$, then the Jordan regions bounded by these curves cover E_{ijk}. Since E_{ijk} is compact, a finite number of the regions cover it. As a result, we obtain a finite number of Jordan curves in $W \cap S_{ijk}$ such that the union of the corresponding Jordan regions covers E_{ijk}. By taking suitable parts of the Jordan curves and a finite number of arcs on $|z|=R$, we obtain an open Jordan arc c_{ijk} joining the end points of A_{ijk} in $W \cap \bar{S}_{ijk} \cap \{z \,|\, |z| \geq R\}$ such that E_{ijk} is in the same component of the complement of c_{ijk} as $z=0$, that is, E_{ijk} is contained in the union of $|z| \leq R$ and the finite region bounded by the curve $c_{ijk} \cup A_{ijk}$.

We construct similarly the curve c'_{ijk} in $W \cap \bar{S}'_{ijk} \cap \{z \,|\, |z| \geq R\}$ with respect to A'_{ijk} and E'_{ijk}.

Now consider the union $\bigcup (c_{ijk} \cup c'_{ijk})$ for all i, j, and k. Together with $\gamma \cap \{z \,|\, |z| \leq R\}$, it divides the plane into two parts. We take the part containing the origin. Its intersection with W is the desired region W_R. In fact, conditions 3 H.(a), 3 H.(b), and 3 H.(d) are derived easily from the construction. To see that 3 H.(c) holds, we may disregard the slits and incisions, since $E = \partial W - \gamma$ and $\gamma \cap \{z \,|\, |z| \leq R\}$ has vanishing area. It follows that the reduced logarithmic area does not exceed the sum of $2\pi \log R$ and the logarithmic area of S_{ijk} and S'_{ijk}. This sum is less than

$$2\pi \log R + \sum_{i,j} \frac{\eta}{2^{i+j+k(i,j)+1}} + \sum_{i,j,k} \frac{|I'_{ijk}|\,\eta}{4\pi} \leq 2\pi \log R + \eta,$$

since

$$\sum_{i,j} \frac{1}{2^{i+j+k(i,j)+1}} < \sum_{i,j} \frac{1}{2^{i+j+1}} \leq \frac{1}{2}, \qquad \sum_{i,j,k} |I'_{ijk}| \leq 2\pi.$$

3 K. Proof of Lemma 3 G. Condition 3 B.(b) trivially implies 3 B.(c). To prove that 3 B.(c) gives 3 B.(b), consider $E_1 = CW \cap \{z \,|\, |z| \leq r_1\}$ and take an arbitrary r' with $r_1 < r' < \infty$. Set $W' = \{z \,|\, |z| < r'\} - E_1$ and $\gamma' = \{z \,|\, |z| = r'\}$. Condition 3 B.(b) is obtained from the equality of $1/r'$ and $c_{0\gamma'}(0)$ for W', since E_1 is then an extremal set of radial slits. Inequality $1/c_{0\gamma'} \geq r'$ is trivial, and it suffices to show that $1/c_{0\gamma'} \leq r'$.

The set $e = \{\theta \,|\, 0 \leq \theta < 2\pi,\ r' e^{i\theta} \in \gamma\}$ is compact and has vanishing measure because of property 3 B.(d), which we showed to be a direct consequence of 3 B.(c) (cf. the beginning of 3 J). Given $\eta > 0$, we cover e by a finite number of disjoint open intervals $I_1, \ldots, I_n$, the sum of whose lengths is less than η. We may assume that their end points are not points on $\partial W - \gamma \cap \{z \,|\, |z| = r'\}$. The set $\{z \,|\, |z| \leq r',\ \arg z \notin I_1 \cup \cdots \cup I_n\}$ can be decomposed into disjoint sectors $S_1, \ldots, S_n$. Since $S_k \cap CW$ contains no incision of W, it coincides with $S_k \cap (\partial W - \gamma)$. By assumption 3 B.(c), it is an extremal set of radial slits; hence so is $S_k \cap E'$, where $E' = CW \cap \{z \,|\, |z| \leq r'\}$.

To estimate $1/c_{0\gamma'}$, we consider the extremal length of the family Γ_ε^* defined for the region W' and denote by Γ_k the collection of curves of Γ_ε^* contained in S_k. Clearly, $\lambda(\Gamma_\varepsilon^*)^{-1} \geq \sum \lambda(\Gamma_k)^{-1}$. Because of the extremality of the set $S_k \cap E'$, we see easily that

$$\lambda(\Gamma_k) = \frac{1}{\theta_k} \log \frac{r'}{\varepsilon},$$

where θ_k is the angle of the sector S_k about the origin (cf. 1 C.(f) and Theorem 1 D). Since $\theta_1 + \cdots + \theta_n \geq 2\pi - \eta$, we obtain

$$\log(r'/\varepsilon) \geq (2\pi - \eta)\lambda(\Gamma_\varepsilon^*) \geq \frac{2\pi - \eta}{2\pi} \log \frac{1}{\varepsilon_2(\varepsilon) c_{0\gamma'}}$$

by Theorem IV.3 G. On letting $\eta \to 0$ and then $\varepsilon \to 0$, we conclude that $r' \geq 1/c_{0\gamma'}$.

4. Tests for Extremal Sets of Slits, with Examples

4 A. Necessary Conditions. For the sake of simplicity, we shall mainly consider extremal sets of vertical slits. Sets of horizontal, circular, and radial slits can be treated in the same manner.

THEOREM. *If E is an extremal set of vertical slits, then*

(a) *the area of E vanishes,*

(b) *any two points $z_1, z_2 \notin E$ with* $\operatorname{Re} z_1 = \operatorname{Re} z_2$ *can be joined in CE by a curve whose length deviates arbitrarily little from $|z_1 - z_2|$.*

Condition (b) is not trivial; in other words, an arbitrary vertical slit plane does not necessarily have this property. A counterexample will be given in Remark 4 F. It is illuminating to compare this with Lemma 2 of VI. 1 E.

The set in 4 G.(a) has vanishing area; yet its complement is a nonextremal vertical slit plane. Thus (a) is not sufficient for extremality.

The example in 4 E has property (b) even if $\prod a_k > 0$. Thus (b) is not sufficient either.

Proof of the Theorem. (a) is a direct consequence of the definition of an extremal vertical slit plane; cf. Theorem VI. 3 B.

The proof of (b) was communicated orally by M. OHTSUKA (see also OIKAWA-SUITA [1]). Suppose there are points z_1 and z_2 with $\operatorname{Re} z_1 = \operatorname{Re} z_2$ and a real number $\varepsilon_1 > 0$ such that any curve in CE joining z_1 and z_2 has length $\geq |z_1 - z_2| + \varepsilon_1$. In order to use Theorem 1 D, consider a rectangle $R=\{z\,|\,a<\operatorname{Re} z<a', b<\operatorname{Im} z<b'\}$ such that $E\subset R$ and $z_1, z_2\in R$. We are going to derive a contradiction by showing that

$$\lambda(\Gamma_R)>\frac{b'-b}{a'-a},$$

where Γ_R stands for the family of locally rectifiable curves joining the horizontal sides of R in $R\cap CE$ (Theorem 1 D).

Let $z_1'=\operatorname{Re} z_1+ib$ and $z_2'=\operatorname{Re} z_2+ib'$. There exists an $\varepsilon_2>0$ with the property that any curve in $R\cap CE$ joining z_1' and z_2' has length $\geq b'-b+\varepsilon_2$. Next let σ be the lower horizontal side of R, that is, $\sigma=\{z\,|\,a\leq\operatorname{Re} z\leq a', \operatorname{Im} z=b\}$. There exists an $\varepsilon_3>0$ such that any curve in $R\cap CE$ joining z_2' with σ has length $\geq b'-b+\varepsilon_3$. We may assume that ε_3 is chosen so small that

$$\{z\,|\,|z-z_2'|\leq\varepsilon_3\}\subset CE \quad\text{and}\quad a<\operatorname{Re} z_2'-\varepsilon_3<\operatorname{Re} z_2'+\varepsilon_3<a'.$$

Then any curve in $R\cap CE$ joining the arc $R\cap\{z\,|\,|z-z_2'|=\varepsilon_3\}$ and σ has length $\geq b'-b$. In order to estimate $\lambda(\Gamma_R)$, consider the density function

$$\rho(z)=\begin{cases}1 & \text{for } z\in R-\{z\,|\,|z-z_2'|\leq\varepsilon_3\},\\ 0 & \text{for } z\in R\cap\{z\,|\,|z-z_2'|\leq\varepsilon_3\}.\end{cases}$$

It gives $\int_c\rho\,|dz|\geq b'-b$, whether or not $c\in\Gamma_R$ meets the disk $\{z\,|\,|z-z_2'|\leq\varepsilon_3\}$. Thus we obtain the desired contradiction:

$$\lambda(\Gamma_R)\geq\frac{(b'-b)^2}{(b'-b)(a'-a)-\frac{1}{2}\pi\,\varepsilon_3^2}>\frac{b'-b}{a'-a}.$$

4 B. A Sufficient Condition. The following result is due to Koebe [7]:

Theorem. *Let E be a closed bounded set. If its projection on the real axis has vanishing linear measure, then E is an extremal set of vertical slits.*

Examples 4 F.(a) and 4 G.(b) will show that this condition is not necessary for extremality.

Proof of the Theorem. The assertion follows immediately from Theorem 1 D. In fact, under the present assumption it is easy to verify that $\lambda(\Gamma_R)=(b'-b)/(a'-a)$.

4 C. Vanishing of the Span. A region W containing ∞ has vanishing span if and only if it is simultaneously an extremal vertical and horizontal slit plane (1 E). Thus from Theorems 4 A and 4 B, we have a sufficient condition and two necessary conditions. If we take into account the conformal invariance of the vanishing of the span, the results can be stated in a somewhat generalized form (AHLFORS-BEURLING [1]):

(a) If W has vanishing span, then CW has vanishing area.

(b) If W has vanishing span, then any two points $z_1, z_2 \in W$ can be joined by a curve in W whose length deviates from $|z_1 - z_2|$ by an arbitrarily small amount.

(c) If the projections of CW onto two orthogonal lines have vanishing linear measures, then W has vanishing span.

As before, none of these conditions is equivalent to the vanishing of the span.

4 D. Generalized Cantor Set. In order to construct counterexamples, we introduce generalized Cantor sets (cf. SARIO [1], AHLFORS-BEURLING [1]).

Take a sequence $\{a_k\}$ of real numbers with $0 < a_k < 1$ and a sequence $\{n_k\}$ of positive integers. Let e_0 be the unit interval $[0, 1]$ on the x-axis, and e_1 the set containing 0 and 1 which consists of $n_1 + 1$ disjoint closed intervals of length $a_1/(n_1+1)$ equally spaced in $[0, 1]$; the total length of e_1 is a_1. On each interval constituting e_1, take n_2+1 intervals of length $a_1 a_2/(n_1+1)(n_2+1)$ and denote their union by e_2. Its total length is $a_1 a_2$. Continue this process to obtain $e_3, e_4, \ldots$. The intersection

$$e(\{a_k\}, \{n_k\}) = \bigcap_{k=0}^{\infty} e_k$$

is called a *generalized Cantor set*. It is a closed bounded set of linear measure $\prod_{k=1}^{\infty} a_k$. We can easily see that it is totally disconnected.

In particular, if $n_1 = n_2 = \cdots = 1$, then we denote the set by $e(\{a_k\})$. The set $e(\{\frac{2}{3}\})$ is the *Cantor ternary set*.

4 E. First Example. Consider $e = e(\{a_k\}, \{n_k\})$ on the real axis and the unit interval $I = [0, i]$ on the imaginary axis. Their Cartesian product $E = e \times I$ is a closed set whose complement $W = CE$ is a vertical slit plane, because e is totally disconnected.

If the length of e vanishes, then W is an extremal vertical slit plane by Theorem 4 B. If it is positive, then so is the area of E and, therefore, by (a) in Theorem 4 A, W is not extremal. We conclude:

$e \times I$ is an extremal set of vertical slits if and only if the length $\prod_{k=1}^{\infty} a_k$ of e is zero.

If I is replaced by a single point, the situation changes (cf. (b) in Theorem 4 G).

4 F. Sets Whose Projections Are Intervals. We assert:

(a) *There is an extremal set of vertical slits whose projection on the real axis is an interval.*

(b) *There is a set whose complement has vanishing span and whose projection on every line is an interval.*

The first example for (a) was exhibited by GRÖTZSCH [4]. The one to be given below is due to TAMURA-OIKAWA-YAMAZAKI [1].

Proof. Consider the Cantor ternary set e on the real and imaginary axes, and let E be the Cartesian product

$$E = e \times e .$$

Since e has vanishing linear measure, the region CE has vanishing span. Rotate E by 45° about its "center" $\frac{1}{2}(1+i)$, and let E' be the resulting set. The region CE' also has vanishing span and, therefore, E' is an extremal set of vertical slits.

The projection of E' onto the real axis is, as we can easily show, an interval of length $\sqrt{2}$. Thus E' is the example for (a).

Consider the union $E \cup E'$. The fact that $C(E \cup E')$ has vanishing span will be proved in XI.3 D. As is readily seen, the projection of $E \cup E'$ on any line is a line segment. Thus the proof of (b) is complete.

Remark. If we consider a suitable $e = e(\{a_k\}, \{n_k\})$ instead of the Cantor ternary set and rotate $e \times e$ by 45°, then the resulting set E' does not have property (b) of Theorem 4 A, yet it is totally disconnected and its complement is therefore a vertical slit plane. Accordingly, condition (b) in Theorem 4 A is not trivial. The choice of $\{a_k\}$ and $\{n_k\}$ will be left to the reader.

4 G. A Totally Disconnected Linear Set. If a totally disconnected closed bounded set E lies on the real axis, then its complement W is simultaneously a vertical and horizontal slit plane. It is an extremal horizontal slit plane by Theorem 4 B. But as a vertical slit plane, it is not necessarily extremal ((a) of the theorem below). If its length vanishes, then it is extremal by Theorem 4 B. Contrary to the case in 4 E, however, this condition is far from necessary ((b) of the following theorem).

Note that, in the present case, the span of the region vanishes if and only if the region is an extremal vertical slit plane.

THEOREM. (a) *There exists a totally disconnected closed bounded set E on the real axis which is not an extremal set of vertical slits.*

(b) *There exists a closed bounded set on the real axis which has positive length yet is an extremal set of vertical slits. Explicitly* $E=e(\{a_k\},\{n_k\})$ *has this property if*

$$\lim_{k\to\infty}\frac{1}{n_k}\log\frac{1}{1-a_k}=0.$$

Remark. (a) was pointed out by KOEBE [7]. The proof below is due to TAMURA [1] (cf. also REICH [1], JENKINS [2], and KÜHNAU [1]). (b) was qualitatively mentioned by GRÖTZSCH [4], whose method we shall follow. AHLFORS-BEURLING [1] showed that E is extremal if

$$\sum_{k=1}^{\infty}(1-a_k)\log\frac{n_k(1-a_{k-1})}{2}=\infty.$$

This will be verified in Remark XI. 3 J.

4 H. Proof of (a). The proof of the theorem extends to the end of this chapter.

We begin with (a). Consider a real-valued function $h(\xi)$ which is defined on $-1\le\xi\le 1$ and has the following properties:

(α) $h(\xi)\ge 0$, $h(\xi)\not\equiv 0$,

(β) $\overline{\lim}_{\xi\to\xi_0+}h(\xi)=h(\xi_0)$ and $\overline{\lim}_{\xi\to\xi_0-}h(\xi)=h(\xi_0)$ for all $-1\le\xi_0\le 1$,

(γ) the support of h is totally disconnected.

A method of constructing such an h will be given in 4 I.

Let e_0 be the support of the function h, and consider in the $(\zeta=\xi+i\eta)$-plane the set

$$E_0=\{\zeta\,|\,\xi\in e_0,|\eta|\le h(\xi)\}.$$

It is closed and bounded since e_0 is closed and h is upper semicontinuous on e_0. The complement $W_0=CE_0$ is a vertical slit plane since e_0 is totally disconnected. By condition (α), E_0 has a component which does not reduce to a point; in other words, W_0 is not a horizontal slit plane. We map W_0 conformally onto an extremal horizontal slit plane W by the function $z=P_0(\zeta;\infty)\not\equiv\zeta$ on W_0. We shall show that $E=CW$ is the desired set.

Since W_0 is symmetric about the real axis, the function $P_0(\zeta;\infty)$ is symmetric and, therefore, W is symmetric about the real axis (cf. Problem 9). Since every boundary component of W_0 meets the real axis, the same is true of W; as a consequence, $E=CW$ lies on the real axis. Let H_z be the upper half of the z-plane. The function $P_0(\zeta;\infty)$ maps the simply connected region $\tilde{W}_0=W_0\cap H_\zeta$ onto $\tilde{W}=W\cap H_z=H_z$. Therefore, by Carathéodory's theory of boundary correspondence, each point z on the real axis is the image of a prime end of W_0.

Property (β) of the function h shows that every upper half-segment of E_0 represents a single prime end. As a consequence, the image of every boundary component of W_0 is a single point; in other words, E is totally disconnected.

The set E is an extremal set of horizontal slits. If it is also an extremal set of vertical slits, the region W must have vanishing span; then the family $\mathscr{V}_\infty(W)$ consists only of the identity mapping (Theorem VI.2 D), a contradiction. We conclude that E is not an extremal set of vertical slits.

41. Proof of (a) (continued). We construct a function h as the limit of a sequence of functions h_n defined inductively and set $h_0(\xi)=1-|\xi|$, $-1\leq\xi\leq 1$. Assuming that we have constructed h_{n-1}, we show how to construct h_n. The graph of the function h_{n-1} consists of a finite number of line segments. Let AB be a segment of the graph, and let CD be the projection of AB onto the ξ-axis. Take the segment $A'B'$ on AB which is at the center of AB and has length $2^{-n}\cdot\overline{AB}$; and the segment $C'D'$ on CD which is at the center of CD and has length $2^{-(n+1)}\cdot\overline{CD}$. Replace the segment AB by the union $AA'C'D'B'B$ of these line segments. Carry out this process for every line segment of the graph of h_{n-1}. The resulting curve, consisting of a finite number of line segments, shall be the graph of h_n.

The functions $h_0, h_1, \ldots$ are continuous and non-negative. They form a monotone decreasing sequence, and, therefore,

$$h(\xi)=\lim_{n\to\infty} h_n(\xi)$$

exists, is non-negative, upper semicontinuous, and satisfies $h(\xi)\not\equiv 0$ (since $h(0)=1$).

The support of h is contained in the intersection of the supports of the h_n. On each of the intervals $-1\leq\xi\leq 0$ and $0\leq\xi\leq 1$, the intersections are similar to the generalized Cantor set $e(\{a_k\})$, though not completely the same. It is not difficult to see that the support is totally disconnected.

Finally, to show that condition (β) holds, it suffices to prove $\overline{\lim}\, h(\xi)\geq h(\xi_0)$, since the inequality in the opposite direction follows from the upper semicontinuity. We shall prove that

$$\overline{\lim_{\xi\to\xi_0+}}\, h(\xi)\geq h(\xi_0). \tag{1}$$

The inequality with respect to the limit $\xi\to\xi_0-$ can be established in exactly the same manner. There is nothing to prove if $h(\xi_0)=0$, so we may assume that $h(\xi_0)>0$.

A vertex of the graph of h_n is also one of h_m for $m\geq n$. The difference between the ξ-coordinates of consecutive vertices of the graph of h_n not on the ξ-axis is less than $2^{-(n-1)}$. The same is true of the η-coordinates.

Since $h(\xi_0)>0$, we have $h_n(\xi_0)>0$ for every n. If $(\xi_0, h_n(\xi_0))$ is a vertex of the graph of h_n, then there is a vertex $(\xi, h_n(\xi))$ such that $0<\xi-\xi_0<2^{-(n-1)}$ and $|h_n(\xi)-h_n(\xi_0)|<2^{-(n-1)}$. The same is true if $(\xi_0, h_n(\xi_0))$ is not a vertex; it is possible to find a vertex $(\xi, h_n(\xi))$ such that $0<\xi-\xi_0<2^{-(n-1)}$ and $h_n(\xi_0)<h_n(\xi)+2^{-(n-1)}$. In either case we obtain $0<\xi-\xi_0<2^{-(n-1)}$ and $h_n(\xi_0)-2^{-(n-1)}<h_n(\xi)$.

We know that $h_n(\xi_0)\geq h(\xi_0)$ and $h_n(\xi)\geq h(\xi)$. Since ξ is the ξ-coordinate of a vertex, we have $h_n(\xi)=h(\xi)$. Consequently,

$$h(\xi_0)-\frac{1}{2^{n-1}}<h(\xi),$$

from which we deduce (1).

4 J. Proof of (b). To prove (b) of Theorem 4 G, recall the definition of $e=e(\{a_k\}, \{n_k\})$ as $\bigcap_{k=0}^{\infty} e_k$, where e_k is obtained inductively from the unit interval $e_0=[0, 1]$ as the union of

$$\nu_k=(n_1+1)\cdots(n_k+1)$$

closed intervals, to be called α_k-intervals, of equal length

$$\alpha_k=\alpha_{k-1}\frac{a_k}{n_k+1}=\frac{a_1\ldots a_k}{(n_1+1)\ldots(n_k+1)}, \qquad \alpha_0=1.$$

These in turn are obtained from each α_{k-1}-interval of e_{k-1} by deleting n_k open intervals, to be referred to as β_k-intervals, of equal length

$$\beta_k=\alpha_{k-1}\frac{1-a_k}{n_k}=\frac{a_1\ldots a_{k-1}}{(n_1+1)\ldots(n_{k-1}+1)}\,\frac{1-a_k}{n_k}.$$

Let R be a square containing e_0 at the center with sides of length $L\geq 3$ parallel to the coordinate axes. We denote by λ and λ_k, $k=1,2,\ldots$, the extremal lengths of the families of curves joining the upper and lower sides of R in $R-e$ and in $R-e_k$, respectively. By Theorem 1 D, e is an extremal set of vertical slits if and only if $\lambda=1$. Since $1\leq\lambda\leq\lambda_k\leq\lambda_{k-1}$, we see that e has this property if

$$\lim_{k\to\infty}\lambda_k=1. \tag{2}$$

The distance between α_k-intervals dominates $\eta_k=\min(\beta_1,\ldots,\beta_k)$. We consider the rectangles of height L and width $\gamma_k=\alpha_k+\eta_k$, each containing one α_k-interval at the center. There are ν_k of them. Each will be denoted by R_k without an additional index. If we remove the R_k from R, we obtain a finite number of rectangles R_{kj} of height L and width γ'_{kj}, say. Clearly

$$L=\nu_k\gamma_k+\sum_j\gamma'_{kj}. \tag{3}$$

Let $\lambda_{(k)}$ be the extremal length of the family of curves joining the upper and lower sides of R_k in the doubly connected region D_k obtained by deleting the α_k-interval from R_k. We obtain

$$\frac{1}{\lambda_k} \geq \frac{\nu_k}{\lambda_{(k)}} + \sum_j \frac{\gamma'_{kj}}{L}. \tag{4}$$

On the other hand, D_k can be mapped conformally onto the region obtained by deleting a vertical line segment from the center of a rectangle $\tilde{R}_k$ of height L whose sides are parallel to the coordinate axes. Let $\tilde{\gamma}_k$ and $\tilde{\alpha}_k$ be the width of $\tilde{R}_k$ and the length of the vertical segment, respectively. We have

$$\frac{1}{\lambda_{(k)}} = \frac{\gamma_k}{L} \cdot \frac{\tilde{\gamma}_k}{\gamma_k} \tag{5}$$

and also

$$\tilde{\gamma}_k \leq \gamma_k, \tag{6}$$

since $\gamma_k/L \geq 1/\lambda_{(k)}$. Furthermore, the extremal length $\lambda^*_{(k)}$ of the family of curves in D_k joining the vertical sides of R_k satisfies $\lambda^*_{(k)} = \gamma_k/L$ and $\lambda^*_{(k)} \leq \tilde{\gamma}_k/(L-\tilde{\alpha}_k)$. Thus

$$\frac{\tilde{\gamma}_k}{\gamma_k} \geq \frac{L-\tilde{\alpha}_k}{L}$$

and, by (3), (4), and (5),

$$\frac{1}{\lambda_k} \geq 1 - \frac{\nu_k \gamma_k \tilde{\alpha}_k}{L^2}.$$

We conclude that (2) holds if

$$\varliminf_{k\to\infty} \nu_k \gamma_k \tilde{\alpha}_k = 0.$$

4 K. Proof of (b) (continued). In order to estimate $\tilde{\alpha}_k$, consider the disk in R_k of diameter γ_k with center at the mid-point of the α_k-interval. Let A_k be the doubly connected region obtained from the disk by deleting the α_k-interval. It is conformally equivalent to the extremal region D_{h_k} of GRÖTZSCH with

$$h_k = \frac{\alpha_k^2 + \gamma_k^2}{2\alpha_k \gamma_k}$$

and $\mu(A_k) = \Phi(h_k)$ (VIII.2 C). On the other hand, consider the image of A_k in $\tilde{R}_k$. The modulus is expressed by the extremal length of the family of curves separating the two boundary components. By using (6) also, we easily obtain $2\pi/\log \mu(A_k) \geq (2\tilde{\alpha}_k)^2/L\tilde{\gamma}_k \geq (2\tilde{\alpha}_k)^2/L\gamma_k$. Consequently, (2) is satisfied if

$$\varliminf_{k\to\infty} \left(\frac{\nu_k^2 \gamma_k^3}{\log \Phi(h_k)}\right) = 0.$$

We remark that $\eta_k \leq \beta_k \leq 2^{-(k-1)} \to 0$ as $k \to \infty$. Accordingly, any sequence $\{k_i\}$ of positive integers contains a subsequence $\{k_j\}$ with $\eta_{k_j} = \beta_{k_j}$. We conclude that (2) is met if

$$\lim_{k \to \infty} \frac{a_1^3 \dots a_{k-1}^3}{(n_1+1) \dots (n_{k-1}+1)} \cdot \frac{1}{n_k} \cdot \frac{1}{\log \Phi(h_k)} = 0, \tag{7}$$

where

$$h_k = 1 + \frac{(n_k+1)^2 (1-a_k)^2}{2 a_k n_k (n_k + 1 - a_k)}.$$

Suppose now that the assumption of the theorem is satisfied. There exists a sequence $\{k_i\}$ such that

$$\frac{a_1^3 \dots a_{k_i-1}^3}{(n_1+1) \dots (n_{k_i-1}+1)} \cdot \frac{1}{n_{k_i}} \log \frac{1}{1-a_{k_i}} \leq \frac{1}{n_{k_i}} \log \frac{1}{1-a_{k_i}} \to 0$$

as $i \to \infty$. If $\underline{\lim}_{i \to \infty} a_{k_i} < 1$, there exists a subsequence $\{k_j\}$ of $\{k_i\}$ such that $a_{k_j} \leq \delta < 1$, $j = 1, 2, \dots$. Then (7) for $k = k_j$ is dominated by $\text{const} \cdot \delta^{3(j-1)}$, which tends to zero as $j \to \infty$. If, on the other hand, $\lim_{i \to \infty} a_{k_i} = 1$, then $\lim h_{k_i} = 1$. By VIII.2 D.(d),

$$\frac{1}{\log \Phi(h_{k_i})} \sim \frac{2}{\pi^2} \log \frac{1}{h_{k_i} - 1} \sim \frac{4}{\pi^2} \log \frac{1}{1 - a_{k_i}}$$

as $i \to \infty$, and we obtain (7). Consequently, (2) is satisfied, and e is an extremal set of vertical slits.

PART THREE

Null Classes

Chapter X

Degeneracy

Roughly speaking, the vanishing of the capacity expresses smallness, or degeneracy, of a boundary or a boundary component (cf. Preface). For example, on a simply connected region, the capacity is the reciprocal of the mapping radius, so that its vanishing implies the degeneracy of the boundary into a single point. Other conformal invariants, such as the span, could also be used to characterize smallness of the boundary, but they do not have the versatility of being applicable to subboundaries.

On an arbitrary Riemann surface, the "smallness" of a subboundary can in turn be expressed in terms of the degeneracy of certain families of functions on a boundary neighborhood or on the entire surface; we discuss these topics in 1 and 2. Intuitively related to such degeneracy are the problems of removability, boundary behavior of functions, and essential extendability, considered in 3, 4, and 5.

Our intuition fails when the genus is infinite: degeneracy of a family of functions does not necessarily imply "smallness" of the boundary (see 2 J). Furthermore, the vanishing of the capacity of a subboundary not only indicates metric smallness but it also gives information concerning the mode of accumulation of other boundary components; this point will be elucidated in Chapter XI.

1. Weak, Semiweak, and Parabolic Subboundaries

1 A. Weak and Semiweak Subboundaries. The terminology introduced for plane regions in VII.2 C can be extended to Riemann surfaces.

DEFINITION. *A subboundary γ of an open Riemann surface W is said to be weak if the capacity $c_{1\gamma}(\zeta)$ vanishes. It is called semiweak if the capacity $c_{0\gamma}(\zeta)$ vanishes.*

The vanishing of a capacity is independent of the reference point $\zeta \in W$ (IV.3 B).

Remark. A weak subboundary is called "parabolic" by JURCHESCU [1]; we shall reserve this word for later usage (1 I). A semiweak boundary is "halb-schwach" in the terminology of CONSTANTINESCU [2].

A weak subboundary is semiweak (III.3 C). That the converse is not true will be apparent from examples to be given in XI.4 B.

Let U be a connected neighborhood of γ, and set $\alpha=\partial U$. Then γ is weak if and only if the modulus $\mu_1(\gamma,\alpha)$ of U is infinite for some, or equivalently every, U. Similarly, γ is semiweak if and only if $\mu_0(\gamma,\alpha)$ is infinite (IV.3 A, 3 B). These properties are also equivalent to the vanishing of the harmonic measures $u_{\nu\gamma}$, $\nu=1,0$, respectively, on U (IV.4 A). From this we see that *weakness and semiweakness are boundary properties.* More precisely, if subboundaries γ and γ' of Riemann surfaces W and W', respectively, have conformally equivalent connected neighborhoods, then the weakness or semiweakness of one implies that of the other.

For an isolated subboundary γ, weakness and semiweakness are equivalent. This is verified by applying the above criteria to a connected neighborhood U with $\beta_U=\gamma$ (IV.3 B). On such a U, we need not distinguish between $\nu=0$ and $\nu=1$; weakness is equivalent to $\mu(\gamma,\alpha)=\infty$, or $u_\gamma=0$. The latter property is expressed in terms of the operator H acting from α into U as follows: γ is weak if and only if $H1=1$ (I.5 D, IV.4 B).

A connected neighborhood U of an isolated subboundary γ with $\beta_U=\gamma$ can be imbedded in a surface W_0 in such a way that W_0-U is compact. Then γ constitutes the entire ideal boundary of W_0. Since weakness is preserved under this process, a considerable part of the study of isolated weak subboundaries reduces to the study of a surface possessing a weak ideal boundary β.

1 B. Classes O_G and O_γ of Riemann Surfaces. An open Riemann surface W will be called *parabolic* if its ideal boundary β, which is an isolated subboundary, is weak. This is equivalent to the fact that W carries no Green's function (III.2 H). For this reason, the class of parabolic Riemann surfaces is denoted by O_G. For convenience, a closed surface will always be classified as a parabolic surface.

A plane region W with $\infty\in W$ is parabolic if and only if the set $E=CW$ has vanishing logarithmic capacity (VII.6 D). In this case, E is totally disconnected (cf. VII.5 G), so that $E=\partial W$.

We shall call the (ideal) boundary of an arbitrarily given Riemann surface *absolutely disconnected* if every boundary component of it is weak (cf. VII.5 C.(b)). We denote by O_γ the class of Riemann surfaces with absolutely disconnected boundaries. Closed surfaces are again included in this class. One could also consider the class of surfaces possessing merely semiweak boundary components, but this class seems to lack sufficiently interesting properties (cf. Problem 22).

A plane region of class O_γ was investigated in VII.5 C; see also 2 L of this chapter.

Every subboundary of a parabolic surface is weak (III.3 C). In particular,

$$O_G \subset O_\gamma.$$

The converse is not true, for in XI.3 J we shall find a plane region in the class $O_\gamma - O_G$.

However, for isolated subboundaries the situation is much simpler. In view of properties I.5 C.(d)–(f) of the operator H, we easily obtain:

Suppose $\beta = \gamma_1 \cup \cdots \cup \gamma_n$ is a partition into a finite number of isolated subboundaries; then the surface is parabolic if and only if $\gamma_1, \ldots, \gamma_n$ are all weak.

Remark. The class O_G of Riemann surfaces without Green's functions was first investigated by Myrberg [1]. In terms of the Schwarz (harmonic) measure it was also discussed by Nevanlinna [1], who developed for it his pioneering theory of square integrable differentials. The class O_γ of Riemann surfaces with absolutely disconnected boundaries was introduced in Sario [10, 13].

1 C. Degeneracy of Families of Functions on a Boundary Neighborhood. Weakness and semiweakness were characterized by the degeneracy of some families of functions on a boundary neighborhood or on the entire surface. Explicitly the weakness of γ is equivalent to any one of the following: $KD_{\zeta\gamma}(W) = \emptyset$ (III.3 B), $\mathscr{H}^1_{\zeta\gamma}(W) = \emptyset$ (III.3 F), or $KD_{\alpha\gamma}(\bar{U}) = \emptyset$ (IV.1 E). Similarly, γ is semiweak if and only if one of the following is true: $KD^0_{\zeta\gamma}(W) = \emptyset$, $\mathscr{H}^0_{\zeta\gamma}(W) = \emptyset$, or $KD^0_{\alpha\gamma}(\bar{U}) = \emptyset$.

In the sequel, we shall consider several other families.

In general, on a Riemann surface W, the families of harmonic functions which are respectively non-negative, bounded, and possess finite Dirichlet integrals will be denoted by $HP(W)$, $HB(W)$, and $HD(W)$. The family $HBD(W) = HB(W) \cap HD(W)$ will also be considered.

If V is a finite union of regularly imbedded subregions with disjoint closures, we shall denote by $HP(\bar{V})$, $HB(\bar{V})$, $HD(\bar{V})$, and $HBD(\bar{V})$ the families of harmonic functions on $\bar{V}$ whose restrictions to each component V_i of V belong to $HP(\bar{V}_i)$, $HB(\bar{V}_i)$, $HD(\bar{V}_i)$, and $HBD(\bar{V}_i)$, respectively. The subfamilies of functions vanishing identically on the relative boundary ∂V will be denoted by $H_0P(\bar{V})$, $H_0B(\bar{V})$, $H_0D(\bar{V})$, and $H_0BD(\bar{V})$.

Let γ be a subboundary of a Riemann surface W, and let U be a neighborhood of γ. In the family $HB(\bar{U})$, we consider functions u such that $u = L_\nu u$ ($\nu = 0, 1$) on $\beta_U - \gamma$, or more precisely, on every end $U' \subset U$ with $\beta_{U'} \cap \gamma = \emptyset$. The operator L_1 here is for **Q** only. We shall denote the family of such u by $HB^\nu_\gamma(\bar{U})$ and set $H_0B^\nu_\gamma(\bar{U}) = H_0B(\bar{U}) \cap HB^\nu_\gamma(\bar{U})$.

In a similar manner, we define $HP_\gamma^\nu(\bar{U})$, $H_0 P_\gamma^\nu(\bar{U})$; $HD_\gamma^\nu(\bar{U})$, $H_0 D_\gamma^\nu(\bar{U})$; and $HBD_\gamma^\nu(\bar{U})$, $H_0 BD_\gamma^\nu(\bar{U})$.

THEOREM. *A subboundary γ is weak ($\nu=1$) or semiweak ($\nu=0$) if and only if one of the equalities*

$$H_0 B_\gamma^\nu(\bar{U})=\{0\}, \quad H_0 D_\gamma^\nu(\bar{U})=\{0\}, \quad H_0 BD_\gamma^\nu(\bar{U})=\{0\}$$

holds for some (or equivalently every) neighborhood U of γ.

This is to be so understood that weakness implies all three relations, and any one of them entails weakness.

Remark 1. If γ is isolated and U satisfies $\beta_U=\gamma$, then the families are replaced by $H_0 B(\bar{U})$, $H_0 D(\bar{U})$, and $H_0 BD(\bar{U})$, respectively. Closed surfaces are included as a trivial case. For $\nu=1$, the above theorem and Theorems 1 F and 1 G below are due to CONSTANTINESCU [2]. For a parabolic W, these theorems as well as Theorem 1 H below are now well known.

Remark 2. Except for closed surfaces, we always have

$$H_0 P_\gamma^\nu(\bar{U}) \neq \{0\}.$$

In fact, a modulus function $q_\nu(z;\gamma,\alpha)\neq$const on U ($\alpha=\partial U$) always exists. It is constant on α and it dominates this value on U; hence $q_\nu-$const$\in H_0 P_\gamma^\nu(\bar{U})$ for some constant. PARREAU [1, p.160] and NAKAI [2] showed that $H_0 P(\bar{V})\neq\{0\}$ for any V which is not relatively compact.

1 D. Proof of the Theorem. We may assume without loss of generality that U is connected (cf. 1 B).

The sufficiency is simple, since each equality implies $u_{\nu\gamma}=0$.

Conversely, suppose γ is weak or semiweak. Exhaust W towards γ by V with $\alpha=\partial U\subset V$. For every $u\in H_0 B_\gamma^\nu(\bar{U})$, set $M=\sup_U u<\infty$. The harmonic measure $u_{\nu\gamma(V)}$ on $U\cap V$ satisfies $u\le M u_{\nu\gamma(V)}$ because of the maximum principle for the operator L_ν. On letting $V\to W$, we conclude that $u\le 0$. Since $-u\in H_0 B_\gamma^\nu(\bar{U})$, we obtain $u\ge 0$ as well. Consequently, $H_0 B_\gamma^\nu(\bar{U})=\{0\}$.

This trivially implies $H_0 BD_\gamma^\nu(\bar{U})=\{0\}$.

Finally, to derive $H_0 D_\gamma^\nu(\bar{U})=\{0\}$, we consider the flux $\int_\alpha * du$ of $u\in H_0 D_\gamma^\nu(\bar{U})$. If it does not vanish, we choose a constant such that $v=\text{const}\cdot u$ has flux $\int_\alpha * dv=2\pi$. Then $v\in KD_{\alpha\gamma}(\bar{U})=\emptyset$ if $\nu=1$, and $v\in KD_{\alpha\gamma}^0(\bar{U})=\emptyset$ if $\nu=0$ (analogue of Theorem III.3 B derived from Theorem IV.1 E); this is a contradiction. Thus $\int_\alpha * du=0$ for every $u\in H_0 D_\gamma^\nu(\bar{U})$. Consequently, from the following lemma we obtain $H_0 D_\gamma^\nu(\bar{U})=\{0\}$, so that the proof of Theorem 1 C will be complete:

LEMMA. *If $H_0 BD_\gamma^\nu(\bar{U})=\{0\}$, then every $u\in H_0 D_\gamma^\nu(\bar{U})$ with $\int_\alpha * du=0$ vanishes identically.*

1 E. Proof of the Lemma. Given a point $\zeta \in U$, we shall consider functions p_ν^0 and p_ν^1, $\nu=0,1$, which are certain generalizations of $p_{\nu\alpha}(z;\zeta)$ on U. In essence, they were introduced in MARDEN-RODIN [1].

Exhaust W towards γ by V with $\alpha=\partial U\subset V$ and $\zeta\in V$. Let $p_{\nu V}^0$ and $p_{\nu V}^1$, $\nu=0,1$, be harmonic functions on $\bar{U}\cap\bar{V}-\{\zeta\}$ which have the singularity $\log|z-\zeta|$ at ζ and possess, for $\mu,\nu=0,1$, the properties $\lim_{z\to\zeta}(p_{\nu V}^\mu-\log|z-\zeta|)=0$, $p_{\nu V}^\mu=\text{const}$ on α, $L_\nu p_{\nu V}^\mu=p_{\nu V}^\mu$ on $\beta_V-\gamma(V)$, $*dp_{\nu V}^0\equiv 0$ along $\gamma(V)$, $p_{\nu V}^1=\text{const}$ on $\gamma(V)$, and $\int_{\gamma(V)}*dp_{\nu V}^1=0$. Denote the constant values of these functions on α by $k_{\nu V}^\mu$. If $V'\supset\bar{V}$, we obtain exactly as in III.2 A (cf. I.2 C, 2 D):

$$D_{U\cap V}[p_{\nu V}^1-p_{\nu V'}^1]\le 2\pi(k_{\nu V'}^1-k_{\nu V}^1),$$
$$D_{U\cap V}[p_{\nu V}^0-p_{\nu V'}^0]\le 2\pi(k_{\nu V}^0-k_{\nu V'}^0).$$

From this we see that the $k_{\nu V}^1$ increase and the $k_{\nu V}^0$ decrease as V increases. We also obtain $k_{\nu V}^1\le k_{\nu V}^0$ using the same process as in III.3 C. We have proved the existence of $\lim k_{\nu V}^\mu$ as $V\to W$ and, therefore, that of p_ν^μ with

$$\lim_{V\to W} D_{U\cap V}[p_{\nu V}^\mu-p_\nu^\mu]=0.$$

It is not difficult to show that, for a suitable constant,

$$p_\nu^1-p_\nu^0-\text{const}\in H_0BD_\gamma^\nu(\bar{U}),$$

where the constant is chosen so that the function vanishes on α. Accordingly, if $H_0BD_\gamma^\nu(\bar{U})=\{0\}$, then $p_\nu^1-p_\nu^0=\text{const}$. On the other hand, we can verify as in III.3 D that every $u\in H_0D_\gamma^\nu(\bar{U})$ with $\int_\alpha *du=0$ satisfies

$$D[u,p_\nu^1-p_\nu^0]=2\pi\,u(\zeta).$$

Since ζ is an arbitrary point of U, we conclude that $u=0$ if $H_0BD_\gamma^\nu(\bar{U})=\{0\}$.

1 F. Characterization by Operators. Consider a neighborhood U of a subboundary γ, and the operators L_0 and L_1 for **Q** acting from $\alpha=\partial U$ into U.

THEOREM. *A subboundary γ is weak ($\nu=1$) or semiweak ($\nu=0$) if and only if one of the relations*

$$u=L_\nu u \quad \textit{for every} \quad u\in HB_\gamma^\nu(\bar{U}),$$
$$u=L_\nu u \quad \textit{for every} \quad u\in HD_\gamma^\nu(\bar{U})$$

holds for some (or equivalently every) U.

Proof. Here and in similar proofs in the sequel, we may assume without loss of generality that U is connected. Sufficiency: Each condition implies that $u_{\nu\gamma}=0$. Necessity: $u-L_\nu u$ belongs to $H_0 B_\gamma^\nu(\bar{U})=\{0\}$ or $H_0 D_\gamma^\nu(\bar{U})=\{0\}$.

Remark. If γ is weak ($\nu=1$) or semiweak ($\nu=0$), then

$$HB_\gamma^\nu(\bar{U})=HD_\gamma^\nu(\bar{U}).$$

This is easily verified using the theorem and the fact that $L_\nu u\in HBD(\bar{U})$. However, as we shall see in 2 F, this is not sufficient for γ to be weak or semiweak.

COROLLARY. *An isolated subboundary γ is weak if and only if one of the following conditions is satisfied on U with $\beta_U=\gamma$:*

(a) $H=L$,

(b) $u=Hu$ *for every* $u\in HB(\bar{U})$,

(b′) $u=Lu$ *for every* $u\in HB(\bar{U})$,

(c) $u=Hu$ *for every* $u\in HD(\bar{U})$.

These relations are to be understood as follows: If γ is weak, then they all hold for every U and every normal operator L; conversely, if one of them holds for H or some L on some U, then γ is weak.

We remark in passing that it is not known whether weakness implies "property (c′)", that is, $u=Lu$ for every $u\in HD(\bar{U})$ for an arbitrary normal operator L.

Proof of the Corollary. Each of the conditions implies that the harmonic measure $u_\gamma=1-H1$ vanishes identically. Conversely, if γ is weak, then (a) holds: $Hf-Lf\in H_0 B(\bar{U})=\{0\}$ for every $f\in C^\omega(\alpha)$. Properties (b), (b′), and (c) are immediate consequences of the theorem.

1 G. Maximum-Minimum Principle. Given an end U of a Riemann surface, consider a family of continuous functions on $\bar{U}$. We say that the *maximum principle* holds for the family if every u in the family satisfies

$$\max_{z\in\alpha} u(z)=\sup_{z\in U} u(z). \tag{1}$$

Similarly, if

$$\min_{z\in\alpha} u(z)=\inf_{z\in U} u(z), \tag{2}$$

the *minimum principle* is said to hold for the family. The validity of both conditions for every u in the family establishes the *maximum-minimum principle.* Finally, if every u in the family satisfies either (1) or (2) (the choice may depend on u), then it will be said that the *maximum-or-minimum principle* holds for the family.

THEOREM. *A subboundary γ of W is weak $(\nu=1)$ or semiweak $(\nu=0)$ if and only if one of the following conditions is satisfied for some (or equivalently every) neighborhood U of γ:*

The maximum-minimum principle holds for $HB_\gamma^\nu(\bar{U})$.

The maximum-minimum principle holds for $HD_\gamma^\nu(\bar{U})$.

Proof. Sufficiency: Each condition implies $u_{\nu\gamma}=0$. Necessity: The conditions are immediate consequences of Theorem 1 F and the maximum principle for the operators L_ν acting from ∂U into U.

From this theorem we obtain the following degeneracy of the family of functions on the entire surface:

COROLLARY. *If γ is weak $(\nu=1)$ or semiweak $(\nu=0)$, then the following families consist only of constants:*

$$\{u \mid u \in HB(W),\ L_\nu u=u \ \text{ on } \ \beta-\gamma\},$$

$$\{u \mid u \in HD(W),\ L_\nu u=u \ \text{ on } \ \beta-\gamma\}.$$

In particular, if W is parabolic, then all functions in $HB(W)\cup HD(W)$ are constant.

In fact, every $u \in HB(W)$ or $HD(W)$ with $L_\nu u=u$ on $\beta-\gamma$ satisfies $u=L_\nu u$ on a neighborhood U of γ. It, therefore, attains its maximum on W at a point of ∂U, so that $u=$const.

That the converse statements are not true will be seen in 2 A.

1 H. Flux Condition. We maintain:

THEOREM. *A subboundary γ of W is weak $(\nu=1)$ or semiweak $(\nu=0)$ if and only if one of the conditions*

$$\int_\alpha * du=0 \quad \textit{for every} \quad u \in HB_\gamma^\nu(\bar{U}),$$

$$\int_\alpha * du=0 \quad \textit{for every} \quad u \in HD_\gamma^\nu(\bar{U})$$

holds on some (or equivalently every) neighborhood U of γ.

Proof. Each condition implies $u_{\nu\gamma}=0$ since $D[u_{\nu\gamma}]=\int_\alpha * du_{\nu\gamma}$; conversely, they are derived from Theorem 1 F and a property of the operator L_ν.

1 I. Parabolic Subboundary. In IV.4 C we introduced the harmonic measure u_γ on a neighborhood U of an arbitrary subboundary γ of an open Riemann surface W. It will be seen in 1 J that the property $u_\gamma=0$ is independent of the neighborhood of γ. Accordingly, we may introduce (CONSTANTINESCU [2]):

DEFINITION. *A subboundary γ is parabolic if $u_\gamma=0$.*

A semiweak γ is parabolic (IV.4 D). That the converse is not true is seen from examples in XI. 5 D.

An isolated γ is parabolic if and only if it is weak. This is apparent if we consider a U with $\beta_U=\gamma$.

1 J. Characterization of Parabolicity. We obtain analogues of Theorems 1 C, 1 F, and 1 G by considering the families

$$HB_\gamma^H(\bar{U})=\{u \mid u\in HB(\bar{U}),\ Hu=u \text{ on } \beta_U-\gamma\}$$

and

$$H_0 B_\gamma^H(\bar{U})=HB_\gamma^H(\bar{U})\cap H_0 B(\bar{U}).$$

THEOREM. *A subboundary γ is parabolic if and only if one of the following conditions is valid:*

(a) $H_0 B_\gamma^H(\bar{U})=\{0\}$.

(b) $Hu=u$ *for every* $u\in HB_\gamma^H(\bar{U})$, *where H acts from ∂U into U.*

(c) $\max_{\partial U} u=\sup_U u$ *for every* $u\in HB_\gamma^H(\bar{U})$ *with* $u\geq 0$.

Remark. The result is due to CONSTANTINESCU [2]. The corresponding properties for the class $HD(U)$ are not available. The main reason for this was pointed out in Remark 3 in IV.4 D.

Proof of the Theorem. Exhaust W towards γ by V with $\alpha=\partial U\subset V$. Consider the harmonic measure $u_{\gamma(V)}$ on $U\cap V$. For $u\in H_0 B_\gamma^H(\bar{U})$, set $M=\sup_U u$; then $u\leq M u_{\gamma(V)}$ on $U\cap V$. Consequently, $u\leq 0$. On considering $-u\in H_0 B_\gamma^H(\bar{U})$ as well, we conclude that (a) holds. (b) and (c) are immediate consequences of (a) and (b), respectively. Conversely, any one of (a), (b), (c) implies $u_\gamma=0$.

COROLLARY. *The property $u_\gamma=0$ is independent of U.*

Proof. Let U, U' be neighborhoods of γ, and u_γ, u'_γ harmonic measures on U and U'. Without loss of generality, we may assume $U'\subset U$. If $u_\gamma=0$, then $u'_\gamma=0$ since $u'_\gamma\leq u_\gamma$. If $u'_\gamma=0$, apply (c) of the theorem to u_γ on U': $\sup_{U'} u_\gamma=\max_{\partial U'} u_\gamma$. Thus u_γ attains its maximum over U at a point on $\partial U'$, so that $u_\gamma=\text{const}=0$.

2. Existence of Functions on Surfaces

2 A. Families of Harmonic Functions. We saw in Corollary 1 G that the weakness or semiweakness of a subboundary implies the degeneracy of certain families of functions on the entire surface. In particular, a parabolic surface carries no non-constant functions in the families

$HB(W)$ and $HD(W)$. To study this in further detail, we begin with a generalization of a part of Theorem 1 G:

LEMMA. *Let V be a regularly imbedded subregion of a parabolic surface W. If a function u which is subharmonic and bounded from above on V satisfies*

$$\overline{\lim_{z\in V,\, z\to\zeta}}\, u(z)\leq M \qquad \textit{at every } \zeta\in\partial V,$$

then $u\leq M$ on V.

Proof. It suffices to consider the case $M=\sup_{\zeta\in\partial V}\overline{\lim}_{z\to\zeta}u(z)$. Take an arbitrary $M'>M$. There exists a $z_0\in V$ with $u(z_0)<M'$. Consider a disk Δ about z_0 such that $\bar{\Delta}\subset V$ and $u\leq M'$ on Δ. Exhaust W by regular subregions Ω with $\bar{\Delta}\subset\Omega$. Let u_Ω be the harmonic measure $u_{\beta(\Omega)}$ on $\Omega-\bar{\Delta}$. Since $M^*=\sup_V u\ (\geq M)$ is finite, the function

$$v(z)=u(z)-M'-(M^*-M)\,u_\Omega(z)$$

is subharmonic. On the boundary of $(\Omega-\bar{\Delta})\cap V$, we have $v\leq 0$, as is easily verified by considering ∂V, $\partial\Delta$, and $\partial\Omega$ separately. Thus $v\leq 0$, so that $u(z)\leq M'+(M^*-M)\,u_\Omega(z)$ on $(\Omega-\bar{\Delta})\cap V$. Since $\lim u_\Omega=0$, we obtain $u\leq M'$ on $V-\bar{\Delta}$ and thus on V. We let $M'\to M$ and conclude that $u\leq M$ on V, as asserted.

In addition to the class O_G, we introduce the classes O_{HP}, O_{HB}, O_{HD}, and O_{HBD}, which consist of surfaces W on which $HP(W)$, $HB(W)$, $HD(W)$, and $HBD(W)$, respectively, contain constants only. In II.2 E, 2 I, we observed that the vanishing of the H-span is a necessary and sufficient condition for a surface to be of class O_{HD}.

In Corollary 1 G we obtained $O_G\subset O_{HB}$ and $O_G\subset O_{HD}$. This result is sharpened as follows:

THEOREM. (a) *W is parabolic if and only if it carries no non-constant subharmonic function which is bounded from above.*

(b) $O_G\subset O_{HP}\subset O_{HB}\subset O_{HD}=O_{HBD}$.

Remark. All inclusion relations in (b) are proper, as was proved by TÔKI [1, 2] and others by counterexamples (cf. also AHLFORS-SARIO [1]). However, for surfaces W of finite genus, it will be seen that $W\in O_G$ if and only if $W\in O_{HD}$ (Theorem 3 D).

Proof. (a) Let u be a subharmonic function bounded from above on W. Apply the lemma to u on $V=U$, a connected neighborhood of the ideal boundary. Then $\max_W u$ is attained at a point on ∂U, so that $u=\text{const.}$ Conversely, if $W\notin O_G$, then the negative of the Green's function is a non-constant subharmonic function bounded from above.

(b) $O_G \subset O_{HP}$ is immediate from (a). Since $O_{HP} \subset O_{HB} \subset O_{HBD}$ and $O_{HD} \subset O_{HBD}$ are trivial, it only remains to prove that $O_{HBD} \subset O_{HD}$. In II.2 I we showed that $W \in O_{HD}$ if and only if $S(\zeta, \zeta') = D[p_3]/\pi$ vanishes. Here $p_3 = \frac{1}{2}(p_0 - p_1)$, and p_0 and $p_1 = p_1(\mathrm{I})$ are principal functions with the singularity $\log|(z-\zeta)/(z-\zeta')|$ (II.2 G). Since $L_0\, p_0 = p_0$ and $L_1\, p_1 = p_1$ on a neighborhood of the ideal boundary, these functions are bounded there. Consequently, $p_3 \in HBD(W)$. If $W \notin O_{HD}$, then $p_3 \neq$ const, and it follows that $W \notin O_{HBD}$.

2 B. Families of Differentials. We shall need the following extension of a part of Theorem 1 H:

LEMMA. *A Riemann surface W is parabolic if and only if*

$$\int_{\partial U} \omega = 0 \qquad \textit{for every } \omega \in \Gamma_c^1(\bar{U})$$

on some (or equivalently every) neighborhood U of the ideal boundary.

Proof. Sufficiency: The condition implies $u_\beta = 0$. To prove the necessity, we exhaust W by regular subregions Ω with $\partial U \subset \Omega$ and consider the function u_Ω which is the harmonic measure $u_{\beta(\Omega)}$ on $\Omega \cap U$. Then

$$\int_{\partial U} \omega = - \int_{\Omega \cap U} du_\Omega \wedge \omega$$

(I.5 D.(2)), and $\|du_\Omega\|^2_{\Omega \cap U} = D_{\Omega \cap U}[u_\Omega] \to 0$ as $\Omega \to W$. The assertion $\int_{\partial U} \omega = 0$ follows.

From this we obtain the following result for the spaces of differentials considered in II.2 B, 3 A:

THEOREM. (a) *If W is parabolic, then*

$$\Gamma_h(W) = \Gamma_{hse}(W), \qquad \Gamma_a(W) = \Gamma_{ase}(W).$$

(b) *If $\Gamma_h(W) = \{0\}$ or, equivalently, $\Gamma_a(W) = \{0\}$, then W is planar and parabolic, and conversely.*

Remark. (a) is due to NEVANLINNA [1]. Its converse does not hold, as we shall see in 2 C.(d).

Proof. (a) Any dividing cycle c is homologous to a finite integral linear combination of the boundaries ∂U of some ends U. Since β_U is weak, every $\omega \in \Gamma_h(W)$ satisfies $\int_c \omega = 0$ by the lemma.

(b) If W is not planar, then the differential $\sigma(c)$ (II.2 A) for a non-dividing cycle c does not vanish, so that $\Gamma_h(W) \neq \{0\}$. If W is planar and not parabolic, then we shall see in 3 D that $W \notin O_{HD}$. Consequently, $\Gamma_h \supset \Gamma_{he} \neq \{0\}$. If W is planar and parabolic, then $\Gamma_h = \Gamma_{hse} = \Gamma_{he}$ by (a). Since $O_G \subset O_{HD}$, $\Gamma_{he} = 0$.

2 C. Existence of Non-Weak Subboundaries. We have observed that $O_G \subset O_\gamma$. If W possesses only one ideal boundary component, then $W \in O_G$ and $W \in O_\gamma$ are trivially equivalent.

(a) *If W has two disjoint subboundaries, one of which is non-weak, then $W \notin O_{HP}$.*

For the proof, let γ be a non-weak subboundary and let γ' be another subboundary such that $\gamma \cap \gamma' = \emptyset$. Consider the capacity functions $p_{1\gamma}$ and $p_{1\gamma'}$ with poles at the same point. The former is bounded from above and the latter is bounded from below outside of a neighborhood of the pole (III.2 F). The difference $p_{1\gamma'} - p_{1\gamma} \neq \text{const}$ is bounded from below.

(b) *If W has two disjoint subboundaries, both of which are non-weak, then $W \notin O_{HD}$.*

In fact, if γ and γ' are such subboundaries, then the modulus function $q_1(z; \gamma, \gamma') \neq \text{const}$ belongs to $HD(W)$.

(c) *Every $W \in O_{HP} - O_G$ possesses only one ideal boundary component and, therefore, $W \notin O_\gamma$.*

This is a direct consequence of (a) and 1 B.

(d) *Suppose W has more than one ideal boundary component. $\Gamma_h(W) = \Gamma_{hse}(W)$ or, equivalently, $\Gamma_a(W) = \Gamma_{ase}(W)$ if and only if at most one subboundary in any partition $\beta = \gamma \cup \gamma'$ into two isolated subboundaries is not weak.*

For example, $W \in O_{HD}$ has this property by (b).

For the proof, suppose neither γ nor γ' is weak. Then the modulus function $q_1(z; \gamma, \gamma')$ gives $\omega = * dq_1 \in \Gamma_h - \Gamma_{hse}$. Conversely, suppose W has the above property. Let c be an arbitrary dividing cycle. It determines a partition $\beta = \gamma \cup \gamma'$, for which a member, say γ, is weak. Then there is a neighborhood U of γ such that $\beta_U = \gamma$ and c is homologous to an integral linear combination of ∂U_i for some components U_i of U. Every $\omega \in \Gamma_h$ satisfies $\int_c \omega = 0$ by Lemma 2 B applied to U imbedded in a surface W_0 with compact $W_0 - U$. Consequently, every $\omega \in \Gamma_h$ belongs to Γ_{hse}.

2 D. Characterization of the Classes O_{HP}, O_{HB}, O_{HD}. Let U be a neighborhood of the ideal boundary of a Riemann surface W. We shall consider a normal operator L and the operator H acting from $\alpha = \partial U$ into U.

THEOREM. *W belongs to the indicated class if and only if one of the following conditions holds for some (or equivalently every) U:*

O_{HP}: (a) $Lu=u$ *for every* $u\in HP(\bar{U})$ *with* $\int_\alpha *du=0$.

O_{HB}: (a) $Lu=u$ *for every* $u\in HB(\bar{U})$ *with* $\int_\alpha *du=0$.

(b) $u=Hu+\text{const}\cdot u_\beta$ *for every* $u\in HB(\bar{U})$.

(b′) $u=Lu+\text{const}\cdot u_\beta$ *for every* $u\in HB(\bar{U})$.

(c) $u=\text{const}\cdot u_\beta$ *for every* $u\in H_0B(\bar{U})$.

(d) *The maximum-or-minimum principle holds for* $HB(\bar{U})$.

O_{HD}: (a) $L_\nu u=u$ *for every* $u\in HD(\bar{U})$ *with* $\int_\alpha *du=0$, $\nu=0,1$; L_1 *is for both* **I** *and* **Q**.

(b) $u=Hu+\text{const}\cdot u_\beta$ *for every* $u\in HD(\bar{U})$.

(b′) $u=L_\nu u+\text{const}\cdot u_\beta$ *for every* $u\in HD(\bar{U})$, $\nu=0,1$; L_1 *is for both* **I** *and* **Q**.

(c) $u=\text{const}\cdot u_\beta$ *for every* $u\in H_0D(\bar{U})$.

(d) *The maximum-or-minimum principle holds for* $HD(\bar{U})$.

(e) L_0 *and* L_1 *for* **I** *coincide.*

With respect to L or L_ν, the theorem is to be understood in the following sense: If W belongs to the class in question, then the corresponding condition is satisfied for all L on all U; if the condition is satisfied by some L on some U, then W belongs to the class.

As for O_{HB}, (b) is due to NEVANLINNA [2, pp. 323ff.].

Proof. (a) Suppose $W\in O_{HP}$ and take an arbitrary $u\in HP(\bar{U})$ with $\int_\alpha *du=0$. By applying Theorem I.1 F with $s=u$, we obtain a function p harmonic on W and such that $L(p-u)=p-u$ on U. The function $p=u+L(p-u)$ is bounded from below on U. Thus $p+\text{const}\in HP(W)$. We conclude that $p=\text{const}$, and therefore $Lu=u$ on U. Conversely, every $u\in HP(W)$ satisfies $u\in HP(\bar{U})$ and $\int_\alpha *du=0$, so that $Lu=u$ on U by the assumption. Therefore u takes its maximum value over W at a point on α; hence $u=\text{const}$.

The same proof applies to (a) for O_{HB} and O_{HD}.

(b), (b′) Suppose $W\in O_{HB}$ and consider an arbitrary $u\in HB(\bar{U})$. If $W\in O_G$, then $\int_\alpha *du=0$ by Theorem 1 H; if $W\notin O_G$, then there exists a constant such that $v=u+\text{const}\cdot u_\beta$ has vanishing flux $\int_\alpha *dv$. In either case, we infer from (a) that $u+\text{const}\cdot u_\beta=L_1(u+\text{const}\cdot u_\beta)=L_1u$. The relation between H and L_1 in I.5 D shows that $u=L_1u+\text{const}\cdot u_\beta=Hu+\text{const}\cdot u_\beta$ with possibly different constants. Conversely, suppose either (b) or (b′) for O_{HB} is satisfied. Any $u\in HB(W)$ belongs to $HB(\bar{U})$, so that $u=Hu+\text{const}\cdot u_\beta$ or $u=Lu+\text{const}\cdot u_\beta$ on U. The maximum principle for L (I.1 D) or for H (I.5 A) implies the following: If the constant is non-positive, then $\max_\alpha u = \sup_U u$; if it is non-negative, then $\min_\alpha u = \inf_U u$. In either case, the harmonic function u attains its maximum or its minimum over W at a point on α. Thus $u=$ const and, therefore, $W \in O_{HB}$.

The same proof can be used for (b) and (b′) with respect to O_{HD}.

(c) This is a direct consequence of (b).

(d) The proof is contained in the proof of (b), (b′).

(e) For every $f \in C^\omega(\alpha)$, we have $L_\nu f \in HD(\bar{U})$ and $\int_\alpha * dL_\nu f = 0$, $\nu = 0, 1$. If $W \in O_{HD}$, then $L_0(L_1 f) = L_1 f$ by (a). Thus $L_0 f = L_1 f$. Conversely, if $L_0 = L_1$, then the principal functions p_0 and $p_1(\mathrm{I})$ in the proof of Theorem 2 A coincide. As we showed there, $W \in O_{HD}$.

2 E. Isolated Subboundaries. Theorem 2 D gives, in particular:

(a) *The relations $W \in O_{HP}$, O_{HB}, and O_{HD} are properties of the ideal boundary.*

This conclusion enables us to make the following observation. Suppose an isolated subboundary γ of a Riemann surface W is given. A connected neighborhood U with $\beta_U = \gamma$ can be imbedded in a Riemann surface W_0 in such a way that $W_0 - U$ is compact. The condition $W_0 \in O_{HP}$ is independent of the choice of U and W_0. If it is satisfied, we shall say that the isolated subboundary γ of W is of *O_{HP}-type.* Similarly, we define subboundaries of *O_{HB}-type* and *O_{HD}-type.*

Let β be the ideal boundary of W. Partition it into a finite number of isolated subboundaries: $\beta = \gamma_1 \cup \cdots \cup \gamma_n$.

(b) *If $W \in O_{HP}$ (O_{HB}, or O_{HD}), then each of $\gamma_1, \ldots, \gamma_n$ is of O_{HP}-type (O_{HB}-type, or O_{HD}-type).*

For the proof, take disjoint neighborhoods U_i of the γ_i such that $\beta_{U_i} = \gamma_i$, $i = 1, \ldots, n$. Let $U = U_1 \cup \cdots \cup U_n$. An arbitrary $u \in HP(\bar{U}_i)$ can be extended to a function u on $\bar{U}$ by defining it to be zero on $\bar{U} - \bar{U}_i$, so that $u \in HP(\bar{U})$. If $\int_{\partial U_i} * du = 0$, then $\int_{\partial U} * du = 0$. By Theorem 2 D we see that $u = L_0 u$ on U and consequently on U_i by a property of L_0 (I.4 B). Hence γ is of O_{HP}-type. The proof is the same for O_{HB} and O_{HD}.

The converse for O_{HP} is readily verified (2 C.(c)). However, for O_{HB} and O_{HD} we have:

(c) *Suppose γ_1 and γ_2 are isolated subboundaries of W with $\beta = \gamma_1 \cup \gamma_2$ and $\gamma_1 \cap \gamma_2 = \emptyset$. Even if both γ_1 and γ_2 are of O_{HB}-type (or O_{HD}-type), it is not necessarily true that $W \in O_{HB}$ (or $\in O_{HD}$).*

In fact, using the counterexamples referred to in Remark 2 A, it is not difficult to construct a W with both γ_1 and γ_2 of O_{HB}-type (or O_{HD}-type), and neither γ_1 nor γ_2 weak. It follows that $W \notin O_{HD}$ and a fortiori $W \notin O_{HB}$ by virtue of proposition 2 C.(b), which we slightly specialize here for the sake of comparison:

(d) *Suppose* γ_1 *and* γ_2 *are isolated subboundaries of* W *with* $\beta = \gamma_1 \cup \gamma_2$ *and* $\gamma_1 \cap \gamma_2 = \emptyset$. *If neither* γ_1 *nor* γ_2 *is weak, then* $W \notin O_{HD}$.

This can also be proved directly from (c) of Theorem 2 D.

Remark. In view of Theorem 1 C, proposition (d) may be stated as follows: *If* W *has two disjoint ends* U_i *with* $H_0 B(\bar{U}_i) \neq \{0\}$ *or, equivalently,* $H_0 D(\bar{U}_i) \neq \{0\}$, $i = 1, 2$, *then* $W \notin O_{HD}$ *and, therefore,* $W \notin O_{HB}$. The converse is not true: a counterexample is the unit disk. A complete characterization of this nature is obtained upon replacing ends by arbitrary subregions. In fact, the following result has been obtained by Bader-Parreau [1]: $W \notin O_{HB}$ *(or* $\notin O_{HD}$*) if and only if* W *has two disjoint regularly imbedded subregions* V_i *with* $H_0 B(\bar{V}_i) \neq \{0\}$ *(or* $H_0 D(\bar{V}_i) \neq \{0\}$, *respectively).* Note again that $H_0 P(\bar{V}) \neq \{0\}$ for any V which is not relatively compact and, therefore, a criterion of this kind is not available for $W \notin O_{HP}$ (cf. Remark 2 in 1 C).

2 F. Relation between $HB(\bar{U})$ and $HD(\bar{U})$. Theorem 2 D enables us to complement the result given in Remark 1 F as follows (the arrow stands for implication):

If U *is a neighborhood of the ideal boundary, then*

$$W \in O_{HB} \Rightarrow HB(\bar{U}) = HD(\bar{U}),$$

$$W \in O_{HD} \Rightarrow HB(\bar{U}) \supset HD(\bar{U}).$$

In fact, if $W \in O_{HB}$, then $u \in HB(\bar{U})$ is expressible as $L_0 u + \text{const} \cdot u_\beta$, which belongs to $HD(\bar{U})$, and therefore $HB(\bar{U}) \subset HD(\bar{U})$. Similarly, if $W \in O_{HD}$, then $HD(\bar{U}) \subset HB(\bar{U})$. Since $O_{HB} \subset O_{HD}$, $W \in O_{HB}$ implies the equality.

The converse statements are not true. The surface W of 2 E.(c) is a counterexample. Its ideal boundary consists of two disjoint isolated subboundaries γ_1 and γ_2 which are of O_{HB}-type but are not weak. Let U_1 and U_2 be disjoint neighborhoods of γ_1 and γ_2 with $\beta_{U_i} = \gamma_i$, $i = 1, 2$. Set $U = U_1 \cup U_2$. Any $u \in HB(\bar{U})$ belongs to $HB(\bar{U}_i) = HD(\bar{U}_i)$, so that $u \in HD(\bar{U})$. Since γ_1 and γ_2 are of O_{HD}-type as well, $u \in HD(\bar{U})$ belongs similarly to $HB(\bar{U})$. Accordingly, $HB(\bar{U}) = HD(\bar{U})$, yet $W \notin O_{HB}$ by 2 E.(d).

Generally speaking, $HB(\bar{U}) = HD(\bar{U})$ is equivalent to $H_0 B(\bar{U}) = H_0 D(\bar{U})$, as is easily seen by using the operator H or L_ν. On the other hand, in terms of the dimension of the vector space $H_0 B(\bar{U})$ or $H_0 D(\bar{U})$, Theorem 1 C and (c) of Theorem 2 D are expressed by the following scheme (where the double arrow stands for "if and only if"):

$$W \in O_G \Leftrightarrow \dim H_0 B(\bar{U}) = 0 \Leftrightarrow \dim H_0 D(\bar{U}) = 0$$

$$W \in O_{HB} \Leftrightarrow \dim H_0 B(\bar{U}) \leq 1, \qquad W \in O_{HD} \Leftrightarrow \dim H_0 D(\bar{U}) \leq 1.$$

The identity of $H_0 B(\bar{U})$ and $H_0 D(\bar{U})$ implies merely the equivalence of dimensions, and not that these dimensions are at most 1. This observation clearly explains why the converse of the above result is not true.

For properties of classes of Riemann surfaces with finite dimensional spaces $H_0 B(\bar{U})$ or $H_0 D(\bar{U})$, the reader is referred to CONSTANTINESCU-CORNEA [1].

2 G. The Classes O_{KB} and O_{KD}. We consider the subfamilies $KB(W)$, $KD(W)$, and $KBD(W)$ of $HB(W)$, $HD(W)$, and $HBD(W)$ consisting of functions u with semiexact $*du$. The classes of Riemann surfaces on which these families consist only of constants will be denoted by O_{KB}, O_{KD}, and O_{KBD}, respectively. The following relations are evident:

$$O_{HB} \subset O_{KB}, \qquad O_{HD} \subset O_{KD} \subset O_{KBD}.$$

For a plane region W, exactness and semiexactness are equivalent. If $u \in KB(W)$, then $\exp(u + i u^*)$ is a bounded analytic function. Conversely, if $\exp(u + i u^*)$ is bounded, then so is u. Accordingly, $W \in O_{KB}$ if and only if W carries no non-constant bounded analytic function. Similarly, we can verify that $W \in O_{KD}$ and $W \in O_{KBD}$ are equivalent to the degeneracy of the corresponding families of analytic functions.

The class O_{KD} was discussed in II. 2 E, 2 G, 2 I by means of principal functions for **Q** in the same manner as O_{HD} was treated using principal functions for **I**. For example, the vanishing of the K-span is a necessary and sufficient condition for a surface to be in the class O_{KD}. Also, the relation

$$O_{KB} \subset O_{KD} = O_{KBD}$$

is obtained in the same way as Theorem 2 A.

Let U be a neighborhood of the ideal boundary. Assume further that each component of $\alpha = \partial U$ is a dividing cycle. Denote by $KB(\bar{U})$ and $KD(\bar{U})$ the subfamilies of $HB(\bar{U})$ and $HD(\bar{U})$ consisting of functions u such that $*du$ is semiexact relative to α. Introduce $K_0 B(\bar{U})$ and $K_0 D(\bar{U})$ similarly. By considering the operators L_0 and L_1 for **Q** acting from α into U, we obtain the following analogue of Theorem 2 D; observe that it is somewhat similar to Theorems 1 C, 1 F, 1 G, even though the proof is parallel to that of Theorem 2 D:

THEOREM. *A Riemann surface W belongs to the indicated classes if and only if one of the following conditions holds for some (or equivalently every) connected neighborhood of the ideal boundary.*

O_{KB}: (a) $u = L_\nu u$ *for every* $u \in KB(\bar{U})$, $\nu = 0, 1$.

(b) $K_0 B(\bar{U}) = \{0\}$.

(c) *The maximum-minimum principle holds for* $KB(\bar{U})$.

O_{KD}: (a) $u = L_\nu u$ *for every* $u \in KD(\bar{U})$, $\nu = 0, 1$.

(b) $K_0 D(\bar{U}) = \{0\}$.

(c) *The maximum-minimum principle holds for* $KD(\bar{U})$.

Condition (a) is to be understood as follows: If W belongs to the class in question, then (a) is true for L_0 and L_1 for $\mathbf{Q}$; conversely, if (a) is satisfied by either L_0 or L_1 for $\mathbf{Q}$, then W belongs to the class.

The proof is left to the reader.

The following consequences are derived as before:

(α) $W \in O_{KB}$ *implies* $KB(\bar{U}) = KD(\bar{U})$. $W \in O_{KD}$ *implies* $KB(\bar{U}) \supset KD(\bar{U})$.

(β) *If* $\Gamma_{hse}(W) = \{0\}$ *or, equivalently,* $\Gamma_{ase}(W) = \{0\}$, *then* W *is planar and* $W \in O_{KD}$, *and conversely.*

2 H. Isolated Subboundaries. From the above theorem, we observe:

(a) *The relations* $W \in O_{KB}$ *and* $W \in O_{KD}$ *are properties of the ideal boundary.*

As a consequence, we can define an isolated subboundary of O_{KB}-*type* and O_{KD}-*type* in exactly the same fashion as in 2 E. Regarding the partition $\beta = \gamma_1 \cup \cdots \cup \gamma_n$ into isolated subboundaries, the conclusion is quite similar to that in 1 B:

(b) $W \in O_{KB}$ *(or* O_{KD}*) if and only if all the* $\gamma_1, \ldots, \gamma_n$ *are of* O_{KB}-*type (or* O_{KD}-*type).*

(c) *Suppose* γ *is a (not necessarily isolated) subboundary of a* W *such that every isolated subboundary of* $\beta - \gamma$ *is of* O_{KB}-*type (or* O_{KD}-*type). Then*

$$\gamma \text{ is weak} \Leftrightarrow \gamma \text{ is semiweak} \Rightarrow W \in O_{KB} \ (\text{or } O_{KD}, \text{ resp.}).$$

Indeed, by assumption we have $H_0 B_\gamma^1(\bar{U}) = H_0 B_\gamma^0(\bar{U}) \supset K_0 B(\bar{U})$ or $H_0 D_\gamma^1(\bar{U}) = H_0 D_\gamma^0(\bar{U}) \supset K_0 D(\bar{U})$, respectively, for any neighborhood U of the ideal boundary.

It is to be noted that, if W is of finite genus and γ is a boundary component, then the three properties in (c) are respectively equivalent (cf. 2 L and 3 A). For an arbitrary surface, however, this is not true; any surface of class $O_{HP} - O_G$ is a counterexample (2 C.(c)).

Remark. The classes O_{KB}, O_{KBD}, and O_{KD} were introduced in SARIO [6] and studied further by ROYDEN [2].

2 I. The Classes O_{AB} and O_{AD}. Denote by $AB(W)$ and $AD(W)$ the families of analytic functions which are bounded and have finite Dirichlet integrals, respectively, on a Riemann surface W. Set $ABD(W)=AB(W)\cap AD(W)$. The classes of Riemann surfaces on which these families consist entirely of constants will be denoted by O_{AB}, O_{AD}, and O_{ABD}. From the definition, we immediately obtain

$$O_{KB}\subset O_{AB}, \qquad O_{KD}\subset O_{AD}\subset O_{ABD}.$$

For planar regions W, we remarked in 2 G that

$$O_{KB}=O_{AB}\subset O_{KD}=O_{AD}=O_{ABD} \qquad (W \text{ planar}).$$

In VII.5 D, we introduced the quantities $c_B(\zeta)$ and $c_D(\zeta)$. We showed that they vanish (independently of the reference point $\zeta\in W$) if and only if $W\in O_{AB}$ and $W\in O_{AD}$, respectively.

We shall say that a plane point set E which is closed and bounded belongs to the *class* N_B (or N_D) if the unbounded component W of CE belongs to the class O_{AB} (or O_{AD}, resp.) (AHLFORS-BEURLING [1]). Sets E in these classes are readily seen to be totally disconnected. Since $W\in O_{KB}$ and $W\in O_{KD}$ are properties of the ideal boundary (2 H), the relations $E\in N_B$ and $E\in N_D$ are conformally invariant in the sense of VII.6 E: if $E\in N_B$ (or N_D) is mapped onto E' by a conformal mapping of a region containing E, then $E'\in N_B$ (or N_D).

We assert that, for Riemann surfaces of arbitrary genus,

$$O_{AB}\subset O_{AD}.$$

In fact, if W carries a non-constant analytic function F with a finite Dirichlet integral, then the image region $F(W)$ in the w-plane has finite area, so that the identity function belongs to $AD(F(W))$. As we observed above, the plane region $F(W)$ then carries a non-constant $F_1\in AB(F(W))$, and we have the non-constant $F_1\circ F\in AB(W)$.

Whether or not O_{AD} coincides with O_{ABD} is unknown.

Remark. The classes O_{AD}, O_{AB} and O_{ABD} were introduced in SARIO [1, p. 76; 2, 6]. The classes N_D and N_B were studied by AHLFORS-BEURLING [1], the class N_B also by KAMETANI [1].

2 J. Myrberg's Example. In general, the theory of the existence of analytic functions on Riemann surfaces is more ramified than that of harmonic functions. In the former case there is, in particular, a striking difference between surfaces of finite genus and of infinite genus. For finite genus we shall show in 3 H that $O_{KB}=O_{AB}$ and $O_{KD}=O_{AD}=O_{ABD}$, and that the boundaries of surfaces in these classes are small; typically, $E\in N_B$ and $E\in N_D$ are both totally disconnected. In contrast, MYRBERG [2] showed:

On a Riemann surface W of infinite genus, there may exist a proper continuum E such that $W-E$ belongs to the class O_{AB} and a fortiori to O_{AD}.

Myrberg's example is constructed as follows. Let W be the 2-sheeted plane $|z|<\infty$, with the sheets connected cross-wise along the cuts $[2n, 2n+1]$, $n=1,2,\dots$, on the real axes. Denote by E the union of the disk $|z-2i|\le 1$ in the upper sheet and the disk $|z+2i|\le 1$ in the lower sheet. To show that $W-E$ belongs to O_{AB}, consider a bounded analytic function F on $W-E$. Given $z\in W$, let $\tilde{z}$ be the corresponding point on the other sheet. The function $(F(z)-F(\tilde{z}))^2=f(z)$ is symmetric, $f(z)=f(\tilde{z})$, so that it can be regarded as a function on the plane with the projection of E removed. Since f is bounded, the point at infinity is removable; and since f vanishes at $z=2,3,\dots$, it is identically zero. Consequently, F is symmetric and can be viewed as a function on the plane. The surface $W-E$ entirely covers the finite plane $|z|<\infty$, where F is analytic and bounded, so that $F=\text{const}$. We conclude that $W\in O_{AB}$.

Remark. KURAMOCHI [2] further obtained the following incisive result: If $W\in O_{HB}-O_G$, then $W-E\in O_{AB}$ for *every* compact set $E\subset W$; similarly, if $W\in O_{HD}-O_G$, then $W-E\in O_{AD}$.

From the above example of MYRBERG we deduce:

$W\in O_{AB}$ *and* $W\in O_{AD}$ *are not properties of the ideal boundary.*

In fact, let E_1 be the union of the disks $|z-2i|\le 1$ on the two sheets. Then $W-E$ and $W-E_1$ permit a conformal mapping between neighborhoods of the ideal boundaries, while the latter evidently does not belong to O_{AD} and a fortiori not to O_{AB}.

2 K. Quantities $c_B(\zeta)$ and $c_D(\zeta)$ on a Riemann Surface. On an open Riemann surface W, consider a point $\zeta\in W$ and choose a local parameter about it. Let $\mathscr{A}_\zeta$ be the family of analytic functions F on W such that $F(\zeta)=0$ and $F'(\zeta)=1$ for the fixed parameter. As in VII.5 D, we introduce the quantities $c_B(\zeta)$ and $c_D(\zeta)$ as follows:

$$\frac{1}{c_B(\zeta)}=\min_{F\in\mathscr{A}_\zeta} M[F], \qquad \frac{1}{c_D(\zeta)}=\min_{F\in\mathscr{A}_\zeta}\left(\frac{1}{\pi}\mathscr{D}[F]\right)^{\frac{1}{2}}.$$

These depend on the parameter about ζ in such a way that $c_B(\zeta)\,|d\zeta|$ and $c_D(\zeta)\,|d\zeta|$ are invariant.

From the definition, it is evident that $W \in O_{AB}$ if and only if $c_B(\zeta)=0$ at *every* $\zeta \in W$. Similarly $W \in O_{AD}$ if and only $c_D(\zeta)=0$ at *every* $\zeta \in W$.

As was anticipated in Remark VIII. 5 E, the word "every" above cannot in general be replaced by "some". This was pointed out by VIRTANEN [2]:

There exists a Riemann surface W such that $W \notin O_{AB}$ and therefore $W \notin O_{AD}$, yet there exists a $\zeta \in W$ where $dF=0$ for every $F \in AB(W) \cup AD(W)$.

For the construction of the example, take two copies of $\{z \mid 1<|z|<\infty\}$. Connect them cross-wise along the cuts $[2n, 2n+1]$, $n=1, 2, \dots$. The resulting surface W clearly does not belong to O_{AD}. On the other hand, every $F \in AB(W)$ is symmetric with respect to the interchange of the sheets, as is seen by an argument similar to that in 2 J. Therefore, $dF=0$ at a branch point ζ; in other words, $c_B(\zeta)=0$. We also conclude that $c_D(\zeta)=0$ at a branch point ζ since $F \in AD(W)$ has a bounded real part in a neighborhood of the ideal boundary component lying over $z=\infty$ (cf. Remark 1 F).

2 L. Families of Univalent Functions. On a plane region we consider the family of univalent analytic functions. We briefly review the results obtained thus far.

In VI. 2 D and VI. 4 B, we observed that $W \in O_{AD}$ is equivalent to each of the following properties:

(a) All univalent analytic functions on W are linear transformations.

(b) All univalent analytic functions on W have image regions whose complements have vanishing area.

In VII. 5 C, we saw that the property $W \in O_\gamma$ is equivalent to each of the following conditions:

(α) All univalent functions on W have image regions whose complements are totally disconnected. Equivalently, all univalent functions on W can be extended to topological mappings of the Riemann sphere onto itself.

(β) W carries no univalent functions in $AB(W)$.

(γ) W carries no univalent functions in $AD(W)$.

As a consequence, we obtain:

$$O_{AD} \subset O_\gamma \quad \text{for plane regions}.$$

Observe the contrast with the case of general Riemann surfaces, where we have $(O_{HP}-O_G) \cap O_\gamma = \emptyset$ (2 C. (c)). Note that a surface in $O_{HP}-O_G$ again has infinite genus (Remark 2 A).

We shall say that a closed bounded plane set E is of class N_{SB} if the unbounded component W of CE carries no SB-function; here "S" stands for "univalent" (schlicht). Clearly E is totally disconnected. By (c), we may also say that E is of class N_{SD}. Using the notation for the corresponding null classes of plane regions, these results (AHLFORS-BEURLING[1], SARIO [9]) can be written

$$O_\gamma = O_{SB} = O_{SD}.$$

3. Removability and Related Topics

3 A. Isolated Subboundary Realized as a Set. Suppose γ_0 is an isolated subboundary of a Riemann surface W, possessing a neighborhood U every component of which is of finite genus. We may assume without loss of generality that U is connected and that $\beta_U = \gamma_0$. It is known that an open Riemann surface U of finite genus can be imbedded in a closed Riemann surface (BOCHNER [2]). Thus it is possible to imbed U into a compact bordered Riemann surface $\bar{U}^*$ in such a way that the relative boundary ∂U of U in W coincides with the border of $\bar{U}^*$. Let U^* be the interior of $\bar{U}^*$ and denote by W^* the Riemann surface obtained in a natural manner from the space $W \cup U^*$ by identifying the corresponding points in U. Then W is a subregion of W^*. The set $S = U^* - U = W^* - W$ is compact and so is the set $E = \partial W$ (the relative boundary of W in W^*). The subboundary γ_0 is realized as E in the sense of I. 1 B; more precisely, by the definition of the subboundary γ_0 as a correspondence $\Omega \to U_\Omega$, we obtain $E = \bigcap \bar{U}_\Omega$, where the closure is taken in W^*.

Note that U and U^* can be taken to be planar, but here we do not make this restriction.

Conversely, suppose W is a subregion of a (closed or open) Riemann surface W^* such that $S = W^* - W$ is compact. Let E be the relative boundary of W in W^*; it is also compact. We obtain a subboundary γ_0 of W as the correspondence $\Omega \to U_\Omega$, where U_Ω is the union of those components of $W - \bar{\Omega}$ whose closures in W^* meet E. Clearly this subboundary is realized as the set E. It is possible to find a regular subregion U^* of W^* such that $S \subset U^*$. Then $U = W \cap U^*$ is a connected boundary neighborhood of γ_0. It satisfies $\beta_U = \gamma_0$, so that γ_0 is isolated. Furthermore, U is of finite genus.

It is to be noted that, if E is totally disconnected (in which case $E = S$), then U^* and U can be taken to be planar.

In general, on a Riemann surface W^*, consider a compact set E which is contained in a planar subregion D. We can imbed D into the Riemann sphere in such a way that $\infty \notin D$. If the set E, thus regarded as a closed bounded plane set, has vanishing logarithmic capacity, we say that E

(on the Riemann surface) has vanishing logarithmic capacity. This property is independent of the choice of D and the imbedding of D in the Riemann sphere, since the vanishing of logarithmic capacity is conformally invariant (VII. 6 E).

Similarly, we can speak of a *compact set of class* N_B, N_D, *or* N_{SB} *on a Riemann surface*, provided it is contained in a planar region.

THEOREM. *An isolated subboundary* γ_0 *is weak (of* O_{KB}*-type,* O_{KD}*-type) if and only if the corresponding set* E *has logarithmic capacity zero (is of class* N_B, N_D, *respectively). Every boundary component of* W *contained in* γ_0 *is weak if and only if* E *is of class* N_{SB}.

Remark. This result and Theorems 3 F, 3 H below are due to A. MORI [1] and ROYDEN [2].

3 B. Proof. The sufficiency is trivial since the requirements for γ_0 are boundary properties. The necessity is also evident for the same reason, if we show that S can be contained in a planar subregion of W^*. We shall do this by proving that S is totally disconnected. If the latter were not the case, we could replace W^* by another surface, also denoted by W^*, with the property that $S=W^*-W=U^*-U$ contains an interior point. Let Δ be a parametric disk contained in S.

Regarding the first part of the theorem, imbed U^* into a closed surface W_0 and view U as a neighborhood of the ideal boundary of the Riemann surface W_0-S. Then we have the principal function $p_{1m}(z;\zeta)$ on W_0 with $\zeta\in\Delta$ (II. 2 C). Its restriction to W_0-S belongs to the family KD. Accordingly, γ_0 is not of O_{KD}-type and a fortiori neither of O_{KB}-type nor weak.

Concerning the second part of the theorem, we shall show that there exists a non-weak ideal boundary component of W contained in γ_0. Let E_1 be the boundary of the component of S containing Δ. It is a proper continuum and a component of E. There is evidently an ideal boundary component γ of W with $\gamma\subset\gamma_0$ which is realized as E_1. We shall show that it is not weak. To this end, we take a parametric disk $\Delta_1\subset U$ and consider the family Γ of curves joining $\partial\Delta_1$ and $\partial\Delta$ in $U^*-(\bar{\Delta}\cup\bar{\Delta}_1)$. Its extremal length $\lambda(\Gamma)$ is clearly finite. The restriction of Γ to $U-\bar{\Delta}_1$ is simply the family $\tilde{\Gamma}^*$ considered in IV. 2 B. Since its extremal length is again finite, we conclude that γ is not weak.

Remark 1. The reasoning in the last paragraph can be applied to a slightly more general situation, and we obtain the following result of JURCHESCU [1], stated in our terminology of I. 1 B: *Suppose a part of an ideal boundary component* γ *of an open Riemann surface is realized as a set* E. *If* E *contains a proper continuum, then* γ *is not weak.*

Remark 2. The above argument remains valid, mutatis mutandis, in all degeneracy phenomena we shall consider in the remainder of 3. Without loss of generality, we shall, therefore, *restrict our attention in* 3 C – 3 I *to surfaces* W *with a totally disconnected* $S = E = W^* - W$.

3 C. Surfaces of Finite Genus. Consider an open Riemann surface W of finite genus. Its ideal boundary satisfies the hypothesis of the first paragraph of 3 A. Furthermore, it is possible to take W^* there as a closed surface. The following is then immediate from Theorem 3 A.

THEOREM. (a) *A surface of finite genus is parabolic if and only if it is obtained from a closed surface by deleting a compact set of logarithmic capacity zero.*

(b) *A surface of finite genus is of class* O_γ *if and only if it is obtained from a closed surface by deleting a compact set of class* N_{SB}.

By the same reasoning, we see that surfaces of finite genus in the classes O_{KB} and O_{KD} are those obtained by removing compact sets of classes N_B and N_D, respectively. We shall further show in 3 H that surfaces of finite genus belonging to O_{AB} and O_{AD} also have these properties.

3 D. Surfaces of Finite Genus (continued). We assert:

THEOREM. *For surfaces of finite genus, the classes* O_G, O_{HP}, O_{HB}, *and* O_{HD} *coincide.*

More generally, we have:

COROLLARY. *For an isolated subboundary* γ *with a neighborhood of finite genus, the following four properties are equivalent:* γ *is weak, of* O_{HP}*-type, of* O_{HB}*-type, or of* O_{HD}*-type.*

Proof of the Theorem. It suffices to show that a $W \notin O_G$ does not belong to O_{HD}. To this end, we imbed W in a closed surface W^* and recall from Remark 2 in 3 B that a totally disconnected $E = W^* - W$ is the only nontrivial case. By assumption, E has positive logarithmic capacity. We can choose two compact subsets E_1 and E_2 such that $E_1 \cap E_2 = \emptyset$, Cap $E_k > 0$, $k = 1, 2$. For the proof, we may consider E in a plane. Then there exists a point $z_1 \in E$ all of whose neighborhoods N_1 satisfy $\mathrm{Cap}(E \cap \bar{N}_1) > 0$, since otherwise we would obtain Cap $E = 0$ (apply XI. 2 B). If N_1 is sufficiently small, then $\mathrm{Cap}(E - N_1)$ is still positive, as is verified using XI. 2 B and the fact that the logarithmic capacity of a disk is equal to its radius (VII. 6 D). Therefore, we can take a $z_2 \in E - N_1$ with neighborhood N_2 such that $\bar{N}_1 \cap \bar{N}_2 = \emptyset$ and $\mathrm{Cap}(E \cap \bar{N}_2) > 0$. The sets $E_k = E \cap \bar{N}_k$, $k = 1, 2$, have the desired properties. The surface $W' = W^* - (E_1 \cup E_2)$ has two disjoint subboundaries realized as E_1 and E_2. They are not weak, so that $W' \notin O_{HD}$ by 2 C. (b). Since $W \subset W'$, we conclude that $W \notin O_{HD}$.

3 E. Removability of a Set of Logarithmic Capacity Zero. Suppose E is a totally disconnected compact set on a Riemann surface W^* and set $W = W^* - E$ (cf. Remark 2 in 3 B). Take a regular subregion U^* of W^* with $E \subset U^*$; the closure $\overline{U}$ of $U = U^* - E$ is understood to be in the topology of W^*.

THEOREM. *E has logarithmic capacity zero if and only if one of the following is true for some (or equivalently every) U^*:*

(a) *Every $u \in HB(\overline{U} \cap W)$ is harmonically extendable to U^*.*

(b) *Every $u \in HD(\overline{U} \cap W)$ is harmonically extentable to U^*.*

In short, Cap E vanishes if and only if E is *removable* with respect to functions in the family HB or HD.

Proof. Consider the subboundary γ of W realized as E. Then U is a boundary neighborhood of γ. If (a) is satisfied, every $u \in H_0 B(\overline{U} \cap W)$ is extendable to U^*, so that $u = 0$. By Theorem 1 C we see that γ is weak, and thus Cap $E = 0$. Conversely, if Cap $E = 0$, then γ is weak, so that $H_0 B(\overline{U} \cap W) = \{0\}$. For any $u \in HB(\overline{U} \cap W)$, let v be the harmonic function on $\overline{U}^*$ such that $u = v$ on $\partial U \cap W$. Since $u - v \in H_0 B(\overline{U} \cap W)$, we obtain $u = v$ on U; hence u is harmonically extendable to U^*. The proof of (b) is analogous.

3 F. Removability with Respect to Analytic Functions. We retain the above notation.

THEOREM. *E is of the indicated class if and only if the given condition is true for some (or equivalently every) U^*:*

N_B: Every $f \in AB(\overline{U} \cap W)$ is analytically extendable to U^*.

N_D: Every $f \in AD(\overline{U} \cap W)$ is analytically extendable to U^*.

N_D: Every $f \in ABD(\overline{U} \cap W)$ is analytically extendable to U^*.

Proof. We shall give the proof for N_B only, since the other cases are analogous. Imbed U^* in a closed surface W_0. Then $W_1 = W_0 - E$ is a Riemann surface of finite genus, and U is a neighborhood of the ideal boundary of W_1. As we remarked immediately after Theorem 3 C, the set E is of class N_B if and only if W_1 is of class O_{KB}. Thus the proof will be complete if we show that (a) $W_1 \in O_{KB}$ if and only if $W_1 \in O_{AB}$, (b) $W_1 \in O_{AB}$ if and only if every $f \in AB(\overline{U} \cap W)$ is analytically extendable to U^*. The former will be proved in 3 H and the latter in 3 G.

3 G. Auxiliary Result. Using the above notation, we state:

LEMMA. *$W_1 \in O_{AB}$ (O_{AD}, or O_{ABD}) if and only if every $f \in AB(\overline{U} \cap W)$ ($AD(\overline{U} \cap W)$, or $ABD(\overline{U} \cap W)$, resp.) is analytically extendable to U^*.*

Proof. We shall give the proof for O_{ABD} and $ABD(\bar{U} \cap W)$. The other cases are analogous and somewhat simpler.

The sufficiency is clear: every $f \in ABD(W_1)$ is extendable to W_0, a closed surface, so that $f \equiv$ const and $W_1 \in O_{ABD}$.

To prove the necessity, we suppose that the condition is not satisfied and show that $W_1 \notin O_{ABD}$.

First consider the case in which $E = W_0 - W_1$ has vanishing area. Let $f \in ABD(\bar{U} \cap W)$ be a function which is not analytically extendable to U^*. We eliminate all removable singularities from E and denote by E' the resulting set, the totality of non-removable singularities. By assumption, it is non-void and totally disconnected (cf. Remark 2 in 3 B). The function f can be extended to $U' = U^* - E'$, where it is bounded. Further, we have assumed that the area of E is zero, so that the Dirichlet integral of f is not affected under this extension, and we have $f \in ABD(\bar{U}' \cap W)$. Let g be the genus of W_1. Since E' is totally disconnected, we can take regions $N_1, \ldots, N_{2g+1} \subset U^*$ such that $\bar{N}_k \cap \bar{N}_l = \emptyset$ if $k \neq l$, $N_k \cap E' \neq \emptyset$, and $\partial N_k \cap E' = \emptyset$. Let f_k be the restriction of f to $N_k - E'$. It belongs to the class $ABD(N_k - E')$ and is not extendable to $E'_k = N_k \cap E'$.

Consider the Riemann surface $W_0 - E'_k$. The set $N_k - E'_k$ is a neighborhood of its ideal boundary on which $u_k = \operatorname{Re} f_k$ has vanishing flux. By Theorem I.1 F, there exists a harmonic function v_k on $W_0 - E'_k$ satisfying $L_0(v_k - u_k) = v_k - u_k$ on $N_k - E'_k$. Clearly $v_k \in KBD(W_0 - E'_k)$. There exist constants $c_1, \ldots, c_{2g+1}$ such that $(c_1, \ldots, c_{2g+1}) \neq (0, \ldots, 0)$ and the differential $*d(\sum c_k v_k)$ has zero periods. Then

$$f_0 = \sum_{k=1}^{2g+1} c_k (v_k + i v_k^*)$$

is a single-valued analytic function on $W_0 - E'$. Clearly its Dirichlet integral is finite and $\operatorname{Re} f_0$ is bounded. Furthermore, $f_0 \neq$ const, since otherwise $-c_k(v_k + i v_k^*) = \text{const} + \sum_{l \neq k} c_l (v_l + i v_l^*)$ with $c_k \neq 0$ would show that f_k is extendable to E'_k, contrary to the assumption. Hence $\exp f_0$ is a non-constant ABD function on W_1 and, therefore, $W_1 \notin O_{ABD}$.

If E has positive area, we consider a planar regular region Ω with $E \subset \Omega$ and $\bar{\Omega} \subset U^*$. Imbed it into the Riemann sphere $\mathbf{S}$ in such a way that $\infty \notin \bar{\Omega}$. The region $\mathbf{S} - E$ has positive span (Corollary VI.4 B), so that it carries a univalent function which is not extendable to E (Theorem VI.2 D). We may assume that it has a pole at ∞ and, therefore, its restriction f to $\Omega - E$ belongs to $ABD(\Omega - E)$. Delete the removable singularities of f from E and denote the remaining set by E'. The extended f is still univalent (cf. VI.1 D), so that it belongs to $ABD(\Omega - E')$. We now carry out the same process as before upon replacing U^* by Ω and conclude again that $W_1 \notin O_{ABD}$.

3 H. Surfaces of Finite Genus. We shall prove:

THEOREM. *For surfaces of finite genus, the classes O_{KB} and O_{AB} coincide, and so do O_{KD}, O_{AD}, and O_{ABD}. These classes consist of precisely those surfaces which are obtained from closed surfaces by deleting compact sets of classes N_B and N_D, respectively.*

Proof. It suffices to prove that $W \notin O_{KB} \Rightarrow W \notin O_{AB}$ and $W \notin O_{KD} \Rightarrow W \notin O_{ABD}$. We shall establish the latter. Imbed W into a closed surface W^* and let $E = W^* - W$. The assumption $W \notin O_{KD}$ implies that $E \notin N_D$. Consider a planar regular region $\Omega \subset W^*$ with $E \subset \Omega$ and imbed it into the Riemann sphere S. Since $E \notin N_D$, the plane region $\mathbf{S} - E$ does not belong to O_{ABD} (2 I). On applying Lemma 3 G to $W_1 = \mathbf{S} - E$, we observe that E is not removable for a function in the family $ABD(\Omega - E)$. By the same lemma with $W_1 = W^* - E$, we then conclude that $W = W^* - E \notin O_{ABD}$.

3 I. Removability of a Set of Class N_{SB}. As in 3 E, let E be a totally disconnected compact set on a Riemann surface W^*. Take a planar regular subregion U^* of W^* with $E \subset U^*$.

THEOREM. *E is of class N_{SB} if and only if every univalent analytic function on U is continuously extendable to U^*.*

Proof. If E is of class N_{SB}, then every ideal boundary component of $W = W^* - E$ realized as a point of E is weak. It is mapped to a single point by an arbitrary univalent function f on U. Thus f is extendable to a continuous function on U^*; for the verification, use Lemma VI. 1 D. This reasoning is reversed to show that $E \in N_{SB}$ if every f is continuously extendable to U^*.

4. Boundary Behavior of Functions

4 A. Boundary Behavior of Harmonic Functions. From Theorems 3 A and 3 E and the fact that a set of logarithmic capacity zero is totally disconnected, we obtain:

Suppose γ_0 is an isolated weak subboundary with a connected boundary neighborhood U of finite genus. Then for an arbitrary boundary component γ contained in γ_0 and for every $u \in HB(\bar{U})$ or $\in HD(\bar{U})$, $\lim_{z \to \gamma} u(z)$ exists.

If no U is of finite genus, this conclusion is not valid, as was shown by a counterexample of HEINS [3] and KURAMOCHI [1].

THEOREM. *A weak isolated subboundary γ_0 does not always have the property that*

$$\lim_{z \to \gamma} u(z)$$

exists for every $u \in HB(\bar{U})$ and every boundary component γ contained in γ_0.

Remark. HEINS [3] introduced the concept of *harmonic dimension* and established the following theorem: The limiting value exists for every $u \in HB(\bar{U})$ if and only if γ has harmonic dimension 1.

4 B. Proof of the Theorem. To construct a counterexample, take sequences $\{a_n\}$ and $\{b_n\}$ of positive real numbers such that $b_{n+1} < a_n < b_n < 1$, $n = 1, 2, \ldots$, and $\lim a_n = \lim b_n = 0$. Delete the closed intervals $[a_n, b_n]$ on the real axis from the punctured extended plane $0 < |z| \leq \infty$, and denote the remaining region by D. Connect two copies of D crosswise along the slits to form a surface W with only one ideal boundary component γ over $z = 0$. It is easily seen that W is parabolic, for example, by using extremal length or the modular test (XI. 1 A). Define a neighborhood U of γ as the part of W lying over $|z| < 1$. Let u be the function in $HB(\bar{U})$ which is equal to 1 on the circle $|z| = 1$ in the upper sheet and equal to -1 on the corresponding circle in the lower sheet. It is obtained by using the linear operator L_0, say, and is uniquely determined. We shall show that if the $b_n - a_n$, $n = 1, 2, \ldots$, are taken sufficiently small, then $\lim_{z \to \gamma} u(z)$ does not exist.

The function u satisfies $u(z) = -u(\tilde{z})$, where z and $\tilde{z}$ have the same projection. The restriction v of u to the upper sheet of U, that is, to the region $D_0 = \{z \mid 0 < |z| < 1\} - \bigcup_{n=1}^{\infty} [a_n, b_n]$, is the bounded harmonic function with the boundary values $v = 1$ on $|z| = 1$ and $v = 0$ on $\bigcup [a_n, b_n]$; because of Lemma 2 A, this suffices to characterize v. For the proof of the theorem, we need only show that

$$\overline{\lim_{x \to 0-}}\, v(x) > 0,$$

since, by symmetry, the approach from the lower sheet gives a negative $\underline{\lim}$ and therefore the non-existence of $\lim_{z \to \gamma} u(z)$.

Let u_n be the harmonic function in $\{z \mid |z| < 1\} - [a_n, b_n]$ with $u_n = 0$ on $|z| = 1$ and $u_n = 1$ on $[a_n, b_n]$. By taking $b_n - a_n$ sufficiently small, $u_n(0)$ can be made as small as we wish (for explicit estimation, cf. XI. 5 G). We obtain $(1 - v(z)) \leq \sum_{n=1}^{\infty} u_n(z)$ on D_0 (cf. Lemma 2 A). In view of $u_n(x) < u_n(0)$ for $x < 0$, it follows that

$$\underline{\lim_{x \to 0-}}\, (1 - v(x)) \leq \overline{\lim} \sum u_n(x) \leq \sum u_n(0).$$

If $\sum u_n(0) < 1$, we conclude that $\underline{\lim}\, v(x) > 0$. This proves the theorem.

4 C. Boundary Behavior of Analytic Functions. Theorem 3 F can be interpreted in analogy with the beginning of 4 A. In contrast with the case of harmonic functions, however, a U of infinite genus will now be permitted under certain circumstances.

Let γ be an ideal boundary component of an open Riemann surface. Let U be its neighborhood, assumed to be connected.

THEOREM. (a) *If γ is weak and if β_U is of O_{KB}-type or O_{KD}-type, then for every $f \in AB(\bar{U})$ or $f \in AD(\bar{U})$, respectively,*

$$\lim_{z \to \gamma} f(z) \tag{1}$$

exists.

(b) *If* (1) *exists for some meromorphic function f on U with finite maximum valence (defined below), then γ is semiweak.*

(c) *If* (1) *exists for some meromorphic function f on U, then γ is parabolic.*

Here the *maximum valence* of a meromorphic function on a Riemann surface is

$$\sup n(w) \qquad (\leq \infty),$$

where $n(w)$ is the number of w-points of the function with due count of multiplicity, and the supremum is taken over all w, $|w| \leq \infty$.

Remark 1. All three statements are due to CONSTANTINESCU [1, 2]. Part (a) for $W \in O_G$ was obtained by HEINS [3] and KURAMOCHI [1].

Remark 2. In (a), the assumption of the weakness of γ is equivalent to that of semiweakness (2 H.(c)). This assumption cannot be omitted in the above general case (a), as is seen by considering an isolated γ and applying (c). The assumption is superfluous, however, if U has finite genus (cf. Theorem 3 F and 3 H).

4 D. Proof of the Theorem. We start with the simplest assertion, namely (c). Without loss of generality, we may assume that $\lim_{z \to \gamma} f(z) = \infty$, so that $|f| \geq 1$ on some $\bar{U}$. For every $n = 1, 2, \ldots$, take neighborhoods $U_n \subset U$ of γ on which $|f(z)| \geq n$. We can choose them such that $\bar{U}_{n+1} \subset U_n$ and $\bigcap U_n = \emptyset$. For $\gamma_n = \beta_{U_n}$, the harmonic measure u_{γ_n} satisfies $\lim u_{\gamma_n} = u_\gamma$ (IV.4 E). On $U - U_n$, we have

$$u_{\gamma_n}(z) \leq \frac{\log |f(z)|}{\log n}.$$

We fix z, let $n \to \infty$, and conclude that $u_\gamma = 0$, so that γ is parabolic.

(b) We estimate the extremal length of the family Γ^* of curves joining γ and ∂U in U. For the normalization, we now choose $\lim_{z \to \gamma} f(z) = 0$. The metric $\rho(z)|dz|$ defined by

$$\rho(z) = \begin{cases} \dfrac{|f'(z)|}{|f(z)| \log(1/|f(z)|)} & \text{for } |f(z)| \leq \dfrac{1}{e}, \\[2ex] \dfrac{|f'(z)|}{e\,|f(z)|^2} & \text{for } |f(z)| \geq \dfrac{1}{e}, \end{cases}$$

is lower semicontinuous on U. Since

$$\int_0^a \frac{dr}{r(\log(1/r))^2} < \infty, \quad \int_0^a \frac{dr}{r\log(1/r)} = \infty$$

for every $0<a<1$, we obtain $\iint_U \rho^2\,dx\,dy<\infty$ and $\int_c \rho\,|dz|=\infty$ for all $c\in\Gamma^*$. Consequently, $\lambda(\Gamma^*)=\infty$, so that γ is semiweak.

To prove (a), we may always assume that $f\in AD(\bar{U})$, for if β_U is of O_{KB}-type, then $AB(\bar{U})\subset AD(\bar{U})$ by 2 G.(α). For an arbitrary neighborhood $U'\subset U$ of γ, we have $\lambda(\Gamma)=0$, where Γ is the family of cycles in U' separating γ from $\partial U'$ (IV.2 C). On considering the metric $\rho\,|dz|=|df|$ with $A(\rho)<\infty$, we can find a sequence of cycles c_n, $n=1,2,\ldots$, such that c_n separates γ from c_1, and c_{n+1} from c_{n-1}, and $\lim \int_{c_n}|df|=0$. We may replace each c_n by a cycle α_n consisting of a finite number of analytic Jordan curves. Consequently, we obtain a sequence of neighborhoods U_n of γ with the properties $\bar{U}_{n+1}\subset U_n\subset U$, $\bigcap U_n=\emptyset$, and

$$\lim_{n\to\infty} \int_{\alpha_n} |df|=0,$$

where $\alpha_n=\partial U_n$. We next apply the lemma below on every U_n and have

$$\lim_{n\to\infty} (\max_{\alpha_n} \operatorname{Re} f-\min_{\alpha_n} \operatorname{Re} f)\le \lim_{n\to\infty} \int_{\alpha_n} |df|=0,$$

$$\lim_{n\to\infty} (\max_{\alpha_n} \operatorname{Im} f-\min_{\alpha_n} \operatorname{Im} f)\le \lim_{n\to\infty} \int_{\alpha_n} |df|=0,$$

so that the proof will be complete, once we have proved:

LEMMA. *Suppose U is a neighborhood of a boundary component γ of a Riemann surface of class O_{KB} or O_{KD}. Then $u=\operatorname{Re} f$ and $u=\operatorname{Im} f$ with f in the class $AB(\bar{U})$ or $AD(\bar{U})$, satisfy*

$$M-m\le \int_\alpha |df|,$$

where $\alpha=\partial U$, $M=\max_\alpha u$, and $m=\min_\alpha u$.

4 E. Proof of the Lemma. Decompose α into components $\alpha_1,\ldots,\alpha_k$ such that $M_1\le M_2\le\cdots\le M_k$, where $M_j=\max_{\alpha_j} u$, $m_j=\min_{\alpha_j} u$. We shall first show that

$$\min(m_{j+1},\ldots,m_k)\le M_j, \qquad j=1,\ldots,k-1. \tag{2}$$

Suppose this were not true and take a j with $M_j<m_l$ for every $l=j+1,\ldots,k$. Let h be the function on α defined by

$$h=\begin{cases} M_j & \text{on} \quad \alpha_1\cup\cdots\cup\alpha_j, \\ u & \text{on} \quad \alpha_{j+1}\cup\cdots\cup\alpha_k. \end{cases}$$

By virtue of $u\le h$ on α, the function $v=L_0 h$ dominates u on U (use the assumption and Theorem 2 G). Since $M_j=\min_{\bar{U}} v$, we have $*dv\le 0$

along $\alpha_1 \cup \cdots \cup \alpha_j$, and since $u \leq v$ on U and $u = v$ on $\alpha_{j+1} \cup \cdots \cup \alpha_k$, we see that $*dv \leq *du$ along $\alpha_{j+1} \cup \cdots \cup \alpha_k$. The function u^* is single-valued on $\bar{U}$; hence $\int *du = 0$ along $\alpha_{j+1} + \cdots + \alpha_k$. Consequently,

$$0 \geq \int_{\alpha_{j+1} + \cdots + \alpha_k} *dv = -\int_{\alpha_1 + \cdots + \alpha_j} *dv \geq 0,$$

which shows that $*dv = 0$ along $\alpha_1 \cup \cdots \cup \alpha_j$. This is sufficient to guarantee that v is constant on U. Then $v = h = M_j$ on α, which contradicts $M_j < m_l = \min_{\alpha_l} u$, $l = j+1, \ldots, k$. Thus we obtain (2).

Now suppose that $m = m_{j_1}$. We have $M - m = M_k - m_{j_1} = M_k - M_{j_1} + M_{j_1} - m_{j_1}$. Since $\min(m_{j_1+1}, \ldots, m_k) = m_{j_2}$ is dominated by M_{j_1}, it follows that $j_2 > j_1$ and $M - m = M_k - M_{j_1} + M_{j_1} - m_{j_1} \leq M_k - M_{j_2} + M_{j_2} - m_{j_2} + M_{j_1} - m_{j_1}$. By repeating the process, we finally obtain

$$M - m \leq M_k - m_k + \cdots + M_{j_1} - m_{j_1} \leq \sum_j \int_{\alpha_j} |df| = \int_\alpha |df|.$$

4 F. Value Distribution and a Covering Property. A topic closely related to boundary behavior is the value distribution of analytic functions on a Riemann surface. For example, the following result is classical:

A non-constant meromorphic function f on a Riemann surface W of class O_{HB} (or O_{AB}) takes on all values except for a compact set of logarithmic capacity zero (or of class N_B, respectively). Moreover, if an f on a W of class O_{AD} has finite maximum valence, then it takes on all values except for a compact set of class N_D.

In fact, if the complement of $f(W)$ has positive logarithmic capacity, then $f(W)$ carries a non-constant bounded harmonic function u. Therefore, $u \circ f$ is non-constant on W, so that $W \notin O_{HB}$. The same reasoning applies to O_{AB} and O_{AD}.

Value distribution of a meromorphic function can also be viewed as a property of a covering surface over the Riemann sphere **S**. Indeed, a non-constant meromorphic function on a Riemann surface W determines W as a covering surface over **S**; conversely, if W is a covering surface of **S**, then the projection map is a non-constant meromorphic function on W. From this point of view, the above result may be stated as follows:

A covering surface W of class O_{HB} (or O_{AB}) over **S** *covers all points except for a compact set of logarithmic capacity zero (or of class N_B, respectively). A covering surface W of class O_{AD} with finite maximum valence covers every point except for a set of class N_D.*

Remark. The result for O_{AB} was first obtained by Kametani [1]; cf. also Besicovitch [1] and Cartwright [1]. The result for O_{HB} has been sharpened as follows: The set of all w such that $n(w)$ is less than the maximum valence has vanishing inner logarithmic capacity. The proof, as well as a detailed discussion of value distribution and covering properties, may be found in SARIO-NOSHIRO [1].

The converse of the above statement does not hold in general. But if we assume the finiteness of the maximum valence, then we obtain the following result, which will be needed in 5 E:

THEOREM. *Suppose W is a covering surface of the Riemann sphere* $\mathbf{S}$ *with finite maximum valence k. Let E be the compact set* $\{w \mid n(w) < k\}$. *Then*

(a) $\operatorname{Cap} E = 0$ *implies* $W \in O_G$.

(b) $E \in N_{SB}$ *implies* $W \in O_\gamma$.

Remark. (a) is classical. (b) is due to TAMURA [3] and JURCHESCU [4]; cf. also ANDREIAN-CAZACU [1].

4 G. Proof of the Theorem. In both cases E is totally disconnected, so that its complement D is a region. Without loss of generality, we may assume that $\infty \in D$.

(a) If W is not parabolic, then it carries a Green's function $g(z; \zeta)$. For every $w \in D$ which is not the projection of a branch point, let $z_1, \ldots, z_k$ be the points of W over w. The function $u(w) = \sum_{j=1}^{k} g(z_j; \zeta)$ can be extended to D, where it is superharmonic and non-constant. Thus $D \notin O_G$ and $\operatorname{Cap} E > 0$.

(b) Let Ω_0 be a parametric disk of W centered at a point lying over ∞, and let D_0 be its projection. Take Ω_0 so small that $\bar{D}_0 \subset D$. Exhaust D by a sequence of regular subregions: $\bar{D}_0 \subset D_1 \subset D_2 \subset \cdots \to D$. Take the part of W lying over D_n, and let Ω_n be its component containing Ω_0. Then $\Omega_n \to W$ is an exhaustion. Suppose there exists a non-weak boundary component γ of W. Then the modulus $\mu_n = \mu_1(\gamma(\Omega_n), \partial\Omega_0)$ of $\Omega_n - \Omega_0$ is bounded:

$$\mu_n \le \mu < \infty .$$

On the other hand, $\gamma(\Omega_n)$ lies over a single component of ∂D_n; with respect to this component and ∂D_0, the modulus m_n of $D_n - \bar{D}_0$ satisfies

$$\lim m_n = \infty .$$

The region Ω_n is an at most k-sheeted covering surface of D_n. Every component of $\partial\Omega_n$ lies over a single component of ∂D_n, and $\partial\Omega_0$ lies over ∂D_0. By means of the extremal length of the family $\tilde{\Gamma}^*$ introduced in IV. 2 B, we easily obtain the inequality

$$k \log \mu_n \ge \log m_n ,$$

which gives a contradiction. Thus $W \in O_\gamma$.

5. Extendability of an Open Riemann Surface

5 A. Extendability. An open Riemann surface W is said to be *extendable* if it can be imbedded in another Riemann surface W^* as a proper subregion; more precisely, if there exists a conformal mapping of W onto a proper subregion of W^*. If W is not extendable, then it is called *maximal.*

A maximal open surface was first exhibited by RADÓ [1]. BOCHNER [2] showed that every open Riemann surface is contained in a maximal one (see also HEINS [1], TSUJI [2], and TAMURA [2]).

The following basic theorem and its alternative expression (5 B) are due, in essence, to DE POSSEL [1, 3] (cf. also TAMURA [1] and RENGGLI [2, 3]):

THEOREM. *An open Riemann surface W is maximal if and only if the following two conditions are satisfied:*

(a) *W has no relatively non-compact connected end of finite genus.*

(b) *W has no regularly imbedded simply connected subregion V which admits a conformal mapping onto a proper subregion of the unit disc $|\zeta|<1$ such that the relative boundary ∂V is mapped into the periphery $|\zeta|=1$.*

Proof. The argument in 3 A implies that (a) is a necessary condition for maximality. If (b) is not satisfied, take a V permitting a conformal mapping described above. The surface obtained from $W\cup\{\zeta\,|\,|\zeta|<1\}$ by identifying the corresponding points in V contains W as a subregion. Thus W is not maximal. Finally, suppose W satisfies (a) and is not maximal. Take a $z_0\in W^*-W$ which is a boundary point of W. By (a), a parametric disk Δ about z_0 has the property that every component V of $\Delta\cap W$ is simply connected. Thus we see that (b) is violated.

5 B. Alternative Form of Theorem 5 A. We shall find alternative expressions for condition (b).

When a regularly imbedded simply connected V is mapped conformally onto the upper half-plane, the relative boundary ∂V corresponds to a relatively open subset A of the real axis. We may assume that $\infty\in A$. Let E be the complement of A with respect to the real axis. Also consider the family $H_0D(\overline{V})$ in the sense of 1 C.

LEMMA. *For a regularly imbedded simply connected subregion V of a Riemann surface, the following three conditions are equivalent:*

(a) *V does not admit a conformal mapping onto a proper subregion of $|\zeta|<1$ such that ∂V is mapped into $|\zeta|=1$.*

(b) *E is of class N_D.*

(c) $H_0D(\overline{V})=\{0\}$.

It may appear plausible that (a) would imply the existence of an interval in E. That this is, however, not true has been demonstrated by Radó [1, p. 2].

Proof. (a) $\Rightarrow$ (b). Consider the horizontal and the vertical slit maps $P_0(z;\infty)$ and $P_1(z;\infty)$ of CE. The former always satisfies $P_0(z;\infty)\equiv z$, and the latter has the property $P_1(\bar{z};\infty)=\overline{P_1(z;\infty)}$ (cf. Problem 9). If $E\notin N_D$, we have $P_1(z;\infty)\not\equiv z$. Then P_1 maps the upper half-plane onto some proper subregion, since otherwise P_1 would reduce to a linear transformation. Thus we obtain a conformal mapping of V onto a proper subregion of the upper half-plane such that ∂V is mapped into the real axis. We conclude immediately that (a) is violated.

(b) $\Rightarrow$ (a). Let φ and f be the conformal mappings of V into $|\zeta|<1$ and onto the upper half of the z-plane, respectively. Extend φ and f to the double $\hat{V}$ of V with respect to ∂V. If $E\in N_D$, then the univalent function $\varphi\circ f^{-1}$ on CE must be linear. Thus it maps the upper half-plane onto $|z|<1$, so that $\varphi(V)$ coincides with $|z|<1$.

(b) $\Leftrightarrow$ (c) (cf. A. Mori [2] and Lokki [2]). If $u\in H_0\,D(\bar{V})$, then $F=u+i\,u^*$ is single-valued and can be continued by the reflection principle to a function in $AD(\hat{V})$. On the other hand, every $F\in AD(\hat{V})$ satisfies $F=F_1+i\,F_2$, where $F_1(z)=\frac{1}{2}(F(z)+\overline{F(\tilde{z})})$ and $F_2(z)=-\frac{1}{2}\,i(F(z)-\overline{F(\tilde{z})})$; here z and $\tilde{z}$ are points in $\hat{V}$ symmetric about ∂V. The imaginary parts of F_1 and F_2 belong to $H_0\,D(\bar{V})$. We conclude that $H_0\,D(\bar{V})=\{0\}$ if and only if $AD(\hat{V})$ consists only of constants, that is, $\hat{V}\in O_{AD}$.

5 C. Essential Extendability. If an open Riemann surface W has an extension W^* such that W^*-W contains an interior point, then W will be called *essentially extendable*. The relation to extendability is as follows:

Lemma. *W is maximal if and only if the following two conditions are satisfied:*

(a) *W has no relatively non-compact connected end of finite genus.*

(b) *W is not essentially extendable.*

Remark. This does *not* mean that (b) is equivalent to (b) of Theorem 5 A.

Proof. We have observed that maximality implies (a). If W is not maximal and satisfies (a), then $S=W^*-W\neq\emptyset$ cannot be totally disconnected. Thus it contains a proper continuum, so that W is essentially extendable.

Essential extendability rules out "smallness" of the ideal boundary of W. In fact, we have the following result (Jurchescu [2–4]), the proof of which is exactly the same as the second and third paragraphs of 3 B:

Theorem. *If $W\in O_{KD}$ or $W\in O_{\gamma}$, then W is not essentially extendable.*

5 D. Essential Extendability (continued). A regularly imbedded subregion V of a Riemann surface will be called *relatively planar* if it is planar and has the property that every chain with end points only on the relative boundary ∂V divides V. Intuitively, V is planar and the components of ∂V are connected by the ideal boundary.

The following result is also due to JURCHESCU [3, 4]:

THEOREM. *W is not essentially extendable if and only if every regularly imbedded relatively planar subregion V has its double $\hat{V}$ about ∂V in the class O_γ.*

Proof. Suppose there is a V such that $\hat{V} \notin O_\gamma$. Since $\hat{V}$ is planar, we may consider it in the plane. It can be mapped conformally onto a circular slit disk of finite radius. By symmetry, this can be done in such a way that the image of ∂V is contained in a diameter of the disk. Thus V is conformally equivalent to a proper subregion of the half-plane H, and the Riemann surface W^* obtained from $W \cup H$ by identifying the corresponding points in V is an essential extension of W.

Conversely, suppose W has an essential extension W^*. Take a $z_0 \in W^* - W$ which is a boundary point of W and is contained in a proper subcontinuum of $W^* - W$. Let Δ be a parametric disk about z_0. Some component V of $\Delta \cap W$ is relatively planar and has the property that $\partial^* V - \partial V$ contains a proper continuum; here $\partial^* V$ means the relative boundary of V in W^* and ∂V means the one in W. By considering $\hat{V}$ in the double of Δ with respect to the periphery, we see that a part of some ideal boundary component of $\hat{V}$ is realized as a continuum. We conclude by Remark 1 in 3 B that $V \notin O_\gamma$.

5 E. Remarks. We observe that, under assumption 5 A.(a), there is no relation between maximality and the properties $W \in O_{AB}$ and $W \in O_{AD}$.

First, Myrberg's example (2 J) is an essentially extendable surface which may be modified so as to satisfy condition 5 A.(a). Thus there is a non-maximal surface of class O_{AB}.

Second, JURCHESCU [4] and TAMURA [3] constructed a maximal surface not in O_{AD} as follows: Two copies of a plane region of class $O_\gamma - O_{AD}$ (XI. 3 J) are connected cross-wise along suitably chosen slits in such a way that the resulting Riemann surface W satisfies condition 5 A.(a). Then W is in $O_\gamma - O_{AD}$ (Theorem 4 F) and is maximal (5 C).

The problem of essential extendability was introduced in SARIO [1], in connection with a study of properties of O_{AD}-surfaces.

Chapter XI

Practical Tests

The characterization of weakness by means of extremal length (Chapter IV) will be modified in 1 into the modular test and also a number of tests with practical applicability. In 2 and 3, the degeneracy of the entire boundary of a plane region is measured by means of concrete metrics. Boundary components consisting of a single point are considered in 4, with weakness tests based on how heavily other components accumulate to it. In 5, semiweakness is discussed on the same basis. The relationship between non-semiweak and strong boundary components is also studied.

1. Tests for Weakness

1 A. Modular Test. Weakness and semiweakness of a subboundary γ can be characterized in terms of extremal length, since the moduli $\mu_1(\gamma,\alpha)$ and $\mu_0(\gamma,\alpha)$ of a connected neighborhood U of γ are determined by it. From this relation we shall derive tests for weakness with more practical applicability.

On a given Riemann surface, consider a finite number of regular subregions A^i, $i=1,\ldots,k$, such that ∂A^i is disconnected and the $\bar{A}^i$ are disjoint by pairs. Together with the set $A=A^1\cup\cdots\cup A^k$, we choose partitions $\partial A^i=\alpha^i\cup\alpha'^i$, $i=1,\ldots,k$, and call this complex an *annular region* $A(\alpha^i,\alpha'^i)$, or simply A if no ambiguity arises. Note that A is not necessarily connected, even though the term "region" is used. Set $\alpha=\bigcup_i\alpha^i$ and $\alpha'=\bigcup_i\alpha'^i$.

In terms of the moduli $\mu(\alpha^i,\alpha'^i)$ of the A^i, we introduce the quantity μ by

$$\frac{1}{\log\mu}=\sum_{i=1}^{k}\frac{1}{\log\mu(\alpha^i,\alpha'^i)}$$

and call it the *modulus* of the annular region A.

Using the harmonic function u_α on A whose restriction to A^i is the harmonic measure u_{α^i} of A^i (IV.1 B), we have

$$\log\mu=\frac{2\pi}{D_A[u_\alpha]}.$$

Let Γ be the family of rectifiable cycles in A separating α from α', and Γ^* the family of locally rectifiable curves joining α and α' in A. Then

$$\log\mu = \frac{2\pi}{\lambda(\Gamma)} = 2\pi\,\lambda(\Gamma^*).$$

Given a subboundary γ of an open Riemann surface W, consider a sequence of disjoint annular regions $A_n(\alpha_n^i, \alpha_n'^i)$, $n=1,2,\ldots$, satisfying the following condition:

(a) For each $n=2,3,\ldots$, every cycle in A_n separating $\alpha_n = \bigcup_i \alpha_n^i$ from $\alpha_n' = \bigcup_i \alpha_n'^i$ separates A_1 from γ and A_{n-1} from A_{n+1}.

Let μ_n be the modulus of $A_n(\alpha_n^i, \alpha_n'^i)$.

THEOREM. *A subboundary γ is weak if and only if there exists a sequence $\{A_n\}$ of annular regions with property* (a) *such that*

$$\prod_{n=1}^{\infty} \mu_n = \infty. \tag{1}$$

Remark. For $\gamma=\beta$, the ideal boundary of W, the sufficiency is due to SARIO [9] and the necessity to NOSHIRO [1]. For a boundary component γ, the sufficiency was proved by SAVAGE [1] and the necessity by OIKAWA [1].

1 B. Proof. Suppose such a sequence $\{A_n\}$ exists. Let U be a connected neighborhood of γ such that $\bigcup A_n \subset U$. Let Γ be the family of cycles in U separating ∂U from γ, and let Γ_n be the family of cycles in A_n separating α_n from α_n'. Then $\lambda(\Gamma)^{-1} \geq \sum \lambda(\Gamma_n)^{-1} = (2\pi)^{-1} \sum \log\mu_n = \infty$ implies $\lambda(\Gamma)=0$, and γ is weak.

To prove the converse, we assume for simplicity that γ is either β or a single boundary component; the general case is analogous.

If γ is the entire ideal boundary β of W, we consider an exhaustion $\Omega_n \to W$ by regular subregions. The harmonic measure $u_{\beta(\Omega_n)}$ on $\Omega_n - \bar{\Omega}_1$ satisfies $\lim D_{\Omega_n - \Omega_1}[u_{\beta(\Omega_n)}] = 0$. The set $\Omega_n - \bar{\Omega}_1$ with the partition of the border into $\alpha_n = \beta(\Omega_n)$ and $\alpha_n' = \beta(\Omega_1)$ is an annular region. Its modulus μ_n' diverges to ∞ as $n \to \infty$, so that it is possible to take an n_1 with $\mu_{n_1}' \geq 2$. We set $A_1 = \Omega_{n_1} - \bar{\Omega}_1$, $\mu_1 = \mu_{n_1}'$. Next for the exhaustion $\Omega_n \to W$, $n = n_1 + 1$, $n_1 + 2, \ldots$, we consider the annular regions $\Omega_n - \bar{\Omega}_{n_1}$ in a similar way. Their moduli μ_n'' diverge to ∞ as $n \to \infty$, so that we can again take an n_2 with $\mu_{n_2}'' \geq 2$. Set $A_2 = \Omega_{n_2} - \bar{\Omega}_{n_1}$, $\mu_2 = \mu_{n_2}''$. By repeating this process, we finally obtain a sequence $\{A_n\}$ of annular regions satisfying condition 1 A.(a), and $\prod \mu_n \geq \prod 2 = \infty$.

Suppose now that γ is a weak boundary component. Consider an exhaustion $\Omega_n \to W$ by *canonical* subregions. On the border $\beta(\Omega_n)$ of Ω_n, the contour $\gamma_n=\gamma(\Omega_n)$ separating Ω_n from γ is well-determined (III.4 A). We may assume that $\Omega_n-\bar{\Omega}_1$ is connected. The moduli $\mu_1(\gamma_n, \gamma_1)$ of the $\Omega_n-\bar{\Omega}_1$, $n=2, 3, \ldots$, diverge to ∞ as $n\to\infty$. Thus, as before, we can take an n_1 with $\mu_1(\gamma_{n_1}, \gamma_1)\geq 2$. Next choose a component Ω_n' of $\Omega_n-\bar{\Omega}_{n_1}$ $(n>n_1)$ such that $\bar{\Omega}_n'\cap\beta(\Omega_{n_1})=\gamma_{n_1}$, and take n_2 so large that the modulus $\mu_1(\gamma_{n_2}, \gamma_{n_1})$ of Ω_{n_2}' is at least 2. We continue this process and denote $\{\Omega_{n_j}\}_{j=1}^{\infty}$ again by $\{\Omega_n\}_{n=1}^{\infty}$. Thus we have obtained a sequence of regions $\tilde{\Omega}_n$, $n=1, 2, \ldots$, which are the components of $\Omega_n-\bar{\Omega}_{n-1}$ determined by $\bar{\tilde{\Omega}}_n\cap\beta(\Omega_{n-1})=\gamma_{n-1}$ and whose moduli $\tilde{\mu}_n=\mu_1(\gamma_n, \gamma_{n-1})$ satisfy $\prod\tilde{\mu}_n=\infty$. Since $\tilde{\Omega}_n, \gamma_n$, and γ_{n-1} do not form an annular subregion, we have one more step to take.

Consider on $\tilde{\Omega}_n$ the modulus function q_1 with respect to γ_n and γ_{n-1}. Normalize it so that it vanishes on γ_{n-1}. It takes the constant value $\log\tilde{\mu}_n$ on γ_n. Suppose $\partial\tilde{\Omega}_n-\gamma_n-\gamma_{n-1}$ consists of contours $\tilde{\gamma}_i$, $i=1, \ldots, m_n$, on which the constant values l_i of q_1 satisfy

$$0=l_0<l_1\leq l_2\leq\cdots\leq l_{m_n}<l_{m_n+1}=\log\tilde{\mu}_n.$$

For simplicity, we assume that these are pairwise distinct; in the general case the pertinent passage below becomes more involved, but the proof is essentially the same. If we take $\varepsilon_n>0$ sufficiently small, then $l_i+\varepsilon_n<l_{i+1}-\varepsilon_n$, $i=0, \ldots, m_n$, and $\sum_{i=0}^{m_n}(l_{i+1}-l_i-2\varepsilon_n)>\log\tilde{\mu}_n-2^{-n}$. The sets

$$A_{ni}=\{z\,|\,l_i+\varepsilon_n<q_1(z)<l_{i+1}-\varepsilon_n\},$$

$i=0, \ldots, m_n$, together with the level lines $q_1=l_i+\varepsilon_n$ and $q_1=l_{i+1}-\varepsilon_n$, are annular subregions whose moduli μ_{ni} are $\exp(l_{i+1}-l_i-2\varepsilon_n)$. It is clear that the sequence $\{A_{ni}\}$ qualifies since

$$\prod_{n,i}\mu_{ni}=\infty.$$

1 C. Remarks Concerning the Modular Test. The necessity of (1) must not be understood to mean that the modular product diverges for *every* sequence $\{A_n\}$ towards a weak γ.

(a) *Every W and every γ have an $\{A_n\}$ with $\prod\mu_n<\infty$.*

In fact, if $\prod\mu_n=\infty$, then we can bisect A_n so that the resulting annular subregions have moduli as close to one as we wish. We shall describe this for $W=\{z\,|\,|z|<\infty\}$ and $A_n=\{z\,|\,n<|z|<n+1\}$, $n=1, 2, \ldots$. Divide A_n into A_n' and A_n'' by the circle $|z-(\frac{1}{2}-\varepsilon)|=n+\frac{1}{2}$. Using extremal length, we can show that the moduli of A_n' and A_n'' tend to 1 as $\varepsilon\to 0$ (cf. Appendix I.K). Thus we obtain a sequence of annular regions A_n', A_n'', whose moduli have converging product.

(b) *In Theorem* 1 A, *we may assume that each component of every* A_n *is a doubly connected (planar) region.*

Indeed, from a given A_n, we can construct desired ones by dividing A_n along suitable level lines of u_{α_n}. The details are left to the reader.

One may ask whether, for an ideal boundary component γ, every A_n can consist of a single doubly connected region. The answer is in the negative:

(c) *There exists a parabolic Riemann surface* W *with only one ideal boundary component yet with* $\prod \mu_n < \infty$ *for every* $\{A_n\}$ *such that each* A_n *is a single doubly connected region.*

A modification of the surface in IV.2 N provides an example of this nature. Let a_k be real numbers with $0 < a_k < 1$, $k = 1, 2, \ldots$, and set $I_k = \{z \mid k \leq \operatorname{Re} z \leq k + a_k, \operatorname{Im} z = 0\}$. Take two replicas of $\{z \mid |z| < \infty\} - \bigcup_1^\infty I_k$ and connect them along $\bigcup_1^\infty I_k$ crosswise. The resulting surface W is parabolic and has only one ideal boundary component γ over $z = \infty$. Denote by U the neighborhood of γ obtained from W by removing a unit disk from each sheet. As in IV.2 N, we can take the a_n so small that the family Γ of closed *curves* in U separating ∂U from γ satisfies $\lambda(\Gamma) > 0$. To prove that $\prod_1^\infty \mu_n < \infty$ for every $\{A_n\}$ such that each A_n is a single doubly connected region, we may assume without loss of generality that $\bigcup_1^\infty A_n \subset U$. From the definition of the moduli μ_n of A_n, we have $\sum_1^\infty \log \mu_n = \sum_1^\infty 2\pi\, \lambda(\Gamma_n)^{-1} \leq 2\pi\, \lambda(\Gamma)^{-1} < \infty$, where Γ_n is the set of members of Γ contained in A_n.

Remark. Heins [3] showed that if $\prod \mu_n = \infty$ for a sequence $\{A_n\}$ of doubly connected regions, then γ has harmonic dimension one (Remark X.4 A).

On a plane region, however, (c) cannot occur. In fact, in the proof of Theorem 1 A, $\exp(q_1 + i\, q_1^*)$ gives a univalent map of $\tilde{\Omega}_n$ onto a circular slit annulus, so that A_{ni} is the inverse image of a concentric circular annulus. We summarize the results (cf. Grötzsch [10]):

Theorem. *A boundary component* γ *of a plane region* W *is weak if and only if there exists a sequence* $\{A_n\}$ *of doubly connected regions with* A_n $(n > 1)$ *separating* γ *from* A_1 *and* A_{n+1} *from* A_{n-1} *and such that*

$$\prod_{n=1}^{\infty} \mu_n = \infty .$$

1 D. Regular Chain Test. We shall derive a number of weakness tests with practical applicability. The results in 1 D – 1 K are due to Savage [1], except for that in 1 E, which was obtained by Constantinescu [2].

Given a subboundary γ of an open Riemann surface W, consider a sequence of disjoint annular regions $A_n(\alpha_n^i, \alpha_n'^i)$, $n=1,2,\ldots$, satisfying condition 1 A. (a). Decompose A_n into components

$$A_n = A_n^1 \cup \cdots \cup A_n^{k(n)}$$

and make the following assumptions:

(a) Each A_n^i is a doubly connected region.

(b) Each A_n^i is the union of a finite number of parametric regions: $A_n^i = \bigcup_{l=1}^{s(n,i)} \Delta_{nil}$.

Set $s_n = \sum_{i=1}^{k(n)} s(n,i)$. Map Δ_{nil} conformally onto the unit disk $|t|<1$ and assume further:

(c) There exists a number N, independent of n and i, such that each Δ_{nil} meets at most N disks Δ constituting A_n^i.

(d) There exists a number δ, $0<\delta<1$, independent of n and i, such that, if $\Delta_{nil} \cap \Delta_{nim} \neq \emptyset$, then the inverse images Δ_{nil}^0 and Δ_{nim}^0 of $|t|<1-\delta$ meet.

(e) For each n, i, l, m, it is possible to find a curve $c(n,i,l,m)$ in Δ_{nil} which is a line segment in the t-plane, joins the center of Δ_{nil} and a point in $\Delta_{nil}^0 \cap \Delta_{nim}^0$ whenever this is non-void, and is such that $\bigcup_{l,m} c(n,i,l,m)$ separates the two boundary components of A_n^i.

THEOREM. *A subboundary γ is weak if $\sum_{n=1}^{\infty} 1/s_n = \infty$.*

Proof. Let μ_n be the modulus of A_n and let u_n be the harmonic function on A_n such that $u_n=0$ on α_n' and $u_n=1$ on α_n. Consider the function $v_n = (\log \mu_n)\, u_n$. For $c_n = \bigcup_{i,l,m} c(n,i,l,m)$, condition (e) implies

$$2\pi \le \int_{c_n} |{*dv_n}| \le \sum_i \sum_{l,m} \int_{c(n,i,l,m)} \left| \frac{d(v_n + i\, v_n^*)}{dt} \right| |dt|. \tag{2}$$

From condition (d) and the area theorem, we obtain

$$\left| \frac{d(v_n + i\, v_n^*)}{dt} \right| \le \frac{1}{\sqrt{\pi}\,\delta} \sqrt{D_{\Delta_{nil}}[v_n]}$$

on $c(n,i,l,m)$. We substitute this into (2). Observe that, by (c), for a given i and l, there are at most N integers m over which the sum is taken. In view of this and the fact that $\int_{c(n,i,l,m)} |dt| \le 1$, we obtain

$$2\pi \le \sum_i \sum_l \frac{N}{\sqrt{\pi}\,\delta} \sqrt{D_{\Delta_{nil}}[v_n]},$$

with s_n terms on the right. By Schwarz's inequality,

$$4\pi^2 \le \frac{N^2}{\pi\delta^2} s_n \sum_i \sum_l D_{\Delta_{nil}}[v_n]$$
$$\le \frac{N^2}{\pi\delta^2} s_n n_0 D_{A_n}[v_n] = \frac{N^3}{\pi\delta^2} s_n \cdot 2\pi \log \mu_n.$$

Thus

$$\sum_n \frac{1}{s_n} \le \text{const} \cdot \sum_n \log \mu_n,$$

and we conclude by Theorem 1 A that γ is weak.

1 E. Poincaré's Metric Test. Suppose the universal covering surface of W is the unit disk $|t|<1$. Introduce the Poincaré metric $ds=\rho\,|dz|$ on W by projecting $|dt|/(1-|t|^2)$.

Given a subboundary γ of W, consider a sequence $\{c_n\}_{n=1}^{\infty}$ of disjoint cycles in W satisfying the following conditions: There exists a regular region Ω_0 such that $(\bigcup c_n)\cap\bar{\Omega}_0=\emptyset$, c_n $(n>1)$ separates γ from Ω_0 and c_{n+1} from c_{n-1}, and for any regular region Ω there is an n_0 such that $c_n \subset W-\bar{\Omega}$ for all $n\ge n_0$.

THEOREM. *A subboundary γ is weak if the lengths $\int_{c_n} ds$ in the Poincaré metric, $n=1, 2, \ldots$, are bounded.*

Remark. In the case where γ is a boundary component, CONSTANTINESCU [2] called γ *k-einsam* if the assumption of the theorem is satisfied for a sequence $\{c_n\}$, each member of which consists of at most k curves. By the above theorem, a k-einsam γ is weak. CONSTANTINESCU showed that a k-einsam γ contains at most k Martin ideal boundary points and proved further (cf. X.4 A, 4 C): If f is a meromorphic function on a neighborhood U of a 1-einsam or k-einsam $(k\ge 2)$ γ omitting at least three or $k+1$ values, respectively, then $\lim_{z\to\gamma} f(z)$ exists. If $W\in O_{KB}$ or O_{KD} and if $u\in HB(\bar{U})$ or $HD(\bar{U})$, respectively, on a neighborhood U of a 1-einsam γ, then $\lim_{z\to\gamma} u(z)$ exists.

In the proof of the theorem, we shall make use of the following auxiliary result:

LEMMA. *Let W_1 be a subregion of W. Consider the Poincaré metrics ds_1 and ds on W_1 and W, respectively. At a point $z\in W_1$ such that the ds distance from $W-W_1$ is at least $\delta>0$, the inequalities*

$$ds \le ds_1 \le \frac{ds}{\tanh\delta}$$

hold.

Proof of the Lemma. The relation $ds \leq ds_1$ is well known (e.g. NEVANLINNA [3, p. 50]). On the other hand, consider the universal covering surface $|t|<1$ of W such that $t=0$ lies over z. Then the component containing $t=0$ of the part over W_1 contains the disk $|t|<\tanh\delta$. From this we easily obtain the right-hand inequality.

Remark. By means of the lemma, we shall obtain a lower bound for the length of c_n in terms of the Poincaré metric of an arbitrary end containing all the c_n for sufficiently large n.

1 F. Proof of Theorem 1 E. We may assume that every c_n is contained in a boundary neighborhood U of γ and that there is a sequence $\{U_n\}$ of boundary neighborhoods of γ such that $\overline{U}_n \subset U$, $\overline{U}_{n+1} \subset U_n$, and $\partial U_n = c_n$. Construct a sequence of integers $0<n(1)<m(1)<n(2)<m(2)<n(3)<\cdots$ as follows. Set $n(1)=1$. Determine $m(1)>1$ such that the distance in the Poincaré metric between $c_{n(1)}$ and $c_{m(1)}$ is at least δ, a positive number given in advance. Then take $n(2)>m(1)$ such that the distance between $c_{n(2)}$ and $c_{m(1)}$ is at least the same δ. Continue the process similarly. Let Γ be the family of cycles in U separating ∂U from γ, and let Γ_k be the family of cyles in $U_{n(k-1)}-\overline{U}_{n(k)}$ separating $c_{n(k)}$ from $c_{n(k-1)}$. We have $\lambda(\Gamma)^{-1} \geq \sum \lambda(\Gamma_k)^{-1}$. Next consider the harmonic function u_k on $U_{n(k-1)}-\overline{U}_{n(k)}$ determined by the boundary behavior $u_k=0$ on $c_{n(k-1)}$, $u_k=1$ on $c_{n(k)}$, and $L_1 u_k = u_k$ on the remaining boundary, where L_1 is the operator for $\mathbf{Q}$. The curve $c_{m(k-1)}$ is contained in $U_{n(k-1)}-\overline{U}_{n(k)}$ and separates $c_{n(k)}$ from $c_{n(k-1)}$. We obtain

$$\begin{aligned}\lambda(\Gamma_k) = D[u_k] = \int_{c_{n(k)}} * du_k = \int_{c_{m(k-1)}} * du_k \\ \leq \int_{c_{m(k-1)}} |* du_k| \leq \int_{c_{m(k-1)}} |du_k + i * du_k| .\end{aligned}$$

The function $u_k + i u_k^*$ is single-valued on the universal covering surface of each component of $U_{n(k-1)}-\overline{U}_{n(k)}$, where the Poincaré metric ds_k satisfies

$$\frac{\pi}{2} \int_{c_{m(k-1)}} |du_k + i * du_k| \leq \int_{c_{m(k-1)}} \frac{\pi\,|du_k + i * du_k|}{2 \sin \pi u_k} \leq \int_{c_{m(k-1)}} ds_k ,$$

since $0<u_k<1$. By the lemma, we obtain

$$\lambda(\Gamma_k) \leq \frac{2}{\pi} \cdot \frac{1}{\tanh\delta} \int_{c_{m(k-1)}} ds ,$$

which is bounded by the assumption. We conclude that $\lambda(\Gamma)=0$ and, therefore, γ is weak.

1 G. Conformal Metric Test. On a Riemann surface W, consider a conformally invariant metric $ds=\rho\,|dz|$ which is complete, i.e. gives infinite distance from every point of W to the ideal boundary. For a fixed point $\zeta\in W$, let $\Omega(r)$ be the set of points whose distance from ζ in the given metric is less than r. Suppose $\rho\,|dz|$ is such that $\Omega(r)$ is a canonical region for every $r>0$. Given a subboundary γ of W, let $\gamma(r)$ be the cycle of smallest ds-length which separates γ from ζ and is a union of some components of the border $\beta(r)$ of $\Omega(r)$. Denote by $L(r)$ the length of $\gamma(r)$ measured in the given metric.

THEOREM. *A subboundary γ is weak if*

$$\int^{\infty}\frac{dr}{L(r)}=\infty.$$

Proof. Clearly $\Omega(r)$ exhausts W as $r\to\infty$. We consider the capacity function $p=p_{1\gamma(r)}$ with pole at ζ and the capacity $c(r)=c_{1\gamma(r)}(\zeta)$ for the region $\Omega(r)$. We are to show that $\lim_{r\to\infty}c(r)=0$. For every τ with $0<\tau<r$, it is evident that $\int_{\gamma(\tau)}{*dp}=2\pi$. Thus

$$\begin{aligned}4\pi^2&\le\left(\int_{\gamma(\tau)}\left|\frac{d(p+i\,p^*)}{ds}\right|ds\right)^2\\&\le\int_{\gamma(\tau)}\left|\frac{d(p+i\,p^*)}{ds}\right|^2 ds\int_{\gamma(\tau)}ds\le\int_{\beta(\tau)}\left|\frac{d(p+i\,p^*)}{ds}\right|^2 ds\cdot L(\tau).\end{aligned}$$

If $r_0>0$ is taken sufficiently small, then

$$\begin{aligned}\int_{r_0}^{r}\frac{d\tau}{L(\tau)}&\le\frac{1}{4\pi^2}\int_{r_0}^{r}d\tau\int_{\beta(\tau)}\left|\frac{d(p+i\,p^*)}{ds}\right|^2 ds\\&=\frac{1}{4\pi^2}D_{\Omega(r)-\Omega(r_0)}[p]=\frac{1}{2\pi}\left(\log\frac{1}{c(r)}-\text{const}\right).\end{aligned}$$

We conclude that $\int^{\infty}L(r)^{-1}\,dr=\infty$ implies $\lim c(r)=0$.

1 H. Ramified Covering Surfaces of a Plane. Suppose a Riemann surface W is realized as a covering surface of the Riemann sphere $\mathbf{S}$ satisfying the following condition: There exists a (smallest) finite set $E_0=\{z_1,\dots,z_k\}\subset\mathbf{S}$ such that the surface W' obtained by deleting all points lying over E_0 is a smooth regular covering surface of $\mathbf{S}-E_0$. Intuitively, E_0 is the set of the "projections" of all algebraic and logarithmic branch points of W. Consider a Jordan curve J passing through $z_1,\dots,z_k$ in this order. Then $\mathbf{S}-J$ consists of two simply connected regions R_1 and R_2. The set of points in W which lie over J divides W

into *faces*, which are 1-sheeted smooth coverings of either R_1 or R_2. Note that the faces in the present case are not necessarily relatively compact. For a face, an *edge* is the subset of its boundary lying over an arc $\widehat{z_j z_{j+1}}$, $j=1, \ldots, k$ $(z_{k+1}=z_1)$, and a *vertex* is an end point of an edge. At a logarithmic branch point, we have an "ideal" vertex not realized as a point of W.

Select one face. Those faces which have an edge in common with the first face are said to constitute the second generation. The n-th generation will consist of those faces which have an edge in common with the $(n-1)$-st generation but no edge in common with the $(n-2)$-nd generation. The union P_n of the first n generations is called the n-th *subpolyhedron.*

The boundary ∂P_n lies over J and consists of a finite number t_n of components β_{ni}. Each β_{ni} is composed of a finite number t_{ni} of edges. Including the ideal vertices at logarithmic branch points, the total number of vertices on β_{ni} is t_{ni}. Around each vertex $V_l \in \beta_{ni}$, real or ideal, there is a finite number σ_l of faces contained in P_n. We set $\sigma_{ni}=\sum_{l=1}^{t_{ni}} \sigma_l$.

The sequence $\{P_n\}$ exhausts W, but this is not an exhaustion by relatively compact subregions. We are going to delete certain subsets from the P_n to obtain such subregions. Let the disks $\Delta_j=\{z \mid |z-z_j| \leq \varepsilon\}$, $j=1, \ldots, k$, be disjoint. Assume that J is a smooth curve consisting of certain diameters of the Δ_j and of connecting arcs exterior to Δ_j. The set of points of P_n which lie over the disk Δ_j consists of a finite number of components which lie interior to P_n or around a vertex V_l, real or ideal, on ∂P_n. We consider the latter regions G_l. Each G_l is the union of σ_l "half-disks". Let H_l be the part of G_l which lies over the disk $\Delta_j^n=\{z \mid |z-z_j| \leq \varepsilon/n\}$ and set $\Omega_n=P_n-\bigcup_l H_l$, where the union is extended over the vertices V_l on ∂P_n. Now the sequence $\{\Omega_n\}$ is an exhaustion of W by relatively compact subregions. The border $\beta(\Omega_n)$ consists of components γ_{ni}, $i=1, \ldots, t_n$, and each γ_{ni} is composed of a part of β_{ni} and parts of H_l with $V_l \in \beta_{ni}$.

Suppose a subboundary γ of W is given. Let $\gamma(\Omega_n)$ be the union of the smallest number of components γ_{ni} which separates γ from Ω_n. For these i we take the sum $\tilde{\sigma}_n=\sum \sigma_{ni}$.

THEOREM. *A subboundary γ is weak if $\sum 1/\tilde{\sigma}_n=\infty$.*

1 I. Proof. We shall apply Theorem 1 D. To this end, we construct a sequence $\{A_n\}$ of annular subregions such that $\gamma(\Omega_n) \subset A_n$ and $s_n \leq 3\tilde{\sigma}_n$, $n=1,2, \ldots$. Consider in **S** the curve J and the disks Δ_j^n, $j=1, \ldots, k$. Recall that $J \cap \Delta_j^n$ is a diameter of Δ_j^n. Delete from $\{z \mid \varepsilon/2n<|z-z_j|<\varepsilon\}$ a radius perpendicular to J. There are two possible radii, so that we obtain two simply connected regions D_j^n and $\tilde{D}_j^n$, the union of which covers $\partial \Delta_j^n$. Next consider a simply connected region D_j which satisfies the following conditions: The arc $\widehat{z_j z_{j+1}}$ of J is contained in D_j except for the end points z_j and z_{j+1}, the boundary ∂D_j is a Jordan curve passing through z_j and z_{j+1}, and $D_j^{(n)} = D_j - \{z \,|\, |z - z_j| \le \varepsilon/2n\} - \{z \,|\, |z - z_{j+1}| \le \varepsilon/2n\}$ is still simply connected for every n. We can further take the D_j such that $D_1^{(n)}, \ldots, D_k^{(n)}$ are disjoint for every n. It is not difficult to see that $\tilde{D}_j^n$, D_j^n, and $D_j^{(n)}$, $j = 1, \ldots, k$, satisfy condition 1 D.(d) for a δ independent of n.

Now lift these regions to W. The union A_n of those intersecting $\beta(\Omega_n)$, $n = 1, 2, \ldots$, is readily seen to be the required annular region: condition 1 D.(c) is satisfied for $N = 3$. We obtain $s_n \le 3\tilde{\sigma}_n$, and $\sum 1/\tilde{\sigma}_n = \infty$ implies the weakness of γ.

1 J. Relative Width Test. In 1 J and 1 K we shall give criteria for the weakness of a boundary *component* of a *plane region* containing ∞. To apply Theorem 1 C, we estimate the modulus of a doubly connected region A_n in the plane. Let δ_n be the distance between the boundary components of A_n and let l_n be the infimum of the lengths of closed rectifiable curves C in A_n separating the components of ∂A_n and such that the distance from ∂A_n is at least $\delta_n/2$. The quantity $w_n = \delta_n / l_n$ will be called the *relative width* of A_n.

THEOREM. *A boundary component γ of a plane region containing ∞ is weak if there exists a sequence $\{A_n\}$ of doubly connected regions such that A_{n+1} separates A_1 from γ and A_n from A_{n+2} and such that*

$$\sum_{n=1}^{\infty} w_n = \infty.$$

Proof. Let $\zeta = f_n(z)$ be a conformal mapping of A_n onto $1 < |\zeta| < \mu_n$. Denote by C' the image of a C. We have $2\pi \le |C'| = \int_C |f_n'| \, |dz|$, where $|C'|$ stands for the length of C'. Every $z \in C$ is the center of a disk of radius $\delta_n/2$ which is contained in A_n. By Koebe's quarter theorem,

$$\frac{1}{4} \cdot \frac{\delta_n}{2} |f_n'(z)| \le \min |f_n - f_n(z)| \le \frac{\mu_n - 1}{2},$$

where the minimum is taken over the boundary of the disk. Then $2\pi \le 4 l_n (\mu_n - 1)/\delta_n$. It follows that $\sum (\mu_n - 1) = \infty$, or equivalently $\prod \mu_n = \infty$. We conclude that γ is weak.

1 K. Square Net Test. Consider a sequence of square nets in the plane, each a refinement of the preceding one and with side 2^{-n}, $n = 1, 2, \ldots$. Given a plane region W containing ∞, let Q_n be the union of those closed squares of side 2^{-n} which have points in common with the complement of W. Denote the components of Q_n by Q_{ni}. A certain subsequence $\{Q_{ni(n)}\}$ tends to γ. Let q_n denote the number of squares in $Q_{ni(n)}$.

THEOREM. *A boundary component γ of a plane region containing ∞ is weak if $\sum 1/q_n=\infty$.*

Proof. Around each $Q_{ni(n)}$ we trace two curves α_n and α'_n at distances 2^{-n-1}, 2^{-n-2}, respectively. The set bounded by α_n and α'_n is a doubly connected subregion A_n of W. Its relative width satisfies

$$w_n \geq \frac{1}{32\, q_n},$$

since $\delta_n=2^{-n-2}$ and, as is easily seen, $l_n \leq 16 q_n 2^{-n-1}$. The sequence of regions A_n meets the requirement of Theorem 1 J, and γ is weak.

Remark. The tests presented thus far, analogues of tests introduced for O_{AD} (SARIO [1]), are for weakness. Interesting tests for hyperbolicity (i.e., non-parabolicity) of open Riemann surfaces were obtained by ROYDEN [1, pp. 85ff.]. Some tests for non-weakness of boundary components of special plane regions will be given in 4.

2. Plane Sets of Logarithmic Capacity Zero

2 A. Logarithmic Capacity. To test for parabolicity of a plane region containing ∞, we can use, in addition to criteria of the types discussed above, the *logarithmic capacity* and the *transfinite diameter* of the complementary compact set (VII. 6 I).

First, from the definition we obtain immediately:

$$E_1 \subset E_2 \Rightarrow \operatorname{Cap}^{(i)} E_1 \leq \operatorname{Cap}^{(i)} E_2. \tag{1}$$

Next, if a set E is mapped onto E' by a function f univalent and analytic on a region $D \supset E$, and if both D and $f(D)$ are convex, then

$$a' \operatorname{Cap}^{(i)} E \leq \operatorname{Cap}^{(i)} E' \leq a \operatorname{Cap}^{(i)} E, \tag{2}$$

where a and a' are the supremum and infimum of $|f'|$ in the region. In particular, if $f(z)=c\,z+b$, then

$$\operatorname{Cap}^{(i)} E' = |c| \operatorname{Cap}^{(i)} E. \tag{2'}$$

Perhaps the easiest proof is by means of transfinite diameters.

From (2), we again obtain the conformal invariance of the vanishing of logarithmic capacity (VII. 6 E).

2 B. Subadditivity. The following theorem will result from potential-theoretic considerations:

THEOREM. *If E is a Borel set contained in the disk $|z|<\frac{1}{2}$ and expressed as the union $E=\bigcup_{n=1}^{\infty} E_n$ of Borel sets E_n, then*

$$\left(\log\frac{1}{\operatorname{Cap}^{(i)} E}\right)^{-1} \le \sum_{n=1}^{\infty}\left(\log\frac{1}{\operatorname{Cap}^{(i)} E_n}\right)^{-1}.$$

COROLLARY. *If a Borel set E is the union of a countable number of Borel sets E_n with $\operatorname{Cap}^{(i)} E_n=0$, then $\operatorname{Cap}^{(i)} E=0$.*

Proof of the Theorem. By assumption, $\log(1/|z-\zeta|)>0$ for every z and ζ in E. Since the result is trivial if $\operatorname{Cap}^{(i)} E=0$, we shall assume in the sequel that $\operatorname{Cap}^{(i)} E>0$.

Take a closed bounded set $E'\subset E$ with $\operatorname{Cap} E'>0$. The conductor potential $u(z)$ of E' satisfies

$$u(z)\le V_{E'}=\log\frac{1}{\operatorname{Cap} E'}$$

at every $z\neq\infty$ (VII.6 C.(2)). Let E'_n be an arbitrary closed bounded subset of $E_n\cap E'$. In terms of the equilibrium distribution μ on E', we consider

$$u_n(z)=\int_{E'_n}\log\frac{1}{|z-\zeta|}\,d\mu(\zeta).$$

If $\mu(E_n)>0$, then take E'_n with $\mu(E'_n)>0$. Since the integrand is non-negative, we have $u(z)\ge u_n(z)$ if $z\in E$. The mass distribution $\mu_n=\mu/\mu(E'_n)$ satisfies $\mu_n(E'_n)=1$, so that

$$\log\frac{1}{\operatorname{Cap} E'_n}=V_{E'_n}\le I(\mu_n)=\frac{1}{\mu(E'_n)^2}\int_{E'_n} u_n\,d\mu$$

$$\le\frac{1}{\mu(E'_n)}\sup_{E'_n} u_n\le\frac{1}{\mu(E'_n)}\sup_{E'} u=\frac{1}{\mu(E'_n)}V_{E'}$$

and, therefore, $\mu(E'_n)\log(1/\operatorname{Cap} E'_n)\le\log(1/\operatorname{Cap} E')$. Since E'_n is arbitrary, $\mu(E_n\cap E')\log\big(1/\operatorname{Cap}^{(i)}(E_n\cap E')\big)\le\log(1/\operatorname{Cap} E')$ and, therefore,

$$\mu(E_n\cap E')\le\left(\log\frac{1}{\operatorname{Cap}^{(i)} E_n}\right)^{-1}\log\frac{1}{\operatorname{Cap} E'}.$$

This is trivially true if $\mu(E_n)=0$. Summing over n gives

$$\left(\log\frac{1}{\operatorname{Cap} E'}\right)^{-1}\le\sum_{n=1}^{\infty}\left(\log\frac{1}{\operatorname{Cap}^{(i)} E_n}\right)^{-1}.$$

Since E' is an arbitrary closed bounded subset of E, the proof is herewith complete.

2 C. Decreasing Sequence of Sets. The following statement is readily obtained by considering the capacity $c_\beta(\infty)$ with respect to the complementary region:

If E and E_n, $n=1,2,\ldots$, are closed bounded sets such that $E_1\supset E_2\supset\cdots$ and $E=\bigcap_{n=1}^{\infty} E_n$, then

$$\operatorname{Cap} E=\lim_{n\to\infty}\operatorname{Cap} E_n. \tag{3}$$

For the proof, let W and W_n be the unbounded components of CE and CE_n. Set $\beta=\partial W$ and $\beta_n=\partial W_n$, and consider $c_\beta(\infty)$ and $c_{\beta_n}(\infty)$ with respect to W and W_n. Since $\beta\subset E\subset CW$, we obtain $\operatorname{Cap} E=c_\beta(\infty)$ by Theorem VII.6 D and similarly $\operatorname{Cap} E_n=c_{\beta_n}(\infty)$. The sets W_n increase with n and $\bigcup_{n=1}^{\infty} W_n=W$. This is not an exhaustion in the ordinary sense, that is, one by regular subregions. But as shown in III.4 B, this is immaterial, and we conclude that $c_\beta(\infty)=\lim c_{\beta_n}(\infty)$.

2 D. Area and Logarithmic Capacity. *If E is a Borel set in the finite plane, then*

$$\operatorname{Cap}^{(i)} E\geq\left(\frac{m\,E}{\pi}\right)^{\frac{1}{2}}, \tag{4}$$

where mE is the area (2-dimensional Lebesgue measure) of E.

In fact, we may assume that E is closed and bounded. For the unbounded component W of CE, $\operatorname{Cap} E=c_\beta(\infty)$. By VII.5 D–5 G, we have $\operatorname{Cap} E\geq\sqrt{S}$, where S is the span of W with respect to $\zeta=\infty$. We obtain $S\geq m(CW)/\pi\geq m\,E/\pi$ on applying Theorem VI.4 B to the function $F(z)\equiv z$. The assertion now follows.

From the definition of logarithmic capacity in VII.6 B, we obtain without using the span,

$$\operatorname{Cap}^{(i)} E\geq\left(\frac{m\,E}{\pi\,e}\right)^{\frac{1}{2}}$$

(see Tsuji [4, p. 58]).

In particular,

$$\operatorname{Cap}^{(i)} E=0\Rightarrow m\,E=0. \tag{5}$$

The converse is not true; a counterexample is obtained by using Theorem 2 E below (see 2 F).

2 E. Length and Logarithmic Capacity. We consider a Borel set E on a rectifiable Jordan curve C. Fix a point $z_0\in C$ and let $s(z)$ be the length of the arc $\widehat{z_0 z}$ along C for any $z\in C$. The length $\int_E ds$ of E will be denoted by $|E|$.

THEOREM. *For a Borel set E on a rectifiable Jordan curve,*

$$\mathrm{Cap}^{(i)} E = 0 \Rightarrow |E| = 0. \tag{6}$$

If there exists a constant K such that $|s(z') - s(z)| \leq K|z' - z|$ *for every* z *and* z' *on E, then*

$$\mathrm{Cap}^{(i)} E \geq \frac{|E|}{4K}. \tag{7}$$

In particular, if E is on a line, then

$$\mathrm{Cap}^{(i)} E \geq \frac{|E|}{4} \tag{8}$$

and $|E|$ is the 1-dimensional Lebesgue measure of E.

These estimates are due to TSUJI [4, pp. 59, 84]. The reader may find it informative to compare them with the estimates involving the quantity c_B to be discussed in 3 G.

That the converse of (6) is false will be seen in 2 F.

Proof of the Theorem. We may assume that E is closed and bounded. Let us start with (7). For $z \in E$, denote by $t(z)$ the length of the set $\{z' \in E \,|\, s(z') \leq s(z)\}$. Then the correspondence $z \to t$ gives a one-to-one mapping of E onto the t-interval $I = [0, |E|]$. For every z_1 and z_2, we have $|t(z_1) - t(z_2)| \leq |s(z_1) - s(z_2)| \leq K|z_1 - z_2|$. Therefore, by comparing the transfinite diameters, we obtain $\mathrm{Cap}\, I \leq K \,\mathrm{Cap}\, E$. The equality $\mathrm{Cap}\, I = |E|/4$ is evident if we consider c_β with respect to the complementary region CI, which is simply connected (VII. 5 H). Thus we have (7).

For the proof of (6), we use the fact that the rectifiable curve C satisfies $\lim_{z' \to z} |s(z') - s(z)|/|z' - z| = 1$ at almost every z. If $|E| > 0$, Egoroff's theorem provides us with a closed subset E' of E such that $|E'| > 0$ and the above limit exists uniformly on E'. Thus $|s(z') - s(z)|/|z' - z| \leq K$ holds for every z and $z' \in E'$, so that $\mathrm{Cap}\, E \geq \mathrm{Cap}\, E' \geq |E'|/4K > 0$, a contradiction. We conclude that $|E| = 0$.

2 F. Logarithmic Capacity of a Cantor Set. In order to show that the converse of (6) does not hold, we consider a generalized Cantor set $e = e(\{a_k\})$ on the real axis (IX. 4 D). NEVANLINNA [3, pp. 154 ff.] and AHLFORS-SARIO [1, pp. 252 ff.] proved:

$$\log \mathrm{Cap}\, e \geq \sum_{k=1}^{\infty} \frac{1}{2^k} \log \frac{a_k(1 - a_k)}{4}. \tag{9}$$

For example, the ternary set (i.e., $a_k = \frac{2}{3}$) satisfies

$$\mathrm{Cap}\, e \geq \tfrac{1}{18}, \qquad |e| = 0$$

and, therefore, serves as the desired counterexample.

2 G. Hausdorff Measure. More delicate metric conditions can be obtained in terms of the Hausdorff measure. Here we shall give its definition and present without proofs the main known relations to the vanishing of the logarithmic capacity.

Let $h(t)$ be a real-valued function which is continuous and strictly increasing on $[0, a)$ for some $0<a<1$, and such that $h(0)=0$. For an arbitrary set E in the finite plane and $0<r<a$, set $\Lambda^{(r)}(E)=\inf\sum_{i=1}^{\infty} h(d(E_i))$, where the infimum is taken over all partitions $E=\bigcup_{i=1}^{\infty} E_i$, $E_i \cap E_j = \emptyset$ $(i \neq j)$ such that the diameter $d(E_i)$ is less than r. The quantity $\Lambda^{(r)}(E)$ does not decrease as r decreases, so that

$$\Lambda(E)=\lim_{r\to 0} \Lambda^{(r)}(E)$$

exists. It is the *Hausdorff measure* of E for the function h (CARATHÉODORY [2], HAUSDORFF [1], SAKS [1, pp. 53ff.]). In particular, for $h(t)=t^{\alpha}$ $(\alpha>0)$, $\Lambda(E)$ is the *α-dimensional measure* $\Lambda_{\alpha}(E)$; for $h(t)=(\log t^{-1})^{-1}$, it is the *logarithmic measure* $\Lambda_{\log}(E)$.

For actual estimation, another approach is sometimes convenient. Instead of partitions, we consider coverings of E by open disks Δ_i with $d(\Delta_i)<r$. Again $\tilde{\Lambda}^{(r)}(E)=\inf\sum_{i=1}^{\infty} h(d(\Delta_i))$ increases as r decreases, so that

$$\tilde{\Lambda}(E)=\lim_{r\to 0} \tilde{\Lambda}^{(r)}(E)$$

exists. In particular, we obtain the measures $\tilde{\Lambda}_{\alpha}(E)$ and $\tilde{\Lambda}_{\log}(E)$.

A relationship between Λ and $\tilde{\Lambda}$ is difficult to find in the general case. We confine ourselves to an h for which

$$k(c)=\lim_{\delta\to 0} \sup_{0<t\le\delta} \frac{h(c\,t)}{h(t)}$$

exists for some $c>0$; for example, $k(c)=c^{\alpha}$ if $h(t)=t^{\alpha}$, and $k(c)=1$ if $h(t)=(\log t^{-1})^{-1}$. It is easily verified that

$$\Lambda(E)\le\tilde{\Lambda}(E)\le k(\sqrt{3})\,\Lambda(E).$$

Accordingly, the vanishing of $\Lambda(E)$ is equivalent to that of $\tilde{\Lambda}(E)$.

The 2-dimensional measure $\Lambda_2(E)$ and the 2-dimensional outer measure $m^* E$ are readily seen to be related by

$$m^* E=\frac{\pi}{4}\tilde{\Lambda}_2(E).$$

If E is contained in a rectifiable curve, then the 1-dimensional measure $\Lambda_1(E)$ and the 1-dimensional outer measure $m^* E$ are known to coincide:

$$m^* E=\Lambda_1(E)$$

(cf. CARATHÉODORY [1]).

For a closed bounded set E, we list the following results:

(a) $\Lambda_{\log}(E)=0 \Rightarrow \operatorname{Cap} E=0$ (LINDEBERG [1]; cf. also NEVANLINNA [3, p. 149]).

(b) $\Lambda_{\log}(E)<\infty \Rightarrow \operatorname{Cap} E=0$ (ERDÖS-GILLIS [1]; see also TSUJI [4, pp. 66 ff.]).

(c) $\operatorname{Cap} E=0 \Rightarrow \tilde{\Lambda}(E)=0$ for all h with the property $\int_0^a h(t)\, t^{-1}\, dt<\infty$ for some $a>0$ (FROSTMANN [1]; NEVANLINNA [3, pp. 151 ff.]). In particular, $\operatorname{Cap} E=0 \Rightarrow \Lambda_1(E)=0$ (cf. Theorem 2 E).

3. Plane Sets of Classes N_B, N_D, and N_{SB}

3 A. Tests by Extremal Length. The condition for a plane region W to be of class O_{AD} has been expressed in terms of extremal length in two ways. First, $W \in O_{AD}$ is characterized by the vanishing of the span $S(\zeta, \zeta')$ (VI. 2 D), which may be given in terms of extremal length (cf. Problem 7). Second, $W \in O_{AD}$ if and only if it is simultaneously an extremal horizontal and vertical slit plane, and the extremality is characterized by extremal length (IX. 1 E).

The condition $W \in O_\gamma$ can be characterized in terms of extremal length, since so can the weakness of a boundary component.

For the class O_{AB}, no such criteria are known.

3 B. Modular Test. In a plane region W, consider a sequence of disjoint annular regions $A_n(\alpha_n^i, \alpha_n'^i)$, $n=1, 2, \ldots$ (cf. 1 A), each A_n consisting of components A_n^i with $\partial A_n^i = \alpha_n^i \cup \alpha_n'^i$, $i=1, \ldots, k(n)$. Instead of the modulus μ_n of A_n, we introduce the *minimum modulus* μ_n^* of A_n as follows:

$$\mu_n^* = \min_{i=1, \ldots, k(n)} \mu(\alpha_n^i, \alpha_n'^i),$$

where $\mu(\alpha_n^i, \alpha_n'^i)$ is the modulus of A_n^i. Evidently $\mu_n \leq \mu_n^*$.

Assume that $\{A_n\}$ satisfies condition 1 A. (a) and that each A_n^i is a doubly connected plane region (cf. 1 C. (b)).

THEOREM. *A plane region W belongs to the class indicated below if it has an $\{A_n\}$ with the above properties and satisfies the corresponding additional condition:*

$$O_{AD}\colon \prod_{n=1}^{\infty} \mu_n^* = \infty,$$

$$O_{AB}\colon \varlimsup_{n\to\infty} k(n)^{-\frac{1}{2}} \prod_{j=1}^{n} \mu_j^* = \infty.$$

Remark 1. These tests were introduced for Riemann surfaces by SARIO [1] and PFLUGER [1], respectively, without the assumption that the A_n^i are doubly connected. If this is supposed, then the original proof is easily modified to show that $\prod \mu_n^* = \infty$ implies $W \in O_{KD}$. No corresponding test is known for O_{KB}.

Remark 2. The proof given below (SUITA [1]) is applicable only to plane regions. The proofs for arbitrary Riemann surfaces are found in the original papers cited above and in the monographs SARIO-NOSHIRO [1, p. 188] and NAKAI-SARIO [2].

It is an open question whether the converse statements are true for plane regions.

3 C. Proof of the Theorem. Without loss of generality, we may assume that $\infty \in W$ and $\infty \notin \bar{A}_n$, $n = 1, 2, \ldots$. Let Ω_n be the unbounded component of $W - \bar{A}_n$, and let $\Omega_n' = \Omega_n \cup \alpha_n' \cup A_n$. Here $\alpha_n' = \bigcup_i \alpha_n'^i$, and $\alpha_n'^i$ is the outer boundary of A_n^i. The regions Ω_n and Ω_n' are regular and of connectivity $k(n)$.

Consider the quantity $c_D(\infty)$ (VII.5 D) for Ω_n as well as for Ω_n'; we use the notation c_n and c_n'. The relation to the span is given by $c_D = \sqrt{S}$. By Theorem VI.4 B, the function $F_n(z) = \frac{1}{2}(P_0(z; \infty) + P_1(z; \infty))$ on Ω_n' satisfies $c_n'^2 = m(CF_n(\Omega_n'))/\pi$. For the doubly connected region $F_n(A_n^i)$, let I_{ni} and I_{ni}' be the areas of the regions bounded by the outer and inner contours, respectively. Clearly $m(CF_n(\Omega_n')) = \sum_i I_{ni}'$. We obtain $\sum_i I_{ni}' = \pi c_n'^2$ by the above, and $\sum_i I_{ni} \le \pi c_n^2$ by Theorem VI.4 B and the fact that $F_n \in \mathscr{V}_\infty(\Omega_n)$. Now apply Golusin's inequality (VIII.2 B): $\mu(\alpha_n^i, \alpha_n'^i)^2 \le I_{ni}/I_{ni}'$. It gives $\mu_n^* c_n' \le c_n$ and, therefore,

$$\mu_n^* c_n' \le c_{n-1}'.$$

We conclude that

$$\left(\prod_{j=1}^{n} \mu_j^*\right) c_n' \le \mu_1^* c_1'. \tag{1}$$

Consequently, $\prod_1^\infty \mu_n^* = \infty$ implies that $c_D(\infty)$ of the region W, which is equal to $\lim c_n'$, vanishes. Thus $W \in O_{AD}$.

Next, consider the quantity $c_B(\infty)$ of the region Ω_n' (VII.5 D); denote it by c_n^*. As mentioned in VII.5 E, a function $G_n \in \mathscr{A}_\infty(\Omega_n')$ with $M[G_n] = 1/c_n^*$ exists and is known to map Ω_n' onto a $k(n)$-sheeted disk of radius $1/c_n^*$. Thus $D[G_n]$ over Ω_n' is equal to $k(n)\, \pi\, c_n^{*-2}$. It follows that $c_n^{*2} \le k(n)\, c_n'^2$. On combining this with (1), we obtain

$$k(n)^{-\frac{1}{2}} \left(\prod_{j=1}^{n} \mu_j^*\right) c_n^* \le \mu_1^* c_1'.$$

Therefore, the assumption of the theorem implies that the quantity $c_B(\infty)$ for the region W, which is equal to $\lim c_n^*$, vanishes, and we have $W \in O_{AB}$.

3 D. Unions of Null Sets. In X.3 F and X.3 I, sets of classes N_B, N_D, and N_{SB} were characterized by removability. By making use of this, we can prove:

THEOREM. *If E_n $(n=1, 2, \ldots)$ are closed bounded sets of class N_B or N_D, then so is every closed bounded subset E of $\bigcup_{n=1}^{\infty} E_n$.*

Proof (NOSHIRO [2, Footnote 2, p. 11]). We may assume that the E_n are uniformly bounded. Consider a regular region Ω containing $\bigcup E_n$. By assumption, $\bigcup E_n$ is totally disconnected; hence so is E. We first suppose that $E \notin N_B$ and derive a contradiction. There exists an $F \in AB(\Omega - E)$ which is not analytically extendable to Ω. We eliminate all removable singularities of F from E and denote the remaining set by E'. Since E' is non-void, it is a perfect subset of E. The extended function on $\Omega - E'$ is single-valued and bounded because of the total disconnectedness of E. By Baire's theorem applied to $E' = \bigcup (E_n \cap E')$ in the relative topology of E', it is possible to find a neighborhood N_0 of a point $z_0 \in E'$ and a number n such that $N_0 \cap E' \subset E_n$. Since E_n is totally disconnected, there is a Jordan curve in $N_0 - E_n$ such that the region N_0' enclosed by it contains z_0 (cf. the last paragraph of VI. 1 E). The compact set $N_0' \cap E_n$ is of class N_B, and F belongs to the family $AB(N_0' - E_n)$. Consequently, F must be analytic at z_0, a contradiction.

For the class N_D, the proof is parallel. In this case, E has vanishing area, so that the extended function on $\Omega - E'$ still has a finite Dirichlet integral. The proof of the theorem is herewith complete.

For sets of class N_{SB} the situation changes. *The union of a finite number of disjoint sets of class N_{SB} is again of class N_{SB}.* It is apparent from the definition that a set is of this class if and only if every point is a weak boundary component of the complementary region. *But if the assumption of finiteness or disjointness is suppressed, then the conclusion does not necessarily hold.* The following counterexample is due to N. SUITA [5].

Take compact sets $e_n \subset A_n = \{z \mid (n+1)^{-1} < |z| < n^{-1}\}$ of class N_{SB} with the property

$$\iint_{A_n - e_n} \frac{dx\,dy}{|z|^2} \leq \frac{1}{n^2}.$$

Their construction is readily carried out by means of the reasoning in 3 J. The union $E = \{0\} \cup \bigcup_{n=1}^{\infty} e_n$ of disjoint sets of class N_{SB} is not of class N_{SB}, for $\{0\}$ is not a weak boundary component of the complementary region of E (see Problem 25).

In the partition $E = E_1 \cup E_2$ into $E_1 = \{0\} \cup \bigcup_{n=1}^{\infty} e_{2n}$ and $E_2 = \{0\} \cup \bigcup_{n=0}^{\infty} e_{2n+1}$, both subsets are of class N_{SB}. In fact, $\{0\}$ is weak by the modular test (Theorem 1 C) and the e_n and hence their components have this property by construction.

3 E. Sets of Classes N_D and N_{SB} on a Curve. We assert:

For a set E on a rectifiable Jordan curve, $E \in N_D$ and $E \in N_{SB}$ are equivalent.

Proof. Every point of $E \in N_{SB}$, regarded as a boundary component of the region CE, is weak. Thus every univalent meromorphic function on CE is continuous on the entire plane including the curve. By a well-known theorem of PAINLEVÉ, the function is meromorphic on the entire plane and therefore a linear transformation. We conclude by Theorem VI.2 D that $E \in N_D$.

Remark 1. STREBEL [4] showed that the properties $E \in N_D$ and $E \in N_{SB}$ are equivalent for an E contained in the union of a countable number of sets with finite 1-dimensional measures (cf. also LOKKI [2]).

Remark 2. In contrast with the above proposition, it is not true that if E is on a rectifiable curve, then $c_D(\infty) = c_1(\infty)$. In fact, if E consists of two line segments, then $c_D(\infty) > c_1(\infty)$ by Theorem VII.5 H.

3 F. Sets of Class N_D on a Circle. Let E be a closed set on a circle c, and let $E' = c \cap CE$. In AHLFORS-BEURLING [1], it was shown that

$$E \in N_D \Leftrightarrow \operatorname{Cap}^{(i)} E' = \operatorname{Cap} c. \tag{2}$$

Proof. We may assume without loss of generality that c is the circle $|z| = 1$, so that $\operatorname{Cap} c = 1$. Consider the span $S(0, \infty)$ of the region $W = CE$. It suffices to verify that

$$S(0, \infty) = -2 \log \operatorname{Cap}^{(i)} E'. \tag{3}$$

The set E' is open relative to c. It can be exhausted by a sequence $\{I_n\}$ of unions of finite numbers of open arcs contained in E'. Clearly, we have $\operatorname{Cap}^{(i)} E' = \lim \operatorname{Cap} \bar{I}_n$. On the other hand, the regions $V_n = C(c - I_n)$ exhaust W. This is not an ordinary exhaustion by regular subregions, but as Theorem II.1 H shows, the principal functions on V_n converge to those on W. Therefore, the span has the corresponding property. We denote by $S_n(0, \infty)$ the span $S(0, \infty)$ for the region V_n and have $S(0, \infty) = \lim S_n(0, \infty)$. Consequently, the proof of (3) reduces to that of

$$S_n(0, \infty) = -2 \log \operatorname{Cap} \bar{I}_n.$$

Let p_0 and p_1 be the principal functions $p_0(z; 0, \infty)$ and $p_1(z; 0, \infty)$ on V_n. Evidently, $p_1(z) = \log |z|$ on V_n, hence

$$S_n(0, \infty) = \tfrac{1}{2} \lim_{z \to \infty} (p_0(z) - \log |z|).$$

On the other hand, let g be the Green's function $g(z, \infty)$ of the region $C(\bar{I}_n)$. We have

$$-\log \operatorname{Cap} \bar{I}_n = \lim_{z\to\infty} (g(z) - \log|z|).$$

Since $g(1/\bar{z}) = g(z) - \log|z|$ by symmetry, the function $2g(z) - \log|z| = g(z) + g(1/\bar{z})$ is symmetric about the unit circle and, therefore, has vanishing normal derivative on $c - \bar{I}_n$. This function vanishes on I_n, so that the function

$$q(z) = \begin{cases} 2g(z) - \log|z| & \text{on } |z| > 1, \\ -2g(z) + \log|z| & \text{on } |z| < 1 \end{cases}$$

can be extended to V_n and has vanishing normal derivative on ∂V_n. Thus

$$p_0(z) = q(z) + k,$$

where the constant k is determined as follows: $0 = \lim_{z\to 0}(p_0(z) - \log|z|) = \lim_{z\to 0}(-2g(z) + k) = \lim_{z\to\infty}(-2g(1/\bar{z}) + k) = \lim_{z\to\infty}(-2g(z) + 2\log|z| + k)$, that is,

$$\tfrac{1}{2}k = -\log \operatorname{Cap} \bar{I}_n.$$

Consequently,

$$\begin{aligned} S_n(0, \infty) &= \tfrac{1}{2} \lim_{z\to\infty} (q(z) - \log|z| + k) \\ &= \lim_{z\to\infty} (g(z) - \log|z| + \tfrac{1}{2}k) = -2\log \operatorname{Cap} \bar{I}_n. \end{aligned}$$

3 G. Metric Estimates. A plane region W with $\infty \in W$ belongs to the classes O_γ, O_{AD}, O_{AB}, and O_G if and only if the quantities $c_1(\infty)$, $c_D(\infty)$, $c_B(\infty)$, and $c_\beta(\infty)$, respectively (considered in VII.5 A – 5 G), vanish. We know that

$$c_1(\infty) \le c_D(\infty) \le c_B(\infty) \le c_\beta(\infty) = \operatorname{Cap} E,$$

where $E = CW$.

Denote the area (=2-dim. Lebesgue measure) of E by mE. Let c be the union of a finite number of rectifiable Jordan curves in $W - \{\infty\}$ such that the union of the regions enclosed by them covers E. We denote by $l(E)$ the infimum of the lengths of all such c. In particular, if E is contained in a line, we have $l(E) = 2|E|$, where $|E|$ is the length (=1-dim. Lebesgue measure) of E.

THEOREM. *For an arbitrary E,*

$$\left(\frac{mE}{\pi}\right)^{\frac{1}{2}} \le c_D(\infty) \le c_B(\infty) \le \frac{l(E)}{2\pi}. \tag{4}$$

In particular, if E is on a line, then

$$\frac{|E|}{4} \le c_B(\infty) \le \frac{|E|}{\pi}. \tag{5}$$

Remark 1. The above result was given in AHLFORS-BEURLING [1]. More recently, POMMERENKE [1] has shown that we actually have

$$c_B(\infty)=\frac{|E|}{4}$$

for an E on a line.

Remark 2. The relation $\sqrt{mE/\pi}\leq \operatorname{Cap} E$ obtained from (4) was previously given in 2 D. The inequality $|E|/4\leq \operatorname{Cap} E$ deduced from (5) was proved in 2 E by a different method. For an E not necessarily on a line, it is impossible to obtain $\Lambda_1(E)\leq \text{const}\cdot \operatorname{Cap} E$, as is easily seen, but VITUŠKIN [1] claims the validity of $l(E)\leq \text{const}\cdot \operatorname{Cap} E$.

Proof of the Theorem. First, the relations $\sqrt{mE/\pi}\leq\sqrt{S(\infty)}=c_D(\infty)$ are direct consequences of Theorem VI.4 B. To prove $c_B(\infty)\leq l(E)/2\pi$, take an arbitrary $F\in\mathscr{A}_\infty(W)$. Since $\int_c F\,dz=-2\pi i$, we have

$$2\pi\leq\int_c |F|\,|dz|\leq M[F]\int_c |dz|,$$

so that $1/M[F]\leq l(E)/2\pi$. Hence $c_B(\infty)\leq l(E)/2\pi$. From this we obtain $c_B(\infty)\leq |E|/\pi$ if E is on a line, which we may assume to be the real axis. Furthermore, the function

$$F(z)=\int_E \frac{d\xi}{z-\xi}$$

satisfies $|\operatorname{Im} F|<\pi$. Therefore,

$$F_0=\frac{4}{|E|}\,\frac{e^{F/2}-1}{e^{F/2}+1}\in\mathscr{A}_\infty$$

and $M[F_0]\leq 4/|E|$. It follows that $|E|/4\leq c_B(\infty)$.

3 H. Explicit Tests. The following relation is an immediate consequence of the above theorem:

(a) *If $l(E)=0$, then $E\in N_B$.*

The 1-dimensional Hausdorff measure satisfies $l(E)\leq\sqrt{3}\,\pi\,\Lambda_1(E)$. There is no relation of the form $\Lambda_1(E)\leq \text{const}\cdot l(E)$, but it is true that $l(E)=0$ implies $\Lambda_1(E)=0$. Consequently, (a) is equivalent to

(a′) *If $\Lambda_1(E)=0$, then $E\in N_B$.*

From the above theorem, we also have the following two corollaries (the latter was previously obtained in Corollary VI.4 B and has been used frequently):

(b) *For an E on a line, $|E|=0$ if and only if $E\in N_B$.*

(c) *If $E\in N_D$, then $mE=0$.*

Result (b) can easily be extended to an E on a rectifiable Jordan arc.

For purposes of comparison, we include here the following two results obtained in IX.4 C (cf. also 3 A):

(d) *If the projections of E onto two orthogonal lines have vanishing lengths, then* $E\in N_D$.

(e) *If* $E\in N_D$, *then any* $z_1, z_2\notin E$ *can be joined by a curve in* CE *whose length deviates arbitrarily little from* $|z_1-z_2|$.

Finally, we remark that (d) gives the following supplement to (c):

(c′) *The Cartesian product* $E=e_1\times e_2$ *of* e_1 *and* e_2 *on a line has the property:*

$$E\in N_D \Leftrightarrow mE=0.$$

3 I. Remarks. (α) Whether the converses of (a) and (a′) are true is not known. (Vituškin [2] claims the existence of an example showing these converses, but the authors were unable to verify the correctness of his proof.) It was shown by Kametani [1] and Ahlfors [1] that, if $E\in N_B$, then its Newtonian capacity vanishes and, therefore, so does the $(1+\varepsilon)$-dimensional Hausdorff measure for every $\varepsilon>0$.

(β) The converse of (c) is trivially false; a line segment is an example of an $E\notin N_D$ with $mE=0$. On the other hand, Lehto [2] exhibited an example showing that a characterization of $E\in N_D$ by vanishing α-dimensional Hausdorff measure is impossible. In fact, for every ε with $0<\varepsilon<2$, the Cartesian product $E=e\times e$ of the generalized Cantor set $e=e(\{a_k\})$ with $a_k=2^{-\varepsilon/(2-\varepsilon)}$, $k=1,2,\ldots$ is the desired set: $E\in N_D$ is immediately seen from $|e|=0$ (cf. 3 H.(c′)) and $\tilde{\Lambda}_{2-\varepsilon}(E)\geq 1$ can be verified by the method found in Hausdorff [1, pp. 177 ff.].

(γ) There exists no relation between $mE=0$ and $E\in N_{SB}$. In fact, a line segment E is a trivial example of an $E\notin N_{SB}$ with $mE=0$, whereas the set E constructed in Theorem 3 J below satisfies $E\in N_{SB}$ and $mE>0$.

(δ) Total disconnectedness of E does not imply $E\in N_{SB}$, as will be seen from a counterexample in 3 L. The same is true even if E lies on a line; a counterexample was given in IX.4 G.(a) (cf. 3 E).

3 J. Examples. An example of a plane region W with

$$W\in O_{AB}-O_G$$

was obtained in 2 F; the complement W of the Cantor ternary set E has the properties Cap $E>0$ and $|E|=0$ (cf. 3 H.(b)).

A plane region W with

$$W \in O_{AD} - O_{AB}$$

was obtained in IX.4 G. A totally disconnected closed bounded set E on the real axis which has a positive $|E|$ and yet is an extremal set of vertical slits belongs to $N_D - N_B$, so that $CE = W$ qualifies. A set in $N_D - N_B$ can also be obtained (cf. 3 F) by constructing a closed set E on the unit circle such that $|E| > 0$ and $\mathrm{Cap}^{(i)} E' = 1$. (For details, see AHLFORS-BEURLING [1] or AHLFORS-SARIO [1, p. 255].)

A plane region W with

$$W \in O_\gamma - O_{AD}$$

will be obtained as the complement of the Cartesian product $E = e \times e$ of a set e on a line with $|e| > 0$ (cf. 3 H.(c')) and $E \in N_{SB}$. Such an E is constructed by using the following result of AHLFORS-BEURLING [1]:

THEOREM. *$E = e \times e$ with $e = e(\{a_k\}, \{n_k\})$ belongs to the class N_{SB} if*

$$\sum_{k=1}^{\infty} (1 - a_k) \log \frac{n_k (1 - a_{k-1})}{2} = \infty. \tag{6}$$

Proof. E is totally disconnected, so that every point of E is a boundary component of $W = CE$. We shall show its weakness by means of the modular test (Theorem 1 C). Given a point x_0 of e, consider all possible concentric circular annuli A_{kj}, $k = 1, 2, \ldots$, $j = 1, \ldots, m_k$, satisfying the following conditions: each A_{kj} has its center at the midpoint of the α_k-interval (cf. IX.4 J) containing x_0; its inner radius is greater than $\alpha_k + \beta_k$ and its outer radius less than β_{k-1}; $A_{kj} \cap e_k = \emptyset$; the inner and the outer peripheries of the A_{kj} pass through the end points of the α_k-intervals; and the A_{kj} are disjoint by pairs. Next, let A'_{kj}, $k = 1, 2, \ldots$, $j = 1, \ldots, m_k$, be the picture-frame-shaped regions defined as follows: their sides are parallel to the coordinate axes; the inner and the outer boundaries of A'_{kj} externally touch the corresponding contours of A_{kj}.

The proof will be complete if we show that the assumption of the theorem implies

$$\prod_{k,j} \mu(A_{kj}) = \infty \tag{7}$$

for an arbitrary $x_0 \in e$. In fact, for every $z_0 \in E = e \times e$, consider the A'_{kj}, $k = 1, 2, \ldots$, $j = 1, \ldots, m_k$, and suitably displace them parallel to the imaginary axis. Then each A'_{kj} is contained in $W = CE$ and separates z_0 from ∞. The modular test shows the weakness of z_0 since (7) implies $\prod_{k,j} \mu(A'_{kj}) = \infty$ by the inequality $\log \mu(A'_{kj}) \geq (\pi/4) \log \mu(A_{kj})$, which is easily verified using extremal length.

Remark. Before proving (7), we insert here the following observation. Since $A_{kj} \subset Ce$, we see similarly that (7) implies $e \in N_{SB}$. Accordingly, (6) gives $e \in N_{SB}$, or equivalently, $e \in N_D$ (3 E), as in Remark IX.4 G.

3 K. Proof (continued). To prove (7), let A_{kj}, $j=1,\dots,m_k$, be so numbered that each A_{kj} is enclosed by those with a larger j. The inner radius of the smallest annulus A_{k1} is $\frac{3}{2}\alpha_k+\beta_k$, and that of A_{k2} is $\frac{3}{2}\alpha_k+\beta_k+(\alpha_k+\beta_k)$. In general, A_{kj} has the inner and the outer radii

$$\tfrac{3}{2}\alpha_k+\beta_k+(j-1)(\alpha_k+\beta_k), \qquad \tfrac{3}{2}\alpha_k+\beta_k+(j-1)(\alpha_k+\beta_k)+\beta_k.$$

Since m_k is the largest j with $\frac{3}{2}\alpha_k+\beta_k+(j-1)(\alpha_k+\beta_k)+\beta_k<\beta_{k-1}$, we see that $\frac{3}{2}\alpha_k+\beta_k+m_k(\alpha_k+\beta_k)+\beta_k \geq \beta_{k-1}$, that is,

$$(m_k+2)(\alpha_k+\beta_k) \geq \beta_{k-1}+\tfrac{1}{2}\alpha_k.$$

We know that $\alpha_{k-1}/n_k>\alpha_k+\beta_k$ and $\beta_{k-1}+\alpha_k/2>\beta_{k-1}>(1-a_{k-1})\,\alpha_{k-1}/a_{k-1}>(1-a_{k-1})\,\alpha_{k-1}$ and, therefore,

$$m_k+2>n_k(1-a_{k-1}).$$

It follows that

$$\log\mu(A_{kj}) = -\log\left(1-\frac{\beta_k}{\alpha_k/2+\beta_k+j(\alpha_k+\beta_k)}\right)$$
$$>\frac{\beta_k}{\alpha_k/2+\beta_k+j(\alpha_k+\beta_k)}>\frac{\beta_k}{(j+1)(\alpha_k+\beta_k)}>\frac{1-a_k}{j+1},$$

which implies

$$\sum_{j=1}^{m_k}\log\mu(A_{kj})>\left(\frac{1}{2}+\cdots+\frac{1}{m_k+1}\right)(1-a_k)$$
$$>\left(\log\frac{m_k+2}{2}\right)(1-a_k)>\left(\log\frac{n_k(1-a_{k-1})}{2}\right)(1-a_k).$$

Consequently, $\sum_{k=1}^{\infty}\sum_{j=1}^{m_k}\log\mu(A_{kj})=\infty$ if the condition of the theorem is satisfied.

3 L. Totally Disconnected $E \notin N_{SB}$. The example in IX.4 G shows that there exists a totally disconnected closed bounded set E with $E \notin N_{SB}$, even if E is on a line. If we suppress the last requirement, the construction of an example is simplified. In fact, we have the following result due to MYRBERG (Appendix to SARIO [1]); cf. $N_{SB}=N_{SD}$ (X.2 L):

There exists a closed totally disconnected set E with $\infty \in E$ whose complement has finite area.

Proof. Consider the following square net in the plane $|z|<\infty$. Each square R_j has sides of unit length parallel to the real and imaginary axes. In a fixed square, say R_1, consider the Cartesian product $E_1=e(\{a_k\})\times e(\{a_k\})$ of generalized Cantor sets (IX.4 D), where $\prod_{k=1}^{\infty} a_k$ is taken so close to 1 that $m(R_1-E_1)\leq\frac{1}{2}$. In each of the eight R_j adjacent to R_1, take $E_j=e(\{a_k\})\times e(\{a_k\})$, so that $m(R_j-E_j)\leq 1/(8\cdot 2^2)$. In each R_j of the n-th generation, we similarly choose an E_j with $m(R_j-E_j)\leq 1/(8(n-1)\cdot 2^n)$. The union E of all E_j and $\{\infty\}$ is the desired set, since the area of CE is $\leq\sum 2^{-n}<\infty$.

4. Weak and Unstable Point Components

4 A. Weak and Unstable Boundary Components. A boundary component γ of a plane region W is weak if and only if, for every univalent function f on W, $f(\gamma)$ consists of a single point (VII.2 C). We called γ strong if $f(\gamma)$ always consists of more than one point (VII.3 D). If γ is neither weak nor strong, then we shall say that it is *unstable* (SARIO [13]).

We shall give several examples of point components whose weakness or instability will depend on the mode of accumulation of the other components.

4 B. First Example. Let E be a compact set on the non-negative real axis such that $0\in E$, $E\subset[0,1)$, and the component of 0 contains no other point. Let h be a real finite-valued function on E which is upper semicontinuous, non-negative, and such that $h(0)=0$. The set

$$W=W(E,h)=C\{z\,|\,\mathrm{Re}\,z\in E,\ |\mathrm{Im}\,z|\leq h(\mathrm{Re}\,z)\}$$

is clearly a region having $\gamma=\{0\}$ as a boundary component. Set $E'=[0,1)-E$. The following criteria are due to OHTSUKA [2], who generalized results of AKAZA-OIKAWA [1]:

THEOREM. *The boundary component* $\gamma=\{0\}$ *of* $W(E,h)$ *is*

(a) *weak if*

$$\int_{E'}\frac{dx}{x+h^*(x)}=\infty,$$

where $h^*(x)=\sup_{t\in E,\,0\leq t\leq x} h(t)$;

(b) *unstable if* $E'\subset\bigcup_{n=1}^{\infty}(a_n,b_n)$ *with* $a_n,b_n\in E\cup\{1\}$ *and*

$$\sum_{n=1}^{\infty}\min\left(\frac{b_n-a_n}{H_n},\ \frac{1}{\log^+\dfrac{1}{(b_n/a_n)-1}}\right)<\infty,$$

where $H_n=\min(h(a_n),h(b_n))$, *and the intervals* (a_n,b_n) *may overlap.*

The proof will be given in 4 C, 4 D.

The range of applicability of the above tests is enlarged by using the following "comparison theorem":

If two functions h_1 and h_2 considered on the same E satisfy

$$\overline{\lim_{x \in E, x \to 0}} \frac{h_2(x)}{h_1(x)} < \infty,$$

then the weakness of the boundary component $\gamma = \{0\}$ of $W(E, h_1)$ implies that of $W(E, h_2)$.

For the proof, we may assume that $h_2(x) \le M h_1(x)$ throughout E for some $M > 0$, since weakness is a boundary property (VII.2 C). The quasiconformal mapping $z \to \operatorname{Re} z + i M \operatorname{Im} z$ maps $W(E, h_1)$ onto $W(E, M h_1)$, a subregion of $W(E, h_2)$. Weakness is characterized by the vanishing of the extremal length of a certain family of curves, a property invariant under quasiconformal mappings (Appendix I.M). Thus the weakness of γ of $W(E, h_1)$ implies that of $W(E, M h_1)$. The weakness of γ of $W(E, h_2)$ is now evident.

4 C. Proof of Theorem 4 B. (a). For $\xi \in E'$, we consider the line segment

$$c'_\xi = \{z \mid \operatorname{Re} z = \xi, |\operatorname{Im} z| \le \xi + h^*(\xi)\}$$

and the arc

$$c''_\xi = \{z \mid |z|^2 = (\xi + h^*(\xi))^2 + \xi^2, \tan^{-1}(1 + h^*(\xi)/\xi) \le |\arg z| \le \pi\},$$

and let $c_\xi = c'_\xi + c''_\xi$. It suffices to show that the extremal length of the family $\Gamma' = \{c_\xi \mid \xi \in E'\}$ vanishes. Let U be a boundary neighborhood of γ such that $c_\xi \subset U$ for every $\xi \in E'$. We shall show that $L(\Gamma', \rho) = \inf_\xi \int_{c_\xi} \rho \, |dz| = 0$ for every ρ with a finite $A(\rho) = \iint_U \rho^2 \, dx \, dy$.

Suppose $L(\Gamma', \rho) = \delta > 0$. For the sets

$$E_1 = \left\{\xi \mid \xi \in E', \int_{c'_\xi} \rho \, |dz| \ge \frac{\delta}{2}\right\}, \quad E_2 = \left\{\xi \mid \xi \in E', \int_{c''_\xi} \rho \, |dz| \ge \frac{\delta}{2}\right\},$$

we have

$$\int_{E_1} \frac{d\xi}{\xi + h^*(\xi)} = \infty \quad \text{or} \quad \int_{E_2} \frac{d\xi}{\xi + h^*(\xi)} = \infty,$$

since $E' = E_1 \cup E_2$. We shall derive a contradiction by showing that neither of these conditions holds. Consider $\Gamma_1 = \{c'_\xi \mid \xi \in E_1\}$ and $\Gamma_2 = \{c''_\xi \mid \xi \in E_2\}$. Then $\lambda(\Gamma_1) \ge (\delta/2)^2 / A(\rho) > 0$ is immediate and $\lambda(\Gamma_1) = 0$ is deduced as follows: for an arbitrary density ρ' on U,

$$L(\Gamma_1,\rho')^2 \le \Big(\int_{c'_\xi} \rho'|dz|\Big)^2 \le 2(\xi+h^*(\xi)) \int_{c'_\xi} \rho'^2\, dy,$$

$$L(\Gamma_1,\rho')^2 \int_{E_1} \frac{d\xi}{\xi+h^*(\xi)} \le 2 \iint \rho'^2\, dx\, dy.$$

Therefore, the first possibility $\int_{E_1}(\xi+h^*(\xi))^{-1}\, d\xi=\infty$ is ruled out.

To exclude the second possibility, we observe that $\lambda(\Gamma_2)\ge(\delta/2)^2/A(\rho)>0$. In order to obtain $\lambda(\Gamma_2)=0$ and derive a contradiction, consider $r(\xi)=((\xi+h^*(\xi))^2+\xi^2)^{\frac{1}{2}}$ and $E_2^*=\{r(\xi)\,|\,\xi\in E_2\}$. For an arbitrary density ρ' on U, extended to $|z|<\infty$ by defining $\rho'=0$ on CU, we obtain

$$L(\Gamma_2,\rho')^2 \le \Big(\int_{c''_\xi} \rho'|dz|\Big)^2 \le 2\pi\, r(\xi) \int_{c''_\xi} \rho'^2\, r(\xi)\, d\theta,$$

$$L(\Gamma_2,\rho')^2 \int_{E_2^*} \frac{dr}{r} \le 2\pi \iint_{|z|<\infty} \rho'^2\, r\, dr\, d\theta.$$

Since h^* is increasing, it is differentiable almost everywhere, and

$$\int_{E_2^*} \frac{dr}{r} \ge \int_{E_2} \frac{(\xi+h^*(\xi))(1+h^{*\prime}(\xi))+\xi}{(\xi+h^*(\xi))^2+\xi^2}\, d\xi$$

$$\ge \int_{E_2} \frac{2\xi+h^*(\xi)}{(\xi+h^*(\xi))^2+\xi^2}\, d\xi \ge \int_{E_2} \frac{d\xi}{\xi+h^*(\xi)}=\infty.$$

We conclude that $\lambda(\Gamma_2)=0$.

4 D. Proof of Theorem 4 B.(b). Let U be a boundary neighborhood of γ such that its intersection with the positive real axis coincides with E'. We consider Γ in U and show that $\lambda(\Gamma)>0$. Let Γ_n be the subfamily of Γ consisting of curves which contain an arc joining the interval (a_n, b_n) and the negative real axis. Since $\Gamma=\bigcup \Gamma_n$, we have $\lambda(\Gamma)^{-1}\le\sum\lambda(\Gamma_n)^{-1}$. Every member of Γ_n contains an arc in the rectangle $\{z\,|\,a_n\le \operatorname{Re} z\le b_n, |\operatorname{Im} z|\le H_n\}$ which joins one of the horizontal sides with the interval $[a_n, b_n]$. Thus

$$\frac{1}{\lambda(\Gamma_n)} \le \frac{b_n-a_n}{H_n}.$$

On the other hand, the assumption in (b) implies $\lim(\max(H_n, a_n)/(b_n-a_n))=\infty$. We choose n_0 such that $\max(H_n, a_n)>b_n-a_n$ for every $n\ge n_0$. If $n\ge n_0$ and $a_n\le H_n$, then

$$\frac{b_n-a_n}{H_n} \le \frac{1}{\log \dfrac{H_n}{b_n-a_n}} \le \frac{1}{\log^+ \dfrac{a_n}{b_n-a_n}}.$$

If $n \geq n_0$ and $a_n \geq H_n$, we consider the doubly connected region $R_n = C([-\infty, 0] \cup [a_n, b_n])$. It is conformally equivalent to the extremal region of TEICHMÜLLER with $h = (b_n/a_n - 1)^{-1}$ (VIII.2 G), and

$$2\pi \lambda(\Gamma_n) \geq \log \Psi\left(\frac{1}{b_n/a_n - 1}\right) \sim \log \frac{1}{b_n/a_n - 1}$$

as $b_n/a_n \to 1$. We choose $n_0' \geq n_0$ so large that, if $n \geq n_0'$ and $a_n \geq H_n$, then

$$\frac{1}{\lambda(\Gamma_n)} \leq \frac{\text{const}}{\log \dfrac{a_n}{b_n - a_n}}.$$

This shows that $\lambda(\Gamma) > 0$.

4 E. Special Cases. In statements (a)–(d) below (OIKAWA [2], AKAZA [1], AKAZA-OIKAWA [1]), γ stands for the boundary component $\{0\}$ of $W(E, h)$, with h defined on the entire interval $(0, 1)$.

(a) *γ is weak if $x^2 + h(x)^2$ is an increasing function satisfying*

$$\int_{E'} \frac{dx}{\sqrt{x^2 + h(x)^2}} = \infty.$$

In fact, $\frac{1}{2}(x + h^*(x)) \leq \max(x, h^*(x)) \leq (x^2 + h(x)^2)^{\frac{1}{2}}$.

(b) *γ is unstable if it is not isolated and if $h(x)$ is an increasing function satisfying*

$$\int_{E'} \frac{dx}{h(x)} < \infty$$

and the following additional condition: there exists a number $K \geq 1$ such that, for every $x \in E - \{0\}$, we can find an $x' \in E$ with

$$x < x', \quad h(x') \leq K\, h(x). \tag{1}$$

For the proof, decompose E' into disjoint components: $E' = \bigcup_{n=1}^{\infty} (a_n, b_n)$. Since γ is not isolated, every a_n is positive. Therefore, $h(b_n) \leq h(a_n') \leq K\, h(a_n)$, so that

$$\sum \frac{b_n - a_n}{H_n} \leq K \sum \frac{b_n - a_n}{h(b_n)} \leq K \int_{E'} \frac{dx}{h(x)} < \infty.$$

(c) *Suppose* $h \equiv 0$. *Then* γ *is weak if*

$$\int_{E'} \frac{dx}{x} = \infty .$$

It is unstable if $E' \subset \bigcup_{n=1}^{\infty} (a_n, b_n)$ *with* $a_n, b_n \in E \cup \{1\}$ *and*

$$\sum_{n=1}^{\infty} \frac{1}{\log^+ \dfrac{1}{(b_n/a_n)-1}} < \infty .$$

This is a special case of Theorem 4 B.

(d) *Suppose* $h \equiv 0$ *and* $E = \{0\} \cup \bigcup_{n=1}^{\infty} [b_{n+1}, a_n]$ *with* $0 < b_{n+1} < a_n < b_n < 1$, $\lim a_n = 0$. *Then* γ *is weak if there exists a subsequence* $\{j(n)\}$ *such that*

$$\sum_{n=n_0}^{\infty} \frac{1}{\log^+ \dfrac{1}{(b_{j(n)}/a_{j(n)})-1}} = \infty$$

for every n_0 *and either* $a_{j(n+1)}/a_{j(n)} \leq 1-\delta$ $(n=1, 2, \ldots)$ *or* $b_{j(n+1)}/b_{j(n)} \leq 1-\delta$ $(n=1, 2, \ldots)$ *for some* δ *with* $0 < \delta < 1$.

4 F. Proof of (d). Suppose $a_{j(n+1)}/a_{j(n)} \leq 1-\delta$, $n=1, 2, \ldots$. If $\overline{\lim}\, b_{j(n)}/a_{j(n)} > 1$, then $\sum (b_n/a_n - 1) = \infty$ and, therefore, γ is weak. We may thus assume that

$$\lim_{n\to\infty} \frac{b_{j(n)}}{a_{j(n)}} = 1 .$$

Consider the doubly connected region

$$A_n = \{z \mid a_{j(n)} < |z| < a_{j(n-1)}\} - [b_{j(n)}, a_{j(n-1)}] .$$

It is conformally equivalent to

$$\{z \mid 1 < |z| < a_{j(n-1)}/a_{j(n)}\} - [b_{j(n)}/a_{j(n)}, a_{j(n-1)}/a_{j(n)}] ,$$

which contains

$$\tilde{A}_n = \{z \mid 1 < |z| < 1/(1-\delta)\} - [b_{j(n)}/a_{j(n)}, 1/(1-\delta)] .$$

The moduli μ_n and $\tilde{\mu}_n$ of A_n and $\tilde{A}_n$ satisfy

$$\log \mu_n \geq \log \tilde{\mu}_n .$$

By the lemma below,

$$\log \tilde{\mu}_n \sim \frac{\pi^2}{2 \log \dfrac{1}{(b_{j(n)}/a_{j(n)})-1}}$$

as $n \to \infty$. Thus $\sum \log \mu_n = \infty$, and we infer that γ is weak.

If $b_{j(n+1)}/b_{j(n)} \leq 1-\delta$, we have only to consider $\{z \mid b_{j(n)} < |z| < b_{j(n-1)}\} - [b_{j(n)}, a_{j(n-1)}]$ instead of A_n.

The proof will be complete once we have shown:

LEMMA. *Suppose $r>1$ is fixed and let $A_\varepsilon = \{z \mid 1<|z|<r\} - [1+\varepsilon, r]$, $0<\varepsilon<r-1$. Denote the modulus of A_ε by μ_ε. Then*

$$\log \mu_\varepsilon \sim \frac{\pi^2}{2\log\dfrac{1}{\varepsilon}} \qquad \textit{as } \varepsilon \to 0.$$

Proof. Map the quadrilateral $A_\varepsilon \cap \{z \mid \operatorname{Im} z>0\}$ joining $[-r, -1]$ with $[1, 1+\varepsilon]$ conformally onto the upper half-plane in such a way that $-r$, -1, 1 correspond to $-\infty$, -1, 0. The image η of $1+\varepsilon$ satisfies $\eta \sim c\,\varepsilon^2$ as $\varepsilon \to 0$, where c is a constant independent of ε. By reflection, A_ε is conformally equivalent to the extremal region D_η of TEICHMÜLLER, and we obtain (cf. VIII. 2 G. (d))

$$\log \mu_\varepsilon = \log \Psi(\eta) \sim \frac{\pi^2}{\log\dfrac{1}{\eta}} \sim \frac{\pi^2}{2\log\dfrac{1}{\varepsilon}}$$

as $\varepsilon \to 0$.

4 G. Second Example. Given sequences $\{a_n\}$, $\{b_n\}$, and $\{\theta_n\}$ with $0<b_{n+1}<a_n<b_n<1$, $0<\theta_n<\pi$, and $\lim a_n = \lim \theta_n = 0$, set $E_n = \{z \mid b_{n+1} \leq |z| \leq a_n, |\arg z| \leq \pi - \theta_n\}$. The region $W = C(\bigcup_{n=1}^{\infty} E_n) - \{0\}$ has a boundary component $\gamma = \{0\}$.

It is immediately verified by using the annular regions $a_n < |z| < b_n$, $n=1, 2, \ldots$, that γ is weak if

$$\sum_{n=1}^{\infty} \left(\frac{b_n}{a_n} - 1\right) = \infty.$$

On comparing W with the region in 4 E. (c), we observe that γ is unstable if

$$\sum_{n=1}^{\infty} \frac{1}{\log\dfrac{1}{(b_n/a_n)-1}} < \infty.$$

These criteria are supplemented by the following (cf. OIKAWA [2]):

THEOREM. *The boundary component $\gamma = \{0\}$ is unstable if*

$$\sum_{n=1}^{\infty} \left(\frac{b_n}{a_n} - 1\right) + \sum_{n=1}^{\infty} \frac{1}{\log\dfrac{1}{\theta_n}} < \infty.$$

Proof. Let Γ be the family of closed curves in $W \cap \{z \mid |z| < b_1\}$ separating γ from $|z| = b_1$. Denote by Γ_n the subfamily of curves which intersect the interval $[a_n, b_n]$. Since $\Gamma = \bigcup_{n=1}^{\infty} \Gamma_n$, we have

$$\frac{1}{\lambda(\Gamma)} \le \sum_{n=1}^{\infty} \frac{1}{\lambda(\Gamma_n)}.$$

Let $\sigma(a_n) = \{z \mid |z| = a_n, |\pi - \arg z| \le \theta_n\}$ and $\sigma(b_n) = \{z \mid |z| = b_n, |\pi - \arg z| \le \theta_n\}$. Consider the following subfamilies of Γ_n: Γ_n^0 consists of the curves in $a_n < |z| < b_n$, Γ_n^a of those intersecting $\sigma(a_n)$, and Γ_n^b of those intersecting $\sigma(b_n)$. Since $\Gamma_n = \Gamma_n^0 \cup \Gamma_n^a \cup \Gamma_n^b$, we have

$$\frac{1}{\lambda(\Gamma_n)} \le \frac{1}{\lambda(\Gamma_n^0)} + \frac{1}{\lambda(\Gamma_n^a)} + \frac{1}{\lambda(\Gamma_n^b)}.$$

It is evident that $\lambda(\Gamma_n^0)^{-1} = (2\pi)^{-1} \log(b_n/a_n)$. Note further that the convergence of $\sum \log(b_n/a_n)$ is equivalent to that of $\sum ((b_n/a_n) - 1)$.

To estimate $\lambda(\Gamma_n^a)$, we map $|z| > a_n$ by $w = z/a_n + a_n/z$ conformally onto the w-plane cut along $[-2, 2]$. Then $\sigma(a_n)$ and $[a_n, b_n]$ correspond to the segments $[-2, -2\cos\theta_n]$ and $[2, a_n/b_n + b_n/a_n]$. By a linear transformation, we further map the w-plane in such a manner that the points $-2\cos\theta_n$, 2, $a_n/b_n + b_n/a_n$ correspond to $0, 1, \infty$. Then the images of $\sigma(a_n)$ and $[a_n, b_n]$ are $[-1/h_n, 0]$ and $[1, \infty]$, where

$$h_n = \frac{1 + \cos\theta_n}{1 - \cos\theta_n} \left(\frac{a_n + b_n}{a_n - b_n} \right)^2. \tag{2}$$

Each member of the image of Γ_n^a contains an arc joining these segments. Thus we see, on considering the extremal region $\tilde{D}_{h_n}$ of TEICHMÜLLER (VIII. 2 G), that

$$\lambda(\Gamma_n^a) \ge \frac{1}{2\pi} \log \Psi(h_n) \sim \frac{1}{2\pi} \log h_n \tag{3}$$

as $h_n \to \infty$. In view of the assumption of the theorem, $\sum \lambda(\Gamma_n^a)^{-1}$ converges, since $h_n \ge (2/\theta_n)^2$.

In the case of Γ_n^b, we consider the mapping of $|z| < b_n$ and obtain the same h_n as in (2). Thus $\sum \lambda(\Gamma_n^b)^{-1}$ converges.

We have shown that $\sum \lambda(\Gamma_n)^{-1}$ converges and, therefore, that γ is unstable.

5. Strong and Unstable Continuum Components

5 A. Boundary Continua. If a boundary component of a plane region consists of more than one point, we shall simply call it a *boundary continuum*. In an illuminating example (5 D – 5 H), we shall give tests for a boundary continuum to be semiweak, parabolic, or strong, depending on the accumulation of the other components. Unstable boundary continua will be discussed in 5 I – 5 M in connection with mappings onto Koebe's circle regions.

Every plane region has at most a countable number of strong boundary continua. For the proof, use Theorem VI. 4 D; the image of a strong boundary continuum γ under the map P_2 has the property that Ξ_γ contains an interior point.

We have already encountered some sufficient conditions for a component to be strong: the properties

(a) γ is not parabolic (X. 1 I),

(b) γ is not semiweak (X. 1 A),

(c) γ is strong,

are interrelated by

$$(a) \Rightarrow (b) \Rightarrow (c).$$

That the converse of (a) ⇒ (b) is false will follow from an example in 5 D. As mentioned in VII. 3 D, we have not succeeded in proving or disproving the converse of (b) ⇒ (c). It also is unknown to us whether "strength" is a boundary property.

Recall the inclined parallel slit mapping $P_{(\theta)}$ discussed in VI. 3 F, where we classified boundary components into three types: 3 F. (a), (b), (c). It is natural to ask whether this classification coincides with that into weak, unstable, and strong components. The example in VI. 4 F reveals that this is not so: the boundary component γ considered there consists of a single point, yet is of type 3 F. (c).

5 B. Boundary Continua with Free Parts. If a boundary continuum γ is isolated, then (a), (b), and (c) of 5 A are always satisfied. In order to generalize this, we shall say that a boundary continuum γ of a plane region W has a *free part* if there exists a disk Δ such that $\Delta \cap \gamma \neq \emptyset$ and the closure of some component of $\Delta \cap W$ is disjoint from $\partial W - \gamma$.

THEOREM. *If γ is a boundary continuum with a free part, then it is not parabolic. Therefore, it is not semiweak, and it is strong.*

Proof. Without loss of generality, we may assume that γ is not isolated. Then we can take a Δ with the property that a component R of $\Delta \cap W$ with $\bar{R} \cap (\partial W - \gamma) = \emptyset$ is simply connected, and $\gamma' = \partial R - \gamma$ contains a proper continuum. Set $\gamma'' = \partial R - \gamma'$. By solving the Dirichlet problem, we construct a bounded harmonic function v on R with boundary values 1 on γ'' and 0 on γ'. Since γ' and γ'' contain proper continua, they contain regular points, and we have $0 < v < 1$ on R. Now let U be a boundary neighborhood of γ with $\bar{R} \cap W \subset U$, on which we consider the harmonic measure u_γ. Recall the property $u_\gamma = \lim u_{\gamma_n}$ in IV.4 E. The function u_{γ_n} is also obtained by solving the Dirichlet problem for the boundary values 0 on $\partial U - \gamma_n$ and 1 on γ_n. Since $\gamma'' \subset \gamma_n$ and $\gamma' \subset U$, we have $u_{\gamma_n} \geq v$, and therefore $u_\gamma \geq v > 0$ on R. Consequently, γ is not parabolic.

5 C. Symmetric Regions. Suppose a plane region W has the axes of symmetry

$$l_j = \{r\, e^{ij\pi/k} \mid -\infty < r < \infty\}, \qquad j = 0, \ldots, k-1,$$

and its boundary component γ contains 0.

THEOREM. *If every boundary component of W meets at least one l_j, then properties* (a), (b), *and* (c) *in* 5 A *are equivalent.*

For example, $\gamma = \{z \mid \operatorname{Re} z = 0,\ |\operatorname{Im} z| \leq 1\}$ is a semiweak boundary component of $W = \{z \mid |z| \leq \infty\} - \gamma - \bigcup_{n=1}^{\infty} \{z \mid \operatorname{Re} z = 1/n,\ |\operatorname{Im} z| \leq 1\} - \bigcup_{n=1}^{\infty} \{z \mid \operatorname{Re} z = -1/n,\ |\operatorname{Im} z| \leq 1\}$. This is verified by using extremal length and the fact that every point of γ except $\pm i$ is inaccessible. Thus γ is unstable and parabolic.

Proof of the Theorem. Suppose γ is strong. We shall show that it is not parabolic. Consider the function $P_{0\gamma}$ with respect to a reference point $\zeta = \infty \in W$. The image $P_{0\gamma}(W)$ is an incised radial slit disk of finite or infinite radius. In the latter case, $P_{0\gamma}(\gamma)$ contains incisions, since $P_{0\gamma}(\gamma)$ cannot degenerate to a point. The region $P_{0\gamma}(W)$ is easily seen to be symmetric about $l_0, \ldots, l_{k-1}$, and the slits are contained in $\bigcup_{j=0}^{k-1} l_j$. It is clear that $P_{0\gamma}(\gamma)$ has a free part. We conclude by the preceding theorem that γ is not parabolic.

5 D. An Example. If a region satisfies the hypothesis of neither 5 B nor 5 C, then we must use condition (a) or (b) to determine the strength of γ.

As an example, consider circular slit disks with certain symmetries. Take a sequence $\{r_n\}$ with $\frac{1}{2} < r_n < r_{n+1} < 1$ and $\lim r_n = 1$, a sequence $\{\theta_n\}$ with $0 < \theta_n < 1$, and a sequence $\{k_n\}$ of positive integers $k_n > 2\theta_n$. On each circle $|z| = r_n$, consider the arcs $\sigma(n, j)$, $j = 1, \ldots, k_n$, which contain both end points, subtend the same central angle $2\pi\, \theta_n / k_n$, and are equally distributed on the circle. The resulting region

$$W = \{z \mid |z| < 1\} - \bigcup_{n,j} \sigma(n, j)$$

has $\gamma = \{z \mid |z| = 1\}$ as a boundary continuum. The relative positions of the $\sigma(n, j)$ for different n's are immaterial.

THEOREM. (a) γ *is semiweak if*

$$\lim_{n\to\infty}\frac{1}{k_n}\log\frac{k_n}{1-\theta_n}=\infty.$$

(b) γ *is semiweak if*

$$1-\theta_n<\frac{k_n}{\pi}\log\frac{r_{n+1}}{r_n},\qquad n=1,2,\ldots,$$

and

$$\sum_{n=1}^{\infty}\frac{1}{k_n}\log\frac{\min(1,(k_n/\pi)\log(r_{n+1}/r_n))}{1-\theta_n}=\infty.$$

(c) γ *is parabolic if*

$$\left(\frac{\pi\,\theta_n}{k_n}\right)^2+\left(\log\frac{1}{r_n}\right)^2<\pi^2,\qquad n=1,2,\ldots,$$

and

$$\lim_{n\to\infty}\sqrt{\frac{\theta_n}{1-\theta_n}}\log\left(1+\frac{k_n}{\pi\,\theta_n}\log\frac{1}{r_n}\right)=\infty.$$

(d) γ *is not semiweak, and it is strong, if the quantities*

$$k_n\log\frac{r_{n+1}}{r_n},\qquad k_n\log\frac{r_n}{r_{n-1}},\qquad 1-\theta_n\qquad (n=1,2,\ldots),$$

with $r_0=\frac{1}{2}$, *are bounded away from zero.*

The proof will be given in 5 E – 5 H.

Case (c) was obtained qualitatively by CONSTANTINESCU [2] in order to exhibit an example of a parabolic γ which is not semiweak. The others were partly obtained by OIKAWA [2].

To illustrate the theorem, consider the case

$$r_n=1-\frac{1}{n},\qquad k_n=n^2.$$

We then have:

(a) γ is semiweak if $\lim n^{-2}\log(n^2/(1-\theta_n))=\infty$.

(b) γ is semiweak if $1-\theta_n<1/\pi$ and $\sum n^{-2}\log(1/(1-\theta_n))=\infty$.

(c) γ is parabolic if $\theta_n\geq\delta$ for some $\delta>0$.

(d) γ is not semiweak, and it is strong, if $\theta_n\leq 1-\delta$ for some $\delta>0$.

Observe that (b) is essentially sharper than (a). In fact, the family Γ^* in IV. 2 B qualifies for (a), as does $\tilde{\Gamma}$ for (b); cf. the last paragraph of IV. 2 C.

5 E. Proof of Theorem 5 D.(a). The set $\{z \mid |z|=r_n\} - \bigcup_j \sigma(n,j)$ consists of open arcs $\sigma'(n,j)$, $j=1,\ldots,k_n$, subtending the same central angle $2\pi(1-\theta_n)/k_n$. Suppose the arcs are numbered so that $\sigma(n,j)$ is succeeded by $\sigma(n,j+1)$, with $\sigma'(n,j)$ between them, $j=1,\ldots,k_n$, and $\sigma(n,k_n+1)=\sigma(n,1)$.

We consider the family Γ^* (IV.2 B) on $U=\{z \mid \frac{1}{2}<|z|\} \cap W$ and shall show that $\lambda(\Gamma^*)=\infty$. Note that condition (a) implies

$$\lim_{n\to\infty} \frac{1-\theta_n}{k_n}=0.$$

Let Γ_{nj}^* be the family of curves in $\frac{1}{2}<|z|<r_n$ which join $|z|=\frac{1}{2}$ and $\sigma'(n,j)$, $j=1,\ldots,k_n$. Then

$$\frac{1}{\lambda(\Gamma^*)} \leq \sum_j \frac{1}{\lambda(\Gamma_{nj}^*)} = \frac{k_n}{\lambda(\Gamma_{n1}^*)}.$$

Here $\lambda(\Gamma_{n1}^*)$ dominates the extremal length of the family of curves in $1/2r_1<|z|<1$ joining $|z|=1/2r_1$ with the arc $\{z \mid |z|=1,\ |\arg z| \leq \pi(1-\theta_n)/k_n\}$. The latter is $\lambda(\Gamma^{(n)})/4$ (cf. Appendix I.I.(a), (b)), where $\Gamma^{(n)}$ is the family of curves in

$$Q=\{z \mid 1/4r_1^2<|z|<1,\ \operatorname{Im} z>0\}$$

joining

$$\sigma_n'=\{z \mid |z|=1/4r_1^2,\ 0 \leq \arg z \leq \pi(1-\theta_n)/k_n\}$$

and

$$\sigma_n''=\{z \mid |z|=1,\ 0 \leq \arg z \leq \pi(1-\theta_n)/k_n\}.$$

If we map Q conformally onto the upper half-plane so that $1/4r_1^2$ and 1 correspond to 0 and 1, then the images $-a$ and $1+b$ of the end points of σ_n' and σ_n'' which are not on the real axis satisfy $a \sim \text{const} \cdot (1-\theta_n)^2 k_n^{-2}$ and $b \sim \text{const} \cdot (1-\theta_n)^2 k_n^{-2}$ as $n \to \infty$. By considering the extremal region $\tilde{D}_h$ of Teichmüller with respect to

$$h=\frac{1+a+b}{a\,b} \sim \text{const} \cdot \frac{k_n^4}{(1-\theta_n)^4},$$

we see that

$$\lambda(\Gamma_{n1}^*) \geq \frac{1}{4}\lambda(\Gamma^{(n)}) \sim \text{const} \cdot \log h \sim \text{const} \cdot \log \frac{k_n}{1-\theta_n}$$

as $n \to \infty$. Therefore, the condition in (a) implies $\lambda(\Gamma^*)=\infty$.

5 F. Proof of Theorem 5 D.(b). Consider the family $\tilde{\Gamma}$ (IV.2 B) on the same U. We shall show that $\lambda(\tilde{\Gamma})=0$. Let $Q(n,j)$ be the quadrilateral joining $\sigma(n,j)$ and $\sigma(n,j+1)$ in the exterior of the closed disk $|z|\le r_n$ as follows:

$$Q(n,j)=\{z\,|\,\pi(1-\theta_n)/k_n<|\log z-\log a_{nj}|<\min(\pi/k_n,\log(r_{n+1}/r_n)),\\ \log|z/a_{nj}|>0\},$$

where a_{nj} is the mid-point of $\sigma'(n,j)$. Since we assumed that $\pi(1-\theta_n)/k_n<\log(r_{n+1}/r_n)$, we have $Q(n,j)\neq\emptyset$. Let $\tilde{\Gamma}_n$ be the family of $c\in\tilde{\Gamma}$ such that $c\subset\bigcup_j Q(n,j)$. We have (cf. Appendix I.L.(b))

$$\lambda(\tilde{\Gamma}_n)=\sum_j 2\pi\left(\log\frac{\min(\pi/k_n,\log(r_{n+1}/r_n))}{\pi(1-\theta_n)/k_n}\right)^{-1}$$

$$=2\pi\,k_n\left(\log\frac{\min(\pi/k_n,\log(r_{n+1}/r_n))}{\pi(1-\theta_n)/k_n}\right)^{-1}.$$

Since $\lambda(\tilde{\Gamma})^{-1}\ge\sum\lambda(\tilde{\Gamma}_n)^{-1}$, we see from the condition in (b) that $\lambda(\tilde{\Gamma})=0$.

5 G. Proof of Theorem 5 D.(c). We consider the harmonic measure u_γ on the same U as above and show that $u_\gamma=0$.

It is readily verified that $u_\gamma=0$ on every $\sigma(n,j)$. If we can find a sequence c_n with

$$0<c_n<1, \qquad \lim c_n=0, \tag{1}$$

and $u_\gamma\le c_n$ on $\bigcup_j\sigma'(n,j)$, then the proof will be complete since $u_\gamma\le c_n$ on $U\cap\{z\,|\,|z|<r_n\}$.

Take a region $D(n,j)$ such that $\sigma(n,j)\subset D(n,j)\subset\{z\,|\,|z|<1\}$. Let v_{nj} be the harmonic function on $D(n,j)-\sigma(n,j)$ with boundary values 0 on $\sigma(n,j)$ and 1 on $\partial D(n,j)$. We have

$$u_\gamma\le v_{nj} \qquad \text{on } U\cap D(n,j).$$

Consequently, if there exists a sequence $\{c_n\}$ satisfying condition (1) and regions $D(n,j)$ with

$$\{z\,|\,|z|=r_n\}\subset\bigcup_j D_{c_n}(n,j), \qquad D_c(n,j)=\{z\in D(n,j)\,|\,v_{nj}(z)<c\} \tag{2}$$

then $u_\gamma\le c_n$ on $\bigcup_j\sigma'(n,j)$, and the proof will be complete.

To prove the existence consider the image of $\sigma(n,j)$ in the ($\zeta=\log z$)-plane. It is a vertical segment on the line $\operatorname{Re}\zeta=\log r_n$ and has length $2\pi\,\theta_n/k_n$. By a suitable parallel displacement, we may assume that its center is at the origin. In the w-plane, we consider the function $z=z(w)$

defined on $|w|>1$ by

$$\log z=\zeta=\frac{\pi\,\theta_n}{2k_n}\left(w-\frac{1}{w}\right).$$

If we determine $b_n>1$ by

$$\log\frac{1}{r_n}=\frac{\pi\,\theta_n}{2k_n}\left(b_n-\frac{1}{b_n}\right), \tag{3}$$

then the image of the circle $|w|=b_n$ is contained in $|z|\le 1$. If b_n also satisfies

$$\frac{\pi\,\theta_n}{2k_n}\left(b_n+\frac{1}{b_n}\right)<\pi, \tag{4}$$

then the image curve is a Jordan curve, and the region enclosed by it is the set $D(n,j)$ which we set out to find.

The function v_{nj} is determined by $v_{nj}(z(w))=(\log|w|)/\log b_n$. Therefore, the level line $v_{nj}(z)=c_n$ is the image of the circle $|w|=\exp(c_n\log b_n)=d_n$. Condition (2) is satisfied, if $D_{c_n}(n,j)$ covers the halves of $\sigma'(n,j-1)$ and $\sigma'(n,j)$ nearer to $\sigma(n,j)$. To accomplish this it suffices to show that

$$\frac{\pi}{k_n}\le\frac{\pi\,\theta_n}{2k_n}\left(d_n+\frac{1}{d_n}\right),$$

that is,

$$1\le\theta_n\cosh(c_n\log b_n).$$

Accordingly, if the $b_n>1$ defined by (3) satisfy (4) and if the $c_n=(\log b_n)^{-1}(\cosh^{-1}(1/\theta_n))$ satisfy (1), then we shall have $u_\gamma=0$. The assumption in (c) is readily seen to be sufficient for these b_n and c_n to meet the requirements. The proof of (c) is herewith complete.

5 H. Proof of Theorem 5 D.(d). Let U be as above. We shall construct a quasiconformal mapping T such that $T(U)$ is a radial slit annulus of inner radius $\frac{1}{2}$ and outer radius 1. Since there are only a countable number of radial slits, the extremal length of the family Γ^* (IV.2 B) is finite on $T(U)$ and hence on U (cf. Appendix I.M). Thus γ cannot be semiweak.

In order to construct T, we shall find simply connected regions $D(n,j)$, $n=1,2,\ldots$, $j=1,\ldots,k_n$, satisfying the following conditions: the regions are bounded by a finite number of analytic Jordan curves; the $\overline{D(n,j)}$ are pairwise disjoint; $\sigma(n,j)\subset D(n,j)$ and $\overline{D(n,j)}\subset U$; and the $D(n,j)-\sigma(n,j)$ are conformally equivalent. It is possible to find radial segments $\tilde{\sigma}(n,j)$ such that $\tilde{\sigma}(n,j)\subset D(n,j)$ and the $D(n,j)-\tilde{\sigma}(n,j)$ are conformally equivalent. Then there exists a quasiconformal mapping T_{nj} which maps $D(n,j)-\sigma(n,j)$ onto $D(n,j)-\tilde{\sigma}(n,j)$, gives the identity mapping on $\partial D(n,j)$, and has a maximal dilatation independent of n and j. We extend T_{nj} by letting it be the identity mapping on $U-\bigcup_{n,j} D(n,j)$, and obtain the desired quasiconformal mapping T of U.

To construct $D(n,j)$, let a_{nj} be the mid-point of $\sigma(n,j)$ and consider the pairwise disjoint quadrilaterals $Q(n,j)=\{z\,|\,\frac{1}{2}\log(r_n r_{n-1})<\log|z|<\frac{1}{2}\log(r_n r_{n+1}),\ |\arg(z/a_{nj})|<\pi/k_n\}$, $n=1,2,\ldots,j=1,\ldots,k_n$. In view of the condition in (d), it is possible to find a δ $(0<\delta<1)$ with $1-\theta_n>\delta$, $k_n\log(r_{n+1}/r_n)>\delta$, $k_n\log(r_n/r_{n-1})>\delta$, so that

$$\frac{\pi\,\theta_n}{k_n(1-\delta)}<\frac{\pi}{k_n}, \qquad \log r_n+\frac{\delta\,\theta_n}{2k_n}<\frac{1}{2}\log(r_n r_{n+1}),$$

$$\frac{1}{2}\log(r_n r_{n-1})<\log r_n-\frac{\delta\,\theta_n}{2k_n}.$$

Then $D(n,j)=\{z\,|\,|\log|z/a_{nj}||<\delta\,\theta_n/2k_n,\ |\arg(z/a_{nj})|<\pi\,\theta_n/k_n(1-\delta)\}$ has its closure in $Q(n,j)$. Since $\sigma(n,j)=\{z\,|\,|z|=r_n,\ |\arg(z/a_{nj})|\le\pi\,\theta_n/k_n\}$, the images of the $D(n,j)-\sigma(n,j)$ in the ($\log z$)-plane are congruent to each other. Thus $D(n,j)$ is the desired region, and we have proved Theorem 5 D.

5 I. Unstable Continua and Koebe's Circle Regions. We now proceed to criteria for the instability of boundary continua. Since we do not know if semiweakness implies instability, we must proceed in a different direction.

We have mentioned previously that a plane region of finite connectivity can be mapped conformally onto a circle region, that is, a region bounded by circles or points only (VI. 5 E). From this we obtain by a simple limiting process:

LEMMA. *If W is a plane region such that all boundary components except one, γ, are isolated, then it can be mapped conformally onto a region such that every boundary component except the image of γ is either a circle or a point.*

If we add the assumption that γ is weak, then the image is a circle region, since the image of γ is a single point. This reasoning was used by DENNEBERG [1], GRÖTSCH [10], and WAGNER [1] to show that certain plane regions with one accumulation boundary component can be mapped conformally onto circle regions (cf. Problems 26, 27). Regions with infinitely many accumulation boundary components were discussed in SARIO [1].

Even if γ is not weak, it is of interest to known when its image reduces to a single point and the region is again mapped onto a circle region. In this case γ is, of course, *unstable*. MESCHKOWSKI [1] has given a sufficient condition for this to occur.

5 J. Meschkowski's Condition. Let γ be a boundary component of a plane region W, $\infty \in W$, such that *every component of* $\partial W - \gamma$ *is isolated.* On a boundary neighborhood U of γ, consider the family $\tilde{\Gamma}$ of IV.2 B. Given a positive integer m, denote by $\tilde{\Gamma}_{(m)}$ the family of chains $c \in \tilde{\Gamma}$ consisting of at most m curves. Meschkowski's result is essentially the following (OIKAWA [2]):

THEOREM. *γ is weak or unstable if*

$$\lambda(\tilde{\Gamma}_{(m)}) = 0$$

for some m. In this case, W can be mapped conformally onto a circle region so that γ corresponds to a single point.

Proof. For simplicity, we assume from the start that every component of $\partial W - \gamma$ is either a circle or a point (cf. Lemma 5 I). We shall show that γ is a single point. Let $\gamma_1, \gamma_2, \ldots$ be the circle components of $\partial W - \gamma$. Since they cluster only at the compact set γ, the sum of the areas of the disks bounded by $\gamma_1, \gamma_2, \ldots$ converges. Hence the radii r_n of the circles γ_n converge to zero.

We may assume that U is bounded. Then the metric $\rho\,|dz| = |dz|$ satisfies $\iint_U \rho^2\, dx\, dy < \infty$, so that we can find

$$c_1, c_2, \ldots \in \tilde{\Gamma}_{(m)} \quad \text{with} \quad \lim_{n \to \infty} \int_{c_n} |dz| = 0.$$

Each c_n consists of curves $c_{n1}, \ldots, c_{nk}$, $k \le m$, and each c_{nj} joins $\gamma_{n(j)}$ and $\gamma_{n(j+1)}$ $(\gamma_{n(k+1)} = \gamma_{n(1)})$, where $\gamma_{n(j)}$ is a component of $\partial W - \gamma$, $j = 1, \ldots, k$. On taking a subsequence if necessary, we may assume from the beginning that the $\gamma_{n(j)}$ for different n or j are different, because each component of $\partial W - \gamma$ is isolated. Then the sum $\sum_{j=1}^{k} r_{n(j)}$ of the radii $r_{n(j)}$ of the $\gamma_{n(j)}$ converges to zero as $n \to \infty$. We join the set $c_{nj} \cap \gamma_{n(j)}$ and the center of $\gamma_{n(j)}$ by a radius, and do the same with $c_{n,j-1} \cap \gamma_{n(j)}$ and the center of $\gamma_{n(j)}$. Together with $c_{n1}, \ldots, c_{nk}$, we obtain a closed curve $\tilde{c}_n$ enclosing γ. Since the length of $\tilde{c}_n$ does not exceed $\int_{c_n} |dz| + 2 \sum_{j=1}^{k} r_{n(j)}$, which converges to zero as $n \to \infty$, we conclude that γ is a single point.

5 K. An Example. Take sequences $\{a_n\}$, $\{b_n\}$, and $\{\theta_n\}$ such that $a_n < b_n$, $a_{n+1} < a_n$, $b_{n+1} < b_n$, $\theta_n < \pi/2$, $\theta_{n+1} < \theta_n$, and $\lim a_n = \lim(1 - b_n) = \lim \theta_n = 0$. Denote by $\sigma_n(\theta)$ the line segment $a_n \le |z| \le b_n$, $\arg z = \theta$. The segment $\gamma=[-1,1]$ on the real axis is a boundary continuum of the region W defined as the complement of the set

$$\gamma\cup\bigcup_{n=1}^{\infty}\left(\sigma_n(\theta_n)\cup\sigma_n(-\theta_n)\cup\sigma_n(\pi+\theta_n)\cup\sigma(\pi-\theta_n)\right).$$

We assert:

The boundary continuum γ is unstable and W can be mapped onto a circle region with γ going to a point if

$$\sum_{n=1}^{\infty}\left(\frac{2\theta_n}{\log(b_n/b_{n+1})}+\frac{\pi}{\log(a_{n-1}/a_n)}\right)^{-1}=\infty.$$

For the proof, consider the quadrilaterals $\{z\,|\,a_n<|z|<a_{n-1},\theta_n<\arg z<\pi-\theta_n\}$ and $\{z\,|\,b_{n+1}<|z|<b_n,-\theta_n<\arg z<\theta_n\}$, as well as their mirror images about the real and imaginary axes, respectively. Let $\tilde{\Gamma}_{(4)}^n$ be the family of members of $\tilde{\Gamma}_{(4)}$ contained in the union of the above four quadrilaterals. We have $\lambda(\tilde{\Gamma}_{(4)})^{-1}\geq\sum_n\lambda(\tilde{\Gamma}_{(4)}^n)^{-1}$ and

$$\lambda(\tilde{\Gamma}_{(4)}^n)\leq 4\theta_n\left(\log(b_n/b_{n+1})\right)^{-1}+2\pi\left(\log(a_{n-1}/a_n)\right)^{-1},$$

so that $\lambda(\tilde{\Gamma}_{(4)})=0$.

Remark 1. This example is also interesting from the viewpoint discussed below. In general, it may seem plausible that even a semiweak boundary component γ would be strong provided its image under some $P_{0\gamma}$ has incisions (VII.4 A). That this is not true, however, is shown by the region W' which is the image of the above W under the transformation $z\to 1/z$. In fact, W' is an extremal radial slit disk of infinite radius with an incision (cf. example in IX.3 G), yet its outer boundary is unstable.

Remark 2. The region W serves as a counterexample referred to at the end of VII.3 C. Consider $F\in\mathcal{S}_{\infty\gamma}$ mapping W onto a circle region with $F(\gamma)=\{\infty\}$. It is evidently not identical with $P_{0\gamma}(z;\infty)=1/z$ although it maximizes m_γ.

5 L. Symmetric Regions. If a region has a symmetry property, as in the case of the above example, we can deduce another criterion for instability.

Suppose W is a region with $\infty\in W$, and γ is a boundary component with $0\in\gamma$. Let W be symmetric about the lines $l_j=\{r\,e^{ij\pi/k}\,|-\infty<r<\infty\}$, $j=0,\dots,k-1$, and assume that every boundary component except γ is isolated. Let U be a boundary neighborhood of γ and denote the components of $(\partial W-\gamma)\cap U$ by $\gamma_1, \gamma_2, \ldots$. Suppose further that one of the following two conditions is satisfied:

(a) For each γ_n, there exists an axis $l_{j(n)}$ such that the family Γ_n of curves in U joining γ_n with γ_n', the symmetric image of γ_n about $l_{j(n)}$, satisfies

$$\lim_{n\to\infty} \lambda(\Gamma_n)=0.$$

(b) There exists a sequence of disjoint quadrilaterals Q_i, $i=1,2,\ldots$, such that

(α) the Q_i cluster only at γ,

(β) each Q_i is symmetric about some axis $l_{j(i)}$,

(γ) each Q_i joins components $\gamma_{n(i)}$ and $\gamma_{n(i)}'$, which are symmetric about $l_{j(i)}$,

(δ) $\bigcup_{i=1}^{\infty}(\gamma_{n(i)}\cup\gamma_{n(i)}')=\bigcup_{n=1}^{\infty}\gamma_n$,

(ε) the family Γ_i of curves joining $\gamma_{n(i)}$ and $\gamma_{n(i)}'$ in Q_i satisfies $\lambda(\Gamma_i)\le M$ for some $M<\infty$ independent of i.

THEOREM. *Suppose that either* (a) *or* (b) *is satisfied. If γ is semiweak, then it is weak or unstable and W can be mapped conformally onto a circle region in such a way that γ corresponds to a single point.*

This result, due to OIKAWA [2], is illustrated by the following:

In example 5 K, *γ is unstable if*

$$\frac{a_n}{a_{n+1}}\ge 1+\delta, \qquad n=1,2,\ldots,$$

for some $\delta>0$.

In fact, the region is symmetric about the real and imaginary axes. By considering the quadrilaterals $\{z\,|\,a_n<|z|<a_{n-1},\ \theta_n<\arg z<\pi-\theta_n\}$ and their mirror images about the real axis, we easily see that condition (b) is satisfied.

5 M. Proof of Theorem 5 L. As in the proof of Theorem 5 J, we may assume in advance that every $\gamma_1, \gamma_2, \ldots$ is a circle or a point and that U is bounded. We shall show that γ consists of the single point 0.

Consider the density $\rho|dz|=|dz|$. If (a) is assumed, then

$$\Big(\inf_{c\in\Gamma_n}\int_c |dz|\Big)^2\le\lambda(\Gamma_n)\iint_U dx\,dy\to 0$$

as $n\to\infty$. If (b) is satisfied, then the Q_i are disjoint and cluster only at γ, their area converges to zero and, therefore,

$$\Big(\inf_{c\in\Gamma_i}\int_c |dz|\Big)^2\le\lambda(\Gamma_i)\iint_{Q_i} dx\,dy\to 0$$

as $i\to\infty$. In both cases, we can thus find, for every γ_n, an axis $l_{j(n)}$ of symmetry and a curve in W joining γ_n and its symmetric image γ'_n about $l_{j(n)}$ such that the length of the curve tends to zero as $n\to\infty$.

The radii of the circles $\gamma_1, \gamma_2, \dots$ tend to zero, as seen in the proof of Theorem 5 J. We extend the above curves to the centers a_n and a'_n of γ_n and γ'_n along suitable radii, thereby joining these centers by curves c_n with $c_n\cap\gamma=\emptyset$ and

$$\lim_{n\to\infty}\int_{c_n}|dz|=0. \tag{5}$$

Let γ^* be the set of accumulation points of the a_n, $n=1,2,\dots$. Clearly $\gamma^*\subset\gamma$. It is further evident that γ and γ^* are symmetric about l_j, $j=0,\dots,k-1$.

We first claim that $\gamma^*\subset l_0\cup\cdots\cup l_{k-1}$. For the proof, suppose there exists a point $z_0\in\gamma^*$ not contained in $l_0\cup\cdots\cup l_{k-1}$. There exists a subsequence $\gamma_{n(m)}$ such that $\gamma_{n(m)}$ and $\gamma'_{n(m)}$ are symmetric about a fixed axis, say l_j, and the center $a_{n(m)}$ of $\gamma_{n(m)}$ converges to z_0 as $m\to\infty$. Then the center $a'_{n(m)}$ of $\gamma'_{n(m)}$ converges to the point z'_0, symmetric with z_0 about l_j. Since $z_0\notin l_j$, we have $z'_0\neq z_0$, in violation of (5). Thus $\gamma^*\subset l_0\cup\cdots\cup l_{k-1}$.

Suppose now that $z_0\in\gamma^*$ is different from 0. We shall show that then no point $z_1\neq 0, z_0$, lying on the line segment $\overline{0\,z_0}$ can belong to γ^*. Let z_0 belong to l_j; then $z_1\in l_j$. If $z_1\in\gamma^*$, we can again find subsequences $\gamma_{n(m)}$, and $\gamma'_{n(m)}$ which are symmetric about a fixed axis l_p and have the property $\lim a_{n(m)}=z_1$. The point $z'_1=\lim a'_{n(m)}$ is symmetric with z_1 about l_p. If $l_p\neq l_j$, then $z_1\neq z'_1$, which contradicts (5). If $l_p=l_j$, then $z_1=z'_1$, but no $c_{n(m)}$ meets γ, a continuum containing 0, z_1, and z_0; therefore,

$$\varliminf_{m\to\infty}\int_{c_{n(m)}}|dz|>0,$$

contrary to (5). Thus $z_1\notin\gamma^*$.

As a consequence, γ^* consists of 0 and a finite number of points on $l_0\cup\cdots\cup l_{k-1}$. The set γ is a continuum containing γ^*. If $\gamma^*\neq\{0\}$, then γ would have a free part in the sense of 5 B, in violation of the semiweakness. Thus $\gamma^*=\{0\}$. If $\gamma\neq\{0\}$, then γ would again have a free part, which is impossible. We conclude that $\gamma=\{0\}$.

Appendices

Appendix I

Extremal Length

The theory of extremal length was introduced by AHLFORS-BEURLING [1]; cf. also AHLFORS-SARIO [1, pp. 219ff.] and AHLFORS [4]. The purpose of Appendix I is to present definitions of extremal length and related notions, together with their properties needed in our book.

I.A. Curves and Chains. By a *curve* on a Riemann surface W we shall mean a continuous mapping $z(t)$, $t \in I$, of an interval I of the real axis into W. The interval I may be $[a, b]$, $(a, b]$, $[a, b)$, or (a, b) with $-\infty < a < b < \infty$. If no ambiguity arises, the set $\{z(t) \mid t \in I\}$ will be called a curve. For example, we shall, in an obvious sense, speak of a curve contained in a set.

A curve is *closed* if $I = [a, b]$ and $z(a) = z(b)$; in all other cases we call it *open*. A closed curve is a *Jordan curve* if it has the additional property $z(t_1) \neq z(t_2)$ for all $t_1, t_2 \in [a, b)$, $t_1 \neq t_2$.

A curve c is *locally rectifiable* if the restriction of c to an arbitrary closed subinterval of I is rectifiable.

Let Ω be an arbitrary subregion of W, and let E_1 and E_2 be subsets of the relative boundary of Ω. A curve with $I = (a, b)$ is said to *join E_1 and E_2 in Ω* if $z(t) \in \Omega$ for all $t \in (a, b)$ and

$$\bigcap_{\varepsilon > 0} \mathrm{Cl}\{z(t) \mid a < t < a + \varepsilon\} \subset E_1,$$

$$\bigcap_{\varepsilon > 0} \mathrm{Cl}\{z(t) \mid b - \varepsilon < t < b\} \subset E_2.$$

Here Cl stands for the closure.

By a *locally rectifiable chain* we mean a formal (finite or infinite) sum $c = \sum k_i c_i$, where the k_i are integers and the c_i locally rectifiable curves. We shall sometimes also use this term to mean the set $\bigcup_i c_i$. If c is a finite sum and if every c_i is a closed curve, then c will be called a *rectifiable cycle.*

I.B. Definition of Extremal Length. Given an open set Ω of a Riemann surface W, an invariant form $\rho\,|dz|$ which is non-negative and Borel measurable will be referred to as a *metric* on Ω. For every locally rectifiable chain $c=\sum k_i c_i$, we set $\int_c \rho|dz|=\sum |k_i| \int_{c_i} \rho|dz|$ ($\le\infty$), where $\int_{c_i} \rho|dz|$ is defined in the usual manner.

Given a family Γ of locally rectifiable chains contained in Ω, set

$$L(\Gamma,\rho)=\inf_{c\in\Gamma}\int_c \rho\,|dz|, \qquad A(\rho,\Omega)=\iint_\Omega \rho^2\,dx\,dy.$$

With the understanding that $0/0=\infty/\infty=0$, we consider

$$\lambda(\Gamma)=\sup\frac{L(\Gamma,\rho)^2}{A(\rho,\Omega)},$$

where the supremum is taken over all metrics on Ω.

This quantity does not depend on the open set Ω containing all the chains in Γ. In fact, for all metrics on W, we obtain

$$\lambda(\Gamma)=\sup\frac{L(\Gamma,\rho)^2}{A(\rho)},$$

where

$$A(\rho)=A(\rho,W).$$

The proof is immediate since every $\rho|dz|$ on Ω can be extended to W by setting $\rho=0$ on $W-\Omega$, and every $\rho\,|dz|$ on W satisfies $A(\rho,\Omega)\le A(\rho)$.

The quantity $\lambda(\Gamma)$, which depends only on Γ, is called the *extremal length* of the family Γ.

In the above definition, we may impose the further restriction that ρ be lower semicontinuous on either Ω or W. However, ρ cannot be assumed to be continuous.

I.C. Extremal Metric. We shall often refer to $\rho\,|dz|$ simply as a metric ρ. Given a family Γ, we denote by $\mathbf{P}(\Gamma)$ the family of metrics on W satisfying

$$L(\Gamma,\rho)\ge 1, \qquad A(\rho)<\infty.$$

Then the extremal length is expressed as follows: $\lambda(\Gamma)=0$ if $\mathbf{P}(\Gamma)=\emptyset$, and otherwise

$$\frac{1}{\lambda(\Gamma)}=\inf_{\rho\in\mathbf{P}(\Gamma)} A(\rho).$$

Under the assumption $\mathbf{P}(\Gamma)\neq\emptyset$, if there exists a $\rho\in\mathbf{P}(\Gamma)$ such that $A(\rho)=\lambda(\Gamma)^{-1}$, it will be called an *extremal metric* for the family Γ; it is a metric minimizing $A(\rho)$ in $\mathbf{P}(\Gamma)$.

In general, an extremal metric may not exist. For this reason, we consider, with STREBEL [3], the completion of $\mathbf{P}(\Gamma)$ with respect to the metric $\sqrt{A(\rho-\rho')}$. Let $\bar{\mathbf{P}}(\Gamma)$ be the family of metrics $\rho|dz|$ on W such that

$$A(\rho)<\infty, \qquad A(\rho-\rho_n)\to 0$$

for a sequence $\{\rho_n\}\subset\mathbf{P}(\Gamma)$. A metric which minimizes $A(\rho)$ in $\bar{\mathbf{P}}(\Gamma)$ will be called a *generalized extremal metric* for the family Γ. An extremal metric is a generalized extremal metric.

If $\lambda(\Gamma)>0$, then a generalized extremal metric ρ_0 exists and is uniquely determined. It satisfies

$$A(\rho_0)=\frac{1}{\lambda(\Gamma)}.$$

For the proof, let $\rho_n\in\bar{\mathbf{P}}(\Gamma)$ be such that $\lim A(\rho_n)=d$, where $d\le\lambda^{-1}<\infty$ is the infimum of $A(\rho)$ in $\bar{\mathbf{P}}(\Gamma)$. We shall make use of the well-known parallelogram law

$$A\left(\frac{\rho_n+\rho_m}{2}\right)+A\left(\frac{\rho_n-\rho_m}{2}\right)=\frac{1}{2}\left(A(\rho_n)+A(\rho_m)\right). \tag{1}$$

Since $(\rho_n+\rho_m)/2\in\bar{\mathbf{P}}(\Gamma)$, we see that if n and m are taken sufficiently large, then $d+A(\rho_n-\rho_m)/4\le d+\varepsilon$. Thus there exists a $\rho_0\in\bar{\mathbf{P}}(\Gamma)$ such that $\lim A(\rho_n-\rho_0)=0$. It satisfies $A(\rho_0)=d$. By definition, $d\le\lambda^{-1}$. On the other hand, there is a sequence $\{\rho_i\}\subset\mathbf{P}(\Gamma)$ with $A(\rho_i-\rho_0)\to 0$. Since $A(\rho_i)\ge\lambda^{-1}$ and $A(\rho_i)\to A(\rho_0)$, we have $d\ge\lambda^{-1}$, and consequently $A(\rho_0)=\lambda^{-1}$.

Let ρ_0' be another generalized extremal metric. Take ρ_i and ρ_i' from $\mathbf{P}(\Gamma)$ such that $A(\rho_i-\rho_0)\to 0$, $A(\rho_i'-\rho_0')\to 0$. By applying (1) we obtain $d+A(\rho_i-\rho_i')/4\le(A(\rho_i)+A(\rho_i'))/2$, so that $d+A(\rho_0-\rho_0')/4\le d$. We conclude that $\rho_0=\rho_0'$ a.e.

I.D. An Inequality Satisfied by the Generalized Extremal Metric. The following relation is crucial in characterizing the extremal metric (SUITA [3]):

Let ρ_0 be the generalized extremal metric for Γ and suppose $\lambda(\Gamma)>0$. Then

$$A(\rho-\rho_0)\le A(\rho)-A(\rho_0)$$

for every $\rho\in\bar{\mathbf{P}}(\Gamma)$.

For the proof, take an ε with $0<\varepsilon<1$ and consider $\rho_\varepsilon=(\rho_0+\varepsilon\,\rho)/(1+\varepsilon)\in\bar{\mathbf{P}}(\Gamma)$. Then $(1+\varepsilon)^2A(\rho_0)\le(1+\varepsilon)^2A(\rho_\varepsilon)=A(\rho_0)+2\varepsilon\iint\rho\,\rho_0\,dx\,dy+\varepsilon^2A(\rho)$. On simplifying and letting $\varepsilon\to 0$, we obtain $A(\rho_0)\le\iint\rho\,\rho_0\,dx\,dy$ and, consequently,

$$A(\rho-\rho_0)=A(\rho)-2\iint_W\rho\,\rho_0\,dx\,dy+A(\rho_0)\le A(\rho)-A(\rho_0).$$

I.E. Another Characterization of the Generalized Extremal Metric. Given a family of rectifiable chains, we say that a property holds for *almost all* chains in the family if it is satisfied by all chains in the family except for those of a subfamily of infinite extremal length. The concept was introduced by OHTSUKA [1] (cf. FUGLEDE [1]).

Given a family Γ with $\lambda(\Gamma)>0$, a metric $\rho_0\,|dz|$ is the generalized extremal metric for Γ if and only if

$$A(\rho_0)=\frac{1}{\lambda(\Gamma)}, \qquad \int_c \rho_0\,|dz|\geq 1$$

for almost all $c\in\Gamma$.

The proof is immediate by the following auxiliary result (SUITA [4]):

LEMMA. *A metric $\rho\,|dz|$ with $A(\rho)<\infty$ belongs to $\bar{\mathbf{P}}(\Gamma)$ if and only if $\int_c \rho\,|dz|\geq 1$ for almost all $c\in\Gamma$.*

Proof. To show the sufficiency set $\Gamma_0=\{c\,|\int_c \rho\,|dz|<1\}\subset\Gamma$. Since $\lambda(\Gamma_0)=\infty$ by assumption, there exists a sequence $\{\rho_n'\}$ such that $\int_c \rho_n'\,|dz|\geq 1$ for every $c\in\Gamma_0$ and $A(\rho_n')\to 0$. Then $\rho_n=\max(\rho,\rho_n')$ belongs to $\mathbf{P}(\Gamma)$. By $(\rho-\rho_n)^2\leq(\rho_n')^2$, we have $A(\rho-\rho_n)\leq A(\rho_n')\to 0$ and, therefore $\rho\in\bar{\mathbf{P}}(\Gamma)$.

Conversely, to show that $\lambda(\Gamma_0)=\infty$, we observe that $\Gamma_0=\bigcup\Gamma_n$, where $\Gamma_n=\{c\,|\int_c \rho\,|dz|<1-1/n\}$, $n=2,3,\ldots$. By I.G.(c) below, it suffices to prove that $\lambda(\Gamma_n)=\infty$. Take $\rho_i\in\mathbf{P}(\Gamma)$ with $A(\rho-\rho_i)\to 0$. Then for any $c\in\Gamma_n$,

$$\int_c n\,|\rho-\rho_i|\,|dz|\geq n\int_c \rho_i\,|dz|-n\int_c \rho\,|dz|\geq n-n\left(1-\frac{1}{n}\right)=1,$$

so that $n\,|\rho-\rho_i|\in\mathbf{P}(\Gamma_n)$. Consequently, $\lambda(\Gamma_n)\geq n^{-2}A(\rho-\rho_i)^{-1}\to\infty$ as $i\to\infty$.

I.F. Conformal Invariance. Let Γ be a family of locally rectifiable chains contained in an open set $\Omega\subset W$. Let f be a direct or indirect conformal mapping of Ω onto an open set Ω_1 of a Riemann surface W_1. Denote by $f(\Gamma)$ the family of images $f(c)$ of $c\in\Gamma$. Then

$$\lambda(\Gamma)=\lambda(f(\Gamma)).$$

For the proof, let f' be the derivative or conjugate derivative of f. For a metric $\rho_1\,|dz_1|$ on Ω_1, the metric $\rho\,|dz|$ on Ω with $\rho(z)=\rho_1(f(z))\,|f'(z)|$ satisfies $L(f(\Gamma),\rho_1)=L(\Gamma,\rho)$ and $A(\rho,\Omega)=A(\rho_1,\Omega_1)$. Consequently, $\lambda(\Gamma)\geq\lambda(f(\Gamma))$. The opposite inequality is obtained by considering f^{-1}.

I.G. Relations between Families. We shall prove the following relations which are frequently used in our book. In (b)–(e) we consider a finite or countably infinite number of families. Result (b) is due to HERSCH [1], (c) to SUITA [4] and ZIEMER [1], the other relations to AHLFORS-BEURLING [1].

(a) *If every* $c\in\Gamma$ *contains a* $c'\in\Gamma'$, *then*

$$\lambda(\Gamma')\leq\lambda(\Gamma).$$

(b) *If* $\Gamma\subset\bigcup_n\Gamma_n$, *then*

$$\frac{1}{\lambda(\Gamma)}\leq\sum_n\frac{1}{\lambda(\Gamma_n)}.$$

(c) *If* $\Gamma_1\subset\Gamma_2\subset\cdots$ *and* $\Gamma=\bigcup_n\Gamma_n$, *then*

$$\lambda(\Gamma)=\lim_{n\to\infty}\lambda(\Gamma_n).$$

(d) *Suppose there exist disjoint open sets* Ω_n *containing the chains in* Γ_n. *If* $\Gamma\supset\bigcup_n\Gamma_n$, *then*

$$\frac{1}{\lambda(\Gamma)}\geq\sum_n\frac{1}{\lambda(\Gamma_n)}.$$

(e) *If, under the above assumptions, there exists, for every* $c\in\Gamma$, *a sequence* $\{c_n\}$ *with* $c_n\in\Gamma_n$ *and* $\bigcup_n c_n\subset c$, *then*

$$\lambda(\Gamma)\geq\sum_n\lambda(\Gamma_n).$$

Proof of (a). Every metric ρ on W satisfies $L(\Gamma',\rho)\leq L(\Gamma,\rho)$, and (a) follows.

Proof of (b). We may assume that $\lambda(\Gamma_n)>0$ for every n, since otherwise the relation is trivial. Given an arbitrary $\varepsilon>0$, there exist metrics ρ_n on W such that $L(\Gamma_n,\rho_n)>0$ and $\lambda(\Gamma_n)^{-1}>A(\rho_n)\,L(\Gamma_n,\rho_n)^{-2}-\varepsilon\,2^{-n}$. Set $\rho_n'(z)=L(\Gamma_n,\rho_n)^{-1}\rho_n(z)$ and $\rho(z)=\sup_n\rho_n'(z)$. Then $L(\Gamma,\rho)\geq 1$ and $\rho^2\leq\sum\rho_n'^2$. Consequently, $\lambda(\Gamma)^{-1}\leq A(\rho)\,L(\Gamma,\rho)^{-2}\leq\sum A(\rho_n)\,L(\Gamma_n,\rho_n)^{-2}\leq\sum\lambda(\Gamma_n)^{-1}+\varepsilon$ for every $\varepsilon>0$.

Proof of (c). It suffices to show that $\lambda(\Gamma)\geq\lim\lambda(\Gamma_n)$, since the opposite inequality is immediate from (a). We may assume that $\lim\lambda(\Gamma_n)>0$. For every n, there exists a unique $\rho_n\in\bar{\mathbf{P}}(\Gamma_n)$ with $A(\rho_n)=\lambda(\Gamma_n)^{-1}$. By the inequality in I.D, we have $A(\rho_n-\rho_m)\leq A(\rho_m)-A(\rho_n)$ if $m>n$. Since $\lim A(\rho_n)<\infty$ by assumption, we obtain a metric ρ_0 such that $A(\rho_n-\rho_0)\to 0$. Clearly $\rho_i\in\bar{\mathbf{P}}(\Gamma_n)$, $i=n,n+1,\ldots$, so that $\rho_0\in\bar{\mathbf{P}}(\Gamma_n)$, $n=1,2,\ldots$. Consequently, $\int_c\rho_0\,|dz|\geq 1$ for almost all $c\in\Gamma_n$ and, therefore, for almost all $c\in\Gamma$, that is, $\rho_0\in\bar{\mathbf{P}}(\Gamma)$ (cf. I.E). We obtain $A(\rho)\geq A(\rho_0)$ for every $\rho\in\bar{\mathbf{P}}(\Gamma)$, since $\rho\in\bar{\mathbf{P}}(\Gamma_n)$ and $A(\rho)\geq A(\rho_n)$. Accordingly, $A(\rho_0)=\lambda(\Gamma)^{-1}$ and it follows that $\lim\lambda(\Gamma_n)=\lim A(\rho_n)^{-1}=A(\rho_0)^{-1}=\lambda(\Gamma)$.

Proof of (d). We may assume that $\lambda(\Gamma)>0$. Given an arbitrary $\varepsilon>0$, there exists a metric ρ on W such that $L(\Gamma,\rho)>0$ and $\lambda(\Gamma)^{-1}+\varepsilon> A(\rho)\,L(\Gamma,\rho)^{-2}$. Since $L(\Gamma,\rho)\leq L(\Gamma_n,\rho)$ and $A(\rho)\geq\sum A(\rho,\Omega_n)$, we obtain $\lambda(\Gamma)^{-1}+\varepsilon>\sum A(\rho,\Omega_n)\,L(\Gamma_n,\rho)^{-2}\geq\sum\lambda(\Gamma_n)^{-1}$ for every $\varepsilon>0$.

Proof of (e). If $\lambda(\Gamma_n)=\infty$ for some n, then $\lambda(\Gamma)=\infty$ by (a) and the relation is trivial. Hence we shall assume that $\lambda(\Gamma_n)<\infty$ for every n. Given $\varepsilon>0$, there exist metrics ρ_n on Ω_n such that $L(\Gamma_n,\rho_n)\geq 1$ and $A(\rho_n,\Omega_n)^{-1}>(1-\varepsilon)\,\lambda(\Gamma_n)$. For any finite m, define a metric ρ on W by $\rho=\lambda(\Gamma_n)(\sum_{k=1}^m\lambda(\Gamma_k))^{-1}\rho_n$ on Ω_n and $\rho=0$ on $W-\bigcup_{k=1}^m\Omega_k$. Then we have $L(\Gamma,\rho)\geq 1$ and

$$\begin{aligned}\lambda(\Gamma)&\geq\frac{L(\Gamma,\rho)^2}{A(\rho)}\\&\geq\Big(\sum_{k=1}^m\lambda(\Gamma_k)\Big)^2\Big(\sum_{n=1}^m\lambda(\Gamma_n)^2A(\rho_n,\Omega_n)\Big)^{-1}\\&\geq\Big(\sum_{k=1}^m\lambda(\Gamma_k)\Big)^2(1-\varepsilon)\Big(\sum_{n=1}^m\lambda(\Gamma_n)\Big)^{-1}\\&=(1-\varepsilon)\sum_{k=1}^m\lambda(\Gamma_k)\end{aligned}$$

for every $\varepsilon>0$ and m.

I.H. Exclusion of Non-Rectifiable Curves. The following is an immediate consequence of (a) and (b):

If $\Gamma_0\subset\Gamma$ *and* $\lambda(\Gamma-\Gamma_0)=\infty$, *then* $\lambda(\Gamma_0)=\lambda(\Gamma)$.

For example, if Γ is considered on a bounded plane region, then $\lambda(\Gamma)=\lambda(\Gamma_0)$ for $\Gamma_0=\{c\,|\,c\in\Gamma,\ \int_c|dz|<\infty\}$.

I.I. Symmetry. We shall only discuss symmetries about a line. Those about a circular arc can be treated similarly.

Denote the upper half-plane $\operatorname{Im} z>0$ by H and the real axis by l. If $c\subset H$ is a curve $z(t)$, $t\in I$, then we designate by $\tilde c$ the curve $\overline{z(t)}$, $t\in I$.

Let Γ be a family of locally rectifiable curves $z(t)$, $t\in(a,b)$, in H such that $z(b)=\lim_{t\to b}z(t)$ exists and is on l. Consider the family $\hat\Gamma=\{\hat c=\bar c-\tilde{\bar c}\,|\,c\in\Gamma\}$, where $\bar c$ is the curve obtained by extending c as follows: $\bar c=z(t)$ for $t\in[a,b]$ if $z(a)=\lim_{t\to a}z(t)$ exists and is on l; otherwise $\bar c=z(t)$ for $t\in(a,b]$.

(a) *Let* Γ, $\hat\Gamma$ *be as above. If* Γ' *is a family such that* $\hat\Gamma\subset\Gamma'$ *and, for all* $c'\in\Gamma'$, *there exist* c_1 *and* c_2 *in* Γ *with* $c_1,\tilde c_2\subset c'$, *then*

$$\lambda(\Gamma')=\lambda(\hat\Gamma)=2\lambda(\Gamma).$$

Proof. One sees without difficulty that $2\lambda(\Gamma)\le\lambda(\Gamma')\le\lambda(\hat{\Gamma})$. We are to show that $\lambda(\hat{\Gamma})\le 2\lambda(\Gamma)$ if $\lambda(\hat{\Gamma}')>0$. Given an arbitrary $\varepsilon>0$, there exists a metric ρ on $W=\{z\,|\,|z|<\infty\}$ such that $L(\hat{\Gamma},\rho)>0$ and $\lambda(\hat{\Gamma})^{-1}+\varepsilon>L(\hat{\Gamma},\rho)^{-2}A(\rho)$. For $\hat{\rho}(z)=\rho(z)+\rho(\bar{z})$, we obtain $4A(\rho)\ge A(\hat{\rho})=2A(\hat{\rho},H)$ and $2L(\hat{\Gamma},\rho)=L(\hat{\Gamma},\hat{\rho})=2L(\Gamma,\hat{\rho})$. Hence $\lambda(\hat{\Gamma})^{-1}+\varepsilon>A(\hat{\rho},H)/2L(\Gamma,\hat{\rho})^2\ge 1/2\lambda(\Gamma)$ for every $\varepsilon>0$. We have proved (a).

Next let Ω be a region in H. Assume the existence of a point on $l\cap\partial\Omega$, a circular neighborhood N of which satisfies $N\cap\Omega=N\cap H$; let l_0 be the set of such points. Given sets E_1 and E_2 on $\partial\Omega$, let Γ be the family of locally rectifiable curves joining E_1 and E_2 in Ω. Consider $\tilde{E}_k=\{\bar{z}\,|\,z\in E_k\}$, $k=1,2$, and $\tilde{\Omega}=\{\bar{z}\,|\,z\in\Omega\}$. For an arbitrary subset l_1 of l_0

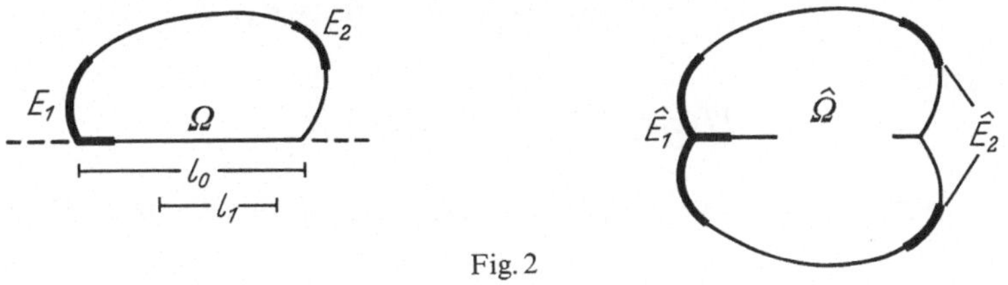

Fig. 2

which is open relative to l and disjoint from $E_1\cup E_2$, the doubles of E_k and Ω are defined as follows (see Fig. 2): $\hat{E}_k=E_k\cup\tilde{E}_k$, $k=1,2$, and $\hat{\Omega}=\Omega\cup l_1\cup\tilde{\Omega}$.

(b) *If Γ is as above and $\hat{\Gamma}$ is the family of curves joining $\hat{E}_1$ and $\hat{E}_2$ in $\hat{\Omega}$, then*

$$\lambda(\hat{\Gamma})=\tfrac{1}{2}\lambda(\Gamma).$$

The result is due to Ohtsuka [1, pp. 184ff.]; cf. also Oikawa [2, p. 268] and Ahlfors [4, p. 16].

Proof. The inequality $\lambda(\hat{\Gamma})^{-1}\ge 2\lambda(\Gamma)^{-1}$ is evident. To prove the opposite inequality, we consider, for every c given by $z(t)$, $t\in(a,b)$, the curve c_0 defined as follows:

$$\begin{cases} z(t) & \text{if } \operatorname{Im} z(t)\ge 0, \\ \overline{z(t)} & \text{if } \operatorname{Im} z(t)\le 0, \end{cases} \qquad t\in(a,b).$$

Set $\hat{\Gamma}_0=\{c_0\,|\,c\in\hat{\Gamma}\}$. The proof will be complete if we show that

$$\lambda(\hat{\Gamma}_0)\le 2\lambda(\hat{\Gamma}), \qquad \lambda(\Gamma)\le\lambda(\hat{\Gamma}_0).$$

The first relation is immediate. Given an arbitrary $\varepsilon>0$, there exists a ρ on $|z|<\infty$ with $\lambda(\hat{\Gamma}_0)-\varepsilon<L(\hat{\Gamma}_0,\rho)^2A(\rho)^{-1}\le L(\hat{\Gamma}_0,\rho)^2A(\rho,H)^{-1}$. The restriction to H of ρ can be extended symmetrically to $\hat{\rho}$ on $|z|<\infty$. Since $L(\hat{\Gamma}_0,\rho)=L(\hat{\Gamma},\hat{\rho})$ and $2A(\rho,H)=A(\hat{\rho})$, we obtain $\lambda(\hat{\Gamma}_0)-\varepsilon<2\lambda(\hat{\Gamma})$ for every ε.

To prove the second inequality, we may assume that $\lambda(\Gamma)>0$, so that for any $\varepsilon>0$, there exists a ρ on Ω such that $L(\Gamma,\rho)=1$ and $\lambda(\Gamma)-\varepsilon<A(\rho)^{-1}$.

The set l_1 consists of disjoint open intervals $l_1=\bigcup_{n=1}^{\infty}(x_n, x'_n)$. It is not difficult to prove the existence of continuously differentiable functions h_n on $[x_n, x'_n]$, $n=1,2,\ldots$, satisfying the following conditions: $h_n(x_n)=h_n(x'_n)=0$; $h_n(x)>0$ for $x_n<x<x'_n$; $|h'_n(x)|<\varepsilon$ in $[x_n, x'_n]$; $H\cap R_n\subset\Omega$, where $R_n=\{z=x+iy|x_n<x<x'_n, |y|<h_n(x)\}$; and $\sum A(\rho, R_n)<\varepsilon$. Define a mapping f of $H\cup\bigcup R_n$ onto H by

$$f(z)=\begin{cases} z & \text{if } z\in H-\bigcup R_n, \\ x+\frac{1}{2}i(y+h_n(x)) & \text{if } z\in R_n. \end{cases}$$

It satisfies

$$|f(z_1)-f(z_2)|\leq(1+\varepsilon)\,|z_1-z_2|$$

for every z_1 and $z_2\in H\cup\bigcup R_n$. For example, if z_1 and z_2 are in R_n, then

$$\begin{aligned}|f(z_1)-f(z_2)|&\leq|x_1-x_2+i(y_1-y_2)/2|+|h_n(x_1)-h_n(x_2)|/2\\ &\leq(1+\varepsilon/2)\,|z_1-z_2|.\end{aligned}$$

The other cases may be verified equally easily.

Consider the family $f(\Gamma_0)=\{f(c)|c\in\hat{\Gamma}_0\}$. It is a subfamily of Γ; that every $f(c)$ is locally rectifiable is verified by means of the above inequality. We have $1=L(\Gamma,\rho)\leq L(f(\hat{\Gamma}_0),\rho)\leq(1+\varepsilon)\,L(\hat{\Gamma}_0,\rho\circ f)$ and $A(\rho)\geq A(\rho, H-\bigcup R_n)=A(\rho\circ f, H\cup\bigcup R_n)-A(\rho\circ f,\bigcup R_n)\geq A(\rho\circ f, H\cup\bigcup R_n)-2\varepsilon$. It follows that

$$\frac{1}{\lambda(\Gamma)-\varepsilon}>A(\rho)\geq\frac{1}{(1+\varepsilon)^2}\,\frac{A(\rho\circ f, H\cup\bigcup R_n)}{L(\hat{\Gamma}_0,\rho\circ f)^2}-2\varepsilon$$

and, therefore,

$$\frac{1}{\lambda(\Gamma)-\varepsilon}\geq\frac{1}{(1+\varepsilon)^2}\,\frac{1}{\lambda(\hat{\Gamma}_0)}-2\varepsilon.$$

Consequently, $\lambda(\Gamma)\leq\lambda(\hat{\Gamma}_0)$.

I.J. Annuli and Rectangular Regions. We shall compute the extremal length of the closed curves separating the two boundary components in an annulus, that of the curves joining the two boundary components in an annulus, and that of the curves joining the two vertical sides in a rectangle.

(a) *In the annulus* $A=\{z\,|\,a<|z|<b\}$, $0\leq a<b\leq\infty$, *consider a closed set E such that* $e=\{r\,|\,r^{i\theta}\in E \text{ for some } \theta\}$ *has vanishing linear measure. Let* Γ *be the family of rectifiable closed curves in* $\Omega=A-E$ *separating* $|z|=a$ *from* $|z|=b$, *and let* Γ_0 *be the family of circles* $c_r=\{r\,e^{i\theta}\,|\,0\leq\theta<2\pi\}$, $r\in(a,b)-e$. *Then*

$$\frac{2\pi}{\lambda(\Gamma)}=\frac{2\pi}{\lambda(\Gamma_0)}=\log\frac{b}{a}.$$

By saying that a curve in a region separates one set from another, we mean that it meets every curve joining the sets in the region.

(b) *In the above annulus A, consider a closed set* E' *such that* $e'=\{\theta\,|\,r\,e^{i\theta}\in E' \text{ for some } r\}$ *has vanishing linear measure. Let* Γ *be the family of locally rectifiable curves joining* $|z|=a$ *and* $|z|=b$ *in* $\Omega=A-E'$, *and let* Γ_0 *be the family of segments* $c_\theta=\{r\,e^{i\theta}\,|\,a<r<b\}$, $\theta\in[0,2\pi)-e'$. *Then*

$$2\pi\,\lambda(\Gamma)=2\pi\,\lambda(\Gamma_0)=\log\frac{b}{a}.$$

(c) *In the rectangle* $R=\{z\,|\,0<\operatorname{Re} z<a,\ 0<\operatorname{Im} z<b\}$, $0<a<\infty$, $0<b<\infty$, *consider a closed set* E^* *such that* $e^*=\{y\,|\,x+i\,y\in E^* \text{ for some } x\}$ *has vanishing linear measure. Let* Γ *be the family of locally rectifiable curves joining* $\{z\,|\operatorname{Re} z=0,\ 0\leq\operatorname{Im} z\leq b\}$ *and* $\{z\,|\operatorname{Re} z=a,\ 0\leq\operatorname{Im} z\leq b\}$ *in* $\Omega=R-E^*$, *and let* Γ_0 *be the family of segments* $c_y=\{x+i\,y\,|\,0<x<a\}$, $y\in(0,b)-e^*$. *Then*

$$\lambda(\Gamma)=\lambda(\Gamma_0)=\frac{a}{b}.$$

Proof of (a). $\Gamma_0\subset\Gamma$ implies that $\lambda(\Gamma)\geq\lambda(\Gamma_0)$. Since the metric $\rho_0(z)\,|dz|=|z|^{-1}\,|dz|$ satisfies $L(\Gamma,\rho_0)=2\pi$ and $A(\rho_0,\Omega)=2\pi\log(b/a)$, we obtain $\lambda(\Gamma)\geq 2\pi(\log(b/a))^{-1}$. Next, for an arbitrary ρ and an $r\in(a,b)-e$,

$$L(\Gamma_0,\rho)^2\leq\Big(\int_{c_r}\rho\,|dz|\Big)^2\leq 2\pi\,r\int_0^{2\pi}\rho(r\,e^{i\theta})^2\,r\,d\theta.$$

On dividing by r and then integrating over $(a,b)-e$, we obtain $L(\Gamma_0,\rho)^2\log(b/a)\leq 2\pi\,A(\rho,\Omega)$, so that $\lambda(\Gamma_0)\leq 2\pi(\log(b/a))^{-1}$.

Proofs of (b) *and* (c). These are obtained in a similar manner by considering $\rho_0(z)\,|dz|=|z|^{-1}\,|dz|$ and $|dz|$, respectively.

The extremal metrics of the families Γ in (a), (b), (c) are

$$\frac{1}{2\pi}\frac{|dz|}{|z|},\qquad \frac{1}{\log\dfrac{b}{a}}\frac{|dz|}{|z|},\qquad \frac{|dz|}{a}.$$

I.K. Punctured Region. Let Ω be a region, z_0 a point of Ω, and E a set on $\partial\Omega$. For every $\varepsilon \geq 0$ with $\{z \mid |z-z_0| \leq \varepsilon\} \subset \Omega$, let Γ_ε be the family of curves joining E and $|z-z_0|=\varepsilon$ in $\Omega - \{z \mid |z-z_0| \leq \varepsilon\}$. Then

$$\lambda(\Gamma_0)=\infty, \qquad \lim_{\varepsilon\to 0} \lambda(\Gamma_\varepsilon)=\infty.$$

In fact, we have $2\pi\,\lambda(\Gamma_\varepsilon) \geq \log(\varepsilon_0/\varepsilon)$ if $0<\varepsilon<\varepsilon_0$ and both ε and ε_0 are sufficiently small.

I.L. Modulus Theorems. For certain families, the "Modulsätze" of Grötzsch [1] (cf. also Rengel [1]) give us more detailed information than the general relations of I.G. This is exemplified by a comparison of I.G.(a) with the following estimate:

(a) *In the annulus* $A=\{z \mid a<|z|<b\}$, *consider a doubly connected region* Ω *such that each component of* $\partial\Omega$ *separates* $|z|=a$ *from* $|z|=b$. *Let* Γ *be the family of rectifiable closed curves in* Ω *separating the components of* $\partial\Omega$. *Then*

$$\frac{2\pi}{\lambda(\Gamma)} \leq \log\frac{b}{a},$$

with equality if and only if $\Omega=A$.

For other modulus theorems, we refer the reader to Grötzsch's original work [1] and to Jenkins [1, pp. 22ff.]. The latter contains an alternative proof of (a). Here we shall add only a very special case of a modulus theorem which is used in our book.

(b) *Suppose* Ω_k, $k=1,\ldots,m$, *are disjoint Jordan regions. Let* E_k *and* E_k' *be disjoint arcs (including end points) on* $\partial\Omega_k$. *Denote by* Γ_k *the family of locally rectifiable curves joining* E_k *and* E_k' *in* Ω_k, *and by* Γ *the family of chains* c *such that* $c\cap\Omega_k$ *contains a curve in* Γ_k, $k=1,\ldots,m$. *Then*

$$\lambda(\Gamma)=\sum_{k=1}^{m} \lambda(\Gamma_k).$$

Proof of (a). The inequality is a simple consequence of I.G.(a) and I.J.(a). To prove the last statement, suppose the equality is satisfied. Map Ω conformally onto $a<|\zeta|<b$ and denote the inverse mapping by $z=f(\zeta)$. Let c_r be the member of Γ which corresponds to $|\zeta|=r$, and let $\Gamma_0=\{c_r \mid a<r<b\}$. With respect to $\rho_0(z)=|z|^{-1}$, we have for every r with $a<r<b$,

$$L(\Gamma_0,\rho_0)^2 \leq \left(\int_0^{2\pi} \rho_0(f(\zeta))\,|f'(\zeta)|\,r\,d\theta\right)^2 \tag{2}$$

$$\leq 2\pi r \int_0^{2\pi} \rho_0(f(\zeta))^2\,|f'(\zeta)|^2\,r\,d\theta, \tag{3}$$

where $\zeta = r\,e^{i\theta}$. From this and from

$$2\pi \leq L(\Gamma_0, \rho_0), \tag{4}$$

we obtain

$$(2\pi)^2 \log\frac{b}{a} \leq L(\Gamma_0, \rho_0)^2 \log\frac{b}{a} \leq 2\pi\, A(\rho_0, \Omega) \leq (2\pi)^2 \log\frac{b}{a}.$$

Consequently, the above inequalities are all equalities. From equality (3), we see that $|f'/f|$ depends only on r; from equalities (2) and (4), we obtain $|f'/f| = 1/r$. Thus $f(z) = \text{const}\cdot z$ with $|\text{const}| = 1$ and, therefore, $\Omega = A$.

The sufficiency of $\Omega = A$ is essentially the same as I.J.(a).

Proof of (b). The inequality $\lambda(\Gamma) \geq \sum \lambda(\Gamma_k)$ is a consequence of I.G.(e). To prove the inequality in the opposite direction, we may assume that Ω_k is the rectangle $\{z \mid a_k < \operatorname{Re} z < a_k', 0 < \operatorname{Im} z < 1\}$, $a_k' < a_{k+1}$, and that E_k and E_k' are the left and right vertical sides. For every η with $0 < \eta < 1$, the chain $c_\eta = \{z \mid \operatorname{Im} z = \eta\} \cap \bigcup \Omega_k$ belongs to Γ. Therefore, for every metric ρ on $\bigcup \Omega_k$,

$$L(\Gamma, \rho)^2 \leq \Big(\int_{c_\eta} \rho\, dx\Big)^2 \leq \Big(\sum_k (a_k' - a_k)\Big) \int_{c_\eta} \rho^2\, dx.$$

On integrating this with respect to η, we obtain $L(\Gamma, \rho)^2 \leq \sum (a_k' - a_k)\cdot A(\rho, \bigcup \Omega_k)$. Since $\lambda(\Gamma_k) = a_k' - a_k$, we conclude that $\lambda(\Gamma) \leq \sum \lambda(\Gamma_k)$.

I.M. Change Under Quasiconformal Maps. We shall extend property I.F to quasiconformal maps under certain strong assumptions which, however, are satisfied in our applications. For the general case, see AHLFORS [4].

Let f be a topological mapping of a region Ω of a Riemann surface onto another such region Ω'. Assume the existence of a set $E \subset \Omega$ with the following properties:

(a) E is closed relative to Ω,

(b) E is the union of at most a countable number of analytic arcs accumulating nowhere in Ω,

(c) f is continuously differentiable and has positive Jacobian in $\Omega - E$.

At a point $z \in \Omega - E$, consider the scalar $K(z) \geq 1$ defined by

$$K(z) + \frac{1}{K(z)} = \frac{u_x^2 + u_y^2 + v_x^2 + v_y^2}{u_x v_y - u_y v_x}$$

for $f = u + i\,v$ and $z = x + i\,y$. If $K(z)$ is bounded in $\Omega - E$, we shall say that f is *quasiconformal*. The quantity

$$\sup_{z \in \Omega - E} K(z) = K < \infty$$

is called the *maximal dilatation* of f.

A conformal map is quasiconformal in our sense with $K=1$.

In terms of complex derivatives, we obtain the simpler expression

$$K(z)=\frac{|f_z|+|f_{\bar z}|}{|f_z|-|f_{\bar z}|}.$$

Note that the Jacobian is $|f_z|^2-|f_{\bar z}|^2$.

Concerning families Γ of chains in Ω, we impose the following restriction: in every chain $c=\sum k_i c_i$, each c_i is the union of at most a countable number of analytic arcs accumulating nowhere in Ω. For example, the families Γ_0 in (a), (b), and (c) of I.J satisfy this condition.

Clearly every $c\in\Gamma$ is locally rectifiable, and so is its image under a quasiconformal map.

If Γ is a family with the above property and if Γ' is its image under a quasiconformal map f in the above sense, then

$$\lambda(\Gamma')\le K\,\lambda(\Gamma),$$

where K is the maximal dilatation of f.

Proof. Let $\rho_1|d\zeta|$ be an arbitrary metric on Ω', with $\zeta=\xi+i\eta$ the local parameter. Consider on Ω the metric $\rho|dz|$ defined by

$$\rho=\begin{cases}(\rho_1\circ f)(|f_z|+|f_{\bar z}|) & \text{on } \Omega-E,\\ 0 & \text{on } E.\end{cases}$$

We have $L(\Gamma',\rho_1)\le L(\Gamma,\rho)$ since, for every $c\in\Gamma$,

$$\int_{f(c)}\rho_1|d\zeta|\le\int_c(\rho_1\circ f)(|f_z|+|f_{\bar z}|)\,|dz|.$$

Moreover, $KA(\rho_1,\Omega')\ge A(\rho,\Omega)$, since

$$\iint_{\Omega'}\rho_1^2\,d\xi\,d\eta=\iint_{\Omega}(\rho_1\circ f)^2(|f_z|^2-|f_{\bar z}|^2)\,dx\,dy$$

$$=\iint_{\Omega}\rho^2\frac{|f_z|-|f_{\bar z}|}{|f_z|+|f_{\bar z}|}\,dx\,dy.$$

Accordingly,

$$\frac{L(\Gamma',\rho_1)^2}{A(\rho_1,\Omega')}\le\frac{KL(\Gamma,\rho)^2}{A(\rho,\Omega)}\le K\,\lambda(\Gamma)$$

for every ρ_1. It follows that $\lambda(\Gamma')\le K\,\lambda(\Gamma)$.

Appendix II

Conductor Potentials

We shall prove the selection theorem for mass distributions and FROSTMAN'S [1] theorem for conductor potentials. Regarding the latter, the proof given here is due to TSUJI [4].

II. A. Mass Distribution. For basic concepts of real analysis, we shall use the terminology adopted by ROYDEN [3].

By a *mass distribution* μ, we mean a Baire measure on the measurable space $(R^2, \mathscr{B})$, where R^2 is the 2-dimensional Euclidean space and $\mathscr{B}$ is the σ-algebra of Baire sets (= Borel sets). The space R^2 will be identified with the complex plane $|z|<\infty$. The *support* S_μ of μ is the smallest closed set whose complement has vanishing μ-value.

Given a compact set E of R^2, denote by $\mathscr{C}(E)$ the family of real-valued continuous functions on E. For every μ with $S_\mu \subset E$,

$$I[f]=\int f\,d\mu$$

is a positive linear functional on $\mathscr{C}(E)$.

Conversely, every positive linear functional I on $\mathscr{C}(E)$ can be expressed as above by means of a uniquely determined μ with $S_\mu \subset E$ (ROYDEN [3, p. 252]).

II. B. Selection Theorem. *Let E be a compact set and let* $\mathbf{M}(E)=\{\mu \mid S_\mu \subset E, \mu(R^2)=1\}$. *For any sequence* $\mu_n \in \mathbf{M}(E)$, $n=1, 2, \ldots$, *there exists a* $\mu \in \mathbf{M}(E)$ *and an* $\{i_n\}$ *such that*

$$\lim_{n\to\infty} \int f\,d\mu_{i_n}=\int f\,d\mu$$

for every $f \in \mathscr{C}(R^2)$. *Furthermore,*

$$\lim_{n\to\infty} \int h\,d(\mu_{i_n}\times\mu_{i_n})=\int h\,d(\mu\times\mu)$$

for every $h \in \mathscr{C}(R^4)$.

Proof. If we use the concept of weak* topology (ROYDEN [3, pp. 173 ff.]), the first half may be verified at once. Here we shall use a direct proof.

The family $\mathscr{C}(E)$ contains a sequence $\{f_j\}$, $j=1,2,\ldots$, which is dense with respect to uniform convergence on E. Indeed, by the Stone-Weierstrass theorem (e.g., ROYDEN [3, p. 150]), it suffices to take polynomials with rational coefficients.

For each j, we have $|\int f_j\,d\mu_n| \le \max_E |f_j|$, $n=1,2,\ldots$, so that we can choose a subsequence $\{k_n\}$ for which $\lim_{n\to\infty}\int f_j\,d\mu_{k_n}$ exists. Carrying this out for every j and using the diagonal process, we obtain a subsequence $\{i_n\}$ for which $\lim_{n\to\infty}\int f_j\,d\mu_{i_n}$ exists for every j. Since the sequence $\{f_j\}$ is dense in $\mathscr{C}(E)$ and $\int d\mu_{i_n}=1$,

$$I[f]=\lim_{n\to\infty}\int f\,d\mu_{i_n}$$

exists for every $f\in\mathscr{C}(E)$. Clearly, $I[f]$ is a positive linear functional on $\mathscr{C}(E)$. We thus obtain a μ such that $S_\mu\subset E$ and $\int f\,d\mu=I[f]$.

The characteristic function f of E belongs to $\mathscr{C}(E)$, so that $\mu(R^2)=\mu(E)=\int f\,d\mu=\lim\int f\,d\mu_{i_n}=1$. We conclude that $\mu\in\mathbf{M}(E)$. An arbitrary $f\in\mathscr{C}(R^2)$ has its restriction to E in $\mathscr{C}(E)$ and, therefore, $\int f\,d\mu=\lim\int f\,d\mu_{i_n}$.

The restriction to $E\times E$ of an arbitrary $h\in\mathscr{C}(R^4)$ belongs to $\mathscr{C}(E\times E)$. Thus it can be approximated uniformly on $E\times E$ by polynomials (of real variables) by the Stone-Weierstrass approximation theorem. Since $\int d(\mu_{i_n}\times\mu_{i_n})=1$, it suffices to show that

$$\lim\int h\,d(\mu_{i_n}\times\mu_{i_n})=\int h\,d(\mu\times\mu)$$

for an arbitrary polynomial h. To do this, it is sufficient to give the proof for the case in which h is a monomial $z^k\bar{z}^l\zeta^p\bar{\zeta}^q$. We have

$$\begin{aligned}\lim\int z^k\bar{z}^l\zeta^p\bar{\zeta}^q\,d(\mu_{i_n}\times\mu_{i_n})&=\lim\left(\int z^k\bar{z}^l\,d\mu_{i_n}\right)\left(\int \zeta^p\bar{\zeta}^q\,d\mu_{i_n}\right)\\&=\left(\int z^k\bar{z}^l\,d\mu\right)\left(\int\zeta^p\bar{\zeta}^q\,d\mu\right)=\int z^k\bar{z}^l\zeta^p\bar{\zeta}^q\,d(\mu\times\mu),\end{aligned}$$

since $z^k\bar{z}^l$ and $\zeta^p\bar{\zeta}^q$ belong to $\mathscr{C}(E)$.

II.C. Frostman's Theorem. Given a μ with compact S_μ, the *logarithmic potential* u_μ and the *energy integral* $I(\mu)$ are defined by

$$u_\mu(z)=\int\log\frac{1}{|z-\zeta|}\,d\mu(\zeta),$$

$$I(\mu)=\iint\log\frac{1}{|z-\zeta|}\,d(\mu\times\mu)=\int u_\mu\,d\mu.$$

For a compact set E, set

$$V_E=\inf_{\nu\in\mathbf{M}(E)} I(\nu)$$

and define the *logarithmic capacity* Cap E as e^{-V_E}. For an arbitrary set E, the *inner logarithmic capacity* $\mathrm{Cap}^{(i)} E$ is defined as the supremum of Cap F over all compact sets $F \subset E$. If E is closed and bounded, then Cap $E = \mathrm{Cap}^{(i)} E$ (cf. (a) below).

If Cap $E > 0$, there exists a mass distribution μ such that

$$\mu \in \mathbf{M}(E), \qquad I(\mu) = V_E < \infty. \tag{1}$$

FROSTMAN'S THEOREM. *If μ satisfies* (1), *then $u_\mu \leq V_E$ on R^2 and $u_\mu = V_E$ on E except for an F_σ-set of inner logarithmic capacity zero.*

Such a μ is uniquely determined.

μ is called the *equilibrium distribution* and u_μ the *equilibrium potential* or *conductor potential.*

The proof, which extends to the end of the appendix, is given in the following steps. Section II.D is devoted to preparatory considerations. In II.E, it is proved that $u_\mu \geq V_E$ on E except for an F_σ-set of inner logarithmic capacity zero. The relation $u_\mu \leq V_E$ is established for S_μ in II.F and for R^2 in II.G. Sections II.H and II.I contain the uniqueness proof.

II.D. Auxiliary Results. We shall derive the following facts needed in the sequel ((a) and (c) are special cases of results contained in XI.2 A, 2 B but here we must prove them directly):

(a) *If $E_1 \subset E_2$, then* $\mathrm{Cap}^{(i)} E_1 \leq \mathrm{Cap}^{(i)} E_2$.

(b) *Let E be a closed bounded set with* Cap $E > 0$, *and let μ be such that $\mu \in \mathbf{M}(E)$ and $I(\mu) < \infty$. Any F_σ-set e with $e \subset E$ and* $\mathrm{Cap}^{(i)} e = 0$ *satisfies*

$$\mu(e) = 0.$$

(c) *Let E be a bounded F_σ-set and let $E_1, E_2, \ldots$ be F_σ-subsets of E with $E = \bigcup_{n=1}^{\infty} E_n$. If* $\mathrm{Cap}^{(i)} E_n = 0$, $n = 1, 2, \ldots$, *then*

$$\mathrm{Cap}^{(i)} E = 0.$$

Proof of (a). If the sets are bounded and closed, then $\mathbf{M}(E_1) \subset \mathbf{M}(E_2)$ gives the inequality; it then immediately extends to arbitrary sets.

Proof of (b). We shall deduce a contradiction from $\mu(e) > 0$. There exists a closed bounded set e_1 with $e_1 \subset e$ and $\mu(e_1) > 0$. Take a constant c such that

$$\frac{c}{|\zeta_1 - \zeta_2|} > 1 \tag{2}$$

for every ζ_1 and $\zeta_2 \in E$. Then the potential u of μ satisfies

$$u(z) + \log c = \int \log \frac{c}{|z - \zeta|}\, d\mu(\zeta) \geq \int_{e_1} \log \frac{c}{|z - \zeta|}\, d\mu(\zeta)$$

and, therefore,

$$\infty > \int u(z)\,d\mu(z) + \log c \geq \int_{e_1 \times e_1} \log \frac{c}{|z-\zeta|}\, d(\mu \times \mu)$$
$$= \int_{e_1 \times e_1} \log \frac{1}{|z-\zeta|}\, d(\mu \times \mu) + \mu(e_1)^2 \log c.$$

It follows that the mass distribution $\mu_1 = \mu/\mu(e_1)$ restricted to e_1 has the property $I(\mu_1) < \infty$. Since $\mu_1 \in \mathbf{M}(e_1)$, we have $V_{e_1} < \infty$, that is, $\operatorname{Cap} e_1 > 0$. In view of (a), this contradicts the fact that $\operatorname{Cap}^{(i)} e = 0$.

Proof of (c). Suppose, contrary to the assertion, that $\operatorname{Cap}^{(i)} E > 0$. Then there exists a closed bounded subset E_0 of E with $\operatorname{Cap} E_0 > 0$ and a mass distribution $\mu \in \mathbf{M}(E_0)$ with $I(\mu) < \infty$. Since $1 = \mu(E_0) \leq \sum \mu(E_0 \cap E_n)$, we can find an n such that $\mu(E_0 \cap E_n) > 0$. We choose a constant c satisfying (2) and obtain

$$\int_{(E_0 \cap E_n) \times (E_0 \cap E_n)} \log \frac{c}{|z-\zeta|}\, d(\mu \times \mu)$$
$$\leq \int_{E_0 \times E_0} \log \frac{c}{|z-\zeta|}\, d(\mu \times \mu) = I(\mu) + \log c < \infty.$$

The set $e = E_0 \cap E_n$ is an F_σ-set with $e \subset E_0$ and $\operatorname{Cap}^{(i)} e = 0$. By (b) it must satisfy $\mu(e) = 0$, contradicting $\mu(E_0 \cap E_n) > 0$.

II.E. Inequality $u \geq V_E$. Write u for u_μ. In this section, we shall show that

$$u \geq V_E$$

on E except for an F_σ-set of inner logarithmic capacity zero.

For E and u given in the theorem, consider $E_0 = \{z \in E \mid u(z) < V_E\}$ and $E_n = \{z \in E \mid u(z) \leq V_E - 1/n\}$, $n = 1, 2, \ldots$. By the lower semicontinuity of u, the latter sets are closed, so that $E_0 = \bigcup_{n=1}^{\infty} E_n$ is an F_σ-set. It suffices to show that

$$\operatorname{Cap}^{(i)} E_0 = 0.$$

We shall derive a contradiction from $\operatorname{Cap}^{(i)} E_0 > 0$. By virtue of this assumption and (c) in II.D, we can find an n with $\operatorname{Cap} E_n > 0$. Since $V_E = \int_{S_\mu} u\, d\mu$, there exists a $\zeta_0 \in S_\mu$ with $u(\zeta_0) > V_E - (2n)^{-1}$. By the lower semicontinuity of u, we can choose a neighborhood N of ζ_0 on which $u > V_E - (2n)^{-1}$. Since $\zeta_0 \notin E_n'$, it is possible to take N so small that $E_n \cap \overline{N} = \emptyset$. Furthermore, since $\zeta_0 \in S_\mu$, we have

$$\mu(N) = m > 0.$$

On the other hand, the assumption Cap $E_n > 0$ implies the existence of a mass distribution ν with

$$S_\nu \subset E_n, \qquad I(\nu) < \infty, \qquad \nu(E_n) = m.$$

We introduce a signed measure σ as follows: for every $e \in \mathscr{B}$, $\sigma(e) = \nu(e \cap E_n) - \mu(e \cap N)$. Clearly, $\sigma(E) = 0$ and

$$I(\sigma) = \int_{E \times E} \log \frac{1}{|z-\zeta|} \, d(\sigma \times \sigma)$$

is well-defined and finite. For every ε with $0 < \varepsilon < 1$, we have $\mu + \varepsilon\sigma \in \mathbf{M}(E)$. If ε is sufficiently small, then

$$\begin{aligned} I(\mu + \varepsilon\sigma) - I(\mu) &= 2\varepsilon \int_E u \, d\sigma + \varepsilon^2 I(\sigma) \\ &= 2\varepsilon \Big(\int_{E_n} u \, d\sigma + \int_N u \, d\sigma \Big) + \varepsilon^2 I(\sigma) \\ &\leq 2\varepsilon \left(m \left(V_E - \frac{1}{n} \right) - m \left(V_E - \frac{1}{2n} \right) \right) + \varepsilon^2 I(\sigma) \\ &= -\varepsilon \left(\frac{m}{n} - \varepsilon I(\sigma) \right) < 0, \end{aligned}$$

which contradicts the fact that $I(\mu)$ is the minimum value of the energy integral over $\mathbf{M}(E)$.

II.F. Inequality $u \leq V_E$ on S_μ. We shall show that

$$u \leq V_E \quad \text{on } S_\mu.$$

We assume the existence of a $\zeta_0 \in S_\mu$ with $u(\zeta_0) > V_E$ and derive a contradiction. By the lower semicontinuity of u, there exists an $\varepsilon > 0$ and a neighborhood N of ζ_0 on which $u > V_E + \varepsilon$. Since $\zeta_0 \in S_\mu$, we have $\mu(N) > 0$. By II.E, we know that $u \geq V_E$ on $S_\mu - N$ except for an F_σ-set of inner logarithmic capacity zero. This exceptional set has μ-measure zero because of II.D.(b). We conclude that

$$\begin{aligned} V_E = \int_{S_\mu} u \, d\mu &= \int_{S_\mu \cap N} u \, d\mu + \int_{S_\mu - N} u \, d\mu > (V_E + \varepsilon)\, \mu(N) + V_E (1 - \mu(N)) \\ &= V_E + \varepsilon\, \mu(N) > V_E, \end{aligned}$$

a contradiction.

II.G. Inequality $u \leq V_E$ on R^2. Next, we shall prove that

$$u \leq V_E \quad \text{on } |z| < \infty.$$

We have shown that $u \leq V_E$ on S_μ. Let G be the complement of S_μ with respect to the Riemann sphere. Since u is harmonic on $G - \{\infty\}$ and $\lim_{z \to \infty} u(z) = -\infty$, it suffices to prove that

$$\overline{\lim_{z \in G,\, z \to \zeta}}\; u(z) \leq V_E \tag{3}$$

at every boundary point ζ of G.

Take an $\varepsilon > 0$. At an arbitrary $\zeta_0 \in \partial G$, we have $\mu(\{\zeta_0\}) = 0$ as $u(\zeta_0) < \infty$. Since $\mu(S_\mu) < \infty$, there exists a circular neighborhood N of ζ_0 such that $\mu(N) < \varepsilon$. For any $z \in N \cap G$, we can find a point $\zeta_1 \in N \cap S_\mu$ with $|z - \zeta_1| \leq |z - \zeta|$ for any $\zeta \in N \cap S_\mu$. Then $|\zeta_1 - \zeta| \leq |z - \zeta_1| + |z - \zeta| \leq 2|z - \zeta|$, so that

$$\begin{aligned}
\int_{N \cap S_\mu} \log \frac{1}{|z-\zeta|}\, d\mu(\zeta) &\leq \mu(N) \log 2 + \int_{N \cap S_\mu} \log \frac{1}{|\zeta_1 - \zeta|}\, d\mu(\zeta) \\
&< \varepsilon \log 2 + \int_{N \cap S_\mu} \log \frac{1}{|\zeta_1 - \zeta|}\, d\mu(\zeta) \\
&= \varepsilon \log 2 + \int_{S_\mu} \log \frac{1}{|\zeta_1 - \zeta|}\, d\mu(\zeta) - \int_{S_\mu - N} \log \frac{1}{|\zeta_1 - \zeta|}\, d\mu(\zeta) \\
&\leq \varepsilon \log 2 + V_E - \int_{S_\mu - N} \log \frac{1}{|\zeta_1 - \zeta|}\, d\mu(\zeta).
\end{aligned}$$

Therefore,

$$\int_{N \cap S_\mu} \log \frac{1}{|z-\zeta|}\, d\mu(\zeta) + \int_{S_\mu - N} \log \frac{1}{|\zeta_1 - \zeta|}\, d\mu(\zeta) \leq \varepsilon \log 2 + V_E.$$

Since $|\zeta_1 - z| \to 0$ as $z \to \zeta_0$, we have, for z sufficiently close to ζ_0,

$$\int_{S_\mu - N} \log \frac{1}{|\zeta_1 - \zeta|}\, d\mu(\zeta) > \int_{S_\mu - N} \log \frac{1}{|z - \zeta|}\, d\mu(\zeta) - \varepsilon.$$

Consequently, for such $z \in G$,

$$u(z) = \int_{N \cap S_\mu} \log \frac{1}{|z-\zeta|}\, d\mu(\zeta) + \int_{S_\mu - N} \log \frac{1}{|z - \zeta|}\, d\mu(\zeta) < \varepsilon \log 2 + V_E + \varepsilon.$$

We have proved (3).

II.H. Uniqueness. We proceed to the uniqueness of the equilibrium distribution.

Suppose the potentials $u_k = u_{\mu_k}$ of $\mu_k \in \mathbf{M}(E)$, $k=1,2$, are such that $u_k \leq V_E$ on R^2 and $u_k = V_E$ on E except for an F_σ-set of inner logarithmic capacity zero.

We shall show in the present section that $u_1 = u_2$ on $|z| < \infty$ except for an F_σ-set of inner logarithmic capacity zero. In the next section, we shall conclude that $\mu_1 = \mu_2$.

For the proof, we need the following extension of Lemma X.2 A:

Let W be a region of the Riemann sphere such that ∂W is bounded, and let v be a harmonic function bounded from below on W. If

$$\underline{\lim}_{z \in W,\, z \to \zeta} v(z) \geq m$$

at every boundary point ζ of W except for an F_σ-set of inner logarithmic capacity zero, then $v \geq m$ on W.

Proof. The results proved thus far in this appendix justify the use of Lemma X.2 A. Let $m_0 = \inf_W v$. We derive a contradiction from $m_0 < m$. For any m_1 with $m_0 < m_1 < m$, the open set $\{z \mid v(z) < m_1\}$ is not void. Let V be one of its components. The set $e = (\partial V) \cap (\partial W)$ is contained in the F_σ-set mentioned in the proposition; hence $\operatorname{Cap} e = 0$. The equality $v = m_1$ holds on $(\partial V) \cap W$. We apply Lemma X.2 A by regarding V as a subregion of a parabolic surface Ce, and obtain $v \geq m_1$ on V, a contradiction.

Now let E_0 be the set of points in E at which $u_1 < V_E$ or $u_2 < V_E$. By assumption, it is an F_σ-set with $\operatorname{Cap}^{(i)} E_0 = 0$. At a point ζ with $u_k(\zeta) = V_E$, the lower semicontinuity of potentials shows that $V_E \leq \underline{\lim}\, u_k(z) \leq \overline{\lim}\, u_k(z) \leq V_E$ as $z \to \zeta$. Therefore,

$$\lim_{z \to \zeta} u_k(z) = u_k(\zeta) = V_E, \qquad k = 1, 2,$$

at every $\zeta \in E - E_0$. Denote the unbounded component of the complement of E by G and set $\beta = \partial G$. On the region G, the function $u_1 - u_2$ is harmonic and bounded, potentials being always bounded from below outside of a neighborhood of ∞. Moreover,

$$\lim_{z \in G,\, z \to \zeta} \bigl(u_1(z) - u_2(z)\bigr) = 0$$

at every point $\zeta \in \beta - E_0$. By the above proposition, we conclude that $u_1 = u_2$ on G.

On a component Ω of the complement of $\bar{G}$, the function u_k is superharmonic and bounded from below. It satisfies

$$\lim_{z \in \Omega,\, z \to \zeta} u_k(z) = V_E$$

at every boundary point ζ of Ω with $\zeta \notin E_0$. By the standard maximum principle, we have $u_k \geq V_E$ on Ω. Since $u_k \leq V_E$ on $|z| < \infty$, we conclude that $u_1 = u_2 = V_E$ on Ω.

Consequently, $u_1 = u_2$ is satisfied on $|z| < \infty$ except for the set $\beta \cap E_0$, which has vanishing inner logarithmic capacity.

II.I. Uniqueness (continued). We complete the proof by showing that $\mu_1 = \mu_2$.

Let f be a function of class C^2 on R^2 with compact support. Apply Green's formula $\iint (v\,\Delta f - f\,\Delta v)\,dx\,dy = \int (f\,\partial v/\partial n - v\,\partial f/\partial n)\,ds$ to $v(z) = \log|z-\zeta|$ on $\varepsilon < |z-\zeta| < r$ for sufficiently large r. On letting $\varepsilon \to 0$, we obtain

$$2\pi f(\zeta) = \iint_{R^2} \log|z-\zeta|\,\Delta f\,dx\,dy, \qquad z = x + iy$$

and, therefore,

$$\int f\,d\mu_k = -\frac{1}{2\pi} \iint_{R^2} u_k\,\Delta f\,dx\,dy, \qquad k = 1, 2.$$

A Borel set of inner logarithmic capacity zero has vanishing area, as was mentioned in XI.2 D. Thus $u_1 = u_2$ almost everywhere. Accordingly,

$$\int f\,d\mu_1 = \int f\,d\mu_2$$

for every f under consideration. By a simple approximation method, we deduce that this relation is valid for every $f \in \mathscr{C}(E)$. In view of the last paragraph of II.A, this result implies that μ_1 and μ_2 coincide.

Problems

These are exercises, whose solutions are known. The numbers in parentheses indicate the topics to which the problems are related.

1. (I. 5 A – 5 D.) Let u be a harmonic function on $\overline{U}$ for which there exists a constant k such that

$$H(u-k)=u-k. \tag{1}$$

Prove the following:

(a) If $u_\gamma=0$, then any k qualifies.

(b) If $u_\gamma \neq 0$, then k is uniquely determined.

(c) Equation (1) is consistent (cf. 5 C).

(d) If U is as in 5 A, then (1) is equivalent to $u=k$ on β_U.

2. (II. 1 A, 1 C.) Show that if p is a principal function with respect to L_1 for **I** and the additive constant is suitably chosen, then

$$Hp=p \quad \text{on } \beta,$$

and conversely. Moreover, the relation fails if the additive constant is arbitrarily chosen.

3. (III. 1 C, 2 B.) Let U be a connected end of a Riemann surface W, ζ a point of U, and L_0, L_1, and H the operators acting from $\alpha=\partial U$ into U. On the other hand, view $\alpha=\partial U$ as a subboundary of the Riemann surface U, and consider the capacity functions $p_{0\alpha}$ and $p_{1\alpha}$ with singularity at ζ. First show that

$$L_\nu f(\zeta)=\frac{1}{2\pi}\int_\alpha f * dp_{\nu\alpha}$$

for $f\in C^\omega(\alpha)$, where L_1 is for **Q**. Next consider the capacity function $p_{1\alpha}$ for the operator L_1 for **I**, and verify the above for $\nu=1$ with respect to this operator. Finally, prove

$$Hf(\zeta)=\frac{1}{2\pi}\int_\alpha f * dp_\beta=-\frac{1}{2\pi}\int_\alpha f * dg,$$

where β is the entire ideal boundary $\alpha\cup\beta_U$, g is the Green's function, and p_β the capacity function for U. (Hint: For the operator H, exhaust U towards β_U and use Theorem III.4 B.)

4. (III.1 F, 3 E.) Prove that III.1 F.(6) for $p^0 = p_{0\gamma}$ can be generalized to an arbitrary subboundary with $c_{0\gamma} > 0$ as follows: for every $p \in HD_{\zeta\gamma}(W)$,

$$D[p - p_{0\gamma}] = \int_\beta p * dp - 2 \lim_{V \to W} \int_{\gamma(V)} p * dp_{0\gamma(V)} + 2\pi k_{0\gamma}.$$

Show also that

$$\lim_{V \to W} \int_{\gamma(V)} p * dp_{0\gamma} = \lim_{V \to W} \int_{\gamma(V)} p * dp_{0\gamma(V)}.$$

Here $V \to W$ is an exhaustion towards γ.

5. (III.2 E, 4 B–4 F.) For exhaustions $\zeta \in \Omega_n \to W$ by canonical subregions, show that

$$\lim_{n \to \infty} (p_{\nu\gamma(\Omega_n)} - p_{\nu\gamma}) = 0 \tag{2}$$

uniformly on every compact set of W in the following sense:

(a) If $c_{1\gamma} > 0$, then (2) with $\nu = 1$ holds for every exhaustion.

(b) If $c_{0\gamma} > 0$, there exists an exhaustion for which (2) with $\nu = 0$ holds.

(c) If $c_{\nu\gamma} = 0$, then for any $p_{\nu\gamma}$, there exists an exhaustion for which (2) holds.

6. (II.1 F, III.3 D.) Under the same assumption as in III.3 D, set $p_{2\gamma} = \frac{1}{2}(p_{0\gamma} + p_{1\gamma})$ and prove that

$$D[p - p_{2\gamma}] = \int_{\beta - \gamma} p * dp - \int_{\beta - \gamma} p_{2\gamma} * dp_{2\gamma}$$

for every $p \in KD_{\zeta\gamma}(\overline{W})$ with $p = \text{const}$ on γ and that

$$\int_{\beta - \gamma} p_{2\gamma} * dp_{2\gamma} = \frac{\pi}{2} (k_{1\gamma} - k_{0\gamma}).$$

7. (II.2 I, IV.3 F, 3 G.) Express $S(\zeta, \zeta')$ in terms of extremal length as follows:

$$S(\zeta, \zeta') = \pi \lim_{\substack{\varepsilon \to 0 \\ \varepsilon' \to 0}} \left(\lambda(\Gamma^*_{\varepsilon\varepsilon'}) - \frac{1}{\lambda(\Gamma_{\varepsilon\varepsilon'})} \right),$$

where the families $\Gamma_{\varepsilon\varepsilon'}$ and $\Gamma^*_{\varepsilon\varepsilon'}$ are the Γ and Γ^* in IV.2 B considered on the surface $W - (\{t \mid |t| \le \varepsilon\} \cup \{t' \mid |t'| \le \varepsilon'\})$ with respect to its subboundaries $|t| = \varepsilon$ and $|t'| = \varepsilon'$; t and t' are local parameters about ζ and ζ' (II.2 G).

8. (V.1 J.) Prove that

$$\mathbf{A}_i^{-1} = \frac{1}{2\pi} (q_{ji}(\gamma_k))_{\substack{j, k = 1, \dots, n \\ j, k \ne i}},$$

where $q_{ji}(\gamma_k)$ is the constant value of $q_1(z; \gamma_j, \gamma_i)$ on γ_k.

9. (VI.2 A.) If W is symmetric about the real axis, show that

$$P_\nu(\bar{z}) = \overline{P_\nu(z)}$$

for $P_\nu(z) = P_\nu(z;\zeta)$ and $P_\nu(z) = P_\nu(z;\zeta,\zeta')$, $\nu = 0, 1$, with real ζ and ζ'. If W is invariant under the transformation $z \to i\,z$, show that

$$P_1(z;\zeta) = -i\,P_0(iz;\zeta)$$

for $\zeta = 0$ or $\zeta = \infty$. If W is symmetric about the circle $|z| = r$, show that

$$P_\nu(r^2/\bar{z};0,\infty) = \text{const} / \overline{P_\nu(z;0,\infty)}.$$

10. (VI.3 E.) If W contains $r < |z| \le \infty$, prove that

$$|a[P_0]|^2 \le r^2 \operatorname{Re} a[P_0]$$

for $P_0 = P_0(z;\infty)$: the point $a[P_0]$ is contained in the circle of radius $r^2/2$ which is tangent to the imaginary axis at the origin. (DE POSSEL [4].)

11. (VI.3 D, 6 C.) Verify that for $-\infty < \alpha < \infty$, the function

$$P_0(z;\zeta,\zeta')^{\frac{1+\lambda}{2}} \cdot P_1(z;\zeta,\zeta')^{\frac{1-\lambda}{2i}}$$

with $\lambda = \exp(2i\tan^{-1}\alpha)$ has a branch in $\mathscr{V}_{\zeta\zeta'}(W)$ which maps W conformally onto a plane slit along logarithmic spirals $\arg w - \alpha \log|w| = \text{const}$. (KOEBE [9], GRÖTZSCH [2], GRUNSKY [1].)

12. (VII.2 G, 3 C, 4 B.) Granted that, for a simply connected region,

$$\inf_{F \in \mathscr{V}_\zeta} (\operatorname{diam} F(\gamma)) = 2\,c_{0\gamma}(\zeta)$$

and that the extremum is taken by the mapping onto the exterior of a disk (JENKINS [3]), prove that the equality remains valid for arbitrary plane regions and that the image under the extremal function, if any, is the exterior of a disk, with radial slits. Show also that the extremal function does not always exist. (GRÖTZSCH [2, 7], KOMATU [3].)

13. (VII.2 G.) Show that the extremal function $1/P_{1\gamma} + \text{const}$ maximizes the area of $X_{F(\gamma)}$ in $\mathscr{V}_\zeta$ (for notation see VI.1 D). Observe the contrast with Theorem VI.4 B. (GRÖTZSCH [2].)

14. (VIII.1 D, 2 B.) Prove that Q_1 maximizes the area of $X_{F(\gamma')}$ among functions $F \in S_{\gamma'\gamma}$ such that $F(\gamma) = \{w \mid |w| = 1\}$. (GRÖTZSCH [3].)

15. (Problem 11.) Discuss conformal mappings onto a disk and an annulus slit along logarithmic spirals. (KOEBE [9].)

16. (VI.3 D, IX.1 A – 1 C.) Define an *extremal parallel slit plane with inclination* θ, and show that it is obtained by rotating an extremal horizontal slit plane about the origin by θ.

17. (IX.1 C, 1 E.) Let E be an extremal set of horizontal slits, and l a line not parallel to the real axis. Prove that $l \cap E \in N_D$ (for the notation, see X.2 I).

18. (IX.3 B.) Without assuming (a), prove directly that (b) is equivalent to (c), (d).

19. (IX.2 B, 3 B.) In (b) of IX.3 B, replace "radial" by "circular", and show that the resulting condition is necessary and sufficient for a region W with $0 \in W$ to be an extremal circular slit disk of radius r ($\leq \infty$).

20. (X.1 F.) Show that γ is weak ($\nu = 1$) or semiweak ($\nu = 0$) if and only if one of the following is true:

$$D_U[u] = \int_\alpha u * du \qquad \text{for every } u \in HB_\gamma^\nu(\bar{U}),$$

$$D_U[u] = \int_\alpha u * du \qquad \text{for every } u \in HD_\gamma^\nu(\bar{U}).$$

21. (X.1 F, 1 H.) Prove that an isolated γ is weak if and only if

$$\int_\alpha * dH f = 0$$

for every $f \in C^\omega(\alpha)$.

22. (X.2 L.) Let N_{SB^*} be the class of compact sets whose complement is connected and possesses only semiweak subboundaries. Prove:

(a) every totally disconnected compact set is of class N_{SB^*},

(b) a compact set on a line is of class N_{SB^*} if and only if it is totally disconnected.

23. (XI.2 D, 2 E.) For a bounded continuum E of the plane, verify that

$$\frac{\operatorname{diam} E}{4} \leq \operatorname{Cap} E \leq \frac{\operatorname{diam} E}{2}.$$

24. (XI.4 A.) Prove that, if γ consists of a single point which is an irregular point for the Dirichlet problem on a plane region W, then it is a weak boundary component of W. (Hint: Use VII.6 G.)

25. (XI.4 A.) Prove that if $\gamma = \{\infty\}$ is a boundary component of a region W with finite area, then it is unstable. Also show that if $\gamma = \{0\}$ is a boundary component of W with $\iint_W |z|^{-2} dx\, dy < \infty$, then it is unstable (cf. XI.3 L).

26. (XI.1 J, 4 A.) Show that the boundary component $\gamma = \{\infty\}$ of a region W is weak if there exist two numbers $0 < c, c' < \infty$ with the following properties: the diameter of every boundary component different from γ is dominated by c; the distance between any two boundary components different from γ dominates c'. (Denneberg [1]; cf. XI.5 I.)

27. (XI.1 J, 4 A.) Verify that the boundary component $\gamma=\{\infty\}$ of a region W is weak if W is invariant under the transformation group $z\to z+n\,\omega+n'\,\omega'$ with $n, n'=0, \pm 1, \pm 2, \ldots$. Here ω and ω' are complex numbers which are linearly independent (over the reals). (Wagner [1]; cf. XI.5 I.)

28. (XI.4 B.) Consider the boundary component γ of $W(E,h)$ with $h\equiv 0$. Let $E_n=\{r\,e^{2\pi n i/k}\,|\,r\in E\}$ for $n=0,\ldots,k-1$, $W^*=C\big(\bigcup_{n=0}^{k-1}E_n\big)$, and $\gamma^*=\{0\}$. Prove that γ^* is a weak boundary component of W^* if and only if the same is true of γ. (Oikawa [2].)

29. (XI.4 E.) For $h(x)=x^p$ $(p\geq 1)$, show that γ of $W(E,h)$ is weak if $\int_{E'}x^{-1}\,dx=\infty$, and unstable if $\int_{E'}x^{-p}\,dx<\infty$. (For the latter, use the method of (b) in XI.4 E after verifying that $(b_n/a_n)^p$ is bounded.) (Akaza-Oikawa [1].)

30. (XI.4 E, 4 F.) Suppose that $h\equiv 0$ and $E=\{0\}\cup\bigcup_{n=1}^{\infty}[b_{n+1},a_n]$ $(0<b_{n+1}<a_n<b_n<1, \lim a_n=0)$. Prove that γ of $W(E,h)$ is weak if there exists a subsequence $\{j(n)\}$ such that

$$\lim_{n\to\infty}\frac{1}{B_n}\log\frac{b_{j(n)}}{a_{j(n)}}$$

exists $(\leq\infty)$ and

$$\sum_{n=1}^{\infty}\frac{B_n}{\left(\log^+\dfrac{1}{(b_{j(n)}/a_{j(n)})^{1/B_n}-1}\right)}=\infty,$$

where $B_n=\log(a_{j(n-1)}/a_{j(n)})$. (Hint: Map A_n in XI.4 F onto $\{z\,|\,1<|z|<e\}-[(1+\varepsilon)^{1/\log r}, e]$ quasiconformally by $z\to\text{const}\cdot|z|^{1/\log r}\,e^{i\arg z}$.) (Oikawa [2].)

31. (XI.5 D, 5 G.) Prove the following by a method similar to that used for Theorem XI.5 D.(c). Let $\frac{1}{2}<a_n<b_n<a_{n+1}<1$, $\lim a_n=1$, $\theta_n=2\pi/k_n$ (k_n an integer),

$$W=\{z\,|\,|z|<1\}-\bigcup_{n=1}^{\infty}\bigcup_{j=1}^{k_n}[a_n\,e^{ij\theta_n}, b_n\,e^{ij\theta_n}],$$

and

$$\gamma=\{z\,|\,|z|=1\}.$$

Then γ is parabolic provided

$$\lim_{n\to\infty}\frac{\theta_n}{\log\dfrac{b_n}{a_n}\log\left(\log\dfrac{1}{a_n\,b_n}\Big/\log\dfrac{b_n}{a_n}\right)}=0.$$

Note that γ is never semiweak.

Open Questions

We collect here the unsolved problems mentioned in the book. Among these, 1 and 2 seem to be fundamental.

1. Is the capacity function $p_{v\gamma}$ unique if $c_{v\gamma}=0$ (III.5 A, VII.2 H)? In the negative case, find relationships between the functions obtained by various limiting processes. (Cf. III.2 E, 4 B, 4 C, 4 F, 4 G, Problem 5.) Consider the corresponding problem for modulus functions.

2. Does a strong boundary component γ of a plane region have positive capacity $c_{0\gamma}$? Is strength a boundary property? (VII.3 D, XI.5 A, Remark 1 in XI.5 K.)

3. Characterize the operators L_0 and L_1 for **Q** by the boundary behavior under compactifications. (Remark I.5 F.)

4. If the span $S_m(\zeta)$ vanishes at some point ζ for some local parameter about it, then does $S_m(\zeta)$ necessarily vanish at every ζ with respect to every local parameter? (Remark 2 in II.2 E.)

5. Does the classification in Theorem VI.3 F depend on ζ?

6. Does $c_D < c_B$ hold for a regular region of connectivity ≥ 2? (VII.5 H.)

7. Find a relation between magnitudes of the quantities c_β, $\sqrt{M}$, and c_B. (VII.5 I.)

8. Characterize the points $z_0 \in \gamma$ which are mapped under $P_{0\gamma}$ to an incision (cf. Theorem VII.4 A). In other words, characterize (in the manner of Theorems 6 F, 6 G, say) the points $z_0 \in \gamma$ for which $\underline{\lim}_{z \to z_0} p_{0\gamma}(z;\zeta) < 1/c_{0\gamma}(\zeta)$. Do the same for modulus functions.

9. Weaken the assumptions of the propositions in IX.1 C. For example, can (b) be extended to a union of countably many sets?

10. In the case of Theorem X.1 F, does the weakness of γ imply $u = L u$ for every $u \in HD(\bar{U})$ and every normal operator L?

Bibliography

ACCOLA, R. D. M.

[1] The bilinear relation on open Riemann surfaces. Trans. Amer. Math. Soc. **96** (1960), 143–161.

AHLFORS, L. V.

[1] Bounded analytic functions. Duke Math. J. **14**, (1947), 1–11.

[2] Open Riemann surfaces and extremal problems on compact subregions. Comment. Math. Helv. **24** (1950), 100–134.

[3] The method of orthogonal decomposition for differentials on open Riemann surfaces. Ann. Acad. Sci. Fenn. Ser. A. I. No. 249/7 (1958), 15 pp.

[4] Lectures on quasiconformal mappings. Princeton, N. J.: Van Nostrand 1966. 146 pp.

[5] Cf. BEURLING, A.; AHLFORS, L. V. [1].

AHLFORS, L. V., and A. BEURLING

[1] Conformal invariants and function-theoretic null-sets. Acta Math. **83** (1950), 101–129.

AHLFORS, L. V., and L. SARIO

[1] Riemann surfaces. Princeton, N.J.: Princeton Univ. Press 1960. 382 pp.

AKAZA, T.

[1] On the weakness of some boundary component. Nagoya Math. J. **17** (1960), 219–223.

AKAZA, T., and K. OIKAWA

[1] Examples of weak boundary components. Nagoya Math. J. **18** (1961), 165–170.

ANDREIAN-CAZACU, C.

[1] Überlagerungseigenschaften Riemannscher Flächen. Rev. Math. Pures Appl. **6** (1961), 685–701.

BADER, R., and M. PARREAU

[1] Domaines non compacts et classification des surfaces de Riemann. C. R. Acad. Sci. Paris **232** (1951), 138–139.

BERGMAN, S.

[1] Über die Entwicklung der harmonischen Funktionen der Ebene und des Raumes nach Orthogonalfunktionen. Math. Ann. **86** (1922), 238–271.

[2] Partial differential equations, advanced topics. Brown University Summer Session for Advanced Instruction and Research in Mechanics, 1941.

[3] A remark on the mapping of multiply-connected domains. Amer. J. Math. **68** (1946), 20–28.

[4] The kernel function and conformal mapping. Math. Surveys 5, Amer. Math. Soc., New York, 1950. 161 pp.

BESICOVITCH, A. S.

[1] On sufficient conditions for a function to be analytic, and on behaviour of analytic functions in the neighborhood of nonisolated singular points. Proc. London Math. Soc. (2) **32** (1931), 1–9.

BEURLING, A., and L. V. AHLFORS

[1] The boundary correspondence under quasiconformal mappings. Acta Math. **96** (1956), 125–142.

[2] Cf. AHLFORS, L. V.; BEURLING, A. [1].

BOCHNER, S.

[1] Über orthogonale Systeme analytischer Funktionen. Math. Z. **14** (1922), 180–207.

[2] Fortsetzung Riemannscher Flächen. Math. Ann. **98** (1927), 406–421.

BOULIGAND, C.

[1] Sur le problème de Dirichlet. Ann. Soc. Polonaise **4** (1926), 59–112.

BRELOT, M.

[1] Familles de Perron et problème de Dirichlet. Acta Szeged **9** (1938), 133–153.

CARATHÉODORY, C.

[1] Über die Begrenzung einfach zusammenhängender Gebiete. Math. Ann. **73** (1913), 323–370.

[2] Über das lineare Maß von Punktmengen. Nachr. Akad. Wiss. Göttingen. Math.-Phys. Kl. (1914), 404–426.

CARLESON, L.

[1] Selected topics on exceptional sets. Van Nostrand Math. Studies No. 13. Princeton, N.J.: Van Nostrand 1968.

CARTWRIGHT, M. L.

[1] The exceptional values of functions with a non-enumerable set of essential singularities. Quart. J. Math. (Oxford Ser.) **8** (1937), 303–307.

CECIONI, F.

[1] Sulla rappresentazione conforme delle aree plane pluriconnesse su un piano in cui siano eseguiti dei tagli paralleli. Rend. Circ. Mat. Palermo **25** (1908), 1–19.

CONSTANTINESCU, C.

[1] Sur le comportement d'une fonction analytique à la frontière idéale d'une surface de Riemann. C. R. Acad. Sci. Paris **245** (1957), 1995–1997.

[2] Ideale Randkomponenten einer Riemannschen Fläche. Rev. Math. Pures Appl. **4** (1959), 43–76.

CONSTANTINESCU, C., and A. CORNEA

[1] Ideale Ränder Riemannscher Flächen. Berlin-Göttingen-Heidelberg: Springer 1963. 244 pp.

CORNEA, A.

[1] Cf. CONSTANTINESCU, C; CORNEA, A [1].

COURANT, R.

[1] Über die Anwendung des Dirichletschen Prinzipes auf die Probleme der konformen Abbildung. Math. Ann. **71** (1912), 145–183.

[2] Cf. HURWITZ, A.; COURANT, R. [1].

DENNEBERG, H.

[1] Konforme Abbildung einer Klasse unendlich-vielfach zusammenhängender schlichter Bereiche auf Kreisbereiche. Ber. Verh. Sächs. Akad. Wiss. Leipzig. Math.-Nat. Kl. **84** (1932), 331–352.

EGGLESTON, H. G.

[1] Convexity. Cambridge Tracts in Mathematics and Mathematical Physics. No. 47. New York: Cambridge Univ. Press 1958. 136 pp.

ERDÖS, P., and J. GILLIS

[1] Note on the transfinite diameter. J. London Math. Soc. **12** (1937), 185–192.

EVANS, G. C.

[1] Potentials and positively infinite singularities of harmonic functions. Monatsh. Math. Phys. **43** (1936), 419–424.

FÉKETE, M.

[1] Über die Verteilung der Wurzeln bei gewissen algebraischen Gleichungen mit ganzzahligen Koeffizienten. Math. Z. **17** (1923), 228–249.

FROSTMAN, O.

[1] Potentiel d'équilibre et capacité des ensembles avec quelques applications à la théorie des fonctions. Medd. Lunds Univ. Math. Sem. **3** (1935), 115 pp.

FUGLEDE, B.

[1] Extremal length and functional completion. Acta Math. **98** (1957), 171–219.

FUJI-I-E, T.

[1] On weak boundary components. Mem. Research Inst. Sci. Engrg., Ritumeikan Univ. **7** (1962), 5–7.

[2] On weak boundary components of a Riemann surface. J. Math. Soc. Japan **15** (1963), 396–403.

GARABEDIAN, P.

[1] Schwarz's lemma and the Szegö kernel function. Trans. Amer. Math. Soc. **67** (1949), 1–35.

GARABEDIAN, P., and M. SCHIFFER

[1] Identities in the theory of conformal mapping. Trans. Amer. Math. Soc. **65** (1949), 187–238.

[2] On existence theorems of potential theory and conformal mapping. Ann. of Math. **52** (1950), 164–187.

GILLIS, J.

[1] Cf. ERDÖS, P; GILLIS, J. [1].

GOLDSTEIN, M.

[1] *K*- and *L*-kernels on an arbitrary Riemann surface. Pacific J. Math. **19** (1966), 445–459.

GOLUSIN, G. M.

[1] Sur la représentation conforme. Rec. Math. **1** (1936), 273–282.

GRÖTZSCH, H.

[1] Über einige Extremalprobleme der konformen Abbildung. I, II. Ber. Verh. Sächs. Akad. Wiss. Leipzig. Math.-Nat. Kl. **80** (1928), 367–376; 497–502.

[2] Über die Verzerrung bei schlichter konformer Abbildung mehrfach zusammenhängender schlichter Bereiche. I–III. Ibid. **81** (1929), 38–47; 217–221; **83** (1931), 283–297.

[3] Über konforme Abbildung unendlich-vielfach zusammenhängender schlichter Bereiche mit endlich vielen Häufungsrandkomponenten. Ibid. **81** (1929), 51–86.

[4] Zum Parallelschlitztheorem der konformen Abbildung schlichter unendlich-vielfach zusammenhängender Bereiche. Ibid. **83** (1931), 185–200.

[5] Das Kreisbogenschlitztheorem der konformen Abbildung schlichter Bereiche. Ibid. **83** (1931), 238 – 253.
[6] Über die Verschiebung bei schlichter konformer Abbildung schlichter Bereiche. Ibid. **83** (1931), 254 – 279.
[7] Über Extremalprobleme bei schlichter konformer Abbildung schlichter Bereiche. Ibid. **84** (1932), 3 – 14.
[8] Über das Parallelschlitztheorem der konformen Abbildung schlichter Bereiche. Ibid. **84** (1932), 15 – 36.
[9] Verallgemeinerung eines Bieberbachschen Satzes. Jber. Deutsch. Math.-Verein. **43** (1933), 143 – 145.
[10] Eine Bemerkung zum Koebeschen Kreisnormierungsprinzip. Ber. Verh. Sächs. Akad. Wiss. Leipzig. Math.-Nat. Kl. **87** (1935), 319 – 324.

GRUNSKY, H.
[1] Neue Abschätzungen zur konformen Abbildung ein- und mehrfach zusammenhängender Bereiche. Schriften Sem. Univ. Berlin **1** (1932), 95 – 140.

GUNNING, R. C., and R. NARASIMHAN
[1] Immersion of open Riemann surfaces. Math. Ann. **174** (1967), 103 – 108.

HAUSDORFF, F.
[1] Dimension und äußeres Maß. Math. Ann. **79** (1918), 157 – 179.

HEINS, M.
[1] On the continuation of a Riemann surface. Ann. of Math. (2) **43** (1942), 280 – 297.
[2] The conformal mappings of simply-connected Riemann surfaces. Ibid. (2) **50** (1949), 686 – 690.
[3] Riemann surfaces of infinite genus. Ibid. (2) **55** (1952), 296 – 317.

HERSCH, J.
[1] Longueurs extrémales et théorie des fonctions. Comment. Math. Helv. **29** (1955), 301 – 337.

HILBERT, D.
[1] Zur Theorie der konformen Abbildung. Nachr. Akad. Wiss. Göttingen. Math.-Phys. Kl. (1909), 314 – 323.

HURWITZ, A., and R. COURANT
[1] Vorlesungen über allgemeine Funktionentheorie und elliptische Funktionen. Berlin-New York: Springer 1964. 706 pp.

JENKINS, J. A.
[1] Univalent functions and conformal mapping. Berlin-Göttingen-Heidelberg: Springer 1958. 169 pp.
[2] On a paper of Reich concerning minimal slit domains. Proc. Amer. Math. Soc. **13** (1962), 358 – 360.
[3] A uniqueness result in conformal mappings. Proc. Amer. Math. Soc. (to appear).

JULIA, G.
[1] Leçons sur la représentation conforme des aires multiplement connexes. Paris: Gauthier-Villars 1934. 94 pp.

JURCHESCU, M.
[1] Modulus of a boundary component. Pacific J. Math. **8** (1958), 791 – 809.
[2] Sur le problème du prolongement des surfaces de Riemann. C. R. Acad. Sci. Paris **249** (1959), 988 – 990.
[3] Bordered Riemann surfaces. Math. Ann. **143** (1961), 264 – 292.
[4] A maximal Riemann surface. Nagoya Math. J. **20** (1962), 91 – 93.

KAMETANI, S.
[1] On Hausdorff's measures and generalized capacities with some of their applications to the theory of functions. Japan. J. Math. **19** (1945), 217–257.

KELLOGG, O. D.
[1] Unicité des fonctions harmoniques. C. R. Acad. Sci. Paris **187** (1928), 526–527.

KERÉKJÁRTÓ, B.
[1] Vorlesungen über Topologie, I. Berlin: Springer 1923. 270pp.

KOEBE, P.
[1] Über die Uniformisierung beliebiger analytischer Kurven, IV. Nachr. Akad. Wiss. Göttingen. Math.-Phys. Kl. (1909), 324–361.
[2] Über ein allgemeines Uniformisierungsprinzip. 4. Internat. Congr. Math. Rome, 1909, II, 25–30.
[3] Über die konforme Abbildung mehrfach zusammenhängender Bereiche. Jber. Deutsch. Math.-Verein. **19** (1910), 339–348.
[4] Über die Uniformisierung der algebraischen Kurven, II. Math. Ann. **69** (1910), 1–81.
[5] Über die Hilbertsche Uniformisierungsmethode. Nachr. Akad. Wiss. Göttingen. Math.-Phys. Kl. (1910), 59–74.
[6] Über die Uniformisierung beliebiger analytischer Kurven, I. Das allgemeine Uniformisierungsprinzip. J. Reine Angew. Math. **138** (1910), 192–253.
[7] Zur konformen Abbildung unendlich-vielfach zusammenhängender schlichter Bereiche auf Schlitzbereiche. Nachr. Akad. Wiss. Göttingen. Math.-Phys. Kl. (1918), 60–71.
[8] Abhandlungen zur Theorie der konformen Abbildung, IV. Abbildung mehrfach zusammenhängender schlichter Bereiche auf Schlitzbereiche. Acta Math. **41** (1918), 305–344.
[9] Abhandlungen zur Theorie der konformen Abbildung, V. Abbildung mehrfach zusammenhängender schlichter Bereiche auf Schlitzbereiche (Erste Fortsetzung). Math. Z. **2** (1918), 198–236.
[10] Abhandlungen zur Theorie der konformen Abbildung, VI. Abbildung mehrfach zusammenhängender schlichter Bereiche auf Kreisbereiche. Uniformisierung hyperelliptischer Kurven (Iterationsmethode). Ibid. **7** (1920), 235–301.

KOMATU, Y.
[1] Untersuchungen über konforme Abbildung von zweifach zusammenhängenden Gebieten. Proc. Phys.-Math. Soc. Japan **25** (1943), 1–42.
[2] Zur konformen Abbildung vielfach zusammenhängender Gebiete. Proc. Japan Acad. **22** (1946), 343–351.
[3] A variational problem in conformal mappings of multiply-connected regions. Rigaku **3** (1948), 3–6. [Japanese.]
[4] Theory of conformal mappings. Kyoritsu Co., Tokyo, Vol. I, 1944, 579 pp. Vol. II, 1949, 409 pp. [Japanese.]

KÜHNAU, H.
[1] Über ein Koebesches Beispiel zur Theorie der minimalen Schlitzbereiche. Wiss. Z. Univ. Halle, Math.-Naturw. **14** (1965), 319–321.

KURAMOCHI, Z.
[1] Potential theory and its applications, I. Osaka Math. J. **3** (1951), 123–174.
[2] On the behaviour of analytic functions on abstract Riemann surfaces. Ibid. **7** (1955), 109–127.

KURODA, T.

[1] On analytic functions on some Riemann surfaces. Nagoya Math. J. **10** (1956), 27–50.

KUSUNOKI, Y.

[1] Theory of Abelian integrals and its applications to conformal mappings. Mem. Coll. Sci. Univ. Kyoto, Ser. A. Math. **32** (1959), 235–258; Supplement and Correction, **33** (1961), 429–433.

[2] Characterizations of canonical differentials. J. Math. Kyoto Univ. **5** (1966), 197–207.

LEHTO, O.

[1] Anwendung orthogonaler Systeme auf gewisse funktionentheoretische Extremal- und Abbildungsprobleme. Ann. Acad. Sci. Fenn. Ser. A. I. No. 59 (1949), 51 pp.

[2] On the existence of analytic functions with a finite Dirichlet integral. Ibid. No. 67 (1949), 7 pp.

LINDEBERG, J.

[1] Sur l'existence de fonctions d'une variable complexe et de fonctions harmoniques bornées. Ann. Acad. Sci. Fenn. Ser. A. I. No. 6 (1918), 27 pp.

LOKKI, O.

[1] Über Existenzbeweise einiger mit Extremaleigenschaft versehenen analytischen Funktionen. Ann. Acad. Sci. Fenn. Ser. A. I. No. 76 (1950), 15 pp.

[2] Beiträge zur Theorie der analytischen und harmonischen Funktionen mit endlichem Dirichletintegral. Ibid. No. 92 (1951), 11 pp.

MARDEN, A., and B. RODIN

[1] Extremal and conjugate extremal distance on open Riemann surfaces with applications to circular-radial slit mappings. Acta Math. **115** (1966), 237–269.

MESCHKOWSKI, H.

[1] Über die konforme Abbildung gewisser Bereiche von unendlich hohem Zusammenhang auf Vollkreisbereiche, I, II. Math. Ann. **123** (1951), 392–405; **124** (1952), 178–181.

MIZUMOTO, H.

[1] On conformal mapping of a multiply-connected domain onto a canonical covering surface. Kōdai Math. Sem. Rep. **10** (1958), 177–188.

[2] On conformal mapping of a multiply-connected domain onto a circular slit covering surface. Ibid. **13** (1961), 127–134.

[3] On extremal properties of circular slit covering surfaces. Math. J. Okayama Univ. **12** (1966), 147–152.

MORI, A.

[1] A remark on the prolongation of Riemann surfaces of finite genus. J. Math. Soc. Japan **4** (1952), 27–30.

[2] A remark on the class O_{HD} of Riemann surfaces. Kōdai Math. Sem. Rep. (1952), 57–58.

MORI, M.

[1] Contributions to the theory of differentials on open Riemann surfaces. J. Math. Kyoto Univ. **4** (1964), 77–97.

MYRBERG, P. J.

[1] Über die Existenz der Greenschen Funktionen auf einer gegebenen Riemannschen Fläche. Acta Math. **61** (1933), 39–79.

[2] Über die analytische Fortsetzung von beschränkten Funktionen. Ann. Acad. Sci. Fenn. Ser. A. I. No. 58 (1949), 7 pp.

NAKAI, M.
[1] On Evans potential. Proc. Japan Acad. **38** (1962), 624–629.
[2] Existence of positive harmonic functions. Proc. Amer. Math. Soc. **17** (1966), 365–367.

NAKAI, M., and L. SARIO
[1] Construction of principal functions by orthogonal projection. Canad. J. Math. **18** (1966), 887–896.
[2] Classification theory (to appear).

NARASIMHAN, R., and R. C. GUNNING
[1] Cf. GUNNING, R. C.; NARASIMHAN, R. [1]

NEHARI, Z.
[1] The kernel function and canonical conformal maps. Duke Math. J. **16** (1949), 165–178.
[2] Bounded analytic functions. Bull. Amer. Math. Soc. **57** (1951), 354–366.
[3] Conformal mapping. New York-Toronto-London: McGraw-Hill 1952. 396 pp.

NEVANLINNA, R.
[1] Quadratisch integrierbare Differentiale auf einer Riemannschen Mannigfaltigkeit. Ann. Acad. Sci. Fenn. Ser. A.I. No. 1 (1941), 34 pp.
[2] Uniformisierung. Berlin-Göttingen-Heidelberg: Springer 1953. 391 pp.
[3] Eindeutige analytische Funktionen. 2. Aufl. Berlin-Göttingen-Heidelberg: Springer 1953. 379 pp.

NICKEL, P.
[1] On extremal properties for annular radial and circular slit mappings of bordered Riemann surfaces. Pacific J. Math. **11** (1961), 1487–1503.

NOSHIRO, K.
[1] Open Riemann surface with null boundary. Nagoya Math. J. **3** (1951), 73–79.
[2] Cluster sets. Berlin-Göttingen-Heidelberg: Springer 1960. 135 pp.
[3] Cf. SARIO, L.; NOSHIRO, K. [1].

OHTSUKA, M.
[1] Dirichlet problem, extremal length and prime ends. Lecture Notes, Washington Univ., St. Louis, 1962. 350 pp.
[2] On weak and unstable components. J. Sci. Hiroshima Univ. Ser. A–I **28** (1964), 53–58.

OIKAWA, K.
[1] On a criterion for the weakness of an ideal boundary component. Pacific J. Math. **9** (1959), 1233–1238.
[2] On the stability of boundary components. Ibid. **10** (1960), 263–294.
[3] Minimal slit regions and linear operator method. Kōdai Math. Sem. Rep. **17** (1965), 187–190.
[4] Remarks to conformal mappings onto radially slit disks. Sci. Papers Coll. Gen. Ed. Univ. Tokyo **15** (1965), 99–109.
[5] Cf. AKAZA, T.; OIKAWA, K. [1].
[6] Cf. TAMURA, J.; OIKAWA, K.; YAMAZAKI, K. [1].

OIKAWA, K., and N. SUITA
[1] On parallel slit mappings. Kōdai Math. Sem. Rep. **16** (1964), 249–254.
[2] Circular slit disk with infinite radius. Nagoya Math. J. **30** (1967), 57–70.

PARREAU, M.
[1] Sur les moyennes des fonctions harmoniques et analytiques et la classification des surfaces de Riemann. Ann. Inst. Fourier (Grenoble) 3 (1951), 103–197.
[2] Cf. BADER, R.; PARREAU, M. [1].

PFLUGER, A.
[1] Sur l'existence de fonctions non constantes, analytiques, uniformes et bornées sur une surface de Riemann ouverte. C. R. Acad. Sci. Paris **230** (1950), 166–168.
[2] Theorie der Riemannschen Flächen. Berlin-Göttingen-Heidelberg: Springer 1957. 248 pp.

POMMERENKE, C.
[1] Über die analytische Kapazität. Arch. Math. **11** (1960), 270–277.

DE POSSEL, R.
[1] Sur le prolongement des surfaces de Riemann. C. R. Acad. Sci. Paris **186** (1927), 1092–1095; **187** (1928), 98–100.
[2] Sur les ensembles du type maximum, et le prolongement des surfaces de Riemann. Ibid. **194** (1932), 585–587.
[3] Zum Parallelschlitztheorem unendlich-vielfach zusammenhängender Gebiete. Nachr. Akad. Wiss. Göttingen. Math.-Phys. Kl. (1931), 199–202.
[4] Sur quelques propriétés de la représentation conforme des domaines multiplement connexes, en relation avec le théorème des fentes parallèles. Math. Ann. **107** (1932/33), 496–504.

RADÓ, T.
[1] Über eine nicht fortsetzbare Riemannsche Mannigfaltigkeit. Math. Z. **20** (1924), 1–6.

REICH, E.
[1] A counterexample of Koebe's for slit mappings. Proc. Amer. Math. Soc. **11** (1960), 970–975.
[2] On radial slit mappings. Ann. Acad. Sci. Fenn. Ser. A.I. No. 296 (1961), 12 pp.

REICH, E., and S. E. WARSCHAWSKI
[1] On canonical conformal maps of regions of arbitrary connectivity. Pacific J. Math. **10** (1960), 965–985.
[2] Canonical conformal maps onto a circular slit annulus. Scripta Math. **25** (1960), 137–146.

RENGEL, E.
[1] Über einige Schlitz-Theoreme der konformen Abbildung. Schrift. Math. Sem. Univ. Berlin **1** (1933), 140–162.
[2] Existenzbeweise für schlichte Abbildungen mehrfach zusammenhängender Bereiche auf gewisse Normalbereiche. Jber. Deutsch. Math.-Verein. **44** (1934), 51–55.

RENGGLI, H.
[1] Zur konformen Abbildung auf Normalgebiete. Comment. Math. Helv. **31** (1956), 5–40.
[2] On point-like boundaries of Riemann surfaces. Arch. Math. **17** (1966), 264–266.
[3] On maximal Riemann surfaces. Amer. J. Math. **88** (1966), 179–186.

RODIN, B.
[1] Reproducing kernels and principal functions. Proc. Amer. Math. Soc. **13** (1962), 982–992.
[2] Cf. MARDEN, A.; RODIN, B. [1].

RODIN, B., and L. SARIO
[1] Principal functions. Princeton, N. J.: Van Nostrand 1968. 347 pp.

ROYDEN, H. L.
[1] Harmonic functions on open Riemann surfaces. Trans. Amer. Math. Soc. **73** (1952), 40–94.
[2] On a class of null-bounded Riemann surfaces. Comment. Math. Helv. **34** (1960), 37–51.
[3] Real analysis. New York: MacMillan 1963. 284 pp.

SAKAI, A.
[1] On minimal slit domains. Proc. Japan Acad. **35** (1959), 128–133.

SAKS, S.
[1] Theory of the integral. Monografie Matematyczne, Vol. 7, Warsaw, 1937. 347 pp.

SARIO, L.
[1] Über Riemannsche Flächen mit hebbarem Rand. Ann. Acad. Sci. Fenn. Ser. A. I. No. 50 (1948), 79 pp.
[2] Sur la classification des surface de Riemann. 11. Scand. Congr. Math. Trondheim, 1949, pp. 229–238.
[3] Existence des fonctions d'allure donnée sur une surface de Riemann arbitraire. C.R. Acad. Sci. Paris **229** (1949), 1293–1295.
[4] Existence des intégrales abéliennes sur les surfaces de Riemann arbitraires. Ibid. **230** (1950), 168–170.
[5] A linear operator method on arbitrary Riemann surfaces. Trans. Amer. Math. Soc. **72** (1952), 281–295.
[6] An extremal method on arbitrary Riemann surfaces. Ibid. **73** (1952), 459–470.
[7] Construction of functions with prescribed properties on Riemann surfaces. Contributions to the theory of Riemann surfaces. Ann. of Math. Studies 30, Princeton Univ. Press, 1953, pp. 63–76.
[8] Minimizing operators on subregions. Proc. Amer. Math. Soc. **4** (1953), 350–355.
[9] Modular criteria on Riemann surfaces. Duke Math. J. **20** (1953), 297–286.
[10] Capacity of the boundary and of a boundary component. Ann. of Math. **59** (1954), 135–144.
[11] Extremal problems and harmonic interpolation on open Riemann surfaces. Trans. Amer. Math. Soc. 79 (1955), 362–377.
[12] Functionals on Riemann surfaces. Lectures on functions of a complex variable. The University of Michigan Press 1955, pp. 245–256.
[13] Strong and weak boundary components. J. Analyse Math. **5** (1956/57), 389–398.
[14] On univalent functions. 13. Scand. Congr. Math. Helsinki, 1957, pp. 202–208.
[15] Stability problems on boundary components. Seminars on analytic functions. Institute for Advanced Study, Princeton, N. J., 1958, II, pp. 55–72.
[16] Cf. AHLFORS, L. V.; SARIO, L. [1].
[17] Cf. NAKAI, M.; SARIO, L. [1].
[18] Cf. NAKAI, M.; SARIO, L. [2].
[19] Cf. RODIN, B.; SARIO, L. [1].

SARIO, L., and K. NOSHIRO
[1] Value distribution theory. Princeton, N. J.: Van Nostrand 1966. 236 pp.

SAVAGE, N.
[1] Weak boundary components of an open Riemann surface. Duke Math. J. **24** (1957), 79–96.

SCHIFFER, M.
[1] Sur un théorème de représentation conforme. C. R. Acad. Sci. Paris **207** (1938), 520–522.
[2] The span of multiply connected domains. Duke Math. J. **10** (1943), 209–216.
[3] The kernel function of an orthonormal system. Ibid. **13** (1946), 529–540.
[4] An application of orthonormal functions in the theory of conformal mapping. Amer. J. Math. **70** (1948), 147–156.
[5] Various types of orthogonalization. Duke Math. J. **17** (1950), 329–366.
[6] Some recent developments in the theory of conformal mappings. Appendix to R. Courant, Dirichlet's principle. New York: Interscience Publishers 1950, pp. 249–323.
[7] Cf. GARABEDIAN, P.; SCHIFFER, M. [1].
[8] Cf. GARABEDIAN, P.; SCHIFFER, M. [2].

SCHIFFER, M., and D. C. SPENCER
[1] Functionals of finite Riemann surfaces. Princeton, N.J.: Princeton Univ. Press 1954. 451 pp.

SCHOTTKY, F.
[1] Über die conforme Abbildung mehrfach zusammenhängender ebener Flächen. J. Reine Angew. Math. **83** (1877), 300–351.
[2] Über die Wertschwankungen der harmonischen Funktionen zweier reeller Veränderlichen und der Funktionen eines komplexen Arguments. Ibid. **117** (1897), 225–253.

SCHWARZ, H. A.
[1] Über die Integration der partiellen Differentialgleichung $\partial^2 u/\partial x^2 + \partial^2 u/\partial y^2 = 0$ unter vorgeschriebenen Grenz- und Unstetigkeitsbedingungen. Berliner Monatsber. (1870), 767–795; Ges. Abh., II. Berlin: Springer. 890 pp., 144–171.

SELBERG, H.
[1] Über die ebenen Punktmengen von der Kapazität Null. Avh. Norske Vid. Acad. Oslo No. 10 (1937), 10 pp.

SPENCER, D. C.
[1] Cf. SCHIFFER, M.; SPENCER, D. C. [1].

STONE, M. H.
[1] Hilbert space methods in conformal mapping. Proc. Intern. Symp. Linear Spaces 1960, Jerusalem (1961), 409–425.
[2] Topological aspects of conformal mapping theory. General topology and its relations to modern analysis and algebra, Proc. Symp. Prague, 1961, pp. 343–346.

STREBEL, K.
[1] Eine Ungleichung für extremale Längen. Ann. Acad. Sci. Fenn. Ser. A. I. No. 90 (1951), 8 pp.
[2] Die extremale Distanz zweier Enden einer Riemannschen Fläche. Ibid. No. 179 (1955), 21 pp.
[3] A remark on the extremal distance of two boundary components. Proc. Nat. Acad. Sci. U.S.A. **40** (1954), 842–844.
[4] On the maximal dilation of quasiconformal mappings. Proc. Amer. Math. Soc. **6** (1955), 903–909.

SUITA, N.
[1] On certain criteria for a set to be of class N_B. Nagoya Math. J. **19** (1961), 189–194.
[2] Minimal slit domains and minimal sets. Kōdai Math. Sem. Rep. **17** (1965), 166–186.
[3] On radial slit disc mappings. Ibid. **18** (1966), 219–228.

[4] On a continuity lemma of extremal length and its applications to conformal mapping. Ibid. **19** (1967), 129–137.
[5] A note on function-theoretic null sets of class N_{SD}. Proc. Japan Acad. **43** (1967), 1013–1015.
[6] On circular and radial slit disc mappings. Ibid. **20** (1968), 127–145.
[7] Cf. Oikawa, K.; Suita, N. [1].
[8] Cf. Oikawa, K.; Suita, N. [2].

Szegö, G.
[1] Über orthogonale Polynome, die zu einer gegebenen Kurve der komplexen Ebene gehören. Math. Z. **9** (1921), 218–270.
[2] Bemerkungen zu einer Arbeit von M. Fékete: „Über die Verteilung der Wurzeln bei gewissen algebraischen Gleichungen mit ganzzahligen Koeffizienten." Ibid. **21** (1924), 203–208.

Tamura, J.
[1] A prolongable Riemann surface. Sci. Papers Coll. Gen. Ed. Univ. Tokyo **6** (1956), 123–127.
[2] On the maximal Riemann surface. Ibid. **7** (1957), 19–22.
[3] On a theorem of Tsuji. Japan. J. Math. **29** (1959), 138–140.

Tamura, J., K. Oikawa, and K. Yamazaki
[1] Examples of minimal parallel slit domains. Proc. Amer. Math. Soc. **17** (1966), 283–284.

Teichmüller, O.
[1] Untersuchungen über konforme und quasikonforme Abbildung. Deutsche Math. **3** (1938), 621–678.

Tôki, Y.
[1] On the classification of open Riemann surfaces. Osaka Math. J. **4** (1952), 191–201.
[2] On the examples in the classification of open Riemann surfaces, I. Ibid. **5** (1953), 267–280.

Tsuji, M.
[1] Theory of conformal mapping of a multiply connected domain. Japan. J. Math. **18** (1943), 759–775.
[2] Maximal continuation of a Riemann surface. Kōdai Math. Sem. Rep. **4** (1952), 55–56.
[3] Existence of a potential function with a prescribed singularity on any Riemann surface. Tôhoku Math. J. **4** (1952), 54–68.
[4] Potential theory in modern function theory. Tokyo: Maruzen 1959. 590 pp.

Virtanen, K. I.
[1] Über die Existenz von beschränkten harmonischen Funktionen auf offenen Riemannschen Flächen. Ann. Acad. Sci. Fenn. Ser. A. I. No. 75 (1950), 8 pp.
[2] Über Extremalfunktionen auf offenen Riemannschen Flächen. Ibid. No. 141 (1952), 7 pp.

Vituškin, A. G.
[1] Some theorems on the possibility of a uniform approximation of continuous functions by analytic functions. Dokl. Akad. Nauk SSSR **123** (1958), 959–962. [Russian.]
[2] Example of a set of positive length but of zero analytic capacity. Ibid. **127** (1959), 246–249. [Russian.]

Wagner, R.
[1] Ein Kontaktproblem der konformen Abbildung. J. Reine Angew. Math. **196** (1956), 99–132.

WARSCHAWSKI, S. E.
[1] Cf. REICH, E.; WARSCHAWSKI, S. E. [1].
[2] Cf. REICH, E.; WARSCHAWSKI, S. E. [2].

WEILL, G. G.
[1] Capacity differentials on open Riemann surfaces. Pacific J. Math. **12** (1962), 769–776.

WEYL, H.
[1] Die Idee der Riemannschen Fläche. Leipzig: Teubner 1913. 169pp. 3rd Ed., Stuttgart: Teubner 1955. 162pp.

WIRTINGER, W.
[1] Über eine Minimalaufgabe im Gebiet der analytischen Funktionen. Monatsh. Math. Phys. **39** (1932), 377–384.

YAMAZAKI, K.
[1] Cf. TAMURA, J.; OIKAWA, K.; YAMAZAKI, K.

ZIEMER, W.
[1] Extremal length and conformal capacity. Trans. Amer. Math. Soc. **126** (1967), 460–473.

Author Index

Subject and Notation Index

Italicized page numbers refer to definitions

Universitätsdruckerei H. Stürtz AG Würzburg

71. Meixner/Schäfke: Mathieusche Funktionen und Sphäroidfunktionen mit Anwendungen auf physikalische und technische Probleme. DM 52,60; US $ 13.15
73. Hermes: Einführung in die Verbandstheorie. DM 46,– ; US $ 11.50
74. Boerner: Darstellungen von Gruppen. DM 58,– ; US $ 14.50
75. Rado/Reichelderfer: Continuous Transformations in Analysis, with an Introduction to Algebraic Topology. DM 59,60; US $ 14.90
76. Tricomi: Vorlesungen über Orthogonalreihen. DM 37,60; US $ 9.40
77. Behnke/Sommer: Theorie der analytischen Funktionen einer komplexen Veränderlichen. DM 79,– ; US $ 19.75
78. Lorenzen: Einführung in die operative Logik und Mathematik. DM 54,– ; US $ 13.50
79. Saxer: Versicherungsmathematik. 1. Teil. DM 39,60; US $ 9.90
80. Pickert: Projektive Ebenen. DM 48,60; US $ 12.15
81. Schneider: Einführung in die transzendenten Zahlen. DM 24,80; US $ 6.20
82. Specht: Gruppentheorie. DM 69,60; US $ 17.40
84. Conforto: Abelsche Funktionen und algebraische Geometrie. DM 41,80; US $ 10.45
85. Siegel: Vorlesungen über Himmelsmechanik. DM 33,– ; US $ 8.25
86. Richter: Wahrscheinlichkeitstheorie. DM 68,– ; US $ 17.00
87. van der Waerden: Mathematische Statistik. DM 49,60; US $ 12.40
88. Müller: Grundprobleme der mathematischen Theorie elektromagnetischer Schwingungen. DM 52,80; US $ 13.20
89. Pfluger: Theorie der Riemannschen Flächen. DM 39,20; US $ 9.80
90. Oberhettinger: Tabellen zur Fourier Transformation. DM 39,50; US $ 9.90
91. Prachar: Primzahlverteilung. DM 58,– ; US $ 14.50
92. Rehbock: Darstellende Geometrie. DM 29,– ; US $ 7.25
93. Hadwiger: Vorlesungen über Inhalt, Oberfläche und Isoperimetrie. DM 49,80; US $ 12.45
94. Funk: Variationsrechnung und ihre Anwendung in Physik und Technik. DM 98,– ; US $ 24.50
95. Maeda: Kontinuierliche Geometrien. DM 39,– ; US $ 9.75
97. Greub: Lineare Algebra. DM 39,20; US $ 9.80
98. Saxer: Versicherungsmathematik. 2. Teil. DM 48,60; US $ 12.15
99. Cassels: An Introduction to the Geometry of Numbers. DM 69,– ; US $ 17.25
100. Koppenfels/Stallmann: Praxis der konformen Abbildung. DM 69,– ; US $ 17.25
101. Rund: The Differential Geometry of Finsler Spaces. DM 59,60; US $ 14.90
103. Schütte: Beweistheorie. DM 48,– ; US $ 12.00
104. Chung: Markov Chains with Stationary Transition Probabilities. DM 56,– ; US $ 14.00
105. Rinow: Die innere Geometrie der metrischen Räume. DM 83,– ; US $ 20.75
106. Scholz/Hasenjaeger: Grundzüge der mathematischen Logik. DM 98,– ; US $ 24.50
107. Köthe: Topologische Lineare Räume I. DM 78,– ; US $ 19.50
108. Dynkin: Die Grundlagen der Theorie der Markoffschen Prozesse. DM 33,80; US $ 8.45
109. Hermes: Aufzählbarkeit, Entscheidbarkeit, Berechenbarkeit. DM 49,80; US $ 12.45
110. Dinghas: Vorlesungen über Funktionentheorie. DM 69,– ; US $ 17.25
111. Lions: Equations différentielles opérationnelles et problèmes aux limites. DM 64,– ; US $ 16.00
112. Morgenstern/Szabó: Vorlesungen über theoretische Mechanik. DM 69,– ; US $ 17.25
113. Meschkowski: Hilbertsche Räume mit Kernfunktion. DM 58,– ; US $ 14.50
114. MacLane: Homology. DM 62,– ; US $ 15.50
115. Hewitt/Ross: Abstract Harmonic Analysis. Vol. 1: Structure of Topological Groups. Integration Theory. Group Representations. DM 76,– ; US $ 19.00

116. Hörmander: Linear Partial Differential Operators. DM 42,– ; US $ 10.50
117. O'Meara: Introduction to Quadratic Forms. DM 48,– ; US $ 12.00
118. Schäfke: Einführung in die Theorie der speziellen Funktionen der mathematischen Physik. DM 49,40; US $ 12.35
119. Harris: The Theory of Branching Processes. DM 36,– ; US $ 9.00
120. Collatz: Funktionalanalysis und numerische Mathematik. DM 58,– ; US $ 14.50
121.
122. Dynkin: Markov Processes. DM 96,– ; US $ 24.00
123. Yosida: Functional Analysis. DM 66,– ; US $ 16.50
124. Morgenstern: Einführung in die Wahrscheinlichkeitsrechnung und mathematische Statistik. DM 38,– ; US $ 9.50
125. Itô/McKean: Diffusion Processes and Their Sample Paths. DM 58,– ; US $ 14.50
126. Letho/Virtanen: Quasikonforme Abbildungen. DM 38,– ; US $ 9.50
127. Hermes: Enumerability, Decidability, Computability. DM 39,– ; US $ 9.75
128. Braun/Koecher: Jordan-Algebren. DM 48,– ; US $ 12.00
129. Nikodým: The Mathematical Apparatus for Quantum-Theories. DM 144,– ; US $ 36.00
130. Morrey: Multiple Integrals in the Calculus of Variations. DM 78,– ; US $ 19.50
131. Hirzebruch: Topological Methods in Algebraic Geometry. DM 38,– ; US $ 9.50
132. Kato: Perturbation Theory for Linear Operators. DM 79,20; US $ 19.80
133. Haupt/Künneth: Geometrische Ordnungen. DM 68,– ; US $ 17.00
134. Huppert: Endliche Gruppen I. DM 156,– ; US $ 39.00
135. Handbook for Automatic Computation. Vol. 1/Part a: Rutishauser: Description of ALGOL 60. DM 58,– ; US $ 14.50
136. Greub: Multilinear Algebra. DM 32,– ; US $ 8.00
137. Handbook for Automatic Computation. Vol. 1/Part b: Grau/Hill/Langmaack: Translation of ALGOL 60. DM 64,– ; US $ 16.00
138. Hahn: Stability of Motion. DM 72,– ; US $ 18.00
139. Mathematische Hilfsmittel des Ingenieurs. Herausgeber: Sauer/Szabó. 1. Teil. DM 88,– ; US $ 22.00
140. Mathematische Hilfsmittel des Ingenieurs. Herausgeber: Sauer/Szabó. 2. Teil. In Vorbereitung
141. Mathematische Hilfsmittel des Ingenieurs. Herausgeber: Sauer/Szabó. 3. Teil. DM 98,– ; US $ 24.50
142. Mathematische Hilfsmittel des Ingenieurs. Herausgeber: Sauer/Szabó. 4. Teil. In Vorbereitung
143. Schur/Grunsky: Vorlesungen über Invariantentheorie. DM 32,– ; US $ 8.00
144. Weil: Basic Number Theory. DM 48,– ; US $ 12.00
145. Butzer/Berens: Semi-Groups of Operators and Approximation. DM 56,– ; US $ 14.00
146. Treves: Locally Convex Spaces and Linear Partial Differential Equations. DM 36,– ; US $ 9.00
147. Lamotke: Semisimpliziale algebraische Topologie. DM 48,– ; US $ 12.00
148. Chandrasekharan: Introduction to Analytic Number Theory. DM 28,– ; US $ 7.00
149. Sario/Oikawa: Capacity Functions. DM 96,– ; US $ 24.00
150. Iosifescu/Theodorescu: Random Processes and Learning. DM 68,– ; US $ 17.00
151. Mandl: Analytical Treatment of One-dimensional Markov Processes. DM 36,– ; US $ 9.00
152. Hewitt/Ross: Abstract Harmonic Analysis. Vol. II. In preparation
153. Federer: Geometric Measure Theory. DM 118,– ; US $ 29.50
154. Singer: Bases in Banach Spaces. In preparation

155. Müller: Foundations of the Mathematical Theory of Electromagnetic Waves. DM 58,– ; US $ 14.50
156. van der Waerden: Mathematical Statistics. In preparation
157. Prohorov/Rozanov: Probability Theory. DM 68,– ; US $ 17.00
158. Constantinescu/Cornea: Potential Theory on Harmonic Spaces. In preparation
159. Köthe: Topological Vector Spaces I. In preparation